Reproduction
in Farm Animals

Reproduction
in Farm Animals

Edited by
E. S. E. HAFEZ

School of Medicine
Wayne State University
Detroit, Michigan

THIRD EDITION

181 Illustrations and 26 Plates

LEA & FEBIGER
Philadelphia

Library of Congress Cataloging in Publication Data

Hafez, E. S. E.   1922–   ed.
    Reproduction in farm animals.

    Includes bibliographies.
    1. Reproduction.   2. Stock and stock-breeding.   3. Veterinary physiology.   I. Title.   [DNLM: 1.
Animals, Domestic—Physiology.   2. Reproduction.   SF768   H138r]
SF678.H2   1974   636.089'26   74–8671
ISBN 0–8121–0295–9

Reprinted, 1975

PRINTED IN THE UNITED STATES OF AMERICA

This volume is dedicated to Dr. Robert Denamur (1925–1973), Head of Laboratoire de Physiologie de la Lactation, Institut National de la Recherche Agronomique, Jouy-en-Josas, France, for his pioneer and outstanding research in the physiology of the corpus luteum and in lactation.

Preface

The objective of the first edition, appearing in 1962, was to present the basic and comparative aspects of reproductive physiology in a simplified manner which would meet the needs of students in veterinary medicine and animal sciences. This aim has not changed. The first and second editions translated into Japanese by Professor Y. Nishikawa were published in Tokyo, 1965 and 1971; and the Spanish editions were published in Mexico City in 1967.

The last five years have been characterized by a gratifying accumulation of data, due to important advances in methodology such as radioimmunoassay, scanning electronmicroscopy and enzymology. The third edition is a complete revision of the second edition in order to include recent progress in gamete ultrastructure and transport, cytogenetics and biochemistry of reproduction, and hypothalamic control of reproduction and sexual behavior. There have been numerous deletions from the second edition, as well as integration of new references and new ideas. Several new chapters have been added dealing with reproductive life cycle, gamete transport, cytogenetics, and mammalian and avian reproduction. The two chapters dealing with reproductive infections and the chapters dealing with behavior were consolidated. Much of the information on laboratory animals presented in the second edition has been replaced by recent data on farm animals. No attempt was made to provide a detailed bibliography, but a selected number of classic papers and review articles are listed at the end of each chapter.

The material in the book is arranged in six parts. Part I deals with the functional anatomy of male and female reproduction with emphasis on physiological mechanism, cell morphology, and physiology. Part II deals with the endocrinology of reproduction; the reproductive life cycle including senility; the biochemistry of semen; the transport of ova and spermatozoa; fertilization, cleavage, and implantation; pregnancy, prenatal development, and parturition; lactation; and reproductive and neonatal behavior. Part III includes the species-specific aspects of the reproductive cycles of farm animals. Part IV deals with cytogenetics of mammalian and avian reproduction. Part V is a general discussion of reproductive failure in the male and female, such as intersexes and reproductive infections. Part VI includes techniques for improving reproductive efficiency, such as artificial insemination and detection and synchronization of estrus, pregnancy diagnosis, and egg transfer.

Detroit, Michigan   E. S. E. Hafez

Acknowledgments

Much of the data presented in Chapters 2, 6, 8, 16, 18 and 20 (Hafez, Jainudeen, and Kanagawa) was obtained through research supported by The Ford Foundation Grant 710-0287.

Much of the data presented in Chapter 3 (Niswender, Nett, and Akbar) was obtained through research supported by The National Institutes of Health (Contract 69–2134), a Program Project in Reproductive Endocrinology (awarded to the University of Michigan), and a grant from G. D. Searle and Company. Dr. T. M. Nett was supported by Postdoctoral Fellowship FO2 HD53980–01. The authors wish to thank Dr. H. J. Brinkley, Department of Zoology, University of Maryland, College Park, Maryland; Dr. H. D. Hafs, Department of Dairy Science, Michigan State University, East Lansing, Michigan; and Dr. D. M. Henricks, Department of Food Science, Animal Science and Dairy Science, Clemson University, Clemson, South Carolina, for providing the data regarding levels of gonadotropic and steroid hormones in pigs and cows shown in Figures 3–10 and 3–11. The authors are also indebted to Drs. M. L. Hopwood, L. C. Faulkner, and M. H. Pineda for assistance and review of this manuscript.

Dr. G. Alexander, contributor to Chapter 10, wishes to acknowledge the cooperation of J. Bareham, P. English, M. Bryant, A. Holmes, and D. Frazer for providing ready access to their unpublished data.

The scanning electron micrograph of the two-cell egg in the frontispiece was kindly provided by Dr. Patricia Calarco from the Department of Anatomy, School of Medicine, University of California, San Francisco.

The sincere thanks of the editor are extended to all contributors and to the staff of Lea & Febiger for their excellent cooperation during the preparation and production of the volume.

Contributors

AKBAR, A. M.: Department of Physiology and Biophysics, Colorado State University, Fort Collins, Colorado 80521

ALEXANDER, G.: CSIRO, Division of Animal Physiology, Ian Clunies Ross Animal Research Laboratory, Prospect, N.S.W., Australia

ANDERSON, L. L.: Department of Animal Science, 11 Kildee Hall, Iowa State University, Ames, Iowa 50010

ASHDOWN, R. R.: Department of Veterinary Anatomy, The Royal Veterinary College, University of London, London N.W. 1, England

BASRUR, P. K.: Department of Biomedical Science, Ontario Veterinary College, University of Guelph, Guelph, Ontario N1G 2W1, Canada

BUTTLE, H. L.: National Institute for Research in Dairying, Shinfield, Reading, RG2 9AT, England

CARROLL, E. J.: Department of Physiology and Biophysics, Colorado State University, Fort Collins, Colorado 80521

COWIE, A. T.: National Institute for Research in Dairying, Shinfield, Reading RG2 9AT, England

FAULKNER, L. C.: Department of Physiology and Biophysics, Colorado State University, Fort Collins, Colorado 80521

FOOTE, R. H.: Department of Animal Science, Cornell University, Ithaca, New York 14850

FOOTE, W. D.: Department of Animal Sciences, University of Nevada, Reno, Nevada 89507

GILBERT, A. B.: Poultry Research Centre, West Mains Road, Edinburgh 9, Scotland

HAFEZ, E. S. E.: Departments of Gynecology-Obstetrics and Physiology, Wayne State University School of Medicine, Detroit, Michigan 48201

HANCOCK, J. L.: Department of Veterinary Anatomy, The Royal Veterinary College, University of London, London N.W. 1, England

HOWARTH, J. A.: Department of Epidemiology and Preventive Medicine, School of Veterinary Medicine, University of California, Davis, California 95616

JAINUDEEN, M. R.: Departments of Gynecology-Obstetrics and Physiology, Wayne State School of Medicine, Detroit, Michigan 48201 (Present address: Faculty of Medicine, P.O. Box 203, Sungei Selangor, Malaysia)

KANAGAWA, H.: Departments of Gynecology-Obstetrics and Physiology, Wayne State School of Medicine, Detroit, Michigan 48201
(Present address: Department of Biomedical Sciences, Ontario Veterinary College, University of Guelph, Guelph, Ontario, N1G 2W1, Canada)

KENDRICK, J. W.: Department of Reproduction, College of Veterinary Medicine, University of California, Davis, California 95616

LAUDERDALE, J. W.: Agricultural Division, The Upjohn Company, Kalamazoo, Michigan 49001

LEVASSEUR, M.-C.: Station de Recherches de Physiologie Animale, I.N.R.A., Jouy-en-Josas, France

McFEELY, R. A.: New Bolton Center, School of Veterinary Medicine, University of Pennsylvania, Kennett Square, R.D. 1, Pennsylvania 19348

McLAREN, A.: Department of Genetics and A.R.C. Unit of Animal Genetics, Institute of Animal Genetics, West Mains Road, Edinburgh EH 9 3 JN, Scotland

NETT, T. M.: Department of Physiology and Biophysics, Colorado State University, Fort Collins, Colorado 80521

NISHIKAWA, Y.: Department of Animal Husbandry, Kyoto University, Kitashirakawa, Kyoto-Shi, Japan

NISWENDER, G. D.: Department of Physiology and Biophysics, Colorado State University, Fort Collins, Colorado 80521

SIGNORET, J. P.: Station de Physiologie de la Reproduction, Centre de Recherches de Tours, B.P. 1 37-Nouzilly, France

TERRILL, C. E.: Sheep, Goat and Fur Animal Research Branch, Agricultural Research Service, U.S.D.A., Beltsville, Maryland 20705

THIBAULT, C.: Station de Recherches de Physiologie Animale, I.N.R.A. Jouy-en-Josas, France

WHITE, I. G.: Department of Veterinary Physiology, The University of Sydney, Sydney, N.S.W., Australia

ZEMJANIS, R.: Department of Veterinary Obstetrics and Gynecology, College of Veterinary Medicine, St. Paul, Minnseota 55101

ZIMBELMAN, R. G.: Agricultural Division, The Upjohn Company, Kalamazoo, Michigan 49001

Contents

Introduction to Animal Reproduction

Birds (*Aves*), fish (*Pisces*), and amphibians (*Amphibia*) are "oviparous." Their eggs are large, produced in abundance, surrounded by an abundant yolk, and in many cases, fertilized outside the body of the female whose external genitalia are poorly developed. Reptiles (*Reptilia*) are "ovoviviparous"; their eggs are covered with a protective shell and have an abundant yolk. Furthermore, the larvae hatch inside the body of the female. Mammals, on the other hand, with the exception of *Monotremata* are "viviparous." They produce fewer eggs which contain a scant yolk, fertilization is internal, fetal development is completed in the uterus, and the external genitalia are well-developed. The echidna (*Tachyglossus aculeata*) and the platypus (*Ornithorhynchus anatinus*) are the only mammals that lay eggs. The eggs of these mammals are relatively small but do contain enough nutriment to support development up to an advanced stage, though not to the stage of hatching.

There are several thousand mammalian species, but reproductive biology has been extensively studied in less than 25: namely, rodents (*Rodentia*), rabbits (*Lagomorpha*), primates (including man), farm animals and a few marsupials. Some of these species are characterized by peculiar reproductive phenomena, such as restricted sexual season, absence of estrus, presence of menstruation, dissociation of ovulation and estrus, nonspontaneous ovulation, spontaneous multiple ovulation with limited implantation, delayed implantation, and ovulation during pregnancy.

The animals which man has domesticated over the centuries to meet his own needs for food, clothing, power, or companionship include cattle, sheep, goats, pigs (*Artiodactyla*); horses and asses (*Perissodactyla*); cats and dogs (*Carnivora*); and poultry (*Galliformes*). These animals vary with respect to sexual season, sexual cycle, gestation period, type of placentation, litter size, lactation period, and susceptibility to reproductive diseases. For example, cattle, pigs and chickens breed throughout the year, horses and asses in the spring, and most sheep and goats in the fall. These seasonal variations are not so evident in tropical zones compared to temperate and frigid zones, where periodicity is very seasonal. Furthermore, these variations are less evident in domesticated than in wild species.

The activity of the gonads and the accessory glands are influenced directly or indirectly by hereditary factors, ambient temperature, photoperiod, and nutrition. The reproductive cycle is regulated by interactions between the central nervous system, the pituitary, and the gonads. The hypothalamus controls the secretion of gonadotropins by releasing a regulatory substance into the portal blood flowing from the median eminence region of the tuber cinereum into the pars distalis of the pituitary gland.

Natural and synthetic LH-RH releasing hormones have similar biological and physiochemical properties. These hormones greatly enhance the release of both LH and FSH in vivo and from pituitaries incubated in vitro in doses of fractions of a nanogram.

Prostaglandins (PG), a group of chemically related 20-carbon chain hydroxy fatty acids, are widely distributed in mammalian tissue. They have been used to induce luteolysis (destruction of the corpus luteum) in early stages of pregnancy, and to synchronize estrus. The potent oxytocic effect of prostaglandins has been used to induce abortion and labor in women.

Recent advances in reproductive endocrinology are primarily due to the development of radioimmunoassay, a standard and highly sensitive method, used for assay of releasing factors, gonadotropins, steroids, and prostaglandins.

There are remarkable species differences in the degree to which the scrotum is held near or away from the abdomen. Optimal temperature for spermatogenesis and storage of epididymal spermatozoa seems to be related to their survival time in the female reproductive tract. In cattle and sheep the scrotum is very pendulous, and the life span of spermatozoa in the female reproductive tract is only about thirty hours. In horses the scrotum is close to the abdominal wall, and the life span of spermatozoa in the female tract extends to three days. The avian testes is located in the abdominal cavity, and the epididymal spermatozoa survive for some thirty days. Turkey spermatozoa survive in the female tract for prolonged periods and eggs can be fertilized for up to thirty days after a single mating. Sperm survive for several months in the female tract of hibernating bats of the families *Vespertilionidae* and *Phinolophidae*. Species differences in sperm survival are due to differences in the anatomy and physiology of the female reproductive tract, sperm concentration and motility, rate of sperm transport, the female's endocrine state at copulation and her inflammatory response to it, and the dilution of sperm by luminal fluids secreted in the female tract.

The mammalian spermatozoon of most species, before being able to penetrate the zona pellucida, undergoes final maturation in the female reproductive tract (capacitation). This phenomenon is followed by the acrosome reaction, involving multiple fusions between the plasma and outer acrosomal membranes, with subsequent vesiculation.

Cleavage and blastulation of the embryo depend on critical oviductal and uterine factors. Quantitative and qualitative changes in the uterine fluids are related to some reproductive process or to specific needs of the embryo during particular stages during pre-implantation development. In some species the implantation of the embryo is delayed, increasing the duration of pregnancy.

Early pregnancy is characterized by active secretion of progesterone from the corpus luteum. In marsupials, such as opossum and kangaroo, the life span of the corpus luteum is similar in pregnant and nonpregnant females, and the length of gestation period is similar to that of the estrous cycle. In farm mammals, the life span of the corpus luteum is prolonged during pregnancy and ovulation is suppressed except in the mare. In general, the duration of gestation increases with the size of the species and with the stage of development at which the young are born.

The neonate in placental mammals is very immature, develops slowly, and depends on maternal care. The stages of development at birth vary greatly in different species and determines the extent to which parental care is required. In the rat and rabbit, neonates are born blind, naked, and with a poorly developed thermoregulatory system; thus, they require a warm maternal nest. In ungulates, the young are born in an advanced stage of development and can fend for themselves in a few days. The extent of mother-young social interactions also varies widely and is necessary for the full development of the physical and behavioral characteristics of the species.

The efficiency of reproduction in a given species depends on the length of the sexual season, frequency of estrus, number of ovulations, duration of pregnancy, litter size, suckling period, puberty age, and duration of the reproductive period in the animal's life. In general, the age at which puberty is attained is earlier in smaller sized species than in large ones, as well as in females compared to males. There is no definite age at which reproductive functions cease abruptly during life, constituting menopause or climacteric in man. Many other female mammals, however, die before arrest of reproductive functions occurs.

The efficiency of reproduction may decline as a result of seasonal, genetic, nutritional, anatomical, hormonal, neural, immunological, humoral, or pathological factors. These factors may result in partial or complete reproductive failure. Those concerned with farm animals have the continuous objective of preventing such failure. Several methods have been used to control and enhance fertility in an attempt to keep some balance between the supplies and demands of exploding population.

E. S. E. Hafez

I. Functional Anatomy of Reproduction

Chapter 1

Functional Anatomy of Male Reproduction

R. R. Ashdown and J. L. Hancock

The male gonads, the testes, lie outside the abdomen within the scrotum, which is a purselike structure derived from the skin and fascia of the abdominal wall. Each testis lies within the vaginal process, a separate extension of the peritoneum, which passes through the abdominal wall at the inguinal canal. The internal and external inguinal rings are the deep and superficial openings of the inguinal canal. Besides providing for the passage of the vaginal process and its contents, the inguinal canal also gives passage to important vessels and nerves supplying the external genitalia. Blood vessels and nerves reach the testis in the spermatic cord which lies within the neck of the vaginal process; the *ductus deferens*, which at first accompanies the vessels, leaves them at the internal inguinal ring to join the urethra.

The spermatozoa produced by the testis leave via a number of efferent ductules which lead into the coiled duct of the epididymis, which becomes the straight ductus deferens. Sets of accessory glands discharge their contents either into the ductus deferens or near its termination in the pelvic portion of the urethra.

The urethra originates at the neck of the bladder. Its pelvic portion, which is enclosed by the striated urethral muscle and receives secretions from various glands at the pelvic outlet, leads into a second, penile, portion where it is joined by two more cavernous bodies to make up the body of the penis which lies beneath the skin of the body wall. Throughout its length the urethra is surrounded by cavernous vascular tissue. A number of muscles grouped around the pelvic outlet contribute to the root of the penis. The apex or free end of the penis is covered by modified skin—the penile integument. In the resting condition it is enclosed within the sheath. The topographic features of the organs of the important farm species are shown in Figure 1–1; their dimensions are listed in Table 1–3. Detailed descriptions of the organs are given by Nickel and associates (1973).

The testis is supplied with blood from the testicular artery which originates from the dorsal aorta near the embryonic site of the testes. The internal pudendal artery supplies the pelvic genitalia and branches leave the pelvis at the ischial arch to supply the penis. The external pudendal artery leaves the abdominal cavity via the inguinal canal to supply the scrotum and sheath.

Afferent and efferent (sympathetic) nerves accompany the testicular artery to the testis. The pelvic plexus supplies anatomic (sympathetic and parasympathetic) fibers to the pelvic genitalia and to the penis. Sacral nerves supply motor fibers to the muscles of the penis and

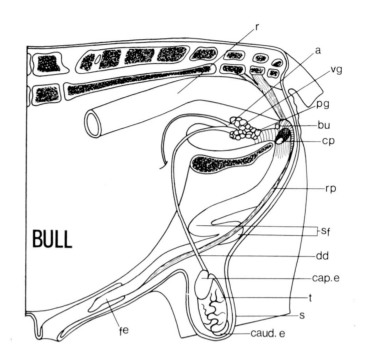

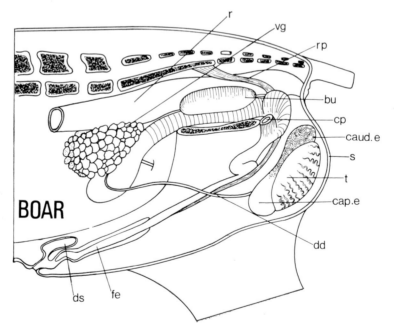

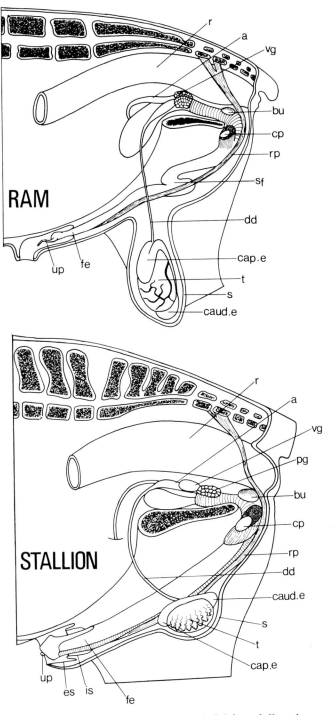

FIG. 1–1. Diagram of the male reproductive tracts as seen in left lateral dissections.

a, Ampulla; *bu*, bulbourethral gland; *cap.e*, caput epididymidis; *caud.e*, cauda epididymidis; *cp*, left crus of penis, severed from the left ischium; *dd*, ductus deferens; *ds*, dorsal diverticulum of sheath; *es*, external sheath; *fe*, free end of penis; *is*, internal sheath; *pg*, prostate gland; *r*, rectum; *rp*, retractor penis muscle; *s*, scrotum; *sf*, sigmoid flexure of penis; *t* testis; *up*, urethral process; *vg*, vesicular gland. (*Adapted from Popesko, 1968. Atlas der topographischen Anatomie der Haustiere. Vol. 3, Jena, Fischer.*)

sensory fibers to the free end of the penis. Afferent fibers from the scrotum and sheath travel to the spinal cord, mainly in the inguinal nerve (Larson and Kitchell, 1958; Hodson, 1970).

I. DEVELOPMENT

A. Prenatal Development

The testes develop in the gonadal ridge which lies medial to the embryonic kidney (*mesonephros*). The gonads of the male embryo differentiate following the arrival of the primordial germ cells; these migrate to the gonadal ridge in the bull fetus about day 26 (Gier and Marion, 1970). The primordial germ cells, carried into the medulla by the primary sex cords, formed from the coelomic epithelium provide the element from which the germinal epithelium of the seminiferous tubule is formed. The *rete testis* develops as a separate mass of cords which establishes connection with the mesonephric (kidney) tubules on one hand and with the future seminiferous tubules on the other; as a result the mesonephric duct becomes the excurrent duct of the testis (Fig. 1–2). In the course of development in the bull, boar, ram and stallion the rete comes to lie centrally

in the testis and not peripherally as in man. The mesonephric tubules which become connected to the rete form the efferent ductules located in the *caput epididymidis*. The rest of the epididymis is formed from the first part of the mesonephric duct; the remainder of the duct forms the ductus deferens and a terminal diverticulum forms the vesicular gland. The paramesonephric duct, which is the primordium of the female duct system, degenerates in the male.

There are two active agents produced by the fetal testis which are responsible for differentiation and development of the duct system. Fetal androgen produced by the testis causes development of the male reproductive tract. The second component (so-called factor X) is responsible for suppression of the paramesonephric (Müllerian) duct from which the uterus and vagina develop in the female.

Early in fetal life the urogenital sinus, into which the mesonephric ducts open, is separated off from the termination of the gut; the male urethra forms from the urogenital sinus. At the urogenital orifice the genital tubercle forms and within it the penile part of the urethra is developed. A separate fold of skin grows

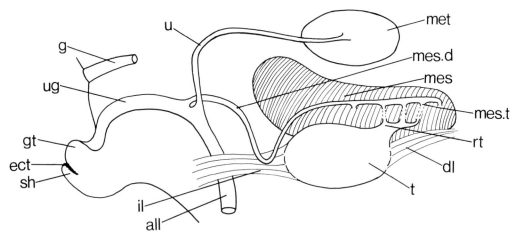

FIG. 1–2. Diagram to show the origins of the male reproductive organs in a mammal.

all, Allantois; *dl*, diaphragmatic ligament; *ect*, ectodermal lamella; *gt*, genital tubercle; *g*, termination of hind gut; *il*, inguinal ligament; *mes*, mesonephros; *mes.d*, mesonephric duct; *mes.t*, mesonephric tubules; *met*, metanephros; *rt*, rete testis; *sh*, developing penile sheath; *t*, testis; *u*, ureter; *ug*, urogenital sinus. (*Adapted from Gier and Marion, 1969. Biol. Reprod. 1, 1.*)

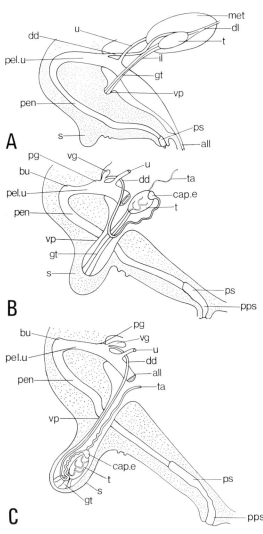

FIG. 1–3. Diagrams to show descent of the testis and development of the reproductive tract in the bovine fetus. *A* at 62 days, *B* at 102 days, *C* at 140 days.

all, Allantois; *bu*, bulbourethral gland; *cap. e*, caput epididymidis; *dd*, ductus deferens; *dl*, diaphragmatic ligament; *gt*, gubernaculum testis; *il*, inguinal ligament; *met*, metanephros; *pel. u*, pelvic urethra; *pen*, penis; *pg*, prostate gland; *ps*, penile sheath; *pps*, prepenile sheath; *s*, scrotum; *t*, testis; *ta*, testicular artery; *u*, ureter; *vg*, vesicular gland; *vp*, vaginal process. (*Adapted from Gier and Marion, 1970. In "Development of the Mammalian Testis" in The Testis. Vol. 1. Johnson, Gomes and VanDemark [eds.], New York, Academic Press.*)

distally over the genital tubercle to form the penile sheath (Fig. 1–2) and remains fused to the penis until after birth.

B. Descent of the Testis

Testicular descent (Fig. 1–3) involves abdominal migration to the internal inguinal ring, inguinal migration through the canal, and finally migration within the scrotum. Descent into the scrotum is preceded by formation of the vaginal process, a peritoneal sac extending toward the scrotum and enclosing the inguinal ligament of the testis which together with the diaphragmatic ligament and the mesorchium, suspends the fetal testis. The inguinal ligament connects the gonad and the mesonephric duct; distal to this point it is often called the *gubernaculum testis* and it terminates in the region of the scrotal rudiments. It enlarges greatly during late fetal life and is closely involved in descent of the testes. The time of descent varies according to the species (Table 1–1). In the horse the epididymis commonly enters the inguinal canal before the testis and that part of the inguinal ligament connecting testis and epididymis (proper ligament of testis) remains extensive until after birth.

Sometimes the testis fails to enter the scrotum. In this condition (*cryptorchidism*) the special thermal needs of the testis are not met and normal spermatogenic function is impossible although the endocrine function of the testis is unimpaired. Cryptorchid males therefore show more or less normal sexual desire but are sterile. Occasionally some of the abdominal viscera pass through the orifice of the vaginal process and enter the scrotum; scrotal hernia is particularly common in pigs. Anomalies involving differentiation of the gonads and duct system can result in varying degrees of intersexuality.

C. Postnatal Development

Each component of the tract of all farm animals grows in size relative to overall

Table 1-1. Chronology of Development of the Male Reproductive
Tract in Farm Animals*

	Bull	Ram	Boar	Stallion
Testicular descent	Enters scrotum halfway through fetal life	Enters scrotum halfway through fetal life	Enters scrotum in last quarter of fetal life	Enters scrotum just before or just after birth
Primary spermatocytes in seminiferous tubules	24 weeks	12 weeks	10 weeks	Very variable throughout seminiferous tubules of each testis
Spermatozoa in seminiferous tubules	32 weeks	16 weeks	20 weeks	56 weeks (very variable)
Spermatozoa in cauda epididymidis	40 weeks	16 weeks	20 weeks	60 weeks (very variable)
Spermatozoa in the ejaculate	42 weeks	18 weeks	22 weeks	—
Completion of separation between penis and penile part of sheath	32 weeks	>10 weeks	20 weeks	4 weeks
Age at which animal can be considered to be sexually "mature"	150 weeks	>24 weeks	30 weeks	90–150 weeks (very variable)

* Compiled from various sources.

Table 1-2. Growth of the Reproductive Tract in Holstein and Holstein-cross
Bulls During the Postpubertal Period*

	Age in Months (Mean)			
Items	37	59	80	133
Number of bulls	7	20	4	7
Body weight (lb)	1864	2081	2046	2006
Testis weight (gm)	259	335	359	395
Epididymis weight (gm)	27	35	38	40
Vesicular gland weight (gm)	55	78	79	81
Bulbourethral gland weight (gm)	5.2	6.5	7.1	6.0
Penis length (cm)	95	97	103	106

* From Almquist and Amann, 1961. J. Dairy Sci., 44.

PLATE 1.

A–D, Penis of a crossbred Galloway calf, protruded to show separation of the free end of the penis from its penile sheath at four different ages. A—278 days; B—306 days; C—319 days; D—326 days. (Ashdown, 1960. J. Agric. Sci. 54, 348.)

E, F, Transverse sections through the urethral process and sheath of a week-old calf (H and E). E—low power view (× 10) to show the urethral process surrounded by the ectodermal lamella which fuses it completely to the penile sheath. F—higher power view (× 200) of part of the ectodermal lamella showing evidence of keratinization. These changes extend to separate penis from sheath.

PLATE 1

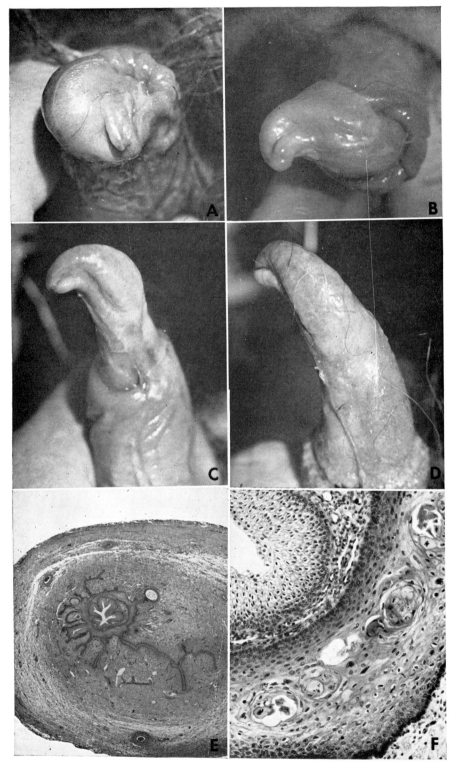

Legend on facing page.

PLATE 2

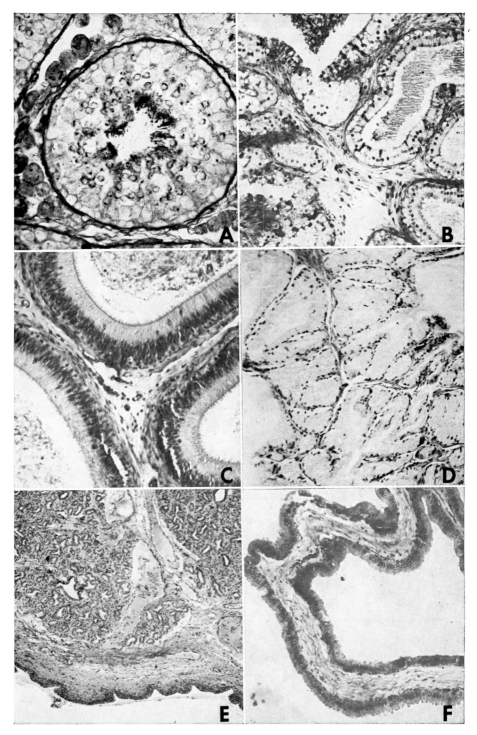

(A) Boar testis, showing tubular epithelium and interstitial tissue. P.A.S.-Methyl green. × 325.
(B) Ductus deferens of bull showing ampullary glands. Iron haematoxylin. × 140.
(C) Boar epididymis showing columnar epithelium with stereocilia. Masson stain. × 140.
(D) Bulbourethral gland of boar. Mallory stain. × 140.
(E) Prostate gland of boar. Haematoxylin and eosin. × 56.
(F) Vesicular gland of boar. Masson stain. × 140.

body size and undergoes histologic differentiation (Abdel-Raouf, 1960; Nishikawa, 1959) but functional competence is not achieved simultaneously in all components of the reproductive system. Thus in the bull the capacity for erection of the penis precedes the appearance of spermatozoa in the ejaculate by several months. At puberty all the components of the male reproductive system have reached a sufficiently advanced stage of development for the system as a whole to be functional. The period of rapid development that precedes puberty is known as the prepubertal period, although this period is itself sometimes referred to as "puberty" (Donovan and Werff ten Bosch, 1965; Skinner and Rowson, 1968). During the postpubertal period, development continues and the reproductive tract reaches full sexual maturity months, or even years, after the age of puberty. Some important anatomic changes occurring during postnatal development are summarized in Tables 1–1 and 1–2, and illustrated in Plate 1 and Figures 1–4 and 1–5.

II. TESTIS

A. Structure

The testes are secured to the wall of the vaginal process along the line of their epididymal attachment (Fig. 1–6). The position in the scrotum and the direction of the long axis of the testis relative to the body differs with the species (Fig. 1–1). The epididymis is closely apposed to the surface of the testis and the point of origin of the efferent ducts from the rete testis lies under the flattened expanded head of the epididymis. The surface of the testis is covered by an extension of the parietal peritoneum of the abdominal cavity. Beneath this lies a tough fibromuscular *tunica albuginea* from which, at the epididymal attachment, extensions penetrate the parenchyma of the organ to join the *mediastinum*, a cord of connective tissue running through the testis (Fig. 1–6). Fibrous septa divide the parenchyma into lobules of coiled seminiferous tubules which lead via straight tubules into the rete testis.

B. Endocrine Function

The interstitial (Leydig) cells which lie between the tubules are the source of the male hormone (testosterone). The epithelium of the tubule consists of spermatogenic cells and supporting sustentacular (Sertoli) cells (Plate 2). The basement membrane contains contractile "myoid" cells. Spermatozoa are produced by differentiation of the last of several generations of cells which result from the division of peripherally situated spermatogonia.

The two important functional roles of the testes are governed by the gonadotrophic hormones of the pituitary gland. Follicle stimulating hormone (FSH) is closely connected with initiating activity in the seminiferous tubules. Luteinizing or interstitial cell stimulating hormone (LH or ICSH) controls the endocrine

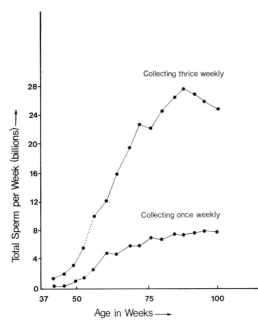

FIG. 1–4. Average sperm output of Holstein bulls during the postpubertal period, up to 100 weeks of age. (*Simplified from Almquist, Amann and Hale, 1963. Ann. Meet. Amer. Soc. Anim. Sci., Morgantown, West Virginia.*)

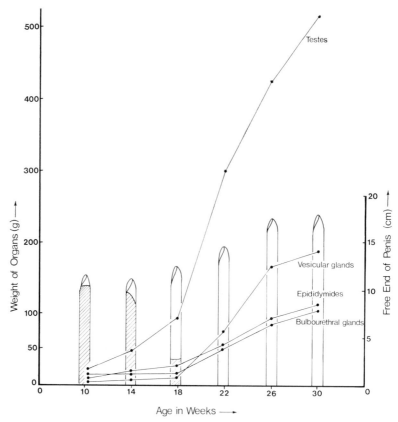

FIG. 1–5.   Development of the reproductive tracts in six littermate Landrace boars from 10 to 30 weeks of age.   Organ weight is the sum of left and right components.   Length of the free end of the penis is shown; the shaded area represents the area still adherent to the penile sheath.

activity of the cells.   Male sex hormone produced by the interstitial cells supports the action of FSH on spermatogenesis and is responsible for the development of the secondary sexual characters and for the growth and functional integrity of the male reproductive tract as a whole. Castration of prepubertal males results in suppression of development (Fig. 1–10).   Regressive changes in behavior and structure take place following castration of adult males.   Castration is a standard procedure in animal husbandry to modify aggressive male behavior and to eliminate undesirable carcass qualities, e.g. boar taint.   Testosterone exerts some sparing effects on protein metabolism and there is now a trend toward the use of intact males for meat production.

C. Exocrine Function

The spermatozoa leave the testis in an important fluid secretion (Setchell, 1970).   This "rete fluid" differs markedly in composition from blood plasma and lymph.   There is thus an important blood-testis barrier which effectively separates the seminiferous epithelium from the general circulation.   This barrier seems to be formed by special cells of the basement membrane of the tubule and by some special features of the sustentacular cells (Dym and Fawcett, 1970). The barrier effectively divides two compartments of the testis, with the parent spermatogonia separated by the sustentacular cells from their progeny.   The integrity of the blood-testis barrier is believed to be important for normal testis

Fig. 1–6.   Diagrammatic horizontal section through the scrotum of the bull to show the relationships of the organs. The inset diagram shows more clearly the layers of the scrotal wall and the vaginal process.

corp.e, Corpus epididymidis; *ct*, loose connective tissue lying between the vaginal process and the wall of the scrotum; *dar*, tunica dartos; *dd*, ductus deferens; *der*, dermis; *e*, epidermis; *med*, mediastinum testis; *pvp*, parietal layer of vaginal process; *ss*, scrotal septum; *t.u.*, tunica albuginea of testis; *ta*, testicular artery; *tp*, testicular parenchyma; *vc*, cavity of vaginal process; *vvp*, visceral layer of vaginal process. (*Adapted from Blom and Christensen, 1947. Skand. Vet. Tidskrift 37, 1.*)

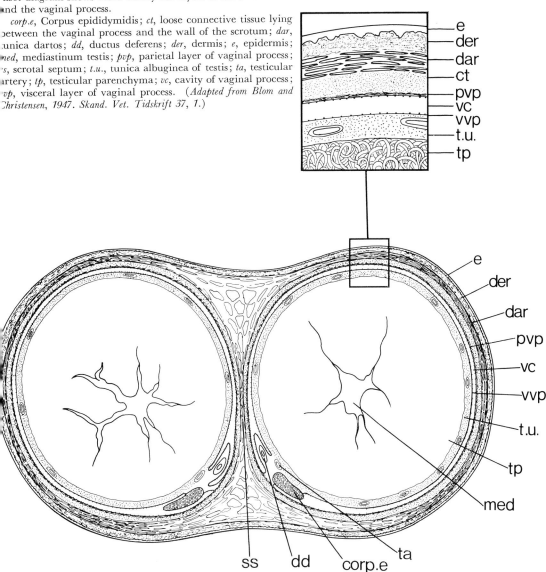

function. The damaging effects of certain heavy metals (e.g. cadmium) on testis function are believed to be due to their effect on this barrier.

D. Sperm Production

Numbers of spermatozoa produced by the bull per gm testes (13×10^6 to 19×10^6) are rather lower than for ram (24×10^6 to 27×10^6) and boar (24×10^6 to 31×10^6) (Amann, 1970). Sperm production in the bull increases with age up to seven years. Production is usually greater than output as measured by the number of spermatozoa recovered at ejaculation.

E. Thermoregulation of the Testis

For effective functioning, the mammalian testes must be maintained at a temperature lower than that of the body. Anatomic features of the testis and scrotum provide for the regulation of testicular temperature. The scrotal skin is noticeably lacking in subcutaneous fat, is very richly endowed with sweat glands, and its muscular (*dartos*) component enables it to alter the thickness of the scrotum and its surface area and to vary the closeness of contact of the testes with the body wall. In the horse this action may be supported by the internal cremaster muscle within the spermatic cord which can lower or raise the testis. In cold conditions the *cremaster* and dartos muscles contract, elevating the testis and wrinkling and thickening the scrotal wall. In hot conditions the muscles relax, lowering the testis within the thin-walled pendulous scrotum. The advantages offered by these mechanisms are enhanced by the special relationship of the veins and arteries.

In all farm animals, the testicular artery is a convoluted structure in the form of a cone the base of which rests on the cranial or dorsal pole of the testis (Fig. 1–7). These arterial coils are enmeshed by the so-called pampiniform plexus of testicular veins. This arrangement pro-

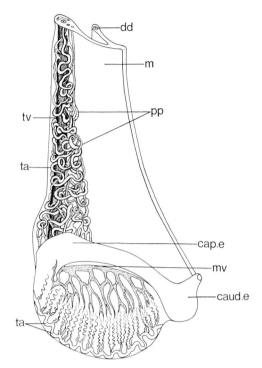

FIG. 1–7. Diagrammatic lateral view of the left testis of a stallion to show arrangement of arteries and veins.

cap.e, Caput epididymidis; *caud.e*, cauda epididymidis; *dd*, ductus deferens; *m*, mesorchium; *mv*, marginal vein of testis; *pp*, pampiniform plexus of veins; *ta*, testicular artery; *tv*, testicular vein. (*Adapted from Tagand and Barone, 1956. Anatomie des Équidés Domestiques. 2, iii, Lyons, École Nat. Vet.*)

vides an effective countercurrent mechanism by which arterial blood entering the testis is cooled by the venous blood leaving the testis. In the ram the temperature of the blood in the testicular artery falls $4°C$ in its course from the external inguinal ring to the surface of the testis; the blood temperature in the veins rises by a similar amount between the testis and the external inguinal ring. The position of the arteries and veins close to the surface of the testis tends to increase direct loss of heat from the testis (Waites and Setchell, 1969).

Temperature receptors in scrotal skin of sheep can elicit responses which tend to lower *whole* body temperature.

The position of the testis relative to the heart, in a region where little muscular activity occurs, makes a rather unusual demand on the mechanism for the return of blood; venous return is promoted by the close contact between arteries and veins. A particular feature of the blood supply to the testis which is more difficult to explain is the absence of a pulsatile flow.

III. EPIDIDYMIS

A. Structure

Three anatomic parts of the epididymis are recognized (Fig. 1–8). The *caput epididymidis* (head), in which a variable number of efferent ductules (6–20) join the duct of the epididymis, forms a flattened structure applied to one pole of the testis. It is continued as the narrow *corpus epididymidis* (body) which terminates at the opposite pole in the expanded *cauda epididymidis* (tail). The contour of the cauda epididymidis is a visible feature in the live animal. The *caput corpus* and cauda epididymidis are less clearly differentiated in the stallion than in other farm species, and in the foal the attachment to the testis is very loose.

The wall of the duct of the epididymis has a prominent layer of circular muscle fibers and a pseudostratified epithelium of columnar cells. Three regions of the duct of the epididymis can be distinguished histologically; these do not coincide with the gross anatomic regions (Nicander, 1957). The initial segment is characterized by a high epithelium with very long straight stereocilia which almost obliterate the lumen. In the middle segment stereocilia are not so straight and the lumen of the duct is wide. In the terminal segment stereocilia are short; the lumen is very wide and packed with spermatozoa.

B. Function

The cytologic features suggest that absorption is an important function of the initial and middle segments but not of the terminal segment. The effects of ligation of the epididymis at different

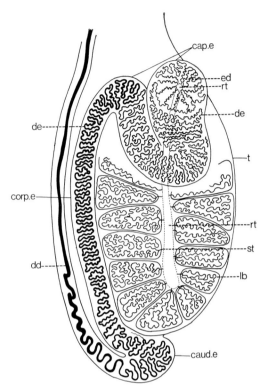

FIG. 1–8. Schematic drawing of the tubular system of the testis and epididymis in the bull (for clarity the duct system of the rete testis is omitted).

cap.e, Caput epididymis; *caud.e*, cauda epididymidis; *corp.e*, corpus epididymidis; *dd*, ductus deferens; *de*, duct of the epididymis; *ed*, efferent ductule; *lb*, lobule with seminiferous tubules; *rt*, rete testis; *st*, straight tubule; *t*, testis. (*Simplified from Blom and Christensen, 1960. Nord. Vet. Med. 12, 453.*)

levels show clearly that active absorption of fluid occurs in the caput epididymidis.

By cannulating the efferent ductules it has been shown that in the ram up to 60 ml fluid leave the testis daily although total ejaculate volume averages about 1.0 ml. Hourly flow rate per 100 gm testis (0.4 ml) is rather less in the bull than in the ram and goat. The epithelium of the efferent ductules is also capable of removing particulate matter including spermatozoa from the lumen.

Secretory activity is a feature of the epithelium of the duct of the epididymis which is suppressed by castration; the secretions may maintain the viability of spermatozoa during storage. The role of the epididymis in the maturation of spermatozoa is less certain. Migration of the cytoplasmic bead from the neck of the spermatozoa to the terminal part of the mid-piece normally occurs in the course of the journey through the epididymis. This change in morphology is associated with important physical and cytochemical changes (Hamilton, 1971) and with an increase in their capacity for motility and their fertilizing ability (Orgebin-Crist, 1969). However, there is evidence from laboratory species that these maturation changes still occur even when spermatozoa are confined to the caput epididymidis by ligatures. Spermatozoa in dilute suspension are transported through the efferent ductules by the action of the ciliated epithelium, supported by contraction of the musculature of the duct wall and by the action of the smooth muscle cells in the tunica albuginea and the myoid cells in the walls of the seminiferous tubules. The time required for the transport of spermatozoa through the epididymis is remarkably constant for individual species. By incorporating a radioactive label into spermatocytes and by sampling different levels of the epididymis at varying intervals it has been found that the duration of the epididymal journey for bull, ram

and boar is 10 days, 13–15 days and 9–12 days respectively (Ortavant et al., 1961; Swierstra, 1968). The epididymis is an important storage organ; although fertilizing capacity of spermatozoa lasts only a few hours outside the body at the temperature of the epididymis, it lasts for several weeks in the isolated epididymis. The life of the spermatozoa in the epididymis is shortened by castration.

The two epididymides of a mature bull can accommodate up to 74.1×10^9 spermatozoa, equal to 3.6 days' production by the testes. Depletion of these reserves occurs following repeated ejaculation but depletion fails to alter significantly the characteristics of the sperm population of the cauda epididymidis. The speed of transport of spermatozoa through the epididymis is altered only slightly by exhaustive depletion.

The ductus deferens leaves the cauda epididymidis to be supported in a separate fold of peritoneum; it is readily separable from the rest of the spermatic cord. It has a thick muscular wall, and its terminal portion is furnished with branched tubular glands. It enters the pelvic urethra at the *colliculus seminalis*. In some species (Table 1–3) this portion forms a distinct ampulla; the ampullae have muscular walls, which expel the

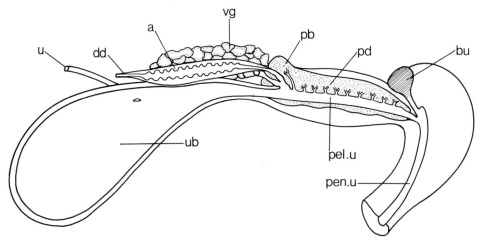

Fig. 1–9. Diagram to show the disposition of the glands which discharge into the pelvic urethra of the bull.

a, Ampulla; *bu*, bulbourethral gland; *dd*, ductus deferens; *pb*, body of prostate gland; *pd*, disseminate part of prostate gland; *pel.u*, pelvic urethra; *pen.u*, penile urethra; *u*, ureter; *ub*, urinary bladder; *vg*, vesicular gland.

Table 1-3. Dimensions and Weights of the Components of the Male Reproductive Tract in Farm Animals*

Organ		Bull	Ram	Boar	Stallion
Testis	Length (cm)	13	10	13	10
	Diameter (cm)	7	6	7	5
	Weight (gm)	350	275	360	200
Epididymis	Length of duct (m)	40	50	18	75
	Weight (gm)	36	—	85	40
Ductus deferens	Length (cm)	102	—	—	70
Ampulla	Length (cm)	15	7	Scattered lobules of gland tissue at termination of the duct	25
	Diameter (cm)	1.2	0.6		2
Vesicular gland	Length (cm)	13	4	13	15
	Breadth (cm)	3	2	7	5
	Thickness (cm)	2	1.5	4	5
	Weight (gm)	75	5	200	—
Prostate gland	Body (cm)	3 × 1 × 1	Scattered lobules of gland tissue	3 × 3 × 1 (20 gm)	Isthmus 2 × 3 × 0.5
	Disseminate part (cm)	12 × 1.5 × 1	—	17 × 1 × 1	Lobe 7 × 4 × 1
Bulbourethral gland	Length (cm)	3	1.5	16	5
	Breadth (cm)	2	1	4	2.5
	Thickness (cm)	1.5	1	4	2.5
	Weight (gm)	6	3	85	—
Penis	Total length (cm)	102	40	55	50
	Length of free end (cm)	9.5	4	18	20
	Urethral process (cm)	0.2	4	Not present	3
Sheath	Length (cm)	30	11	23 (Preputial diverticulum volume about 100 ml)	External 25 / Internal 15

* Average values are given from various sources. It is especially difficult to give figures for the horse because of breed variation in size.

semen from the ductus deferens into the urethra; this process of emission is only one component of the process of ejaculation.

IV. ACCESSORY GLANDS

The prostate, vesicular and bulbourethral glands pour their secretions into the urethra where, at the time of ejaculation, they are mixed with the fluid suspension of spermatozoa and ampullary secretions from the ductus deferens (Fig. 1–9). All the accessory glands are essentially lobular branched tubular glands, with smooth muscle prominent in the interstitial tissue (Plate 2). Anatomic differences between farm species are summarized in Table 1–3 and Figure 1–10.

A. Comparative Anatomy

The Vesicular Glands. These lie lateral to the terminal parts of each ductus deferens. In ruminants and swine they are compact lobulated glands; in the stallion they are true vesicles consisting of large pyriform glandular sacs. The duct of the vesicular gland and the ductus deferens may have a common ejaculatory orifice into the urethra.

Prostate Gland. Two components are distinguished. There is a distinct lobulated external part or body which lies outside the thick urethral muscle surrounding the urethra and a second internal or disseminate part distributed along the length of the pelvic urethra below the urethral muscle. The body of the prostate is small in the bull and large in the boar, while in the ram no body is visible. In the stallion the prostate gland is wholly external and consists of two lateral lobes joined by an isthmus.

The Bulbourethral Glands. These are paired bodies lying dorsal to the urethra near the termination of its pelvic portion. In the bull they are almost hidden by the *bulbospongiosus* muscle and in all species are covered by a thick layer of striated muscle. They are especially large in the boar and contribute the prominent gel-like component of boar semen.

Urethral Glands. The bull lacks urethral glands comparable with those found in man: earlier conclusions about their contribution to bull semen need to be re-examined (Kainer et al., 1969). Glands of this name in the horse have been considered comparable with the disseminate prostate of ruminants, but in the boar the disseminate prostate and the urethral glands are histologically distinct (McKenzie et al., 1938).

B. Function

Apart from providing a liquid vehicle for the transport of spermatozoa, the function of the accessory glands is obscure although much is known of the specific chemical agents contributed by the glands to the ejaculate (Mann, 1964).

These agents serve as markers of the contribution made to the semen by individual glands and as indicators of gland function. Fructose and citric acid are important components of vesicular glands of domestic ruminants. Citric acid alone is found in stallion vesicular gland: boar vesicular gland also contains little fructose and is characterized by a high content of ergothionine and inositol. Glyceryl phosphoryl choline is a distinctive component of the epididymal secretion. Ergothionine is found in the ampullary glands of horse and jackass.

In all species studied, spermatozoa from the cauda epididymidis are capable of fertilization when inseminated without the addition of accessory gland secretions. The gel-like fraction of the boar ejaculate forms a plug in the vagina of mated females but in commercial insemination practice this fraction is removed from the semen by filtration.

In large animals it is possible to palpate per rectum some of the above structures. The positions of the accessory glands relative to the bony pelvis are shown in Figure 1–10. The vesicular

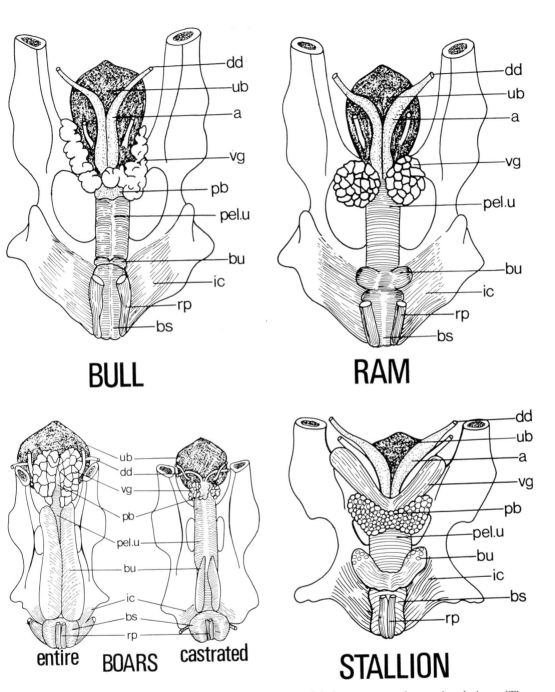

BULL

RAM

entire   BOARS   castrated

STALLION

Fig. 1–10.   Diagrams of the pelvic genitalia, within the pelvic bones, as seen from a dorsal view.   (The cranial parts of the ilium have been removed.)

a, Ampulla; *bs,* bulbospongiosus muscle; *bu,* bulbourethral gland; *dd,* ductus deferens; *ic,* ischiocavernosus muscle; *pb,* body of prostate gland; *pel.u,* pelvic urethra; *rp,* retractor penis muscle; *ub,* urinary bladder; *vg,* vesicular gland.   (*Diagrams of bull, boar and stallion redrawn from Nickel, 1954. Tierärztl. Umschau 9, 386.*)

glands of the bull are readily detectable lying one on each side of the ampullae and the body of the prostate gland is detectable as a hard smooth prominence caudal to the neck of the bladder. The bulbourethral glands, because of their covering of muscle, are not identifiable in the bull.

Rectal palpation of the bulbourethral glands of the boar can be used to identify abdominal cryptorchids; the glands are much smaller in castrated males than in cryptorchid males (Fig. 1–10). Pathologic changes occur in the accessory glands of castrated male sheep grazing certain species of clover. This is due to estrogenic components of the clover which induce feminizing changes in the glandular epithelium.

V. PENIS AND SHEATH

A. Structure of the Penis

In the mammalian penis, three cavernous bodies are aggregated around the penile urethra (Fig. 1–11). The *corpus spongiosum penis* which surrounds the urethra is enlarged at the ischial arch to form the penile bulb. This bulb is covered by the striated bulbospongiosus muscle. The *corpus cavernosum penis* arises as a pair of *crura* from the ischial arch under cover of the striated *ischiocavernosus* muscle. The corpus cavernosum penis continues to the apex of the penis as a more or less paired dorsal cavernous body. A thick collagenous covering (tunica albuginea) covers the cavernous bodies and from it numerous trabeculae enter the corpus

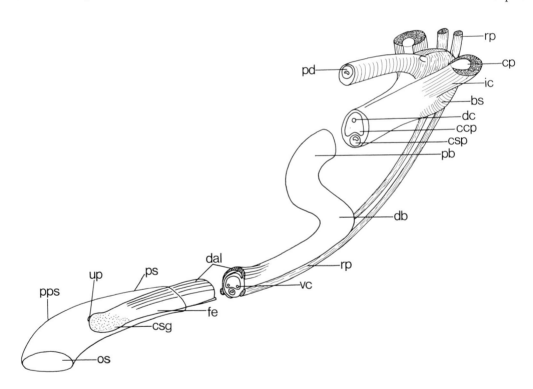

FIG. 1–11.   Diagram to show the anatomy of penis and sheath in the bull.

bs, Bulbospongiosus muscle; *ccp*, corpus cavernosum penis; *csp*, corpus spongiosum penis, surrounding the penile urethra; *cp*, left crus penis; *csg*, corpus spongiosum glandis (a thin mantle of cavernous tissue covers the fibrous tissue of the enlargement at the tip of the penis); *dal*, dorsal apical ligament; *dc*, dorsal erection canal; *db*, distal bend of the sigmoid flexure; *fe*, free end of the penis, covered by penile integument; *ic*, ischiocavernosus muscle; *os*, orifice of the sheath; *pd*, disseminate part of the prostate gland; *pps*, prepenile sheath; *ps*, penile sheath; *pb*, proximal bend of sigmoid flexure; *rp*, left retractor penis muscle; *up*, urethral process; *vc*, left ventrolateral erection canal.

cavernosum penis to support its cavernous tissue. A pair of smooth *retractor penis* muscles arise from the sacral or coccygeal regions of the vertebral column and are especially large in ruminants and swine. In these animals the retractor penis muscles are able to control the size of the sigmoid flexure.

In the stallion the cavernous bodies contain large cavernous spaces; during erection considerable increases in size result from accumulation of blood in these spaces. In bull, ram and boar, the cavernous spaces of the corpus cavernosum penis are small, except in the crura and at the distal bend of the sigmoid flexure. The cavernous spaces of the corpus spongiosum penis are large, but distension is limited by the tunica albuginea. In ruminants and swine, erection results from the inflow of a relatively small volume of blood.

The subcutaneous tissues of the free end of the penis in some species form a well-developed cavernous body, the *corpus spongiosum glandis*. It is poorly developed in the bull and indistinct in the boar. Figure 1–12 illustrates the features of the free end of the penis of bull, ram, boar and stallion.

B. Structure of the Sheath

In ruminants and swine the orifice of the sheath is controlled by a special striated muscle (cranial muscle of the sheath); a second (caudal) muscle may also be present. The sheath can be divided into penile and prepenile parts. The epithelia of the penile part of the sheath and the penile integument are derived from a single ectodermal lamella; during postnatal development this lamella is split by keratinization and the free end of the penis is liberated from the penile sheath (Plate 1, Table 1–1). In the boar, the prepenile sheath consists of a short vestibule with a narrow external orifice and there is a large dorsal diverticulum in which urine and epithelial debris accumulate. The penile (inner) sheath of the stallion is enclosed in a voluminous

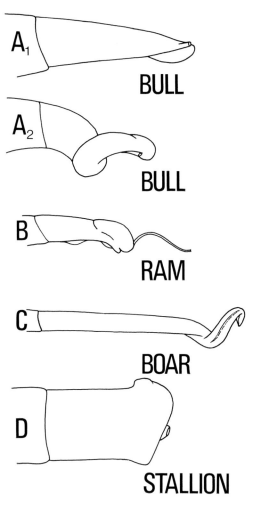

Fig. 1–12. Diagrams to show the shape of the free end of the penis. A_1 shows the shape of the penis just before intromission and A_2 shows the shape after intromission when spiral deviation has occurred. *B* shows the shape of the penis during natural service. *C* does not show the full degree of spiralling that occurs during service. *D* was drawn after injection and shows enlargement of the erectile bodies. (A_1, A_2 *and B from photographs. C and D from fixed specimens. Not drawn to scale.*)

prepenile (outer) sheath. Eversion of the lining of the sheath can expose the epithelium to injury and infection; in cattle of European origin, it occurs most commonly in polled beef breeds (Long, 1969) and prolapse of the sheath is an important condition in some breeds of Asian origin.

C. Erection and Protrusion

Sexual stimulation produces marked dilation of the arteries supplying the cavernous bodies of the penis (especially the crura) (Gilanpour, 1972). Stiffening of the penis in ruminants is mainly brought about by the ischiocavernosus muscle, which pumps blood from the cavernous spaces of the crura into the rest of the corpus cavernosum penis by way of special longitudinal cavernous spaces (erection canals) (Watson, 1964). Pressures of up to 7000 mm Hg have been recorded in the cavernous spaces during natural service (Beckett et al., 1972). The *bulbocavernosus* muscle simultaneously pumps blood from the penile bulb into the rest of the corpus spongiosum penis (Gilanpour, 1972). Protrusion of the penis results from several causes. Rising pressure in the cavernous spaces of the corpus cavernosum penis, particularly the large cavernous spaces in the distal bend of the sigmoid flexure (Ashdown, 1970), eliminates the sigmoid flexure and this is facilitated by relaxation of the retractor penis muscle. When the penis of the bull is protruded, penile and prepenile parts of the sheath are everted over the protruded organ. The spiral arrangement of the fibrous architecture of the penile integument causes the penis to spiral when the integument is stretched; the urethral orifice turns in a counter-clockwise direction through about 300° as ejaculation occurs (Ashdown and Smith, 1969; Seidel and Foote, 1969). Intromission in the bull lasts for about two seconds and straightening of the penis after withdrawal often occurs abruptly as the dorsal apical ligament reasserts its action in keeping the penis straight. Withdrawal into the sheath follows as the pressure in the cavernous spaces subsides. The fibrous architecture of the corpus cavernosum penis in the region of the sigmoid flexure tends to re-form the flexure and is assisted in this by shortening of the retractor penis muscle. The events in the ram are probably similar to those in the bull. Spiral deviation of the penis is a pronounced feature of the boar's penis during copulation; intromission lasts for up to seven minutes, during which time a large volume of semen is ejaculated. Spiral deviation does not occur in the horse and intromission lasts for several minutes.

D. Emission and Ejaculation

Understanding of the physiology of ejaculation is still imperfect. Emission of semen from the ductus deferens is controlled by sympathetic fibers of the hypogastric nerves. The stored secretions of some if not all of the accessory glands are probably released by muscular contraction controlled by autonomic nerves; secretory activity of the glandular epithelium may be under similar control. Ejaculation involves contraction of striated muscles which are innervated by sacral nerves. One of these, the bulbospongiosus muscle, compresses the penile bulb during ejaculation and the resultant waves of pressure within the penile urethra may help to transport the ejaculate (Watson, 1964). The fully integrated sequence of erection, emission and ejaculation is under complex nervous control; electrical stimulation of ejaculation in farm animals is a very crude imitation of the natural mechanisms. Reflux of semen into the bladder has been shown radiographically to occur in rams during electrical stimulation (Hovell et al., 1969). Although spermatozoa can be recovered from the urine of sexually rested rams, retro-ejaculation is probably normally limited by the action of the urethral muscle.

VI. OTHER SPECIES

Anatomic features of the male reproductive tracts have been described for the camel (Elwishy et al., 1972), the llama (Casas Peréz et al., 1967), the buffalo (Joshi et al., 1967), the goat (Yao and Eaton, 1954) and the jackass (Tagand and Barone, 1956).

REFERENCES

Abdel-Raouf, M. (1960). The postnatal development of the reproductive organs in bulls with especial reference to puberty. *Acta Endocr. Copnh.* Suppl. 49, *34*, 1–109.

Amann, R. P. (1970). "Sperm Production Rates." In *The Testis.* Vol. I. A. D. Johnson, W. R. Gomes and N. L. VanDemark (eds.), New York, Academic Press.

Ashdown, R. R. and Smith, J. A. (1969). The anatomy of the corpus cavernosum penis of the bull and its relationship to spiral deviation of the penis. *J. Anat. 104*, 153–159.

Ashdown, R. R. (1970). Angioarchitecture of the sigmoid flexure of the bovine corpus cavernosum penis and its significance in erection. *J. Anat. 106*, 403–404.

Beckett, S. D., Hudson, R. S., Walker, D. F., Vachon, R. I. and Reynolds, T. M. (1972). Corpus cavernosum penis pressure and external penile muscle activity during erection in the goat. *Biol. Reprod. 7*, 359–364.

Casas Peréz, J. H., San Martin, M. and Copaira, M. (1967). Histology of the testis in the alpaca (*Lama pacos*). *Revta Fac. Méd. Vét. Univ. Nac. Lima 18–20* (1963/1966), 223–238.

Donovan, B. T. and van der Werff ten Bosch, J. J. (1965). *Physiology of Puberty.* London, Arnold.

Dym, M. and Fawcett, D. W. (1970). The blood testis barrier in the rat and the physiological compartmentation of the seminiferous epithelium. *Biol. Reprod. 3*, 308–326.

Elwishy, A. B., Mobarak, A. M. and Fouad, S. M. (1972). The accessory genital organs of the one humped male camel (*Camelus dromedarius*). *Anat. Anz. 131*, 1–12.

Gier, H. T. and Marion, G. B. (1970). "Development of the Mammalian Testis." In *The Testis.* Vol. I. A. D. Johnson, W. R. Gomes and N. L. VanDemark (eds.), New York, Academic Press.

Gilanpour, H. (1972). *Angioarchitecture and Functional Anatomy of the Penis in Ruminants.* Ph.D. Thesis, University of London.

Hamilton, D. W. (1971). "The Mammalian Epididymis." In *Reproductive Biology.* H. Balm and S. Glasser (eds.), Basel, S. Karger, Excerpta Med. Fdn.

Hodson, N. P. (1970). "The Nerves of the Testis, Epididymis and Scrotum." In *The Testis.* Vol. I. A. D. Johnson, W. R. Gomes and N. L. Van-Demark (eds.), New York, Academic Press.

Hovell, G. J. R., Ardran, G. M., Essenhigh, D. M. and Smith, J. C. (1969). Radiological observations on electrically induced ejaculation in the ram. *J. Reprod. Fert. 20*, 383–388.

Joshi, N. H., Luktuke, S. N. and Chatterjee, S. N. (1967). Studies on the biometry of the reproductive tract and some endocrine glands of the buffalo male. *Indian Vet. J. 44*, 137–145.

Kainer, R. A., Faulkner, L. C. and Abdel-Raouf, M. (1969). Glands associated with the urethra of the bull. *Am. J. Vet. Res. 30*, 963–974.

Larson, L. L. and Kitchell, R. L. (1958). Neural mechanisms in sexual behavior. *Am. J. Vet. Res. 19*, 853–865.

Long, S. E. (1969). Eversion of the preputial epithelium in bulls at artificial insemination centres. *Vet. Rec. 84*, 495–499.

Mann, T. (1964). *Biochemistry of Semen and of the Male Reproductive Tract.* London, Methuen.

McKenzie, F. F., Miller, J. C. and Bauguess, L. C. (1938). The reproductive organs and semen of the boar. *Res. Bull. Mo. Agric. Exp. Sta. No. 279.*

Nicander, L. (1957). Studies on the regional histology and cytochemistry of the ductus epididymidis in stallions, rams and bulls. *Act. Morph. Neerl.-Scand. 1*, 337–362.

Nickel, R., Schummer, A. and Seiferle, E. (1973). *The Viscera of the Domestic Mammals* translated and revised by W. O. Sack, Berlin, Parey.

Nishikawa, Y. (1959). *Studies on Reproduction in Horses.* Tokyo, Japan Racing Association.

Orgebin-Crist, M.-C. (1969). Studies on the function of the epididymis. *Biol. Reprod.* Suppl. 1, *1*:155-175.

Ortavant, R., Orgebin, M. C. and Singh, G. (1961). "Étude Comparative de la Durée des Phénomènes Spermatogénétiques chez les Animaux Domestiques." In *The Use of Radioisotopes in Animal Biology and the Medical Sciences* (symposium) Mexico City. London and New York, Academic Press.

Seidel, G. E. Jr. and Foote, R. H. (1969). Motion picture analysis of ejaculation in the bull. *J. Reprod. Fert. 20*, 313–317.

Setchell, B. P. (1970). "Testicular Blood Supply, Lymphatic Drainage and the Secretion of Fluid." In *The Testis.* Vol. I. A. D. Johnson, W. R. Gomes and N. L. VanDemark (eds.), New York, Academic Press.

Skinner, J. D. and Rowson, L. E. A. (1968). Puberty in Suffolk and cross-bred rams. *J. Reprod. Fert. 16*, 479–488.

Swierstra, E. E. (1968). Cytology and duration of the cycle of the seminiferous epithelium of the boar. Duration of spermatozoan transit through the epididymis. *Anat. Rec. 161*, 171–185.

Tagand, R. and Barone, R. (1956). *Anatomie des Équidés Domestiques.* 2, iii. Lyons, École Nat. Vet.

Watson, J. W. (1964). Mechanisms of erection and ejaculation in the bull and ram. *Nature* (Lond.) *204*, 95–96.

Waites, G. M. H. and Setchell, B. P. (1969). "Physiology of the Testis, Epididymis and Scrotum." In *Advances in Reproductive Physiology.* Vol. 4. A. McLaren (ed.), London, Logos, pp. 1–21.

Yao, T. S. and Eaton, O. N. (1954). Postnatal growth and histological development of reproductive organs in male goats. *Am. J. Anat. 95*, 401–431.

Chapter 2

Functional Anatomy of Female Reproduction

E. S. E. Hafez

The female reproductive organs are composed of ovaries, oviducts, uterus, cervix uteri, vagina and external genitalia. The internal genital organs (the first of four components) are supported by the *broad ligament*. This ligament consists of the *mesovarium* which supports the ovary; the *mesosalpinx*, which supports the oviduct; and the *mesometrium*, which supports the uterus. In cattle and sheep, the attachment of the broad ligament is dorsolateral, in the region of the ilium, resulting in the uterus being arranged like a ram's horns with the convexity dorsal and the ovaries located near the pelvis (Fig. 2–1). The ovary, oviduct

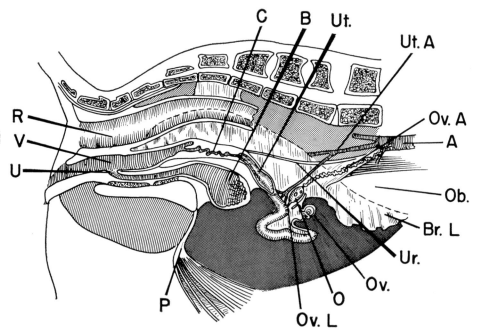

Fig. 2–1. Sagittal section through pelvic region (view of left side) showing attachments of rectum and urogenital tract, the pouches of the pelvic peritoneum, and attachments of the abdominal muscles to the prepubic tendon in the ewe.

A, Aorta; *B*, bladder; *Br.L*, broad ligament of uterus; *C*, cervix; *O*, ovary; *Ob.*, internal oblique muscle; *Ov.*, oviduct; *Ov. A*, ovarian artery; *Ov. L*, ovarian ligament; *P*, prepubic tendon; *R*, rectum; *U*, urethra; *Ur.*, ureter; *Ut. A*, uterine artery; *V*, vagina. (*Redrawn from Bassett, 1965. Aust. J. Zool. 13, 201.*)

and uterus are supplied primarily by autonomic nerves. The pudic nerve supplies sensory fibers and parasympathetic fibers to the vagina-vulva and clitoris.

The following discussion deals with the embryology, morphology, anatomy, physiology and biochemistry of the female reproductive organs in farm mammals.

I. EMBRYOLOGY

The rudimentary reproductive system of the mammal consists of two sexually undifferentiated gonads, two pairs of ducts, a urogenital sinus, a genital tubercle and vestibular folds (Fig. 2–2). It arises primarily from two germinal ridges on the dorsal side of the abdominal cavity and can potentially differentiate into a male or female system (this is

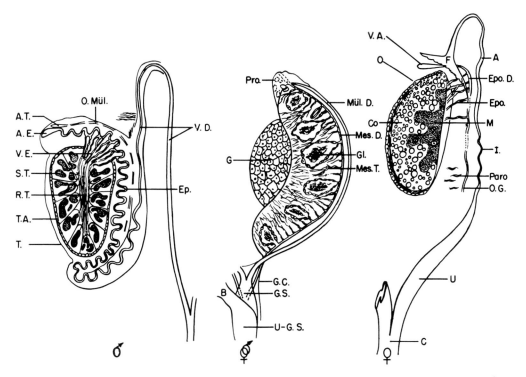

FIG. 2–2. Diagram representing embryonic differentiation of male and female genital systems. *Center:* the undifferentiated system with its large mesonephros, mesonephric duct, Müllerian duct and undifferentiated gonad. Note that the Müllerian and mesonephric ducts cross before they enter the genital cord. *Right:* the female system, in which the ovary and Müllerian ducts differentiate while the remnants of the mesonephros and mesonephric ducts atrophy into the epoophoron, paroophoron and Gartner's duct. *Left:* the male system in which the testes and mesonephric (Wolffian) ducts differentiate; the sole remnants of the Müllerian ducts are the testicular appendix and prostatic utricle (vagina masculinus). (*Moustafa and Hafez, unpublished data, 1972.*)

A Ampulla	*G. C.* Genital cord	*Paro* Paroophoron
A.E. Appendage of epididymis	*Gl.* Glomerulus	*Pro.* Pronephros
A.T. Appendage of testis	*G.S.* Genital sinus	*R.T.* Rete tubules
B Bladder	*I.* Isthmus	*S.T.* Seminiferous tubules
C Cervix	*M* Ovarian medulla	*T.* Testis
Co Ovarian cortex	*Mes. D.* Mesonephric duct	*T.A.* Testicle artery
Ep. Epididymis	*Mes. T.* Mesonephric tubules	*U* Uterus
Epo. Epoophoron	*Mul. D.* Müllerian duct	*U-G.S* Urogenital sinus
Epo. D. Duct of epoophoron	*O* Ovary	*V.A.* Vesicular appendage
F Fimbriae	*O.G.* Obliterated Gartner's duct	*V.D.* Vas deferens
G Gonad (undifferentiated)	*O. Mul.* Obliterated Müllerian duct	*V.E.* Vasa efferentia

Table 2-1. Developmental Fate of the Sexual Rudiments in the Male and the Female Mammalian Fetus*

Sexual Rudiment	Male	Female
Gonad		
Cortex	Regresses	Ovary
Medulla	Testis	Regresses
Müllerian ducts	Vestiges	Uterus, oviducts, parts of vagina
Wolffian ducts	Epididymis, vas deferens	Vestiges
Urogenital sinus	Urethra, prostate Bulbourethral glands	Part of vagina, urethra
Genital tubercle (phallus)	Penis	Clitoris
Vestibular folds	Scrotum	Labia

* From Frye, 1967. *Hormonal Control in Vertebrates.* New York, Macmillan.

referred to as *embryonic bisexuality*). The developmental fate of the sexual rudiments in the male and female fetus is shown in Table 2–1.

The sex of an animal depends on three distinct but related factors: (*a*) inherited genes; (*b*) gonadogenesis; and (*c*) the formation and maturation of accessory reproductive organs. The sex of an embryo is basically determined by its genes, but the expression of the genetic sex is a developmental process depending on the function of the fetal gonads and, occasionally, the adrenal cortex.

Estrogen and androgen cause sex reversal in male and female embryos, respectively, during only a brief period early in sexual differentiation. In contrast, the accessory reproductive organs remain sexually labile for a much longer time and hormone treatment late in development can induce considerable reversal (Frye, 1967). The age at which this bisexual potential is completely lost varies with the species.

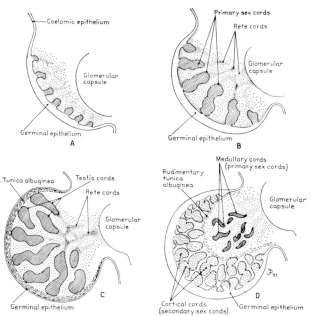

Fig. 2–3. Differentiation of the undifferentiated gonad of higher vertebrates into testis and ovary.
A, The primary sex cords arise from the germinal epithelium. *B*, Primary sex cords have developed, but the gonad is still undifferentiated. *C*, Differentiation of the testis is taking place: the primary sex cords continue to proliferate, while the germinal epithelium diminishes in size; the tunica albuginea also develops. *D*, Differentiation into an ovary involves development of secondary sex cords from the cortex and reduction of the primary sex cords and tunica albuginea. (*Burns, 1961. In Sex and Internal Secretions. Vol. 1. 3rd ed. W. C. Young [ed.], Baltimore, Williams & Wilkins.*)

Gonads. The gonads form from a group of large granulated yolk sac cells which invade the germinal ridges. Two invasions occur in the female. The initial one is abortive, but the second results in the formation of sex cords which later break up into primordial germ cells (oogonia). The sex cords of the female are called *medullary cords;* those of the male are the *seminiferous tubules.*

The testis develops predominantly from the medulla of the sexually undifferentiated gonad, whereas the ovary arises primarily from its cortex. The primordial germ cells congregate in the developing gonad and proliferate the oogonia of the female. They and the secondary sex cords which are forming concurrently move into the cortex (Fig. 2–3) while the primary sex cords and the medulla regress in size.

Reproductive Ducts. Wolffian and Müllerian ducts are both present in the sexually undifferentiated embryo. In the female, the Müllerian ducts develop into a gonaduct system and the Wolffian ducts atrophy. The opposite is true for the male. The female Müllerian ducts fuse caudally to form a uterus, a cervix and the anterior part of a vagina. The oviduct becomes coiled and acquires differentiated epithelia and fimbriae just before birth. Signs of epithelial differentiation include increased mucosal height, pseudostratification, the presence of peg cells and evidence of secretory activity. The onset of these developmental events varies with the species.

In the male fetus, testicular androgen plays a role in the persistence and development of the Wolffian ducts and the atrophy of the Müllerian ducts. However, the growth of the female Müllerian ducts beyond the ambisexual stage is apparently hormonally independent and the duct is capable of considerable autonomous growth, coiling and epithelial differentiation. Adrenal hormones are thought to stimulate oviductal growth in the fetal guinea pig (Price et al., 1969).

Urogenital Sinus. The urogenital sinus gives rise to the vestibule. The folds of skin which border the sinus form the lips of the vulva. The female phallus or clitoris, homologous to the male penis, grows little in size.

II. THE OVARY

The ovary, unlike the testis, remains in the abdominal cavity. It performs both an exocrine (egg release) and endocrine (estrogen and progesterone secretion) function.

The shape and size vary both with the species and the stage of the estrous cycle (Table 2–2). In cattle and sheep the ovary is almond-shaped; whereas in the horse it is bean-shaped due to the presence of a definite ovulation fossa, an indentation in the attached border of the ovary. In swine the ovary resembles a cluster of grapes, the markedly protruding follicles and corpora lutea obscuring the underlying ovarian tissue.

The part of the ovary which is not attached to the mesovarium is exposed and bulges into the abdominal cavity. It is on this surface that ovarian follicles protrude, as found in the case of cattle, sheep and swine.

The ovary is composed of the medulla and cortex; it is surrounded by the germinal epithelium and in general increases four to seven times the birth weight by the onset of puberty.

The ovarian medulla consists of irregularly arranged fibroelastic connective tissue and an extensive nerve and blood vessel system which enters the ovary by way of the *hilus* (the attachment between the ovary and mesovarium). The arteries are arranged in a definitive spiral shape. The ovarian cortex contains the ovarian follicles, their precursors and end products. Furthermore, it is the site of both egg formation and hormone production. Thus ovaries may possess different component structures (ovarian follicles or corpora lutea) at various stages of development or regression.

The connective tissue of the cortex contains many fibroblasts, some collagen and reticular fibers, blood vessels, lymphatic vessels, nerves and smooth muscle fibers.

Table 2-2. Comparative Anatomy of the Ovary in the
Adult Female of Farm Mammals

Organ	Animal			
	Cow	Ewe	Sow	Mare
Ovary				
Shape	Almond-shaped	Almond-shaped	Berry-shaped (cluster of grapes)	Kidney-shaped; with ovulation fossa
Weight of one ovary (gm)	10–20	3–4	3–7	40–80
Mature graafian follicles				
Number	1–2	1–4	10–25	1–2
Diameter of follicle (mm)	12–19	5–10	8–12	25–70
Diameter of egg without zona pellucida (μ)	120–160	140–185	120–170	120–180
Mature corpus luteum				
Shape	Spheroid or ovoid	Spheroid or ovoid	Spheroid or ovoid	Pear-shaped
Diameter (mm)	20–25	9	10–15	10–25
Maximum size attained (days from ovulation)	10	7–9	14	14
Regression starts (days from ovulation)	14–15	12–14	13	17

The measurements included in this table vary with age, breed, parity, plane of nutrition and reproductive cycle.

The connective tissue cells near the surface are arranged roughly parallel to the ovarian surface and are somewhat denser than the cells lying toward the medulla. This dense layer is known as the tunica albuginea. On the surface of the ovary is a layer of flattened cells known as the *germinal epithelium*. Although in the newborn foal the surface of the ovary is covered by these cells, the ventral border of the organ becomes concave shortly afterwards and the germinal tissue sinks below the surface to form the *ovulation fossa*.

A. Development of Ovarian Follicles

Unlike the male in which the germinal epithelium is located deep within the seminiferous tubules, this tissue lies on the surface of the ovary. For many years, development of this tissue has been a center of controversy and has not yet been adequately resolved. The view generally held is that, with few exceptions, primordial follicles are not formed during postnatal life (Brambell, 1952; Zuckerman, 1962).

The ovary is a dynamic organ in which primordial follicles are constantly developing (Fig. 2–4). The development of ovarian follicles is often classified for convenience by histologic criteria, e.g. (a) size; (b) number of layers of granulosa cells; (c) development of the theca; and (d) position of the oocyte within its surrounding cumulus oophorus (Fig. 2–5). Primordial follicles are oocytes (a potential egg) surrounded by one layer of flattened cells. These cells increase in number by mitosis and become cuboidal; the follicle is then called "secondary follicle." Vesicular follicles are then formed by accumulation of fluids in spaces (antra) between the epithelial cells. With further division of granulosa cells, the *membrana granulosa* has two, three and then four layers; blood capillaries invade the fibrous layer of cells surrounding the follicle and form a vascular layer, the *theca interna*. The granulosa cells within the wall of the graafian follicle are deprived of a blood supply by the basement membrane (Fig. 2–6).

With the progressive development of the cortical blood vessels in the neighborhood of the growing follicle and the formation of the two layers of the theca, a basketlike vascular meshwork develops around the follicle, particularly in the theca interna.

Steroids Secreted by Follicle. The ovaries of farm animals lack the relatively large amounts of steroid-secreting interstitial tissue so prominent in the ovaries of rodents and rabbits. The theca cells of farm animals appear to differentiate almost completely during the atretic processes, and thus these animals lack an important source of certain steroids present in those animals that have large amounts of interstitial tissue (Hansel et al., 1973). All cell types in the ovary have the capacity to make steroid hormones. The steroid hormones secreted by a particular cell type are determined by the stage of the estrous cycle. However, estrogens are largely secreted by the theca cells of the follicle. Some progesterone produced by the granulosa cells may be used by the theca cells for the synthesis of androgens and estrogens

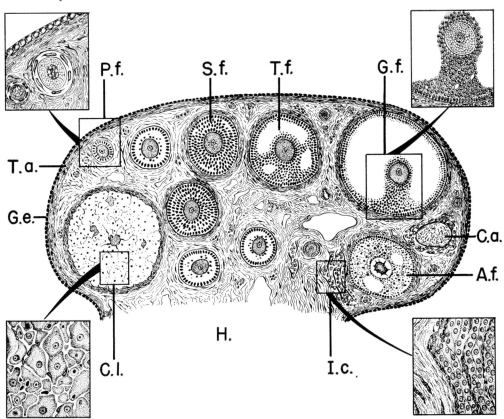

Fig. 2–4. A composite diagram of the mammalian ovary. Progressive stages in the differentiation of a graafian follicle are indicated (*upper left to upper right*). The mature follicle may become atretic (*lower right*) or ovulate and form a corpus luteum (*lower left*).

A.f., Atretic follicle; *C.a.,* corpus albicans; *C.l.,* corpus luteum; *G.e.,* germinal epithelium; *G.f.,* graafian follicle; *H,* hilus; *I.c.,* interstitial cells; *P.f.,* primary follicle; *S.f.,* secondary follicle; *T.a.,* tunica albuginea; *T.f.,* tertiary follicle. (*Partly adapted from Turner, 1948. General Endocrinology. Philadelphia, courtesy of W. B. Saunders Co.*)

Surface

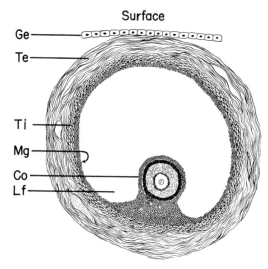

FIG. 2–5.   Illustration of a graafian follicle. *Co*, Cumulus oophorus; *Ge*, germinal epithelium; *Lf*, liquor folliculi; *Mg*, membrana granulosa; *Te*, theca externa; *Ti*, theca interna.

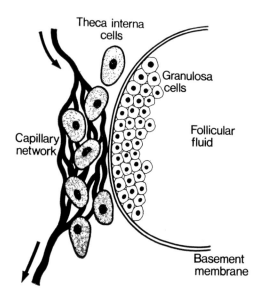

FIG. 2–6.   The structure of the wall of the graafian follicle showing how the granulosa cells are deprived of a blood supply by the basement membrane. (*Baird, 1972. In Reproduction of Mammals. C. R. Austin and R. V. Short [eds.], Cambridge, Cambridge University Press.*)

(Baker, 1972).   Following ovulation, the luteinized granulosa cells become vascularized and secrete large amounts of progesterone into the ovarian vein.

Number of Follicles to Ripen.   The number of graafian follicles which develop per estrous cycle depends on hereditary and environmental factors.   In cattle and horses, one follicle usually develops more rapidly than the others, so that at each estrus only one egg is released.   The remaining follicles then regress and become atrophied.   In swine 10–25 follicles ripen at each estrus.   In sheep, one to three follicles may reach maturity depending on the breed, age and stage of sexual season.

Hormonal Mechanism.   Both the rate of development of follicles and the number to ripen per estrus depend on the pituitary gonadotropins.   In the adult female, a limited amount is available to the ovary.   During each estrous cycle as a result of the release of sufficient FSH by the pituitary gland, a crop of "growing" follicles is stimulated to undergo further growth and maturation.

Many follicles grow during the first stages of the estrous cycle, but few mature completely.   Perhaps less FSH is required to initiate the growth of small follicles than to maintain larger follicles and bring them to ovulatory size (Fig. 2–7), since the number of follicles which mature can be greatly increased (superovulation) by injecting animals with large doses of gonadotropins.   Why then do so many follicles begin to grow?   They may supply substances, like estrogen, which are essential for the ovulation of other larger follicles.

B. Rupture of Follicles (Ovulation)

At ovulation, the egg is released while embedded in a solid mass of follicular cells, the *cumulus oophorus*, which protrudes into the fluid filled antrum.   The cumulus oophorus is usually attached to the granulosa cells opposite the side that will eventually rupture.

The rupture per se occurs at the apex of the follicle. The outermost layer is the first to part. The inner layers protrude through the gap to form a papilla or *stigma*. The stigma thins out, bulges on the surface of the ovary, and becomes completely avascular. The bulging stigma soon ruptures releasing some of the thin follicular fluid. After a short interval the egg mass moves toward the opening becoming elongated as it progresses. More fluid moves through the opening carrying the egg, the connections of which with the cumulus oophorus are dissociated during the latter stages of follicular development into the peritoneal cavity. It is then picked up by the fimbriated opening of the oviduct.

The length of time required to complete ovulation depends upon the location of the egg within the follicle. This time is shorter when the egg is at the base of the follicle than when the egg lies in close proximity to the bulging stigma. After ovulation is completed, the follicle collapses.

Follicular development in the mare progresses toward the ovulation fossa.

Thus ovulation is followed by fairly extensive hemorrhaging in the cavity. Hemorrhage is also noted in the cow, sow and only rarely in the ewe.

Many views have been advanced to explain why the follicle wall ruptures, but the precise mechanism involved remains uncertain. For example, it may be facilitated by muscular contraction of the fibers encircling the follicle. It apparently is not dependent on any ultimate size or internal pressure of the follicular fluid.

C. Formation of Eggs

An egg (ovum) is a highly differentiated cell that is capable of being fertilized and subsequently undergoes embryonic development. Mammalian eggs were first recognized by de Graaf in 1672 and then described and identified by Cruickshank in 1797 and von Baer in 1827. The following discussion is concerned with the formation, structure, ultrastructure, biochemistry, transport and manipulation of mammalian eggs.

The precursor cells of either male or

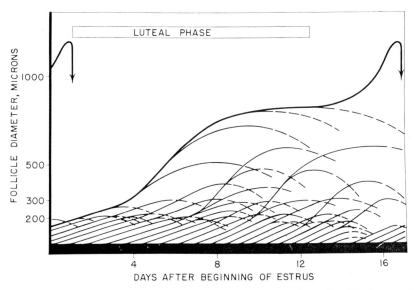

Fig. 2–7. A schematic representation of the follicular cycle in the guinea pig. The heavy solid line indicates the average diameter of the largest follicles. Ovulation occurs at the arrow. The other solid and broken lines represent the concomitant growth and atresia, respectively, of other groups of follicles that do not normally ovulate. (*From Evertt, 1961. In Sex and Internal Secretions. Vol. I. 3rd ed. W. C. Young [ed.], Baltimore, Williams & Wilkins.*)

female gametes, called *gonocytes*, originate probably from extra-embryonic endodermal tissue (extragonadally). They migrate to the presumptive intra-embryonic gonadal zone where they differentiate either into oogonia or spermatogonia. In the female fetus, the germinal epithelium forms into clusters in which one gonocyte differentiates into an *oogonium* containing typical cell constituents, e.g. Golgi apparatus, mitochondria, nucleus and one or more nucleoli. The oogonia then undergo proliferation prior to or shortly after birth resulting in the fetal ovaries containing the sole reservoir of all future ova called oocytes.

The growth of the oocyte is characterized by: (*a*) the enlargement of the cytoplasm by accumulation of different sizes of granules of deutoplasm (yolk); (*b*) the development of an egg membrane (zona pellucida); and (*c*) the mitotic

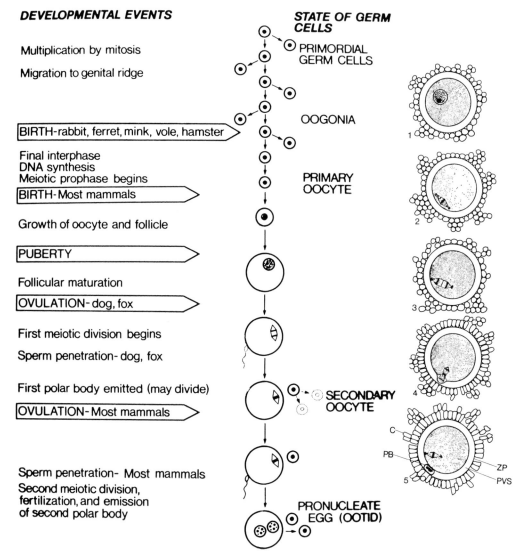

Fig. 2–8. *Left,* Diagrammatic representation of development events in the life cycle of the female germ cell. *Right,* Preovulatory maturation of the egg. (*Baker, 1972. In Reproduction in Mammals. C. R. Austin and R. V. Short* [eds.], *Cambridge, Cambridge University Press.*)

proliferation of follicular epithelium and adjacent tissue. These follicular cells may serve as nurse cells by providing the deutoplasm of the oocyte. By maturity, the egg has accumulated reserves of material to provide an energy source for subsequent development. Factors determining which of the ovarian oocytes are destined to begin their growth or to complete their growth during a reproductive cycle are unknown.

There are two stages in the growth of the oocyte (Fig. 2–8). During the first phase, growth is rapid and intimately associated with the development of the ovarian follicle. Attainment of its mature size occurs at about the time antrum formation begins in the follicle. During the second phase, the oocyte does not grow in size, while the ovarian follicle responding to pituitary hormones increases very rapidly in diameter. This growth is in fact confined primarily to follicles in which the egg has attained its full dimensions.

During the latter phase of follicular growth, the oocyte undergoes maturation. The nucleus which had entered into the prophase of the meiotic division during the growth of the oocyte prepares to undergo reductive divisions. The nucleoli and the nuclear membrane disappear and the chromosomes condense in a compact form. The centrosome (specialized area of dense cytoplasm) then divides into two centrioles around which *asters* appear (groups of radiations at both poles of the oocyte). These asters move apart and a *spindle* is formed between them. The chromosomes in diploid pairs are set free in the cytoplasm and become arranged on the equatorial plate of the spindle (metaphase I). The primary oocyte now undergoes two *meiotic* divisions. In the first division two daughter cells arise, each containing one-half of the chromosome complement (2n). However, unlike the divisions occurring in spermatogenesis, one acquires almost all of the cytoplasm. This cell is known as the *secondary oocyte;* the other much smaller cell is known as the *first polar*

body. At the second maturation division, the secondary oocyte divides into the *ootid* (n) and a *second polar body* (n). The two polar bodies, containing very little cytoplasm, are entrapped within the zona pellucida of the oocyte, and here they degenerate. The first polar body may also divide, thus the zona pellucida may contain one, two or three polar bodies.

The time at which the two reductive divisions occur is not necessarily coincident to the time of ovulation. The oocyte is usually in the pachytene or diplotene stage of prophase I during diestrus. Shortly before ovulation the oocyte may undergo the first meiotic division. The second division begins but is not completed until or unless fertilization takes place. Thus, the second polar body and the female pronucleus are formed at fertilization. The ova of cattle, sheep and swine contain one polar body at ovulation, whereas horse, dog and fox ova are in the first maturation division at ovulation.

It should be pointed out that it is the secondary oocyte which is liberated at ovulation (primary oocyte in the case of the horse). The oocyte continues the process of maturation until fertilization when it becomes a "zygote." In the process of oogenesis, one primary oocyte gives rise to *one egg;* in spermatogenesis, *one* primary spermatocyte gives rise to *four* sperm.

Atresia and Degeneration. The estimated number of oocytes in both ovaries at the time of birth varies from 60,000 to 100,-000 according to the species and breed. However, not all of these oocytes develop to the mature stage; for every egg that matures and is ovulated several start to develop but never reach maturity. Consequently every normal ovary contains some degenerating oocytes in follicles that fail to rupture (*atretic* follicles). The degenerating oocyte is characterized by hyalinization, thickening of the zona pellucida and/or fragmentation of the cytoplasm (Plate 10A). It is engulfed by the ovarian fibrocytes by phagocytosis and eventually disappears into a scar.

D. Structure of Eggs

Differences in the size of eggs depend almost entirely on the amount of accumulated deutoplasm. With the exception of monotremes' eggs, the diameter of the intrazonal vitellus at the time of ovulation ranges between 80 and 200μ. In farm mammals, it is usually less than 185μ. Even though the egg is larger than most of the somatic cells, there is little relationship between the size of the egg and that of the adult animal.

Corona Radiata. Before ovulation, the egg lies at one side of the ovarian follicle, embedded in a solid mass of follicular cells called the *cumulus oophorus*. Recently ovulated eggs are usually surrounded by a variable number of granulosa cell layers (*corona radiata*) and a matrix of follicular fluid. The connection of the egg with the granulosa cells is loosened by the development of new, liquid filled, intercellular spaces in the cumulus. Both cumulus and corona cells present on cattle and sheep eggs persist for only a few hours after ovulation. Protoplasmic extensions of these cells penetrate the zona pellucida in oblique or irregular directions, and intertwine with thin projections (microvilli) present in the oocyte itself (Fig. 2–9). Soon after ovulation, however, these projections are withdrawn. In vitro exposure of recently ovulated eggs to oviductal fluids containing fibrinolytic enzymes also results in retraction and degeneration of these projections. This is followed by regression of the main part of the cell. The ensuing necrosis leads directly to the denudation of the egg, which is slower in the eggs of pigs and rabbits than of cattle, sheep and horses.

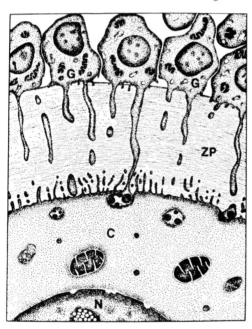

Fig. 2–9. Structure of fully formed zona pellucida (*ZP*) around an oocyte in a graafian follicle. Microvilli arising from the oocyte interdigitate with processes from the granulosa cells (*G*). These processes penetrate into the cytoplasm of the oocyte (*C*) and may provide nutrients and maternal protein (*N*) oocyte nucleus. (*Baker, 1972. In Reproduction in Mammals. C. R. Austin and R. V. Short [eds.], Cambridge, Cambridge University Press.*)

Egg Membranes. The egg has two distinct membranes: the *vitelline membrane* and the *zona pellucida*. The vitelline membrane is a cortical differentiation of the oocyte and can be considered to have essentially the same structure and properties of the plasma membrane of somatic cells (diffusion and active transport). The zona pellucida is a homogeneous and apparently semipermeable membrane which is made of a conjugated protein that can be dissolved by proteolytic enzymes, such as trypsin and chymotrypsin.

In certain species, another "membrane" is present; such membranes accumulate during the passage through the oviduct, which secretes a great variety of materials for the protection and nutrition of the eggs. For example, fish and amphibian eggs are provided with jelly envelopes, membranes and shells. A layer of mucin is deposited by the oviductal epithelium around the zona pellucida of rabbit eggs. However, cattle, sheep and swine oviductal eggs are not surrounded by such a coat.

The egg membranes probably are important for protection of the egg as well as for the selective absorption of inorganic ions and metabolic substances as exemplified by the physiochemical changes that occur at the time of ovulation, fertilization, cleavage and expansion of the blastocyst.

Vitellus. The vitellus comprises most of the volume within the zona pellucida at the time of ovulation. After fertilization, it shrinks and a *perivitelline space* is formed between the zona pellucida and the vitelline membrane in which the polar bodies are situated. The vitellus shows marked species-dependent morphology, mainly due to varying amounts of yolk

and fat droplets. In goat and rabbit eggs, the yolk granules are finely divided and uniformly distributed; thus, the various nuclear changes that occur during meiosis and fertilization are readily visible. Horse and cow eggs, however, are filled with fatty and highly refractile droplets, resulting in the nucleus being obscured by the dark mass of vitellus. If eggs fail to become fertilized, the vitellus fragments into a number of units of unequal size, each containing one or more abortive nuclei.

Structural Abnormalities. Ova structural abnormalities may appear as small or giant eggs, oval or flattened eggs, ruptured zona pellucida, large polar bodies or vacuoles within the vitellus. Such abnormalities can be the result of faulty or incomplete maturation of the oocyte, genetic factors or environmental stress.

The increase in the weight of the corpus luteum is initially very rapid. In general the period of growth is slightly longer than half of the estrous cycle. In the cow, the weight and progesterone content of the corpus luteum

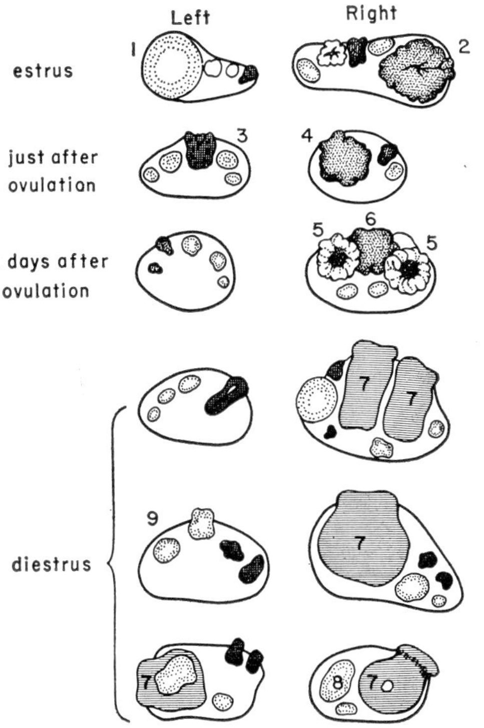

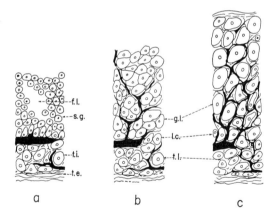

Fig. 2–10. Diagrammatic illustration of the morphologic changes in the estrous cycle of bovine ovary.

1, Ripe follicle; *2,* regressing corpus luteum (brick brown); *3,* collapsed follicle—surface wrinkled and bloodstained walls; *4,* regressing corpus luteum (bright yellow); *5,* twin corpus luteum—some hemorrhage; *6,* regressing corpus luteum (bright yellow); *7,* corpus luteum of diestrus; *8,* largest follicle; *9,* corpus albicans. (*Redrawn from Arthur, 1964. Wright's Obstetrics. London, Bailliere, Tindall & Cox.*)

Fig. 2–11. Diagram showing the organization of cells in the corpus luteum of the ewe. *a,* Corpus haemorrhagicum; *b,* corpus luteum of the second day following estrus; *c,* corpus luteum of the fourth day following estrus. Blood vessels are shown with heavy black lines.

f.l., Lake of follicular fluid; *g.l.,* lutein cells from the membrana granulosa; *t.i.,* theca interna; *t.e.,* theca externa. (*Adapted from Warbritton, 1934. J. Morphol. 56, 181.*)

increase rapidly between days 3 and 12 of the cycle (Figs. 2–10 and 2–11) and remain relatively constant until day 16 when regression begins (Erb et al., 1971). In the ewe and sow, corpora lutea increases rapidly in weight and progesterone content from day 2 to day 8, and remain relatively constant until day 15 when regression begins (Erb et al., 1971). The diameter of the mature corpus luteum is larger than that of a mature graafian follicle, except in the mare where it is smaller (Table 2–2).

The corpus luteum has a possible stimulatory effect on follicular development and ovulation through a local intra-ovarian mechanism. Robinson demonstrated that the presence of a previously formed corpus luteum increased the efficacy of pregnant mare serum gonadotropin (PMSG) in inducing ovulation in sheep. Rowson reported that the presence of a corpus luteum greatly facilitated the ovulatory response to PMSG in cows if given in the presence of the corpus luteum rather than after enucleation of the corpus luteum.

If pregnancy occurs, the corpus luteum retains its size. Furthermore, it functions through pregnancy in all farm mammals except the horse. The corpus luteum of pregnancy is known as the *corpus luteum verum* and may be larger than the *corpus luteum spurium* (false yellow body) of the estrous cycle. In cattle it increases in size for two to three months, regresses up to four to six months of gestation and thereafter remains relatively constant until calving.

Regression of Corpus Luteum. If fertilization does not occur, the corpus luteum regresses, allowing other graafian follicles to mature. As these cells degenerate, the whole organ decreases in size, becomes white or pale brown and is known as the *corpus albicans.* Regressive changes include thickening of the walls of the arteries in the corpus luteum, a decrease in cytoplasmic granulation, a rounding of the cell outline, and peripheral vacuolation of the large luteal cells. After two

or three cycles a barely visible scar of connective tissue remains. Remnants of the bovine corpus albicans persist during several successive cycles as small red patches of lipochromic pigments. In contrast, the corpus albicans of the preceding estrus is large (about 1 cm in diameter) and whitish and has a very rough fibrous consistency.

Regression of the corpus luteum in the cow which is not pregnant commences 14 to 15 days after estrus. It proceeds rapidly and the size may be halved within 36 hours. The corpus luteum of pregnancy is quite degenerate by seven days after calving.

III. THE OVIDUCT

A. Anatomy of the Oviduct

An intimate anatomic relationship exists between the ovary and oviduct. In farm mammals, the ovary lies in an open *ovarian bursa* in contrast to some species (e.g. rat, mouse) where it is in a closed sac. This bursa in farm animals is a pouch consisting of a thin peritoneal fold of mesosalpinx, which is attached to a suspended loop at the upper portion of the oviduct (Fig. 2–12). In cattle and sheep the ovarian bursa is wide and open. In swine it is well-developed and, although open, it largely encloses the ovary. In horses it is narrow and cleftlike and encloses only the ovulation fossa.

The oviducts are suspended in the mesosalpinx, a peritoneal fold that is derived from the lateral layer of the broad ligament. The oviduct can be subdivided into the infundibulum with its fimbriae, the ampulla and the isthmus. The length and degree of coiling vary in farm mammals (Table 2–3).

The oviductal end adjacent to the ovary is expanded to form a funnel-like structure called the *infundibulum.* Its size varies with the species and age of animals; the surface area is 6 cm^2 to 10 cm^2 in sheep, and 20 cm^2 to 30 cm^2 in cattle. The opening of the infundibulum, *ostium abdominale,* lies in the center of a fringe of irregular processes, forming the ex-

Table 2-3. Comparative Anatomy of the Reproductive Tract in the
Adult Nonpregnant Female of Farm Mammals

Organ	Animal			
	Cow	Ewe	Sow	Mare
Oviduct				
Length (cm)	25	15–19	15–30	20–30
Uterus				
Type	Bipartite	Bipartite	Bicornuate	Bipartite
Length of horn (cm)	35–40	10–12	40–65	15–25
Length of body (cm)	2–4	1–2	5	15–20
Surface lining of endometrium	70–120 caruncles	88–96 caruncles	Slight longitudinal folds	Conspicuous longitudinal folds
Cervix				
Length (cm)	8–10	4–10	10	7–8
Outside diameter (cm)	3–4	2–3	2–3	3.5–4
Cervical lumen				
Shape	2–5 annular rings	Annular rings	Corkscrewlike	Conspicuous folds
Os Uteri				
Shape	Small and protruding	Small and protruding	Ill-defined	Clearly-defined
Anterior Vagina				
Length (cm)	25–30	10–14	10–15	20–35
Hymen	Ill-defined	Well-developed	Ill-defined	Well-developed
Vestibule				
Length (cm)	10–12	2.5–3	6–8	10–12

The dimensions included in this table vary with age, breed, parity and plane of nutrition.

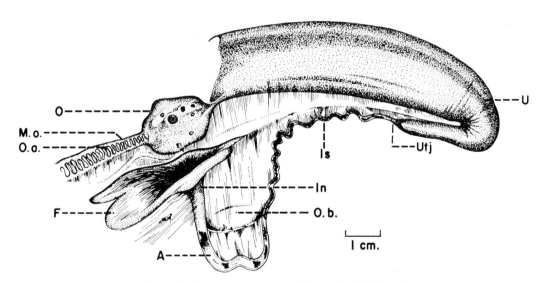

FIG. 2–12. Anatomic relationship between the ovary and the oviduct in the ewe.
A, Ampulla; F, fimbriae; In, infundibulum; Is, isthmus; M.o., mesovarium; O, ovary; O.a., ovarian artery; O.b., ovarian bursa; U, uterus; Utj, uterotubal junction. Note the suspended loop to which the ovarian bursa is attached. The oviduct in the ewe is pigmented.

tremity of the oviduct, the *fimbriae*. The fimbriae are unattached except for one point at the upper pole of the ovary. This assures close approximation of the fimbriae and ovarian surface.

The *ampulla*, comprising about half of the oviductal length, merges with the constructed section known as the *isthmus*. The anatomic and physiologic significance of this *ampullary-isthmic junction* is still unknown. The isthmus is connected directly to the uterine horn (it enters the horn in the form of a small papilla in the mare). No well-defined sphincter muscle is present at this point, the *uterotubal junction*. In the sow, however, this junction is guarded by long fingerlike mucosal processes. In the cow and ewe there is a marked flexure at the uterotubal junction, especially during estrus.

1. MUCOSA AND SUBMUCOSA

The tunica mucosa lines large primary and small secondary folds of the oviductal wall. The folds are particularly complex in the ampulla. The mucosa consists of one layer of columnar epithelial cells. The underlying submucosa of smooth muscle fibers and connective tissue are permeated by fine blood and lymph vessels. The epithelium contains ciliated and nonciliated cells, together with "peg cells" (Stiftzellen or intercalary cells) which are presumably depleted secretory cells. At the ovarian end of the oviduct, the wall is lined mostly with ciliated cells (Plates 3, 4); such cells are less common at the uterine end of the oviduct.

Ciliated Cells. The ciliated cells of the oviductal mucosa have a slender motile cilia (kinocilia) that extend into the lumen. The cilia conform to the standard structural plan of kinocilia, with nine pairs of fibrils arranged concentrically around another central pair. The fimbriae are made mostly of ciliated cells. The percentage of ciliated cells decreases gradually in the ampulla toward the isthmus. In the isthmus there are a few ciliated cells, located deep between the tall secretory ones. The tips of their cilia hardly reach the oviductal lumen and thus exert no effect on egg transport. The cilia beat toward the uterus. Their activity, coupled with oviductal contractions, keeps oviductal eggs in constant rotation which is essential for bringing egg and sperm together (fertilization) and preventing oviductal implantation.

Ciliation of the oviduct is hormonally controlled in the rhesus monkey: cilia disappear almost completely after hypophysectomy (Brenner, 1969) and develop in response to exogenous estrogens. This phenomenon, however, is subject to species differentiation.

Secretory Cells and Fluid. The secretory cells of the oviductal mucosa are nonciliated and characteristically contain secretory granules, the size and number of which vary widely among species and during different phases of the estrous cycle. In general, they become increasingly abundant as one moves from the infundibulum toward the isthmus; this is also true of the nonciliated cells which produce them. Like cilia, the presence and number of secretory granules depend on ovarian activity. Ovariectomy causes them to disappear in all regions of the oviduct except the isthmus. They reappear if the animal is subsequently treated with estrogen.

The metabolic activity of the epithelium of the oviduct appears to be most pronounced in the vicinity of the egg. In the rabbit, for example, the uptake of radioactive sulfate is always highest in the segment which contains the egg as it passes along toward the uterus.

By means of extra- and intra-abdominal devices which are used to cannulate the oviduct and collect its fluids, it has been possible to show that ovarian hormones also regulate the secretory activity of the epithelium of the oviduct. The secretions are composed predominantly of mucoproteins and mucopolysaccharides.

2. MUCULARIS AND SEROSA

The tunica muscularis consists of an inner circular layer and an outer longitudinal layer of smooth (unstriated)

PLATE 3

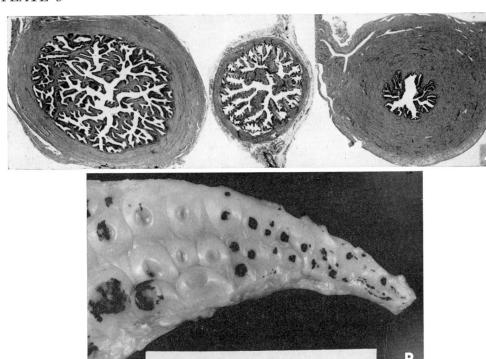

A, Cross section of ampulla in the sow (*left*) and cow (*center*). Note differences in the diameter of the lumen and complexity of mucosal primary and secondary folds. Cross section in isthmus in the sow (*right*); note the thickness of muscular coat ($\times$ 4.5). (*Photographs by E. Schilling*.)

B, Mucosa of uterine horn of the nonpregnant ewe. Note caruncles and pigmentation of the endometrium.

C, Maternal caruncle from a pregnant cow. Note the spongy-like crypts to which the chorionic villi were embedded.

PLATE 4

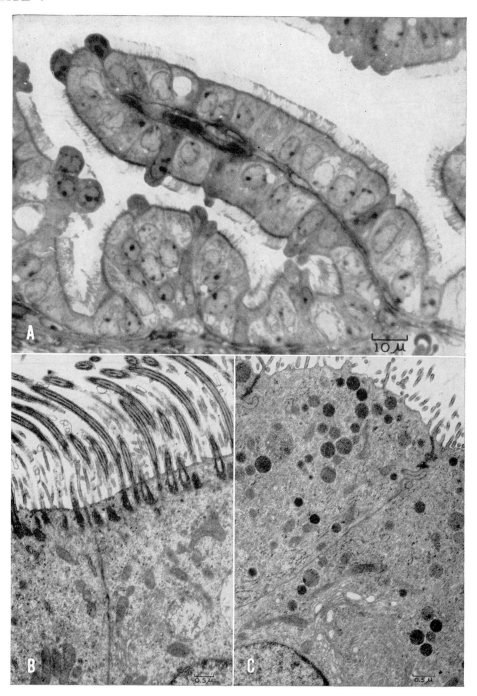

Structure and ultrastructure of mammalian oviduct. (*Scale in microns.*)

A, Infundibulum of mouse oviduct, day 1 of pregnancy. Ciliary cells dominate and among them secretory cells ("peg cells") can be seen. Toluidine blue staining of a plastic section.

B, Ciliary cells in the infundibulum of mouse oviduct, day 1 of pregnancy (time of egg passage).

C, Secretory cells in the isthmus of mouse oviduct, day 2 of pregnancy (time of egg passage). Secretory granules of varying density are observed in the apical part of the cells. (*Photographs by S. Reinius.*)

PLATE 5

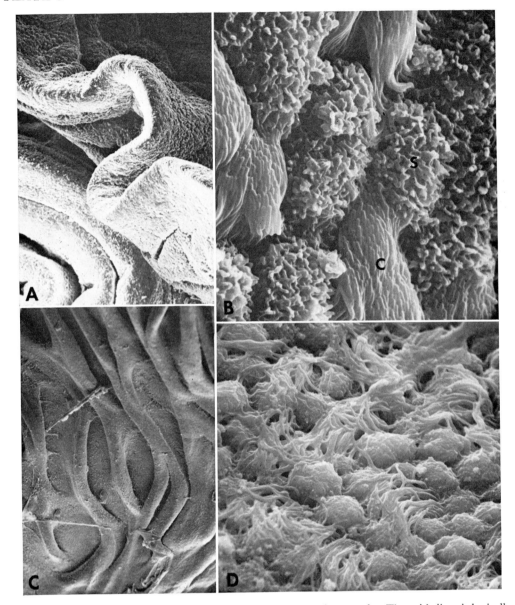

Scanning electron micrographs of the female reproductive tract of mammals. The epithelium is basically made of two types: nonciliated secretory cells with microvilli, and ciliated cells with kinocilia which beat toward the vagina. The topography of the epithelium and the morphology of the kinocilia and microvilli vary in different segments of the reproductive tract and with the stage of the estrous cycle.

A, Fimbriae of the oviduct. Note the complicated mucosal folds.

B, Ampulla of the oviduct showing secretory cells (S) and ciliated cells (C) × 25.

C, Bovine cervical canal when the cervix was cut open. Note the complexity of cervical crypts. × 59.

D, Bovine cervical epithelium showing the kinocilia overlapping on secretory cells. Compare with B. × 440.

muscle (Fig. 2–13). Muscle also extends from these layers into the connective tissue of the mucosal folds, permitting coordinated contractions of the entire wall. The thickness of the musculature increases as one advances from the ovarian to the uterine end of the oviduct.

B. Physiology of the Oviduct

The oviduct has the unique function of conveying the eggs and spermatozoa in opposite directions, almost simultaneously. The structure of the oviduct is well-adapted to its multiple function. The fringelike fimbriae transport ovulated eggs from the ovarian surface to the infundibulum. The eggs are transported through the mucosal folds of the ampulla to the ampullary-isthmic junction where fertilization and early cleavage of fertilized eggs take place. The embryos are kept in the oviduct for three days before they are transported to the uterus. The mesosalpinx and oviductal musculature coordinate ovarian hormones, estrogen and progesterone.

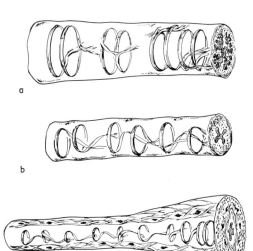

Fig. 2–13. Diagrammatic illustration of the musculature in the oviduct of ungulates. *A*, Ampulla: the musculature consists of spiral fibers arranged almost circularly. *B*, Isthmus: note differences in morphology of muscle fibers. *C*, Uterotubal junction: note the longitudinal muscle coat of uterine origin, as well as peritoneal fibers. (*Schilling, 1962. Zentralbatt Veter. 9, 805.*)

The oviductal fluid provides a suitable environment for fertilization, cleavage of fertilized eggs and capacitation of spermatozoa. These fluids are actively secreted from the epithelial lining of the oviduct, the rate being regulated by ovarian hormones. By utilization of different methods of cannulation for continuous collection of oviductal secretions, the volume of fluid secreted by both oviducts has been shown consistently to be variable during the estrous cycle (Fig. 2–20). The volume is low during the luteal phase, increases at the onset of estrus, reaches a maximum a day later and then declines to characteristic luteal phase levels.

The uterotubal junction controls, in part, the transport of sperm from the uterus to the oviduct and the transport of embryos from the oviduct into the uterus.

Contractions of the Oviduct. The contractile activities of the oviduct are independent of those which occur in the uterus. They appear to occur in specific peristaltic and antiperistaltic patterns which propel the egg at a definite rate and in a defined rotational manner. The latter is reflected in the remarkably uniform mucin coat of rabbit ova.

In part, these patterns of contraction arise because of the architecture of the oviduct. The ampulla, for example, has a wide lumen surrounded by a wall with a thin layer of circular muscle and a mucous membrane with highly branched folds. Transport of an ovum through the ampulla depends mainly on the rate and amplitude of these segmenting contractions, which originate in the infundibulum and pass progressively toward the isthmus.

Norepinephrine and acetylcholine may also initiate oviductal contractions. The oviduct and the uterine horns have opposite sensitivities to these two compounds; the ampulla is more sensitive to acetylcholine and less sensitive to norepinephrine than the uterus.

The contractile activities of the fimbriae, oviduct and oviductal ligaments are also partly coordinated by the ratio of estrogen to progesterone in the blood. The oviduct is most efficient at engulfing ovulated eggs during estrus, less efficient during other periods of the sexual cycle. In some species, neurohormonal mechanisms also increase the muscular activity of the fimbriae at copulation.

IV. THE UTERUS

The uterus consists of two uterine horns (cornua), a body and a cervix (neck). The relative proportions of each, as well as the shape and arrangement of the horns, vary considerably from species to species (Fig. 2–16). In swine, the uterus is of the bicornuate type (*uterus bicornis*). The horns are folded or convoluted and may be as long as four to five feet, while the body of the uterus is very short (Fig. 2–14). This length is an anatomic adaptation for successful litter bearing. In cattle, sheep and horses, the uterus is of the bipartite type (*uterus bipartitus*). These animals have a septum which separates the two horns and a prominent uterine body (the horse has the largest). Superficially the body of the uterus in cattle and sheep appears larger than it actually is because the caudal parts of the horns are bound together by the intercornual ligament.

Both sides of the uterus are attached to the pelvic and abdominal walls by the broad ligament (Fig. 2–15). In multiparous animals the uterine ligaments stretch allowing the uterus to drop into the pelvic cavity. In the mare this may hinder the removal of endometrial fluids or even allow small amounts of urine to flow through the cervix during estrus resulting in mild catarrhal inflammation.

The uterus receives its blood and nerve supply through the broad ligament. The blood vessels are very numerous, thick-walled and tortuous (Reuber and Emmerson, 1959). The *middle uterine artery*, a branch of either the *internal iliac artery* or the *external iliac artery*, is the chief blood supply to the uterus in the

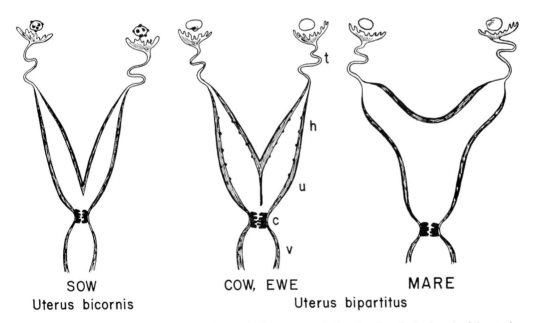

SOW
Uterus bicornis

COW, EWE
MARE
Uterus bipartitus

Fig. 2–14. Diagram of different types of uterus in farm mammals showing the relative length of the uterine horns and the uterine body. In the sow the uterine horns are elongated and the uterine body is very small. In the cow and ewe a septum separates the uterine horns and in the mare the uterine body is especially prominent.

region of the developing fetus, and thus it enlarges greatly with advancing gestation. The *cranial uterine artery*, a branch of the *utero-ovarian artery*, supplies blood to the ovary by the ovarian artery and to the anterior extremity of the uterine horn by the cranial uterine artery. The utero-ovarian artery supplies blood to the oviduct. The *caudal uterine artery* is a forward continuation of a branch of the internal pudic artery which supplies blood to the vagina, vulva and anus. The sympathetic nerves are supplied through the uterine and pelvic plexuses,

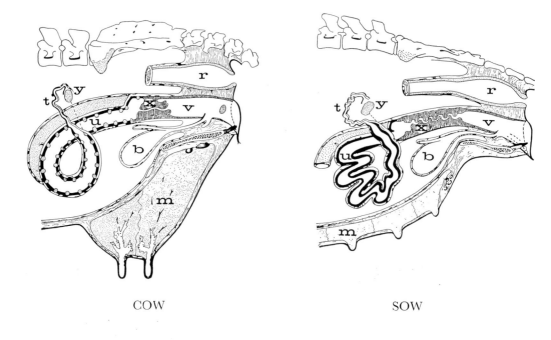

COW   SOW

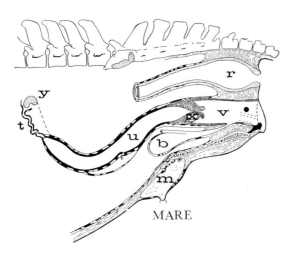

MARE

Fig. 2–15.   Comparative anatomy of the reproductive organs in the female.
b, Bladder; *m*, mammary gland; *r*, rectum; *t*, oviduct; *u*, uterus; *v*, vagina; *x*, cervix; *y*, ovary. Note species differences in anatomy of cervix, uterus and mammary gland. (*Redrawn from Ellenberger and Baum, 1943. Handbuch der vergleichenden Anatomi der Haustiere. 18th ed. Zietszchmann, Ackernecht and Grau [eds.], Berlin, Springer*).

terminating partly in the muscle fibers and partly in the mucosa.

A. Uterine Anatomy

Like most other hollow internal organs the walls of the uterus consist of a mucous membrane lining, an intermediate smooth

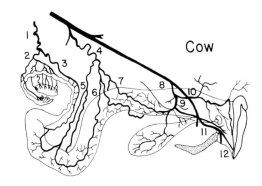

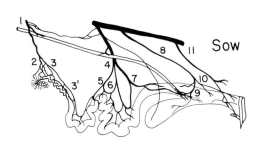

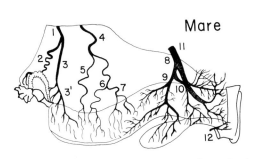

Fig. 2–16. Diagrammatic illustration of the arterial blood supply in the female reproductive tract. *1,* utero-ovarian artery; *2,* ovarian artery; *3,* uterotubal artery; *3′,* anterior cornual artery or anterior artery of uterine horn; *4,* uterine artery; *5,* middle cornual artery; *6,* posterior cornual artery; *7,* artery of corpus uterii; *8,* vaginal artery; *9* cervicouterine artery; *10,* vagino-rectal artery; *11,* internal pudic artery; *12,* artery of the clitoris. (*Barone and Pavaux,* 1962. *Bull. Soc. Sci. Vet. Lyon No. 1,* 33-52.)

muscle layer and an outer serous layer, the peritoneum. From the physiologic standpoint, only two layers are recognized—the endometrium and the myometrium.

1. The Endometrium

The endometrium is a highly glandular structure consisting of an epithelial lining of the lumen, a glandular layer and connective tissue. It varies in thickness and vascularity with ovarian hormonal changes and pregnancy.

Caruncles. The inner surface of the uterus in ruminants contains nonglandular projections, the *caruncles* (Plate 3). They are arranged in four rows, extending from the uterine body into the two uterine horns and consisting of connective tissue comparable to that found in the cortical stroma of the ovary. The deeper areas between the prominences are rich in blood vessels but contain no glands.

The uterus of the nonpregnant cow has 70 to 120 caruncles, each measuring approximately 15 mm in diameter. During pregnancy, they may attain a diameter of 10 cm and appear spongy due to the numerous crypts that receive the placental chorionic villi. These villi develop in localized areas, the *cotyledons* which invade the caruncles. The cotyledons and caruncles are referred to together as the *placentome.*

The uterus of horses and swine have no caruncles. Instead, the mucosa is characterized by conspicuous longitudinal folds which pass into the cervix to form the internal and external orifices.

Uterine Glands. These glands are scattered throughout the endometrium with the exception of the caruncles. They are branched and tubular and are considerably coiled especially toward the ends. Their density varies with the species, breed, parity and estrous cycle. The variation in proximity to one another during estrous cycle is largely a result of diameter changes and amount of stromal ground substance. The number of glands is higher in the horns than in the mucosa adjacent to the cervix. The glands may be rapidly increased by budding and outward growth from the basal zone. Buds that do not reach the surface become dilated and cystic. The glands are thought to serve an important function in nutrition during the entire gestation period in the pig which has a simple diffuse placentation.

Cyclic Changes. The maturation and growth of the endometrium are under the control of ovarian hormones estrogen and progesterone, which must be secreted cyclically and with an optimal quantitative ratio (Odell and Moyer, 1971). During proestrus, when the endometrium is under the influence of follicle-secreted estrogen (follicular phase), vascularity is increased, the surface epithelium is a short columnar and the glands undergo some growth although they remain straight with few branches (Fig. 2–14). As a result of changes in the ovary at estrus, the endometrium becomes controlled by the corpus luteum-produced progesterone. In early diestrus (luteal phase), the endometrium increases in thickness, the surface epithelium becomes highly columnar and the uterine glands reach maximal development (they are larger and more coiled and branched). During this period the glands are actively secretory. In late diestrus, the thickened endometrium shrinks, the glands become much smaller and their secretory activity ceases. These cyclic changes occur whether fertilization takes place or not.

The cyclic shedding of the endometrial surface epithelium followed by regeneration causes no extensive bleeding, such as that observed in primates where a substantial portion of the endometrium is shed. During "metestrus bleeding" in the cow, the caruncles show a pronounced capillary distension, but the epithelium remains intact.

2. THE MYOMETRIUM

The myometrium is the muscular portion of the uterine wall. It consists of two layers of smooth muscle: a thick inner circular layer and a thinner outer longitudinal layer. Between them lies a vascular layer made of blood and lymph vessels, nerves and connective tissue. This layer is not very distinct in the sow and mare, and may even occur within the circular layer in ruminants. During pregnancy the amount of muscle tissue in the uterine wall increases markedly both by cell enlargement and cell number.

B. Physiology of the Uterus

The uterus serves a number of functions. The endometrium and its fluids play a major role in the reproductive process. As spermatozoa are transported through the uterine lumen to the oviducts they undergo "capacitation" in endometrial secretions. Before implantation the developing blastocyst is nourished by the uterine secretions. After implantation, the embryo depends upon an adequate vascular supply within the endometrium. Throughout gestation the physiologic properties of the endometrium

FIG. 2–17. Diagrammatic illustration of the changes taking place in size and shape of the ruminant uterus during pregnancy. Three uteri are shown in the diagram; the inner one represents a nonpregnant uterus; the outer one represents a gravid uterus prior to delivery, and the middle one represents a uterus after delivery in the process of involution.

and its blood supply are important for the survival and development of the fetus.

At mating, the contractions of the myometrium are essential for the transport of spermatozoa from the site of ejaculation to the site of fertilization. Large numbers of spermatozoa aggregate in the endometrial glands; the physiologic and/or immunologic significance of this phenomenon is not known. The uterus is capable of undergoing tremendous changes in size, structure and position in order to accommodate the needs of the growing conceptus. Its great contractile power remains dormant until the time of parturition, when it plays the major role in fetal expulsion. Following parturition, the uterus almost regains its former size and condition by a process called *involution* (Fig. 2–17). In the sow the uterus continuously declines in both weight and length for 28 days after parturition. Thereafter it remains relatively unchanged during the lactation period. Immediately after the young are weaned, however, the uterus increases in both weight and length for four days.

1. Uterus and Luteolytic Mechanisms

There is a local utero-ovarian cycle, whereby the corpus luteum stimulates

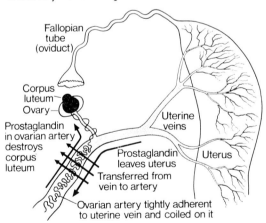

Corpus luteum
Ovary
Prostaglandin in ovarian artery destroys corpus luteum
Fallopian tube (oviduct)
Uterine veins
Prostaglandin leaves uterus
Transferred from vein to artery
Uterus
Ovarian artery tightly adherent to uterine vein and coiled on it

Fig. 2–18. Postulated route by which prostaglandin manufactured by the progesterone-primed uterus is able to enter the ovarian artery and destroy the corpus luteum in sheep. (*Short, 1972. In "Hormones in Reproduction," in Reproduction of Mammals. C. R. Austin and R. V. Short [eds.], Cambridge, Cambridge University Press.*)

3

the uterus to produce a substance which in turn destroys the corpus luteum. This lytic substance, formed by the endometrium, seems to be prostaglandin F_2 which diffuses from the uterine vein directly into the ovarian artery (Fig. 2–18). Short has reported that infusions of prostaglandin F_2 into either the uterine vein or the ovarian artery stop progesterone secretion more effectively than if they are given into the peripheral circulation, and prostaglandin is present in uterine venous blood in high concentrations on day 15 of the cycle. Thus, the presence of the uterus is required for normal regression of the corpus luteum (Fig. 2–19). Hysterectomy greatly prolongs the functional life span of the cyclic corpus luteum and thus causes prolonged anestrus (Anderson, 1966). The removal of the corpus luteum of the sow and cow, either at the time of hysterectomy or months after hysterectomy is followed by ovulation and formation of a new corpus luteum which in turn is maintained for an extended period of time. Thus, the ability of the hypophysis to secrete gonadotropins and the ability of the ovary to respond to gonadotropins do not seem to be markedly impaired by hysterectomy (Hansel et al., 1973).

The stimulation of the uterus during the early stages of the estrous cycle hastens regression of the corpus luteum and causes precocious estrus. Uterine stimulation can be initiated by placing a small foreign body in the lumen. The subsequent estrous cycle will be either shortened or prolonged depending on when the foreign body is inserted and on the nature and size of the introduced material. The nervous system is implicated by the fact that the estrous cycle is unaffected when the uterine segment containing the foreign body is denervated.

2. Uterine Secretions

The uterine luminal fluid is believed to be a combination of a blood plasma and uterine gland secretions. Fritz and May have demonstrated that metabolites present in the maternal circulation can be excreted into the female reproductive duct system and eventually absorbed by embryo. Upon administration of S^{35} to mated rabbits, they found isotope in both the uterine fluid and the zona pellucida of the developing embryo.

Differences in concentration as well as distribution of components in the uterine fluids compared to the blood provides evidence that secretion as well as transudations occur. For example, the potassium content in rat uterine fluids is 10 times that in the blood, whereas the sodium concentration is slightly lower in uterine fluids than in blood. Furthermore, at least two types of proteins found in the uterine fluid are not present in the blood.

The volume and biochemical composition (Olds and VanDemark, 1957) of the uterine fluid show consistent variation during the estrous cycle (Fig. 2–20). In

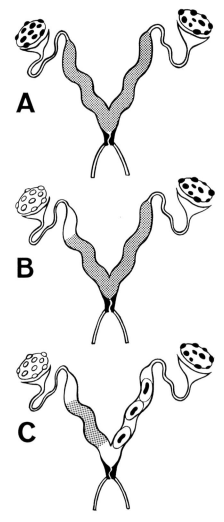

 PART OF UTERUS REMOVED

 REGRESSING CORPORA LUTEA

PERSISTENT CORPORA LUTEA

FIG. 2–19. Diagram showing the effect of partial hysterectomy on persistence of corpora lutea in the pig.

A, Total hysterectomy during the luteal phase causes retention of the corpus luteum for a period similar to gestation. B, Unilateral hysterectomy and partial removal of the other horn, retaining only a fragment 20 cm in length, causes asymmetric functioning of the two ovaries. C, The corpora lutea on the intact side are normally maintained whereas those on the other side regress before the 22nd day of gestation. (*Data from Du Mesnil du Buisson, 1961. Ann. Biol. Anim. Bioch. Biophys. 1, 105.*)

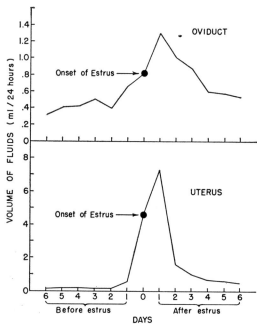

FIG. 2–20. Fluctuations in the volume of secretions of the oviduct and the uterus during the estrous cycle of sheep. Maximum secretion rates occur one day following the onset of estrus; this period coincides with the time of ovulation and reception of the ovum by the fimbriae. During estrus the volume of uterine fluids exceeds that of the oviduct, whereas the reverse is true during the luteal phase. (*Perkins et al., 1965. J. Anim. Sci. 24, 383.*)

sheep the volume of the fluid in the uterus exceeds that of the oviduct during estrus, whereas during the luteal phase the reverse is true (Perkins et al., 1965).

The uterine fluid has two important functions: namely, to provide a favorable environment for sperm capacitation and to provide nutrition for the blastocyst until implantation is complete.

3. Uterine Motility

The motility of the uterus is coordinated with the rhythmic movements of the oviduct and ovary. The degree and frequency vary with the estrous cycle. During diestrus, the uterus shows extremely slow, feeble, uncoordinated movements. The contractions may arise in any part and extend in any direction. At estrus, however, the contractions become rhythmic and move in a wave starting at the oviductal end. The amplitude and rate increase in both the uterus and oviduct. However, as ovulation takes place, ovarian changes cause a modification of these patterns. Thus long before the eggs reach the uterine lumen, the uterine muscles have become quiescent; moreover, they remain so throughout pregnancy. This sort of myometrial activity can also be induced experimentally by the injection of estrogen and progesterone.

4. Uterine Metabolism

The endometrium metabolizes carbohydrates, lipids and proteins to supply the necessary requirements for cell nutrition, rapid proliferation of the uterine tissue and conceptus development. Cyclic metabolic variations in this tissue consist of changes in the rate of nucleic acid synthesis, the availability of glucose, and the amount of glycogen reserves. These reactions depend on four phenomena: (a) the enzymatic reactions involved in glucose metabolism; (b) the increase in circulation through the spiral arterioles; (c) the morphologic changes which occur in the endometrium and myometrium; and (d) the stimulating action of the ovarian and other hormones (Jacquot and Kretchmer, 1964).

Two compounds that are of special significance in endometrial metabolism are glucose and glycogen. Glucose is converted to glucose-6-phosphate by hexokinase, perhaps as it passes through the cell wall. Once in the form of glucose-6-phosphate it may undergo a number of further conversions. For example glucose-6-phosphatase may merely return it to the glucose pool. It may, however, be metabolized through the Embden-Meyerhof-Parnas pathways or hexose monophosphate shunt. The former is primarily involved in energy production; the latter supplies pentoses for nucleic synthesis and reducing agents ($NADPH_2$) required for anabolic reactions (Hughes et al., 1964). Another metabolic fate is the conversion to glycogen which then is stored in the glands.

The role of specific enzymes for implantation and fetal development still awaits investigation. When available, it may be of academic and clinical importance in birth control and manipulation of fetal size.

Ovarian hormones play a substantial role in regulating uterine metabolism (Mounib and Chang, 1965). Growth of the uterus (both protein synthesis and cell division) is induced by estrogen; in the process it utilizes a considerable amount of energy in the form of ATP. Estrogen causes hyperemia followed by an increase in amino acid incorporation, nucleic acid synthesis and nitrogen retention. This hormone also stimulates phosphorus incorporation, oxidative metabolism, aerobic and anaerobic glycolysis and glycogen deposition.

Progestational responses in the endometrium involve major growth, a striking increase in DNA and RNA and a loss in water. A rapid change occurs in the metabolism of the endometrium at about the time the egg passes through the uterotubal junction.

5. Postpartum Involution of Uterus

During the postpartum interval the destruction of endometrial tissue is accompanied by the presence of large numbers of leukocytes and the reduction of the endometrial vascular bed. The cells of the myometrium are reduced in number and size. These rapid and disproportional changes in the uterine tissue are a possible cause of low postpartum conception rate. Neither the presence of suckling calves nor anemia delays uterine involution. Caruncular tissues are sloughed off and expelled from the uterus 12 days after calving. Regeneration of the surface epithelium over the caruncles occurs by growth from the surrounding tissue and is completed 30 days after calving.

V. THE CERVIX UTERI

A. Anatomy of the Cervix

The cervix is a sphincterlike structure which projects caudally into the vagina. It is characterized by a thick wall and constricted lumen. Although the structure of the cervix differs in detail among farm mammals, the cervical canal is characterized by various prominences (Fig. 2–21). In ruminants these are in the form of transverse or spirally interlocking ridges known as annular rings, which develop to varying degrees in the different species. They are especially prominent in the cow (usually four rings) and in the ewe, where they fit into each other to close the cervix securely. In the sow, the rings are in a corkscrew arrangement which are adapted to the spiral twisting of the tip of the boar's penis. Distinguishing features of the mare's cervix are the conspicuous folds in the mucosa and the projecting folds into the vagina.

The cervix does not contain any glands as erroneously assumed. The cervical mucosa is thrown into primary and secondary mucosal crypts (Fig. 2–21) which provide an extensive secretory surface. These intricacies give the cervix its typical "fern leaf" microscopic appearance. The cervical mucosa is made of two types of columnar epithelial cells: ciliated cells with kinocilia and nonciliated secretory cells (Fig. 2–22, Plate 5). The cytologic characteristics of the cervical epithelium resemble those of the oviductal epithelium in that the kinocilia are beating towards the vagina and the nonciliated cells contain massive numbers of secretory granules. The greatest secretory activity of these cells occurs at estrus; furthermore, this is when the mucus is least viscid.

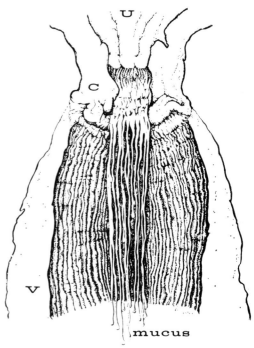

Fig. 2–21. (*Top*) Tracing of a longitudinal section of the bovine cervix showing the complexity of the cervical crypts which attract massive numbers of spermatozoa. *E*, external, or *I*, internal, or *M*, mucus-secreting mucosa; *S*, cervical stroma.

(*Bottom*) Diagrammatic illustration showing how the strands of cervical mucus flow from the crypts of the cervix (*C*) to the epithelium of the vagina (*V*). The biophysical characteristics of cervical mucus and arrangement of the macromolecules of mucus facilitate sperm transport from the vagina to the uterus (*U*).

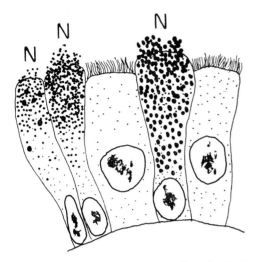

Fig. 2–22.   Diagrammatic illustration of epithelial cells of the cervix showing two types of cells—ciliated cells with kinocilia and nonciliated cells with abundant secretory granules (N). (*Hafez et al., 1972. Acta Biochem. 39, 195.*)

The cervical wall consists primarily of fibral, elastic and collagenous tissue and a small amount of muscle. The connective tissue is made of fibrous constituents, cellular components and ground substance. Collagen, the principal fibrous element, has a very high tensile strength. Collagenase, a specific degenerative enzyme for collagen, may be involved in the process of labor. The reticular network is an intimate part of the collagen system. The mass of the cervix increases as gestation advances and, although involution occurs after parturition, part of the gain in mass is maintained.

B. Physiology of the Cervix

The primary function of the cervix is to prevent intruders from entering the uterine lumen. It is tightly closed except during estrus at which time it relaxes slightly permitting sperm to enter the uterus. Mucus, discharged from the cervix, is expelled from the vulva. In swine and horses semen is ejaculated in the cervix near the internal os.

After mating or artificial insemination massive numbers of spermatozoa are lodged in the complicated cervical crypts. It is possible that the cervix acts as a reservoir of spermatozoa, thus providing the upper reproductive tract with subsequent releases of sperm. It is also possible that spermatozoa which are trapped in the cervical crypts are never released, thus preventing excessive numbers of spermatozoa from reaching the site of fertilization. It is also believed that spermatozoa in the cervix or uterus may stimulate the infiltration of leukocytes into the uterus, with subsequent phagocytosis of excessive spermatozoa by these leukocytes. During pregnancy an increased amount of mucus is secreted to occlude the cervical canal and thus to prevent uterine infections. The only other time the cervix is open is prior to parturition. At this time the cervical plug liquefies and the cervix dilates to permit the expulsion of the fetus and fetal membranes.

C. Cervical Mucus

Cervical mucus consists of macromolecules of mucin of epithelial origin, which are composed of glycoproteins (particularly sialomucin type) which contain some 25% amino acids and 75% carbohydrates. The mucin is made of a long, continuous polypeptide chain with numerous oligosaccharide side chains. The carbohydrate portion is made of galactose, glucosamine, fucose and sialic acid. The proteins of cervical mucus include prealbumin, lipoprotein, albumin, and beta and gamma globulins. The cervical mucus contains several enzymes including glucuronidase, amylase, phosphorylase, esterase and phosphatases. Cervical mucus which accumulates in the vaginal pool may also contain endometrial, oviductal, follicular and peritoneal fluids as well as leukocytes and cellular debris from uterine, cervical and vaginal epithelia.

Due to its unique biophysical characteristics, the cervical mucus has several rheologic properties such as ferning (crystallization in fern leaf-like shape upon drying), elasticity, viscosity, thixotropy and tack (stickiness). The secretion of cervical mucus is stimulated by ovarian estrogen and inhibited by progesterone. Cyclic qualitative changes in the cervical mucus throughout the estrous cycle, and cyclic variations in arrangement and viscosity of these macromolecules cause periodic changes in the penetrability of spermatozoa in the cervical canal. Optimal changes of cervical mucus properties, such as an increase in quantity, viscosity, ferning and pH and decrease in viscosity and cell content occur during estrus and ovulation, and are reversed during the luteal phase when sperm penetration in the cervix is inhibited. Under the influence of estrogens the macromolecules of glycoprotein of the mucus are oriented so that the spaces between them measure 2μ–5μ. In the luteal phase the spaces of the meshwork of macromolecules becomes increasingly smaller. Thus, at the time of estrus and ovulation the large size of the meshes allows for the transport of spermatozoa through the meshwork of filaments and through cervical canal. During pregnancy, a highly viscid, nonferning thick and turbid mucus occludes the cervical canal, acting as an effective barrier against sperm transport and invasion of bacteria in the uterine lumen.

Sperm Physiology in Cervical Mucus. After mating, spermatozoa are oriented toward the internal os. As the flagellum beats and vibrates the sperm head is propelled forward in the channels of least resistance. The macro- and microrheologic properties of cervical mucus play a major role in sperm migration. Sperm penetrability increases with the cleanliness of mucus, since cellular debris and leukocytes delay sperm migration. The aqueous spaces between the micelles allow the passage of sperm as well as diffusion of soluble substances. Proteolytic enzymes may hydrolyze the backbone protein or some of the cross-linkages of the mucin and reduce the network to a less resistant mesh with more open channels for the passage of sperm migration. When cervical mucus and semen are placed in apposition in vitro, phase lines immediately occur between the two fluids. Sperm phalanges soon appear and develop marked degrees of arborization, the terminal aspects of which consist of canals through which one or two spermatozoa can pass.

VI. THE VAGINA

The vagina is a dilatable organ for copulation and the passageway for fetal and placental delivery. In cattle and sheep, semen is ejaculated into the vaginal fornix, an arch formed by the projecting cervix.

The wall of the vagina consists of mucosa, muscularis and serosa. The mucous membrane is composed of glandless, stratified, squamous epithelial cells, except in the cow. Here some mucous cells are present in the cranial part next to the cervix and the epithelial surface fails to cornify, probably because of low levels of circulating estrogens. Cyclic changes occurring in the vaginal epithelium can be followed to some extent by the vaginal smear technique. This is however most accurate in animals with short estrous cycles, e.g. rats and mice.

The muscular coat of the vagina is not as well-developed as the outer parts of the uterus. It consists of a thick inner circular layer and a thin outer longitudinal layer; the latter continues for some distance into the uterus. The muscularis is well-supplied with blood vessels, nerve bundles, small groups of nerve cells and loose and dense connective tissue. The cow is peculiar in possessing an anterior sphincter muscle in addition to the posterior sphincter (at the junction of the vagina and vestibule) found in the other farm mammals.

VII. THE EXTERNAL GENITALIA

The external genitalia are comprised of the vestibule, the labia majora, the labia minora, the clitoris and the vestibular glands.

Vestibule. The junction of the vagina and vestibule is marked by the external *urethral orifice* and frequently by a ridge (the vestigial hymen). In some cattle the hymen may be so prominent that it interferes with copulation.

The vestibule of the cow extends inward for approximately 10 cm, where the external urethral orifice opens into its ventral surface. Just posterior to this opening lies the suburethral diverticulum, a blind sac (Fig. 2–15). Gartner's tubes (remnants of the Wolffian ducts) open into the vestibule posteriorly and laterally to Gartner's ducts. The glands of Bartholin which secrete a viscid fluid, most actively at estrus, have a tubo-alveolar structure similar to the bulbourethral glands in the male.

Labia Majora and Labia Minora. The integument of the labia majora is richly endowed with sebaceous and tubular glands. It contains fat deposits, elastic tissue and a thin layer of smooth muscle, and has the same outer surface structure as the external skin.

The labia minora are small and have a core of spongy connective tissue. The surface contains many large sebaceous glands.

Clitoris. The ventral commissure of the vestibule conceals the clitoris, which has the same embryonic origin as the male penis. It is made up of erectile tissue covered by stratified squamous epithelium and is well-supplied with sensory nerve endings. In the cow, the greater part of the gland is buried in the mucosa of the vestibule. However, in the mare it is well-developed, whereas in the sow it is long and sinuous, terminating in a small point or cone.

REFERENCES

Anderson, L. L. (1966). Pituitary-ovarian-uterine relationships. In Ovarian regulatory mechanisms, *J. Reprod. Fert.* Suppl. 1, 21–32.

Baker, T. (1972). "Oogenesis and Ovulation." In *Reproduction in Mammals.* C. R. Austin and R. V. Short (eds.), Cambridge, Cambridge University Press.

Brambell, F. W. R. (1952). *Marshall's Physiology of Reproduction.* 3rd Ed., Vol. 1, Chapter 5, London, Longmans.

Brenner, R. M. (1969). "The Biology of Oviductal Cilia." Chap. 8 in *The Mammalian Oviduct.* E. S. E. Hafez and R. J. Blandau (eds.), Chicago, The University of Chicago Press.

Erb, R. E., Randel, R. D. and Callahan, C. J. (1971). Female sex steroid changes during the reproductive cycle. *J. Anim. Sci.* Suppl. 1, *32*, 80–106 (IX Biennial Symp. on Anim. Reprod.).

Frye, B. E. (1967). *Hormonal Control in Vertebrates.* New York, Macmillan.

Hansel, W., Concannon, P. W. and Lukaszewska, J. H. (1973). Corpora lutea of the large domestic animals. *Biol. Reprod.* 8, 222–245.

Hughes, C. E., Jacobs, R. D. and Rubulis, A. (1964). Effect of treatment for sterility and abortion upon the carbohydrate pathways of the endometrium. *Amer. J. Obst. Gynec.* 89, 59–69.

Jacquot, R. and Kretchmer, N. (1964). Effect of fetal decapitation on enzymes of glycogen metabolism (rats). *J. Biol. Chem.* 239, 1301–1304.

Mounib, M. S. and Chang, M. C. (1965). Metabolism of endometrium and fallopian tube in the estrous and the pseudopregnant rabbit. *Endocrinology* 76, 542–546.

Odell, W. D. and Moyer, W. D., (1971). *Physiology of Reproduction.* St. Louis, Mosby.

Olds, D. and VanDemark, N. L. (1957). Composition of luminal fluids in bovine female genitalia. *Fertil. Steril.* 8, 345–354.

Perkins, J. L., Goode, L., Wilder, W. A., Jr. and Henson, D. B. (1965). Collection of secretions from the oviduct and uterus of the ewe. *J. Anim. Sci.* 24, 383–387.

Price, D., Zaaijer, J. and Ortiz, E. (1969). "Prenatal Development of the Oviduct *In Vivo* and *In Vitro.*" Chap. 1 in *The Mammalian Oviduct.* E. S. E. Hafez and R. J. Blandau (eds.), Chicago, The University of Chicago Press.

Reuber, H. W. and Emmerson, M. A. (1959). Arteriography of the internal genitalia of the cow. *J. Amer. Vet. Med. Assn.* 134, 101–109.

Zuckerman, S. (ed.) (1962). *The Ovary.* Vols. I and II, New York, Academic Press.

II. Physiology of Reproduction

Chapter 3

The Hormones of Reproduction

G. D. Niswender, T. M. Nett and A. M. Akbar

Hormones are chemical agents synthesized and secreted by specialized ductless glands and carried by the blood stream to other parts of the body where they evoke systemic adjustments by acting on specific target tissues or organs. There are several types of hormones based upon their site of origin, chemical properties and the responses which they evoke in target tissues. In reproductive endocrinology we are concerned primarily with three different types of hormones. *Releasing factors* or *releasing hormones* originate in the *hypothalamus* and control the synthesis and/or release of hormones from the anterior portion of the pituitary gland (*adenohypophysis*). The *gonadotropic hormones* are produced in the hypophysis and are directly involved in maturation and release of the gametes and stimulate the secretion of *sex steroid hormones* from the *gonads*. Sex steroid hormones play important roles in the behavioral aspects of reproduction, in development and maintenance of the secondary sex characters, in maintenance of the reproductive organs and in regulation of the reproductive cycle and pregnancy.

I. REPRODUCTIVE HORMONES

A. Hypothalamic Releasing Hormones

1. Anatomic Relationships

The anatomic relationships between the hypothalamus and the pituitary gland are shown in Figure 3–1. There does not appear to be direct innervation of the pituitary gland. Efficient, rapid transfer of releasing hormones between the hypothalamus and the adenohypophysis occurs via an elaborate system of portal vessels. These *hypothalamo-hypophyseal portal vessels* have been confirmed in numerous species, and the direction of blood flow (from hypothalamus to anterior pituitary gland) has been established by direct observation of injected materials. The importance of intact portal vessels for normal hypophyseal function has been demonstrated in several species, including cattle, pigs and sheep. In these studies, a physical barrier was inserted between the hypothalamus and the adenohypophysis which led to decreased production of adenohypophyseal hormones. Content as well as the volume of portal blood is important, since lesions of the hypothalamus which do not limit blood flow may result in decreased pituitary function. Lesions in selected portions of the hypothalamus may influence only one of the hormones produced in the anterior pituitary gland.

Secretion of most adenohypophyseal hormones is decreased following transplantation of the pituitary gland from the *sella turcica* to the kidney capsule. This further exemplifies the importance of the anatomic relationships between the hypothalamus and pituitary gland and suggests that the releasing hormones need

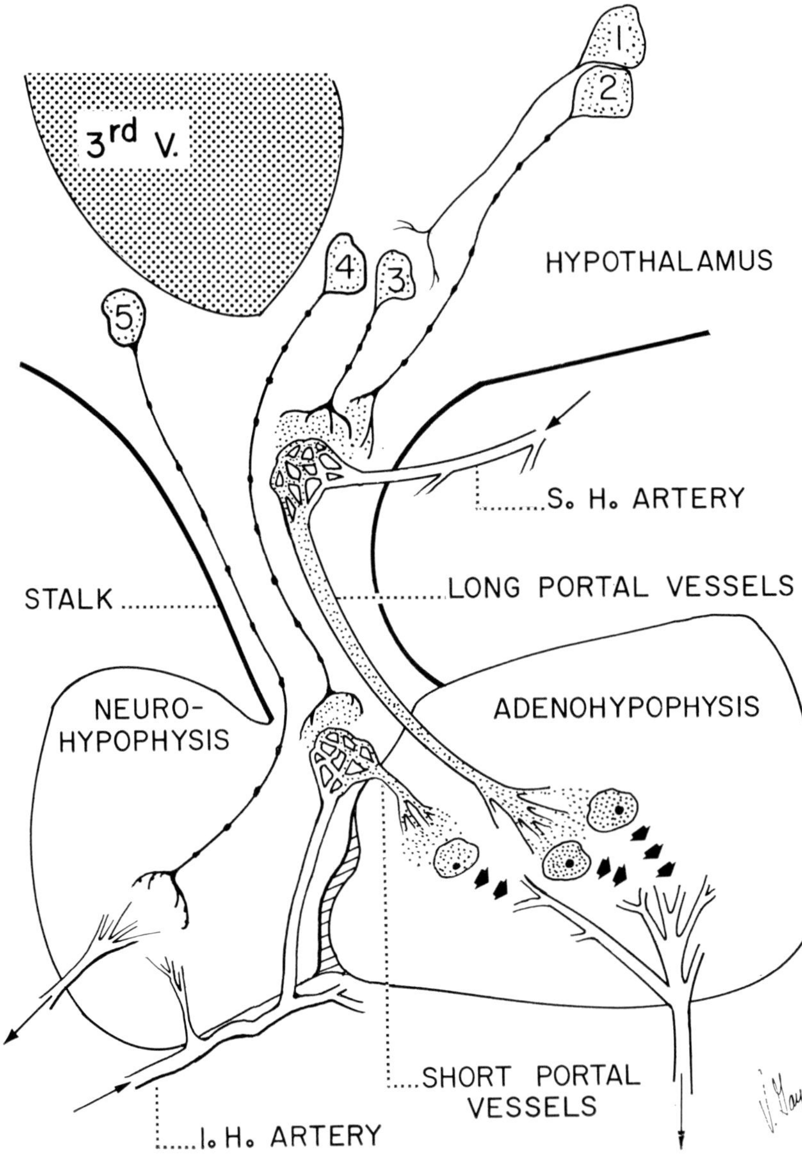

Fig. 3–1. A schematic representation of the anatomic relationships between the hypothalamic area and the pituitary gland. Both long portal vessels (originating from the superior hypothalamic artery) and short portal vessels (originating from the inferior hypothalamic artery) provide means for communication between hypothalamic neurons and the hormone secreting cells of the adenohypophysis. Three types of neural stimuli have been proposed to bring about secretion of releasing hormones. Extrahypothalamic neurons (neuron 1) may stimulate hypothalamic neurons (neuron 3) resulting in the secretion of releasing hormones. Neurons, the cell bodies of which lie outside the hypothalamus, may also secrete releasing hormones. In both cases the releasing hormones are absorbed by the capillary bed and transported to the adenohypophysis via the long portal vessels. The pathway depicted by neuron 4 suggests transport of releasing hormones to the short portal vessels which vascularize the periphery of the adenohypophysis. Neuron 5 is included to illustrate the neurosecretory processes whereby oxytocin and vasopressin are synthesized in the brain and transported to the neurohypophysis. (*From Gay, 1972. Fertil. Steril. 23, 50.*)

to be present in locally high concentrations. The decreased secretion by the transplanted pituitary gland is presumably due to the greatly reduced quantities of releasing hormones which reach it via the general circulation (Gay, 1972).

2. Chemistry

Substances have been isolated in varying degrees of purity from hypothalamic tissues which stimulate the release of hypophyseal hormones. By purifying the material obtained from thousands of ovine or porcine hypothalami, the chemical structure of a molecule which releases both *luteinizing hormone* (LH) and *follicle stimulating hormone* (FSH), termed *gonadotropin releasing hormone* (Gn-RH), has been determined. A substance which releases thyroid stimulating hormone (TSH) called thyrotropin releasing hormone (TRH) has also been characterized and synthesized. Gn-RH is a peptide which contains 10 amino acids while TRH is a tripeptide. TRH also causes release of *prolactin* in several species, including sheep, man and rats. Both a *prolactin-inhibiting factor* and a *prolactin-releasing factor*, other than TRH, have also been demonstrated in hypothalamic tissues. However, there is very little known regarding the exact chemical nature of these factors.

Release of other adenohypophyseal hormones is also controlled by the hypothalamus. Corticotropin releasing factor (CRF), responsible for the release of adrenocorticotropin (ACTH), was the first hypothalamic hormone to be demonstrated, but its structure has not been elucidated. The secretion of growth hormone is controlled by a growth hormone releasing factor (GH-RF) of hypothalamic origin, but to date the exact chemical nature of GH-RF has not been unequivocally determined. Secretion of melanocyte stimulating hormone (MSH) appears to be controlled both by a releasing factor (MRF) and an inhibiting hormone (MIH) (Schally et al., 1973).

3. Biologic Effects

The anatomic relationships illustrated in Figure 3–1 form the basis for current concepts of hypothalamic regulation of hypophyseal hormones. Releasing and inhibiting factors are: (*a*) produced within neurons of the hypothalamus; (*b*) transported down the axons of these neurons; (*c*) released from the terminal nerve endings in response to appropriate neural stimuli; and (*d*) transported via specialized portal vessels directly to the cells of the adenohypophysis. The ability of both crude and purified preparations of hypothalamic releasing factors to release pituitary hormones has been demonstrated repeatedly in many species.

Minute quantities of the synthetic releasing hormones presently available are effective in stimulating secretion of adenohypophyseal hormones; however, they appear to lack specificity. The Gn-RH molecule, initially characterized by its ability to release LH, also stimulates the release of FSH. TRH is effective in releasing TSH but also releases prolactin. Additional data regarding sites of production, transport, specificity and mechanism of action will be necessary before we obtain a complete understanding of the role of hypothalamic releasing hormones in the reproductive processes of domestic animals.

B. Hypophyseal Hormones

1. Chemistry

The hypophyseal hormones which appear to play the most important role in reproduction are luteinizing hormone, follicle stimulating hormone and prolactin. LH and FSH are glycoproteins containing 15%–25% carbohydrate, including small quantities of fucose, mannose, galactose, glucosamine and galactosamine. The carbohydrate is covalently bound to the protein portion of the molecule in the form of oligosaccharide chains, some of which terminate in sialic acid residues. Ovine and bovine LH have a molecular weight of 30,000 and contain 216 amino acids with a known sequence. Ovine FSH has a molecular weight of approximately 32,000, but the amino acid sequence of FSH has not been determined. LH and FSH, and the other hypophyseal glycoprotein hormone—TSH—contain two nonidentical subunits designated as alpha and beta which are held together by noncovalent forces. The alpha subunits of the three hormones in a given species appear to be identical, while the beta subunit is unique for each hormone and dictates the biologic activity of the intact molecule. Isolated subunits appear to be devoid of biologic activity. However, under certain experimental conditions, alpha and beta subunits may be combined to form a biologically active hormone. Isolated subunits apparently are not released into the circulatory system.

Prolactin is a protein with a molecular weight of 23,300 and a known sequence of 198 amino acids. There is no carbohydrate associated with the prolactin molecule (Li, 1972).

2. Secretion

The cells of the adenohypophysis have been classified on the basis of their staining characteristics. Cells devoid of granules and therefore not taking up stain are termed *chromophobes*. Cells containing granules which can be stained specifically are classified as *chromophils*. The chromophils may be further divided into acidophils and basophils. Acidophils show a strong affinity for acid dyes, and basophils are characterized by the presence of granules that take a basic stain. Both LH and FSH are produced in basophilic cells. In fact, both LH and FSH have been demonstrated within the same cell (Phifer et al., 1973). Conversely, single cells in culture develop clones of cells which produce a single hormone, either LH or FSH (Steinberger et al., 1973). Prolactin has been localized in acidophilic cells. Pituitary hormones appear to be biosynthesized primarily by rough-surfaced endoplasmic reticulum and concentrated by the Golgi apparatus into secretory granules for storage. In the process of secretion the storage granules which contain the hormone migrate to the cell membrane. The contents of the granule are released into the intercellular or pericapillary spaces by exocytosis following fusion of the granular and cellular membranes.

3. Functions

Table 3–1 lists some of the functions of the gonadotropic hormones. The relative biologic activities of the hormones appear to be related to their disappearance rate from the circulatory system. Protein hormones, such as prolactin and Gn-RH which contain no sialic acid, disappear from the circulation much more rapidly than do glycoprotein hormones (Fig. 3–2). The disappearance rate of the various glycoprotein hormones also appears to be directly influenced by their native sialic acid content (Fig. 3–2). The enzymatic removal of sialic acid residues from glycoprotein hormones results in their rapid clearance from the blood stream and a loss of biologic activity (Van Hall et al., 1971).

The biologic effects of LH and FSH appear to be limited to stimulation of

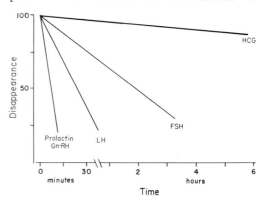

Fig. 3–2. The relative disappearance rates of some protein (prolactin, Gn-RH) and glycoprotein hormones (LH, FSH and human chorionic gonadotropin [HCG]). Prolactin and Gn-RH have no sialic acid residues whereas the LH, FSH and HCG molecules have 1–2, 5 and 12 sialic residues, respectively. If one defines half life ($T\frac{1}{2}$) as the time required to reduce blood levels of a hormone by 50%, assuming no further secretion, then $T\frac{1}{2}$ for Gn-RH and prolactin is approximately seven minutes. The $T\frac{1}{2}$ for LH, FSH and HCG is approximately 0.5, 2 and 24 hours respectively.

Table 3-1.   **Functions of the Polypeptide and Protein**
Hormones of Reproduction

Source	Hormone	Some Functions
Hypothalamus	Gonadotropin releasing hormone (Gn-RH)	Release of LH
		Release of FSH
	Thyrotropin releasing hormone (TRH)	Release of TSH
		Release of prolactin
	Prolactin releasing factor (PRF)	Release of prolactin
	Prolactin inhibiting factor (PIF)	Inhibition of prolactin release
Adenohypophysis	Luteinizing hormone (LH)	Ovulation; formation of corpus luteum; progesterone secretion; estrogen secretion; androgen secretion
	Follicle stimulating hormone (FSH)	Follicular growth; stimulation of Sertoli's cells, spermiogenesis
	Prolactin	Progesterone secretion; lactation; testosterone secretion; stimulation of male accessory organs
Neurohypophysis	Oxytocin	Parturition; milk ejection; sperm transport; ovum transport
Placenta	Pregnant mare serum gonadotropin (PMSG)	Formation of accessory corpora lutea; progesterone synthesis
	Placental luteotropin	Maintenance of corpus luteum of gestation

various physiologic processes within the gonads.   Most of the research regarding the actions of LH and FSH on the testis has been performed in rats but similar mechanisms probably exist in domestic animals.   LH acts to stimulate testosterone secretion from interstitial cells (cells of Leydig).   Prolactin also appears to be necessary to restore testosterone secretion to normal levels following hypophysectomy.   Prolactin may also act synergistically with androgens to stimulate the male accessory structures such as the prostate, seminal vesicles and preputial gland.   FSH appears to stimulate Sertoli's cells and/or the primary spermatogonia and does not appear to stimulate steroidogenesis.

In the female, LH is thought to be primarily responsible for ovulation, although FSH alone or in combination with LH also causes ovulation under experimental conditions.   Follicular growth, including proliferation of the granulosa cells is stimulated by FSH, but FSH alone will not promote estrogen secretion.   LH acts synergistically with FSH to promote estrogen secretion and to increase ovarian weight.   There is no evidence that prolactin plays a role in follicular development or ovulation, although blood levels of prolactin increase at estrus.   Prolactin does appear to play a role in maintenance of the corpus luteum and in progesterone synthesis by luteal tissue (Schroff et al., 1971; Denamur et al., 1973).   Prolactin also has a stimulatory effect on the development of the mammary gland and milk secretion.

4. Mechanism of Action of Gonadotropins

The net result of stimulation of a target tissue by a gonadotropic hormone is a function of the specific hormone and the ability of the stimulated cell to respond. The factors which influence the ability of a cell to respond include its membrane structure and potential, the net energy charge of the cell, the complement of intracellular enzymes and the ability for genetic expression within the cell.   In addition, cooperative effects may occur between stimulated cells and other adjacent cells within the target tissue which might influence its response.

In general, cells in target tissues are characterized by receptor molecules which have a high binding affinity for the hormone and are present in relatively low numbers per cell, so physiologic concentrations of hormone are capable of saturating them. Figure 3–3 summarizes the current concepts regarding the mechanism whereby LH could stimulate progesterone synthesis within the luteal cell (Niswender et al., 1972). The receptor molecule for protein hormones is one of two components of the adenyl cyclase system associated with the cell membrane. Besides the receptor unit which binds the hormone, this system contains a catalytic unit with the adenyl cyclase enzymatic activity. Adenyl cyclase catalyzes the reaction in which adenosine triphosphate is converted to *cyclic 3'5' adenosine monophosphate* (cyclic AMP). LH stimulates cyclic AMP formation in ovine and bovine luteal tissue. Activation of a prostaglandin receptor may be a necessary intermediate step for LH to stimulate formation of cyclic AMP.

Several mechanisms can be envisioned for increased steroid production following elevation of intracellular concentrations of cyclic AMP. For example, cyclic AMP may activate the phosphorylase

LH ACTION – HYPOTHETICAL MODEL

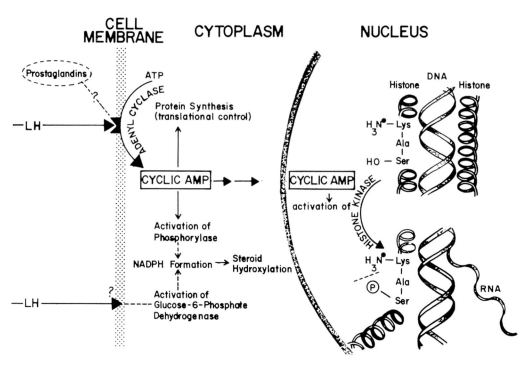

FIG. 3–3. Hypothetical scheme of the action of LH on target cells. The primary event in LH mediated progesterone synthesis may be the activation of membrane bound adenyl cyclase which causes an increase in the intracellular concentration of cyclic AMP. Cyclic AMP may act at the cytoplasmic level to stimulate protein synthesis. Alternatively, cyclic AMP may enter the nucleus causing phosphorylation of histone or other proteins tightly bound to DNA which causes a conformational change and, consequently, unmasks a part of the DNA molecule for transcription by RNA polymerase. The RNA that is synthesized is transported to the site of protein synthesis. The protein synthesized may be a rate-limiting enzyme involved in steroid synthesis. The scheme also depicts the alternate possibility that LH may directly activate glucose-6-phosphate dehydrogenase, resulting in increased formation of NADPH which would satisfy the cofactor requirement for steroid hydroxylations. (*From Niswender et al., 1972. Fertil. Steril. 23, 432.*)

enzyme, which facilitates glycogenolysis. Glucose-6-phosphate, resulting from the breakdown of glycogen, may then be metabolized in the hexose monophosphate shunt, resulting in the generation of reduced nicotinamide adenine dinucleotide phosphate (NADPH). NADPH and molecular oxygen are required for the conversion of cholesterol to pregnenolone, a precursor of progesterone. It has also been suggested that LH can stimulate NADPH formation directly via the activation of glucose-6-phosphate dehydrogenase. This reaction does not require elevation of intracellular levels of cyclic AMP as an intermediate step.

A third alternative suggests that increased intracellular levels of cyclic AMP result in the direct activation of enzymes which lead to the synthesis of new or additional quantities of the specific enzymes involved in biosynthesis of steroid hormones. A final alternative suggests that levels of cyclic AMP may be increased within the nucleus, leading to the activation of a specific kinase which results in the phosphorylation of histones or other nuclear proteins which leads to the unmasking of DNA. The unmasked portion of the chromosomal DNA which contains the genetic information for production of a specific enzyme allows transcription by RNA polymerase. This results in the synthesis of new RNA molecules, which are responsible for increased protein synthesis. The new protein may be a rate-limiting enzyme in steroid biosynthesis.

For low concentrations of hormones to effect major responses in target tissues, secondary catalytic events capable of triggering and amplifying subsequent intracellular activities must occur. Therefore, it is possible that the stimulation of progesterone synthesis by LH involves all of the suggested mechanisms. In fact, stimulation of a target cell by LH may result in a myriad of membrane and intracellular responses, all of which are aimed at increasing the synthesis of progesterone.

Very few data are available regarding the mechanism of action of FSH and prolactin. These hormones presumably act via mechanisms similar to those described for LH, since similar mechanisms have been described for a number of other protein hormones, including ACTH and insulin.

5. EFFECTS OF OTHER PITUITARY HORMONES

Pituitary hormones which do not influence reproduction directly also play an important role in the maintenance of optimal reproductive performance in domestic animals. Growth hormone (GH), or somatotropic hormone, is concerned with the growth of all body tissue and is needed throughout the reproductive life of the animal. GH is a protein, containing a known sequence of 188 amino acids, which has been synthesized. The effects of GH are to increase amino acid uptake and stimulate protein synthesis in all cells of the body, to increase the mobilization of fats and utilization of fats for energy and decrease the utilization of carbohydrates. These metabolic actions of growth hormone are necessary for all body functions, including the normal reproductive processes.

Adrenocorticotropic hormone (ACTH) is a peptide hormone having 39 amino acids. The sequence of amino acids is known, and the hormone has been synthesized. ACTH stimulates the synthesis and secretion of steroidal hormones from the adrenal cortex.

The *posterior pituitary* is responsible for storage and release of two hormones—*oxytocin* and *vasopressin*. Both are synthesized in the supraoptic and paraventricular hypothalamic nuclei and transported as secretory droplets along nerve fibers to endings in the neurohypophysis (posterior pituitary), where they are stored (Fig. 3–1). Release of the hormone occurs as a result of stimulation of the nerve cell bodies in the supraoptic and paraventricular nuclei. Oxytocin is an octapeptide, with a molecular weight of 1000, which stimulates the transport of sperm and ova in the uterus and oviduct by stimulating the contraction of smooth muscle in these tissues. In addition, stimulation of uterine contractions by oxytocin at parturition plays a role in expulsion of the fetus. During lactation, oxytocin stimulates contraction of myoepithelial cells in the mammary gland to facilitate milk ejection. Vasopressin, although not directly involved in reproduction, acts in conjunction with the mineralocorticoids to control the water and electrolyte balance of the animal.

Thyroid stimulating hormone influences reproduction primarily by its influence on the thyroid gland and the production of thyroxine and/or triiodothyronine. These hormones affect the metabolic pools of nitrogen and available energy which are necessary for the reproductive system and the developing embryo. Abnormally low levels of thyroxine delay the onset of puberty and maturation of the reproductive system, interfere with normal pregnancy and may directly influence spermatogenesis in the adult.

C. Gonadal Steroid Hormones

1. CHEMISTRY

Steroid hormones are derived from a common precursor molecule via the

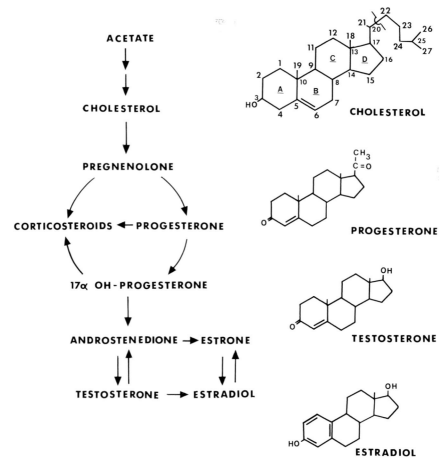

FIG. 3–4. Metabolic pathway for the synthesis of steroid hormones from acetate. The number sequence for the carbon atoms and the lettering sequence for the four rings are shown for cholesterol at the upper right. The structures of the three most important sex steroid hormones—progesterone, testosterone and estradiol—are also depicted.

metabolic pathway in Figure 3–4. It appears that all steroid-producing tissues possess the complete complement of enzymes necessary for the synthesis of all steroid molecules. The ability of different tissues to secrete specific steroids depends upon the quantity of individual enzymes contained within the cell.

Cholesterol is synthesized from acetate primarily in the smooth endoplasmic reticulum of steroid-secreting cells. It is transported to the mitochondria for cleavage of the side chain between carbons 20 and 22, which results in the formation of pregnenolone. *Progesterone* is formed from pregnenolone in the endoplasmic reticulum. The ketone group at carbon three (C3) and the double bond between C4 and C5 appear to be necessary for the biologic activity of progesterone. Changes in other rings (B, C, D) may also result in loss of biologic activity.

Progesterone serves as a precursor for the synthesis of male reproductive steroids, which contain 19 carbon atoms. *Testosterone* is the principal male steroid hormone. The ketone group at C3 and the hydroxyl group at C17 are important substituents for biologic activity. Other androgens, including androstenedione and dihydrotestosterone, may also be synthesized in the testis but in lesser amounts than testosterone.

Estrogens are derived from androgens by elimination of the C19 methyl group and aromatization of the A ring. The phenolic A ring and the substituents associated with C17 determine the biologic activity of the estrogens. *Estradiol-17β* is the principal estrogenic hormone in the ovary and has the greatest biologic activity.

It has not been possible to separate synthesis from release of steroid hormones, since there does not appear to be significant storage of these substances within steroid producing cells. Progesterone is synthesized associated with protein molecules in luteal tissue. These proteins may play a role in the transport of the steroid molecule within the cell and may be involved in release of the steroid into the circulatory system.

2. FUNCTIONS

The maintenance of the structural integrity and secretory activity of the secondary sexual organs in the male is dependent upon the production of testosterone. Secretory products from these organs serve as a source of energy for sperm cells following ejaculation and buffer them from changes in pH. Testosterone stimulates the germinal epithelium to produce spermatogonia. It also stimulates sexual activity (*libido*) in the male and may play a role in aggressive behavior. Androgens induce nitrogen retention and are commonly used for their anabolic effects.

One of the most important biologic effects of estrogen is to induce sexual receptivity (*estrus*) in the female. Small quantities of progesterone have a synergistic influence on the ability of estrogens to induce estrus. Estrogens are necessary for the maintenance of secondary sexual characteristics, they induce ductal development in the mammary gland and are necessary for normal development and function of the female reproductive tract. At puberty, increased levels of estrogens play a role in the calcification of the epiphyseal cartilages and limit bone growth. Estrogens also appear to potentiate the ability of oxytocin and prostaglandins to cause uterine contractions. Based on studies in rodents, it seems probable that estrogens are involved in attachment of the embryo to the uterine wall. In addition to these direct effects on reproduction, estrogens, like androgens, have been shown to exhibit anabolic effects.

Progesterone is the principal progestational hormone. It has synergistic action with estrogens to bring about estrous behavior and is necessary for the maintenance of pregnancy, probably through the ability to inhibit uterine contractility and to promote uterine glandular development. High levels of progesterone can inhibit the ovulatory surge of LH, which

Table 3-2. Functions of Gonadal Steroid Hormones

Source	Hormone	Functions
Ovary Placenta	Estrogens	Estrous behavior Feedback control of gonadotropins Maintenance of secondary sexual characters Duct development in mammary gland Growth of female reproductive tract Increased uterine contractility Anabolic effect
Ovary	Progesterone	Synergistic action with estrogen for estrous behavior Maintenance of pregnancy Inhibits uterine contractility Promotes glandular development in endometrium Promotes mammary alveolar growth Negative feedback on gonadotropins
Testis	Androgens	Secondary sexual characters Maintenance of secondary sexual organs Stimulate spermatogenesis Anabolic effect Libido; aggressive behavior

is the basis for most treatments for synchronization of estrus.

In both the male and female, steroid hormones may exert a *negative feedback* to inhibit the release of gonadotropins. This may result from decreased output of releasing hormones from the hypothalamus or a reduction in the ability of the pituitary gland to respond to the releasing hormones. In addition, appropriate concentrations of estrogens may result in a *positive feedback* and greatly increase the levels of gonadotropins in the blood stream. This may be the result of increased output of releasing hormones or increased ability of the adenohypophysis to respond (Table 3–2).

3. Mechanism of Action

An extensive review of the mechanism of action of steroid hormones is beyond the limits of this chapter (see O'Malley, 1971). Specific steroid binding proteins in blood are important for efficient transport of the steroid hormone from the site of production to target tissues. Binding proteins also impede the clearance of the steroid from the blood stream and, therefore, effectively extend the biologic use-fulness of these molecules. In the case of estradiol and testosterone, *sex hormone binding globulin* is the transport protein. Adrenocorticosteroids and progesterone are associated with corticosteroid-binding globulin. The binding globulins are characterized by a high affinity and a limited capacity for steroids. Albumin has essentially unlimited capacity but a low affinity for the steroid molecule. Free steroids in blood represent an extremely small proportion of the total.

Target tissues for steroids contain specific receptor molecules which have a higher affinity for the steroid molecule than does albumin or binding globulin. Specific receptors for estrogens and progesterone are found in the reproductive tract, the mammary gland, the hypothalamus and the adenohypophysis. Similarly, testosterone receptors are found in the cells of male accessory organs and hypothalamus, and receptors may be present in the adenohypophysis. These receptors are present in relatively low numbers within the target tissue cells and therefore can be saturated with low biologic concentration of circulating steroids.

The mechanism by which steroid hormones enter the target cells has not been completely elucidated (Fig. 3–5). It is known that specific receptors within the cytoplasm of target cells bind with the steroid molecule and form a stable complex. Testosterone is converted to dihydrotestosterone before it may be bound to specific receptor protein. The stable complex is activated through a mechanism which probably involves dissociation of a portion of the receptor protein. This activated complex then enters the nucleus and associates with deoxyribonucleic acid (DNA). The DNA probably comprises a single genome within the chromosome. The association of the steroid receptor complex with the DNA allows the synthesis of specific ribonucleic acid (RNA) molecules by DNA-dependent RNA polymerase. This RNA is transported to cytoplasm which results in the synthesis of specific protein molecules. These protein molecules are responsible for the intracellular changes seen in target tissues following stimulation with steroid hormones.

II. ENDOCRINE REGULATION OF REPRODUCTION

A. Females

1. PUBERTY

Very little is known concerning the endocrine factors which bring about the onset of reproductive cycles in the maturing female. Maturation of the hypothalamus, adenohypophysis and/or ovary may be required for cyclic reproductive performance to begin. Prepuberal pigs, calves and lambs can respond to stimulation with exogenous gonadotropins by ovulation. Therefore, the ovaries are capable of responding to gonadotropic stimulation prior to *puberty*. The adenohypophysis is also capable of secreting gonadotropic hormones within the first two weeks of life (Fig. 3–6). In the immature ewe there are peaks of LH release which increase in both frequency and amplitude until puberty (Foster et al., in press). Serum levels of FSH also appear to elevate as the animal approaches puberty. High constant levels

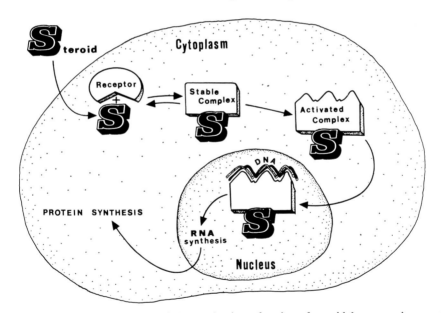

FIG. 3–5. Schematic representation of the mechanism of action of steroid hormones in target tissues. The steroid molecule enters the cell where it associates with a specific receptor molecule to form a stable complex. The stable complex becomes activated, probably by dissociation of a portion of the receptor molecule, and enters the nucleus where it associates with a portion of the chromosome (DNA). The synthesis of RNA occurs and the RNA is transported to the cytoplasm where it stimulates the synthesis of protein. The synthesized protein is responsible for stimulating the biologic activities of the cell.

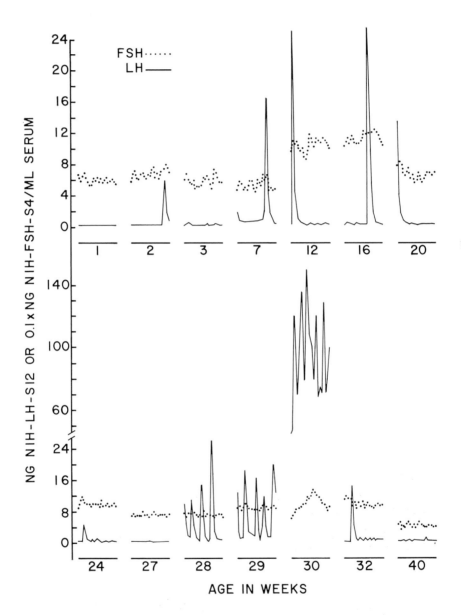

FIG. 3–6. Blood levels of LH and FSH throughout the life of a sheep. In this study 20 samples were obtained over a six-hour period once each week beginning at birth and continuing until the ewe became pregnant. First estrus and ovulation occurred during week 30 in this ewe and the samples were collected during the ovulatory LH peak. This ewe was bred during week 32 and was approximately eight weeks pregnant by week 40. (*From Foster et al., Endocrinology, in press.*)

of exogenous estradiol will suppress the peaks of LH and FSH which occur prior to puberty suggesting that all components of the negative feedback system are functional (Foster, 1970). A single injection of estradiol will also induce an LH release suggesting that the positive feedback systems are also functional. There is some evidence that prior to the first estrus, there is a short-term increase in serum levels of progesterone, however, the source of this progesterone has not been ascertained. (Gonzales-Padilla et al., unpublished data). Induction of specific hormone receptors in target tissues and synthesis of key intracellular enzymes for hormone production are the factors most likely involved in the onset of puberty.

2. Reproductive Cycle

Hormonal interactions are responsible for regulation of the reproductive processes (Fig. 3–7). The hypothalamus secretes releasing hormones into the hypothalamic-hypophyseal portal system.

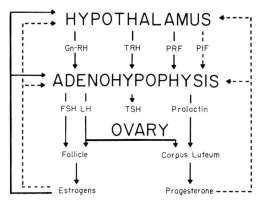

Fig. 3–7. Diagrammatic scheme of the hormonal interrelationships which regulate reproduction in domestic animals. The hypothalamus secretes releasing factors which influence the release of gonadotropic hormones from the adenohypophysis. FSH and LH stimulate follicle growth and estrogen secretion. Estrogens can have both positive (solid line) and negative (broken line) feedback effects on the hypothalamus or adenohypophysis. LH causes ovulation and acts synergistically with prolactin for development and maintenance of the corpus luteum and secretion of progesterone. Progesterone may exert a negative feedback on gonadotropin secretion at the hypothalamic or adenohypophyseal level.

These releasing hormones act on specific cells in the adenohypophysis and result in the release of gonadotropic hormones, FSH, LH and prolactin. FSH is secreted into the circulation and transported to the ovary, where it stimulates follicular development. LH is also secreted into the blood stream by the anterior lobe of the pituitary gland and acts synergistically with FSH to stimulate the secretion of estrogens by the follicle. Estrogens may exert both a positive and a negative feedback on the hypothalamus and/or pituitary gland to control the release of LH, FSH and prolactin.

The peak levels of LH and FSH secreted at estrus are responsible for rupture of the follicle and release of the ovum (*ovulation*). Following ovulation the secretion of estrogen and gonadotropins is greatly reduced. Even at these reduced levels LH is at least partially responsible for the transformation of *granulosa cells* to *luteal cells* which secrete progesterone. This transformation does not occur in hypophysectomized animals unless they are constantly infused with LH (Kaltenbach et al., 1968). Prolactin and LH have a synergistic effect in promoting progesterone secretion by the corpus luteum. Progesterone exerts a negative feedback on the hypothalamus and/or adenohypophysis to decrease gonadotropin secretion.

The events which occur during the estrous cycle of different domestic animals are similar although the length of the cycle and the duration of estrus are unique for each species. The ewe will be discussed in depth because more information is available concerning hormonal interrelationships in this species than for other domestic animals.

Follicular Phase. The relative circulating levels of gonadotropins and steroids during the estrous cycle of the ewe are plotted in Figure 3–8. The relative blood flow to both ovaries is also shown in this figure. A few days prior to estrus several follicles begin to develop. These follicles grow slowly until the follicular phase of the cycle which begins approximately 48 hours prior to estrus. As the follicle matures, granulosa cells increase in number and layers of theca cells grow and encompass the periphery of the follicle. Epithelioid theca cells differentiate into the theca interna, which synthesizes and secretes estrogens under the influence of LH and FSH, and the theca externa, a connective tissue capsule which surrounds the entire follicle. One or two follicles undergo a rapid maturation during the follicular phase of the cycle. FSH is the hormone primarily responsible for follicular growth, however its concentration in the circulation does not increase when the follicles are growing most rapidly. In fact, the levels tend to decrease during the follicular phase of the cycle.

The highest level of estrogen secretion occurs during the late follicular phase of the cycle. Although the theca interna secretes primarily estradiol-17β into the circulation, it may be converted to estrone by enzymes in the red blood cells. These two estrogens are in equilibrium in the general circulation. Estrogen secretion does not begin to increase rapidly until progesterone levels have begun to decline rapidly. During the period of most rapid follicular growth the concentrations of estrogens in the blood stream reach a maximum and the ewe shows estrous behavior (sexual receptivity).

Estrus. Estrus is due to a local effect of estrogens on the central nervous system. Implantation of minute amounts of estradiol-17β into the hypothalamus results in estrous behavior whereas identical quantities administered systemically have no effect. Systemic administration of much higher levels of estradiol are efficacious in inducing estrus. Progesterone has an important synergistic action with estradiol in inducing sexual receptivity. Increasing levels of estrogen during the follicular phase of the cycle also exert a positive feedback on the hypothalamo-hypophyseal axis resulting in the release of peak levels of LH, FSH, and prolactin during estrus. The increased secretion of LH and FSH may be due to increased release of Gn-RH from the hypothalamus or to increased sensitivity of LH and FSH secreting cells within the adenohypophysis to Gn-RH.

High levels of estrogen reduce the release of PIF from the hypothalamus, resulting in increased prolactin secretion. At estrus the concentration of LH in the circulation may increase as much as 200- to 300-fold. The concentration of FSH rarely more than doubles at this time. This is, however, the maximum release of FSH seen at any time during the estrous cycle. The release of LH during the preovulatory surge does not occur as one massive dumping of the pituitary nor as a steady secretion over an extended period of time. The LH is released in short bursts followed by

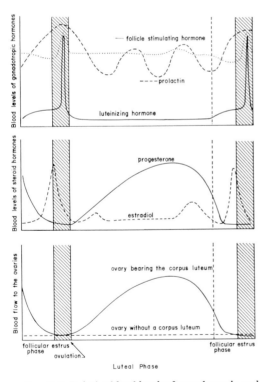

Fig. 3–8. Relative blood levels of gonadotropic and steroid hormones throughout the estrous cycle of the ewe. The data represent mean daily levels from several animals. The relative blood flow to both ovaries was determined using Doppler ultrasonic blood flow equipment.

periods when circulating levels are declining (Fig. 3–9). Many of these bursts at relatively frequent intervals make up the pre-ovulatory LH peak at estrus. Blood levels of LH are maintained by bursts of LH release in all reproductive states examined including times during the estrous cycle (Fig. 3–9), in pre-puberal animals (Fig. 3–6), in castrated animals and in males.

The high concentrations of LH in the circulation at estrus appear to activate

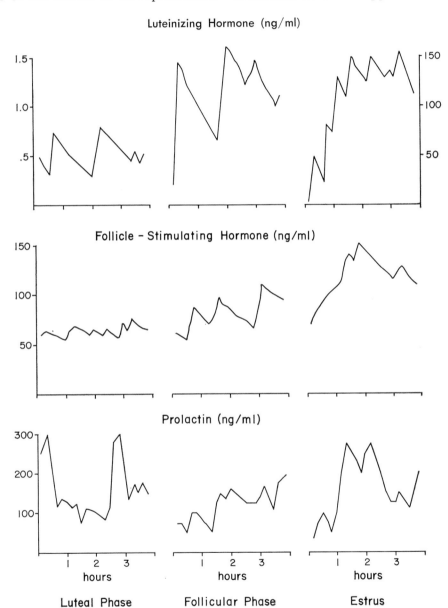

Luteinizing Hormone (ng/ml)

Follicle – Stimulating Hormone (ng/ml)

Prolactin (ng/ml)

hours

hours

hours

Luteal Phase   Follicular Phase   Estrus

Fig. 3–9. Patterns of blood gonadotropins at different stages of the estrous cycle. In all cases blood was collected every six minutes for four hours and assayed for all three gonadotropic hormones. Note the similarity of the pattern of LH release at different stages of the cycle even though the absolute quantity of LH on the day of estrus was 100 times higher than during the luteal or follicular phases of the cycle. As can be seen, FSH and prolactin are also released in bursts at all times during the cycle.

degradatory enzymes including collagenase which digests connective tissues in the wall of the follicle ultimately resulting in ovulation. Fluid pressure within the follicle plays a minimal or no role in the ovulatory mechanism since intrafollicular pressure decreases prior to ovulation (Rondell, 1970). Ovulation occurs approximately 20 hours following the LH peak in most domestic animals.

Luteal Phase. Following ovulation the granulosa cells in the cavity left by rupture of the follicle grow and divide under the influence of gonadotropins. The growing cells accumulate lipid and become large, polygonal cells capable of synthesizing and secreting progesterone and are collectively termed the *corpus luteum.* In some species thecal cells may also luteinize and become part of the corpus luteum. The secretion of progesterone increases as the corpus luteum grows, until maximum size is attained on approximately day 10 of the estrous cycle. The corpus luteum is a highly vascularized structure and when fully developed receives approximately 97% of the blood which supplies the ovary of the ewe (Niswender et al., 1973). The large quantity of blood which reaches the corpus luteum is probably a reflection of its metabolic activity and is necessary to supply this organ with the necessary nutrients for synthesis of progesterone. In addition, the increased blood supply is probably necessary to transport progesterone to its target organs. Although somewhat controversial, both LH and prolactin seem to play a role in regulating luteal function in sheep (Schroff et al., 1971; Denamur et al., 1973).

Once the corpus luteum has developed completely and the secretion of progesterone is at a maximum, the function of the corpus luteum continues only a few days unless the ewe becomes pregnant. In the nonpregnant ewe 13 to 15 days following ovulation the production of progesterone diminishes rapidly, and the corpus luteum begins to degenerate (regress). During this period the concentrations of LH, FSH and prolactin in the circulation are no lower than those found during the period of maximum luteal function. Thus it does not appear that a lack of gonadotropin secretion from the pituitary is responsible for regression of the corpus luteum.

In several species, including the sheep, the uterus is involved in regression of the corpus luteum, since hysterectomy during the luteal phase of the cycle extends luteal function for a period approximating the length of gestation. In addition, the influence of the uterus appears to be a local effect since removal of the uterine horn ipsilateral to the corpus luteum prolongs luteal function while removal of the contralateral horn has no effect. The exact mechanism whereby the uterus influences the corpus luteum remains obscure. It may be mediated via the secretion of prostaglandin (PG) $F_2\alpha$, a lipid derived from fatty acids normally found in the circulation. Levels of PG $F_2\alpha$ increase in the endometrium during the late luteal phase of the cycle concurrent with elevated concentrations in the uterine vein (Bland et al., 1971; McCracken et al., 1971; Wilson et al., 1972). Infusions or injections of massive quantities of PG $F_2\alpha$ may cause premature luteal regression. Since PG $F_2\alpha$ is cleared very rapidly from the circulatory system high local concentrations would be required to influence luteal activity which may explain how the uterus exerts its local effect on the ovary. However, conclusive proof that PG $F_2\alpha$ plays a physiologic role in regression of the corpus luteum is not available.

Once luteal regression begins it occurs rather rapidly. The corpus luteum essentially ceases to function within 12 to 24 hours following the onset of luteal regression. Once luteal regression has occurred the ewe is ready to repeat the estrous cycle in its entirety.

FSH is responsible for growth and maturation of the follicle. However, as seen in Figure 3–8 blood levels of FSH tend to decrease during the follicular phase of the cycle. This may be ex-

Table 3-3. Metabolic Clearance Rates (MCR) and Secretion Rates (SR) of Gonadotropic Hormones During the Estrous Cycle in Sheep

Phase of Estrous Cycle	LH		FSH		Prolactin	
	MCR ml/min	SR ng/min	MCR ml/min	SR μg/min	MCR ml/min	SR μg/min
Luteal[1]	41	10	33	3	96	10
Follicular[2]	48	58	24	5	103	13
Estrus[3]	44	1340	30	10	112	9

[1]Day 8 was used to represent the luteal phase of the cycle.
[2]The day prior to estrus, when follicles were growing most rapidly, was used to represent the follicular phase of the cycle.
[3]The day of the major LH peak was used to represent estrus.

plained by a negative feedback of the rising levels of estrogen on FSH secretion. However, were this true one would expect LH levels to decline as well and this clearly is not the case. An alternative explanation may be that the developing follicles utilize significant quantities of FSH, thereby reducing circulating levels. A similar situation exists between blood levels of LH and prolactin and luteal function. When blood levels of progesterone are maximal, LH and prolactin are at their lowest levels during the cycle. This may reflect a negative feedback of progesterone on LH and prolactin secretion or utilization of these gonadotropins by the highly functional corpus luteum. If utilization of gonadotropins by the ovary is significant this should be reflected in the metabolic clearance rate (MCR) of these hormones (Table 3-3). Since the MCR for LH and prolactin are lowest during the luteal phase of the cycle and are lowest for FSH during the follicular phase of the cycle it does not seem likely that the ovaries play a significant role in the metabolism of gonadotropins. Blood levels of these hormones are indicative of their secretion rates (Akbar et al., 1973a).

Blood flow to the ovary bearing the corpus luteum is increased three- to ten-fold during periods of active luteal function (Moore et al., 1972). Therefore, the quantities of gonadotropic hormones reaching the corpus luteum are increased proportionately to the increased blood flow (Fig. 3–8). These observations probably explain the mechanism whereby LH and prolactin are responsible for increased functioning of the corpus luteum during the luteal phase of the cycle.

3. PREGNANCY

The discussion of pregnancy will be general with respect to the cow, the ewe and the sow, since the events appear to be very similar. The mare will be considered separately.

There are no endocrine changes which are known to occur at the time of fertilization. Likewise, there do not appear to be endocrine changes during the first few days following conception that do not also occur in the animal that has not conceived. Although there are no specific endocrine changes during the first few days of pregnancy which are not also seen in the cycling animal, estradiol and progesterone influence the transport of sperm through the reproductive tract, transport of the ovum through the oviduct and into the uterus and preparation of the uterus for pregnancy. On day 12 or 13 in the sheep there is a signal from the pregnant uterus which results in transformation of the corpus luteum of the cycle into the corpus luteum of gestation, which is maintained until just prior to parturition.

The nature of the signal from the pregnant uterus is not known, however it has been proposed that the embryo may prevent the production of luteolytic substances and thus prevent regression of the corpus luteum. Alternatively, the embryo may be responsible for production of a luteotropic substance which may override the luteolytic effect of the uterus. The signal for maintenance of the corpus luteum does not require attachment of the embryo to the uterus since this signal is given prior to day 18, the day nidation occurs in the ewe. Secretion of progesterone increases slightly following nidation and continues to rise until a few days prior to parturition, at which time progesterone levels in the peripheral circulation begin to decline. Progesterone comes predominantly from the corpus luteum in the cow, ewe and sow, but the placenta may produce small quantities in some cases. During early pregnancy the concentration of estrogens in the peripheral circulation remains very low, comparable to that found during the luteal phase of the estrous cycle. Estrogen levels begin to increase near mid-pregnancy. At first, the increase is gradual but estrogen levels rise sharply during the last few days prior to parturition. It is likely that the placenta is the primary source of estrogen production during pregnancy, since corpora lutea of domestic animals produce limited amounts of estrogens. Ovarian follicles may produce estrogens during gestation. During the very early stages of pregnancy blood concentrations of all three gonadotropins are essentially the same as during the luteal phase of the estrous cycle. Subsequently, the levels decrease and remain depressed until just prior to parturition. At parturition the concentrations of FSH and LH increase slightly, whereas circulating prolactin may increase as much as 200-fold in the ewe and cow, presumably to prepare the mammary gland for the onset of lactation.

The hormones produced by the adrenal cortex (the adrenal corticosteroids) do not fluctuate greatly during pregnancy. These hormones, primarily corticosterone and cortisol, are synthesized and secreted due to stimulation by ACTH from the adenohypophysis. The importance of the adrenal corticosteroids in the pregnant animal is not known, however they are involved in the regulation of metabolic processes necessary for normal pregnancy. The concentrations of these hormones rise just prior to parturition.

The endocrine interrelationships in the pregnant mare are quite different from those in other domestic animals. The corpus luteum of pregnancy is the main source of progesterone the first 30 to 40 days following conception. At this time fetal trophoblastic cells begin to secrete large amounts of pregnant mare serum gonadotropin (PMSG) into the circulation. PMSG stimulates follicular growth and luteinization of these follicles may occur with or without ovulation resulting in the formation of accessory corpora lutea. These accessory corpora lutea secrete large amounts of progesterone. The maximum secretion of PMSG occurs approximately 60 days following conception and the maximum concentration of circulating progesterone occurs a few days later. PMSG levels decline rapidly and become nondetectable at approximately 170 days of gestation.

The concentration of progesterone decreases coincidentally with PMSG and disappears from the circulation about mid-pregnancy; it remains nondetectable until a few days prior to parturition when small amounts reappear in the circulation. The concentration of LH in serum is low during early pregnancy in the mare and is nondetectable from 200 days of gestation until slightly before parturition, when levels are similar to those found during the luteal phase of the estrous cycle.

Estrogenic hormones in the mare are low for the first 80 days following conception, comparable to levels found during the luteal phase of the estrous cycle. Nidation of the embryo occurs about day 60 in the mare and once the attached embryo is fully established the concentration of estrogens in the maternal circulation increase rapidly and are elevated 50-fold by day 210 of gestation.

After day 210 the production of estrogens by the placenta decreases until parturition when the levels are approximately 20% of those found at midpregnancy. Not only do estrogen levels decline during the latter stages of gestation but there is a shift in the type of estrogens produced. Immediately following placentation estrone and estradiol are the major estrogens which appear in the maternal circulation. However, as pregnancy progresses the ratio of equilin and equilenin to estrone increases until more equilin and equilenin are produced during late pregnancy than estrone. Similarly, the ratio of dihydroequilin and dihydroequilenin to estradiol increases throughout gestation (Savard, 1961). The function of these estrogens, unique to the mare, in the maintenance of normal pregnancy is not known.

4. PARTURITION

The concentration of circulating progestins decreases the last few days before parturition. This decrease may remove a progesterone block responsible for maintaining the uterus in a quiescent state. In contrast, the concentrations of estrogens increase dramatically the last few days prior to parturition. Estrogens increase the spontaneous activity of the myometrium, which is normally suppressed by progesterone. In sufficient doses estrogens will terminate pregnancy in the sow, cow and the ewe. These facts suggest a sensitizing action of estrogens on the myometrium of the uterus during late pregnancy. Oxytocin is also involved in the normal process of parturition. This hormone stimulates contractions of the smooth muscle of the uterus. An increase in the estrogen to progesterone ratio increases the response of the uterus to oxytocin, further facilitating uterine contractions.

Adrenal corticosteroids rise just prior to parturition in the cow and ewe. Synthetic analogues of corticosteroids will initiate parturition during the third trimester of pregnancy in these species.

This suggests a role for these hormones in parturition but it has not been clearly defined.

Relaxin is a polypeptide hormone isolated from the ovary and/or placenta of several species. Its role is to cause a relaxation of the pelvic ligaments resulting in a widening of the birth canal so the fetus may be expelled. It acts on tissues already prepared by estrogen and progesterone but is effective to a lesser extent when given alone. In addition, relaxin softens the connective tissue in the uterine musculature during the latter stages of pregnancy to allow the uterus to expand and accommodate the growing fetus.

The role of the fetus in the process of parturition must not be overlooked. The fetal pituitary can secrete gonadotropins which may stimulate placental production of estrogen near term (Foster et al., 1972). Likewise, the fetal adrenals secrete corticosteroids that may initiate parturition. This is indeed likely in the cow since removal of the adrenal glands from the fetus results in prolonged gestation. However, in other species if the fetus is removed from the placenta and uterus, the placental membranes will be expelled at approximately the time when normal parturition would have been expected.

B. Males

1. PUBERTY

LH and FSH are released in prepuberal rams and bulls in a pattern similar to that seen in prepuberal ewes (Fig. 3–6). Exogenous testosterone or estradiol will suppress the release of LH and FSH suggesting that the components of the negative feedback system are functional. However, additional information will be required before we completely understand the endocrine factors which regulate the onset of puberty in males.

2. SPERMATOGENESIS

The formation of spermatogonia from the germinal epithelium in the seminiferous tubules appears to be stimulated by testosterone. There does not appear to be any hormonal requirement for mitotic division of spermatogonia and the formation of primary spermatocytes. Reduction division of the primary spermatocyte to secondary spermatocytes appears to be testosterone-dependent. Further divison of the secondary spermatocytes to spermatids may also require testosterone. The maturation of spermatids into fully developed sperm cells requires the presence of FSH. Normal functioning of the male accessory organs requires testosterone. Both LH and prolactin appear to be important for normal testosterone secretion (Steinberger, 1971).

III. COMPARATIVE REPRODUCTIVE ENDOCRINOLOGY

A. Blood Levels of Gonadotropic Hormones

Until recently, reliable methods for quantifying blood levels of gonadotropic and steroid hormones were not available. However the development of radioimmunoassay procedures has provided very sensitive and reliable techniques for measuring these hormones. Radioimmunoassays are presently available for all three gonadotropic hormones in sheep, cattle and pigs and for luteinizing hormone in the mare. In addition, all major steroid hormones can now be quantified by sensitive radioimmunoassays.

In general, the patterns of gonadotropic hormones in blood are similar in all domestic animals during the estrous cycle (Fig. 3–10). In fact, the patterns of LH release in sheep and cattle are essentially identical. Luteal phase levels range from 0.2 ng/ml to 2 ng/ml and are elevated two- to three-fold during the follicular phase of the cycle. In both of these species, the peak of LH at estrus may increase more than 200-fold above luteal phase levels and this increase is of relatively short duration, i.e. LH can reach peak levels and return to baseline

within 12 hours. In the pig, luteal phase levels are similar to those observed in the ewe and cow, however peak levels at estrus rarely exceed a four-fold increase, although they may remain elevated for up to 20 hours. Luteal phase levels of LH in the mare are somewhat higher (3 ng/ml–8 ng/ml) than those observed in other domestic animals. In addition, LH levels are elevated for the duration of estrus with peak levels (75 ng/ml–100 ng/ml) occurring 12 to 24 hours prior to ovulation. In all four species the blood levels of LH are somewhat lower during pregnancy than those observed during the luteal phase of the cycle.

The cow and ewe also show patterns of FSH release which are very similar (L'Hermite et al., 1972; Akbar et al., 1973b). In both cases FSH tends to decline during the follicular phase of the cycle, peak at estrus coincidentally with the LH peak and then decrease and remain low throughout the luteal phase of the cycle. In the very limited data available in the pig there does not appear to be a cyclic pattern of FSH release (Wilfinger et al., 1973). The general patterns of prolactin secretion in cattle and sheep are quite similar, although prolactin levels in the ewe appear more variable. Prolactin levels fluctuate considerably during the luteal phase of the cycle but are consistently elevated during estrus. During pregnancy, prolactin levels are relatively low (2 ng/ml– 5 ng/ml) until a few days prior to parturition at which time they rise to 300 ng/ml–1000 ng/ml (Davis et al., 1971; Swanson and Hafs, 1971). Prolactin in pigs tends to be low except at estrus when peak levels occur (Brinkley et al., 1972).

B. Blood Levels of Steroid Hormones

In general, estrogen levels are high when progesterone levels are low and vice versa (Fig. 3–11). The levels of these steroid hormones are higher in the sow. This is probably due to the increased

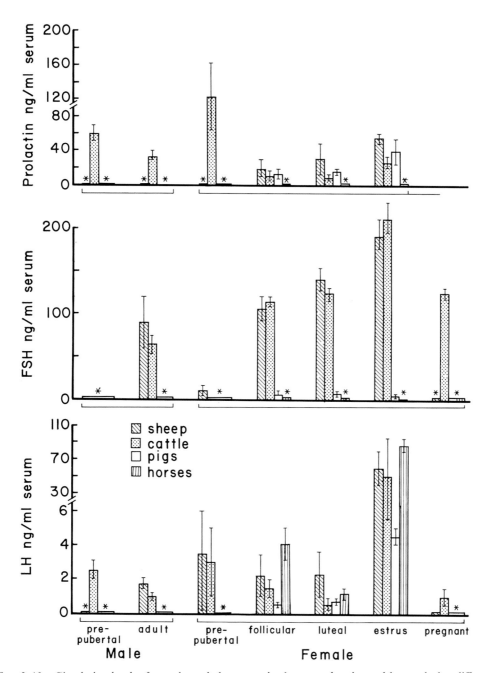

Fig. 3–10. Circulating levels of gonadotropic hormones in sheep, cattle, pigs and horses during different reproductive states. The columns depict average levels of the hormones and the bars (I) encompass the normal physiologic range. An asterisk (*) indicates data are not available.

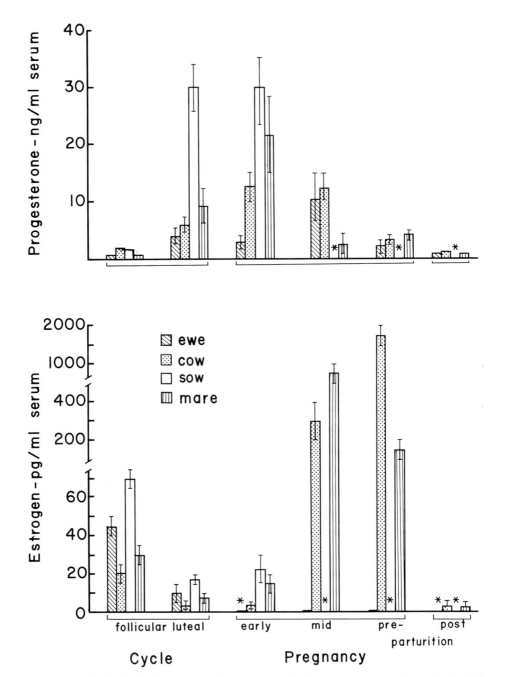

F<small>IG</small>. 3–11. Circulating levels of estrogens and progesterone in the ewe, cow, sow and mare during different reproductive states. The columns depict average levels of the hormone and the bars (I) encompass the normal physiologic range. An asterisk (*) indicates data are not available.

numbers of follicles which mature and ovulate in the sow which leads to increased numbers of corpora lutea. In all species, estrogen levels are higher during the follicular phase than at other times during the estrous cycle. During this period blood concentrations of estrogens in the cow average 20 pg/ml whereas those in the sow average 70 pg/ml (Wettemann et al., 1972; Henricks et al., 1972a). Estrogen levels in the ewe and the mare during the follicular phase are intermediate to those of the cow and sow. The levels of progesterone in all domestic animals are less than 1 ng/ml of serum during the follicular phase of the estrous cycle.

During the luteal phase of the estrous cycle the concentrations of estrogens in serum range from 4 pg/ml to 17 pg/ml in the cow and sow, respectively. During this period progesterone ranges from 4 ng/ml to 30 ng/ml.

The levels of estrogens and progesterone during early pregnancy are similar to those found during the luteal phase of the estrous cycle (Henricks et al., 1971; Guthrie et al., 1973). However, as pregnancy progresses the levels of steroid hormones in the circulation tend to increase. In the ewe and cow the blood concentrations of progesterone rise to 15 ng/ml one to two weeks prior to parturition, at which time they begin to decline. The concentration of estrogens begins to rise about mid-pregnancy in the cow and continues to rise until just prior to parturition, at which time concentrations of 2000 pg/ml of serum or greater may be obtained (Henricks et al., 1972b). Data concerning the blood levels of estrogens and progesterone during mid and late pregnancy in the sow are not available, however urinary excretion patterns of these steroids would indicate a similar pattern of secretion.

The concentrations of steroid hormones in the mare follow a somewhat different pattern during pregnancy. Progesterone is maximal (30 ng/ml) during the period when the accessory corpora lutea of pregnancy are in their most active state

at approximately day 80. The level of progesterone falls as PMSG disappears from the circulation and becomes non-detectable near mid-pregnancy (Holtan et al., 1973). Progesterone reappears in the blood stream of the mare at approximately 300 days of gestation and reaches a maximum level of 4 ng/ml a few days prior to parturition. During the period when progesterone levels are falling the concentrations of estrogens are rapidly increasing. The maximum concentration of estrogen in the serum is found between 210 and 240 days of gestation. Estrogen levels decline for the remainder of gestation to a concentration of 250 pg/ml just prior to parturition (Nett et al., 1972). Following parturition the concentration of steroid hormones in the blood stream declines rapidly and by two or three days postpartum the levels of both estrogen and progesterone are lower than those found at any time during the estrous cycle.

C. Placental and/or Embryonic Hormones

Hormones originating in the pregnant uterus which influence the function of the ovaries have been clearly demonstrated in three species—the horse, man and the rat. Pregnant mare serum gonadotropin and the role it plays in stimulating accessory corpora lutea and progesterone synthesis in mares has already been discussed. Human chorionic gonadotropin (HCG) is a glycoprotein hormone which has biologic activity similar to LH and appears to be responsible for maintenance of the corpus luteum of gestation in women (Niswender et al., 1972). The placenta of the rat secretes a luteotropic hormone with biologic activity similar to both prolactin and LH, which is responsible for maintenance of the corpus luteum during the last half of gestation (Linkie and Niswender, 1973). Rat placental gonadotropin is first detectable in maternal serum and placentae on day 11 of gestation, peaks on day 12 and is no longer detectable by day 15.

The role played by placental or embryonic hormones in the regulation of ovarian function in domestic species other than the horse has not been well-defined. In the ewe, the signal for maintenance of the corpus luteum of gestation occurs on day 13. This conclusion is based on the fact that if the embryo is removed from the uterus on day 12 the ewe will return to estrus at the normal time on day 16. However, if the embryo is removed from the uterus on day 13 the functional life of the corpus luteum is extended and the ewe does not return to estrus until approximately day 25 (Moor and Rowson, 1966). Direct evidence supporting the existence of an embryonic or placental luteotropic hormone in the ewe was obtained when Rowson and Moor (1967) demonstrated that homogenized day 25 sheep embryos would extend the life span of the corpus luteum in cycling ewes when infused into the uterus.

There is no direct evidence for existence of an embryonic or placental luteotropic hormone in sows or cows. However it seems very probable that luteotropic hormones do arise from the pregnant uterus in these species.

As discussed previously, an alternative explanation for maintenance of the corpus luteum of gestation is that the embryo inhibits PG $F_2\alpha$ secretion by the uterus. However, it has not been demonstrated conclusively that PG $F_2\alpha$ plays a physiologic role in luteal regression during the cycle or that the presence of an embryo inhibits uterine secretion of this substance.

CONCLUDING REMARKS

The hormonal interrelationships which are responsible for regulation of the reproductive processes in domestic animals are very complex. The effects of the central nervous system are mediated via releasing hormones secreted by the hypothalamus. Releasing hormones are responsible for the synthesis and release of gonadotropic hormones from the adenohypophysis. The gonadotropic hormones stimulate secretion of the sex steroid hormones and play an important role in the maturation and release of gametes from the gonad. The sex steroid hormones secreted by the gonads are involved in regulating sexual behavior, maintenance of pregnancy, spermatogenesis, maintenance of the secondary sex characteristics and feedback on the hypothalamo-hypophyseal axis to control secretion of gonadotropins.

REFERENCES

Akbar, A. M., Nett, T. M. and Niswender, G. D. (1973a). Metabolic clearance and secretion rates of gonadotropins at different stages of the estrous cycle in ewes. *Endocrinology* (In press).

Akbar, A. M., Reichert, L. E., Jr., Dunn, T. G., Kaltenbach, C. C. and Niswender, G. D. (1973b). Bovine FSH in serum measured by radioimmunoassay. *J. Anim. Sci.* 37, 299.

Bland, K. D., Horton, E. W. and Poysen, N. L. (1971). Levels of prostaglandin $F_2\alpha$ in the uterine venous blood of sheep during oestrous cycle. *Life Sci. 10*, 509–517.

Brinkley, H. J., Rayford, D. L. and Young, E. D. (1972). Porcine prolactin radioimmunoassay. *J. Anim. Sci. 35*, 237.

Davis, S. L., Reichert, L. E., Jr. and Niswender, G. D. (1971). Serum levels of prolactin in sheep as measured by radioimmunoassay. *Biol. Reprod. 4*, 145–153.

Denamur, R., Martinet, J. and Short, R. V. (1973). Pituitary control of the ovine corpus luteum. *J. Reprod. Fert. 32*, 207–220.

Foster, D. L. (1970). *Regulation of Luteinizing Hormone in the Fetal and Neonatal Lamb.* Ph. D. Thesis, University of Illinois.

Foster, D. L., Roche, J. F., Karsch, F. J., Norton, H. W., Cook, B. and Nalbandov, A. V. (1972). Regulation of luteinizing hormone in the fetal and neonatal lamb. I. LH concentrations in blood and pituitary. *Endocrinology 90*, 102–111.

Foster, D. L., Lemons, J. L., Jaffe, R. B. and Niswender, G. D. (In press). Regulation of gonadotropic and gonadal hormones in the immature sheep. I. Sequential patterns of LH and FSH in individual females in the presence or absence of ovaries. *Endocrinology.*

Gay, V. L. (1972). The hypothalamus: physiology and clinical use of releasing factors. *Fertil. Steril. 23*, 50–63.

Gonzales-Padilla, E., Wiltbank, J. N. and Niswender, G. D. (1973). Hormonal interrelationships prior to and during puberty in beef heifers. Unpublished data.

Guthrie, H. D., Henricks, D. M. and Handlin, D. L. (1973). Plasma estrogen, progesterone and luteinizing hormone prior to estrus and during early pregnancy in pigs. *Endocrinology* (In press).

Henricks, D. M., Lamond, D. R., Hill, J. R. and Dickey, J. F. (1971). Plasma progesterone concentrations before mating and in early pregnancy in the beef heifer. *J. Anim. Sci. 33*, 450–454.

Henricks, D. M., Guthrie, H. D. and Handlin, D. L. (1972a). Plasma estrogen, progesterone and luteinizing hormone levels during the estrous cycle in pigs. *Biol. Reprod. 6*, 210–218.

Henricks, D. M., Dickey, F. J., Hill, J. R. and Johnston, W. E. (1972b). Plasma estrogen and progesterone levels after mating, and during late pregnancy and postpartum in cows. *Endocrinology 90*, 1336–1342.

Holtan, D. W., Nett, T. M. and Estergreen, V. L. (1973). Plasma progestins in pregnant and cycling mares. *J. Anim. Sci. 37*, 315–325.

Kaltenbach, C. C., Graber, J. W., Niswender, G. D. and Nalbandov, A. V. (1968). Luteotropic properties of some pituitary hormones in nonpregnant or pregnant hypophysectomized ewes. *Endocrinology 82*, 818–824.

L'Hermite, M., Niswender, G. D., Reichert, L. E., Jr. and Midgley, A. R., Jr. (1972). Serum follicle stimulating hormone in sheep as measured by radioimmunoassay. *Biol. Reprod. 6*, 325–332.

Li, C. H. (1972). Hormones of the adenohypophysis. *Proc. Amer. Philos. Soc. 116*, 365–382.

Linkie, D. M. and Niswender, G. D. (1973). Characterization of rat placental luteotropin: physiological and physiocochemical properties. *Biol. Reprod. 8*, 48–57.

McCracken, J. A., Baird, D. T. and Goding, J. R. (1971). Factors affecting the secretion of steroids by the transplanted ovary in the sheep. *Rec. Prog. Hormone Res. 27*, 537–582.

Moor, R. M. and Rowson, L. E. A. (1966). Local maintenance of the corpus luteum in sheep with embryos transferred to various isolated portions of the uterus. *J. Reprod. Fert. 12*, 539–550.

Moore, R. T., Mohrman, D. E. and Niswender, G. D. (1972). Variations in blood flow to the ovaries of sheep during the estrous cycle. *Proc. 4th Internat. Congr. Endocrinology* p. 64.

Nett, T. M., Holtan, D. W. and Estergreen, V. L. (1972). Plasma estrogens in pregnant mares. *J. Anim. Sci. 34*, 914.

Niswender, G. D., Menon, K. M. J. and Jaffe, R. B. (1972). Regulation of the corpus luteum during the menstrual cycle and early pregnancy. *Fertil. Steril. 23*, 432–442.

Niswender, G. D., Diekman, M. A., Nett, T. M. and Akbar, A. M. (1973). Relative blood flow to the ovaries of cycling and pregnant ewes. *Proc. Soc. Study Reprod. 6, 69.*

O'Malley, B. W. (1971). "Steroid Hormones and Synthesis of Specific Proteins." pp. 455–458. In *Hormonal Steroids* (Proc. 3rd Internat. Congr. Hormonal Steroids. Hamburg, 1970.) Amsterdam, Excerpta Medica.

Phifer, R. F., Midgley, A. R. and Spicer, S. S. (1973). Immunologic and histologic evidence that follicle stimulating hormone and luteinizing hormone are present in the same cell type in the human pars distalis. *J. Clin. Endocr. Metab. 36*, 125–141.

Rondell, P. (1970). Follicular process in ovulation. *Fed. Proc. 29*, 1875–1879.

Rowson, L. E. A. and Moor, R. M. (1967). The influence of embryonic tissue homogenates infused into the uterus, on the life-span of the corpus luteum in the sheep. *J. Reprod. Fert. 13*, 511–516.

Savard, K. (1961). The estrogens of the pregnant mare. *Endocrinology 68*, 411–416.

Schally, A. V., Arimura, A. and Kastin, A. J. (1973). Hypothalamic regulatory hormones. *Science 179*, 341–350.

Schroff, C., Kaltenbach, C. C., Graber, J. W. and Niswender, G. D. (1971). Maintenance of corpora lutea in hypophysectomized ewes. *J. Anim. Sci. 33*, 268.

Steinberger, E. (1971). Hormonal control of mammalian spermatogenesis. *Physio. Rev. 51*, 1–22.

Steinberger, A., Chowdhury, M. and Steinberger, E. (1973). Cultures of rat anterior pituitary cells with differential gonadotropin secretion and tinctorial properties. *Endocrinology 92*, 18–21.

Swanson, L. V. and Hafs, H. D. (1971). LH and prolactin in blood serum from estrus to ovulation in holstein heifers. *J. Anim. Sci. 33*, 1038–1041.

Van Hall, E. V., Vaitukaitis, J. L., Ross, G. T., Hickman, J. W. and Ashwell, G. (1971). Effects of progressive desialylation on the rate of disappearance of immunoreactive HCG from plasma in rat. *Endocrinology 89*, 11–15.

Wettemann, R. D., Hafs, H. D., Edgerton, L. A. and, Swanson, L. V. (1972). Estradiol and progesterone in blood serum during the bovine estrous cycle. *J. Anim. Sci. 34*, 1020–1024.

Wilfinger, W. W., Brinkley, H. J. and Young, E. P. (1973). Plasma FSH in the estrous cycle of the pig. *J. Anim. Sci. 37*, 333–341.

Wilson, L., Cenedella, R. J., Butcher, R. L. and Inskeep, E. K. (1972). Levels of prostaglandins in the uterine endometrium during the ovine estrous cycle. *J. Anim. Sci. 34*, 93–99.

Chapter 4

Reproductive Life Cycle

CHARLES THIBAULT and MARIE-CLAIRE LEVASSEUR

Although reproduction is closely related to fertility, fertility is not an all-or-none phenomenon; environmental factors are able to stimulate, inhibit and modulate reproduction in the female. Both physical and psychologic factors are involved in reproduction, and reproductive efficiency can be improved by manipulating these factors or using exogenous hormones.

After a period of full fertility, there is a progressive decrease of reproductive efficiency in all mammals due to general aging in the male and primarily to uterine aging in the female. At that time farm animals are usually slaughtered since the aging process cannot be retarded or prevented.

I. PUBERTY

Puberty is the period of adolescence when a male or a female is first able to release gametes. The first ovulation or estrus indicates that the female has reached puberty. It is much more difficult to determine the exact time of puberty in the male because the first differentiation of spermatogonia precedes release of the first spermatozoa from the seminiferous tubules by a month or more, and sperm transport from testes to vas deferens needs at least two weeks.

There is a delay of several days in the rat, several weeks in domestic animals

Table 4-1. Age at Puberty in European Sheep (Born in March)

Breed	Age at First Estrus
Romanov	179 ± 5 days
Finnoise	205 ± 6 days
Clun Forest	229 ± 4 days
Limousine	250 ± 5 days

and several years in primates between the first ovulation or ejaculation and the ability to produce offspring, at which point the animal reaches sexual maturity. The mean number of inseminations needed to obtain pregnancy in heifers is 2.1 for the first gestation, and 1.8, 1.7 and 1.6 respectively, for the next three gestations (Ahmed and Tantawy, 1959).

Age at puberty differs with the breed (Table 4–1). Climatic factors modify age at puberty. Social factors do the same; for example, the presence of individuals of the opposite sex generally hastens puberty, whereas the presence of individuals of the same sex delays onset of puberty (see Chapter 10).

Nutrition also plays an important role in age at puberty. There is a correlation between body weight and testicular weight or age at the first estrus. If nutrition is maintained at a normal level, puberty occurs when body weight reaches 60% of the adult body weight in sheep, 45% of the adult weight in cattle. A similar relation does not exist in pigs.

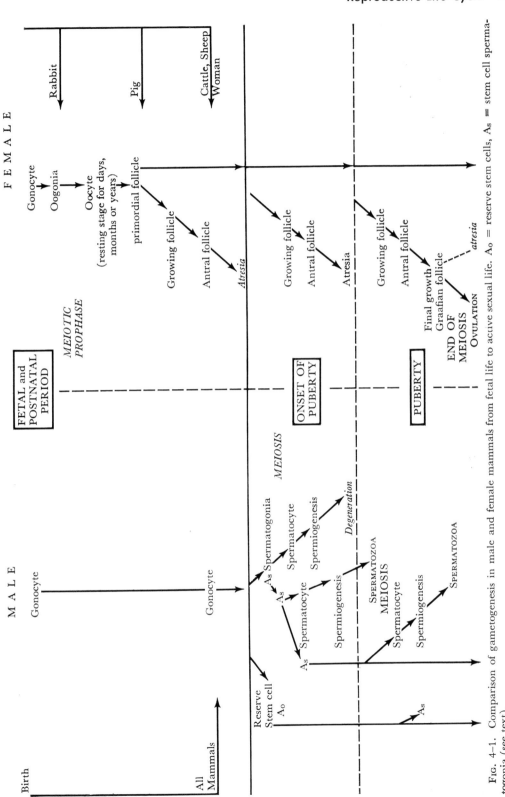

FIG. 4–1. Comparison of gametogenesis in male and female mammals from fetal life to active sexual life. A₀ = reserve stem cells, A$_s$ = stem cell spermatogonia (see text).

If growth is accelerated by overfeeding, body weight at time of puberty is higher than normal and the animal reaches maturity at a younger age. On the other hand, if growth is slowed down by underfeeding, puberty is delayed and body weight does not reach the level of that of normally fed animals.

There are two prerequisites for attaining puberty. First, during the fetal period or immediately after birth, gonads and steroid target organs such as hypothalamus, sexual organs and liver become either masculine or feminine in type. Secondly, progressive maturation of gonads and progressive attainment of steroid sensitivity of hypothalamus and related brain structures lead to sexual activity at puberty.

A. Gametogenesis

MALE

The structure of the testis remains unchanged from sexual differentiation to the onset of puberty; seminiferous tubules are lined with supporting cells, whereas undifferentiated germ cells or gonocytes occupy the central part of the tubules (Fig. 4-1). The only modification during this period is a slow increase in the relative number of germ cells. This increase is not gonadotropin-dependent; in hypophysectomized lamb the increase in gonocyte number is similar to that of the control. On the other hand, supporting cells need gonadotropin to divide during infancy. They are fewer in number in hypophysectomized lambs and can only be transformed into Sertoli's cells under the influence of gonadotropins.

At the onset of puberty, gonocytes migrate to the periphery of the tubule and produce spermatogonia, whereas supporting cells produce Sertoli's cells. When the first stages of meiotic prophase appear, a lumen is visible at the center of the seminiferous tubule. Two types of spermatogonia are differentiated from gonocytes: stem cells or A_S spermatogonia which give rise to A_1 spermatogonia and succeeding spermatocytes, spermatids and spermatozoa, and A_0 or reserve stem cells. The A_0 cells contribute to the

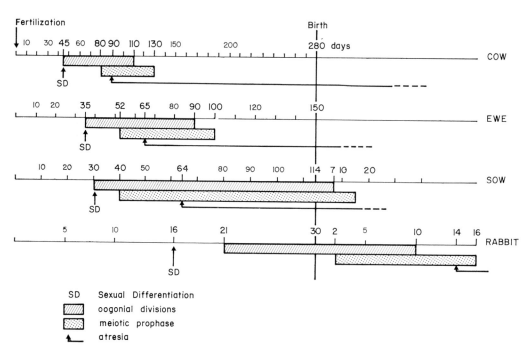

FIG. 4-2. Oogenesis in some mammals.

increase of the A_S population between puberty and adulthood. Such stem cell multiplication involves an increase in the weight of the testes, which is three-fold in the bull between one to five years. Spermatogonial differentiation, meiotic prophase and the transformation of spermatids into spermatozoa are gonadotropin-dependent.

FEMALE

Oogonia and oocytes are formed during the first half of fetal life in the ewe and cow. Evolution is slower in the sow, oogenesis being completed during the first days following birth (Figs. 4–1, 4–2). Around the oocytes forms a layer of somatic cells from which arise the three layers of the follicle—the external and internal theca and the granulosa. Contrary to that which occurs in the male at puberty, no fetal or maternal gonadotropin is necessary to the formation or multiplication of oogonia or to meiotic prophase development. This is demonstrated by the ineffectiveness of hypophysectomy performed on the ovine female fetus before 50 days (Mauleon, 1973).

Well before puberty (and as soon as the end of fetal life in ewe and cow), follicle growth cycles succeed one another, causing constant formation of graafian follicles which disappear by atresia. Thus, during fetal life the definitive stock of oocytes is constituted which will be used during the entire sexual life. As soon as it is formed, this stock diminishes rapidly by atresia; a cow fetus having 2,700,000 oocytes at day 110 of gestation, only has 68,000 at birth.

B. Sexualization

Castration at different times of fetal life (or just after birth in rat), followed or not by injection of testosterone or estradiol, has shown that sexualization in the direction of the male type is under the early influence of steroids. In the absence of steroids, sexual organs and mechanisms involved in all aspects of sexuality are of the female type, whatever the genetic sex is.

Analysis of fetal or neonatal steroid secretions supports these conclusions. The testes produce testosterone almost immediately after sexual differentiation

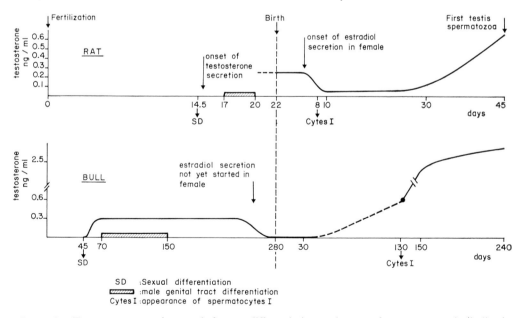

SD   :Sexual differentiation

▨ :male genital tract differentiation

Cytes I :appearance of spermatocytes I

FIG. 4–3. Testosterone secretion, genital tract differentiation and onset of spermatogenesis (indication is given for onset of estradiol secretion in the female).

of the gonads up to the end of fetal life (cattle) or the first days after birth (rat). It is during this period that masculinization of the genital tract, central nervous system and liver occurs. In the fetus or young female, ovarian secretion begins when this period of sexualization is achieved (Figure 4–3).

Particular attention has been given to sexualization of the central nervous system after Pfeiffer discovered this phenomenon in the rat under the influence of androgens. Other experimental approaches have given a complete description of the phenomenon. The rhythmic surge of LH–FSH release in the adult female is a basic property of the preoptic area and hypothalamus, when development of these parts of the central nervous system occurs in the absence of testosterone or estradiol. When one of these steroids is present, the preoptic area loses its rhythmic capacity leading to a male type of LH–FSH release.

The aptitude for the male or female to develop normal sexual behavior after puberty also depends on the presence or absence of sexual steroids during this critical period of differentiation of the CNS. It is probable that the sexualization of reproductive function in domestic mammals is the result of the same mechanism, but little is known about the period of fetal life in which sexualization of the CNS occurs.

C. Neuroendocrine Mechanisms Involved in Puberty

It appears that gonad and pituitary glands can respond to gonadotropins or hypothalamic releasing hormones long before puberty. Thus puberty involves modification at the brain level.

Prepuberal Gonadal Response

Complete spermatogenesis has been obtained by exogenous gonadotropin. For instance, in the hypophysectomized lamb, injection of both ovine LH and FSH induces spermatogenesis (Courot, 1967; 1971). In the female, follicle differentia-

tion and enlargement in response to gonadotropins is obtained from day 18 after birth in rats, and from day 30 in sheep. Thus gonads of both sexes are able to respond to gonadotropins a long time before puberty.

Pituitary Response

The existence of functional pituitary activity in the fetus seems to indicate that the hypophysis is able to respond before puberty to gonadotropin releasing hormones. Direct evidence is given by increase in LH plasma level after injection of LH–RH (L-RF) in 140-day-old lamb fetuses (Foster et al., 1972) and in young children.

Hypothalamic Function

In the rat, man, lamb and calf, the negative feedback regulation of basal gonadotropic secretion by sexual steroids works very early. Castration of female rat on day 13, when steroidogenesis has begun, enhances FSH plasma content. In calves of both sexes, LH plasma level increases when animals over one month of age are castrated (Odell et al., 1970). The blood level of LH in female lambs, ovariectomized when five days old, increases from day 12 onward (Liefer et al., 1972).

Thus, the regulation of gonadotropin level by steroids occurs almost immediately after birth. However, steroid sensitivity is higher before puberty than after, as shown by the results of experimental variation of blood steroid levels.

Central Nervous System Function

In all mammals studied, the mechanisms of the gonadotropin ovulatory surge became potentially functional before the onset of puberty: in 21-day-old rats (about three weeks before puberty), growth of the follicle induced by PMSG is followed by a spontaneous surge of LH and ovulation. In lambs as young as 30–40 days of age, injection of PMSG is also followed by LH discharge. As in the adult ewe, the LH surge in ewe lambs can be obtained by estradiol injection.

Prepuberal discharge of LH can be induced in three- to four-month-old calves after follicular stimulation.

D. Tentative Explanation of Puberty

Pituitary gonadotropic function acts very early in fetal life. Gametogenesis and steroidogenesis of both testis and ovary can be stimulated by exogenous gonadotropins a long time before puberty. Finally, the ability of preoptic and hypothalamic areas to induce cyclic discharge of gonadotropins is also functional before puberty in the female.

Negative steroid feedback mechanisms of gonadotropin regulation differentiates under the influence of the first secretion of either testosterone or estradiol. Such a secretion occurs in fetal life in the male and just after birth in the female. Up to the onset of puberty, the sensitivity of these mechanisms is very high so that the gonadotropin level, after a burst during differentiation of the negative steroid feedback, remains very low and is insufficient to stimulate gametogenesis. The onset of puberty may be explained as a decrease of brain steroid receptor sensitivity with an increase in gonadotropin secretion, subsequent activation of spermatogenesis and preovulatory maturation of graafian follicles.

The reason for this decrease in sensitivity is unknown. Precocious puberty after pinealectomy of young animals shows that the pineal gland inhibits onset of puberty (Wurtman et al., 1968). As the indisputable role of the pineal gland is limited to this action and to counteracting the unfavorable effect of photoperiodism on sexual function (which is related to increase of steroid sensitivity), it is postulated that pineal secretion (melatonin) can modulate the sensitivity of brain steroid receptors.

An inhibitory role of the rhinencephalon (amygdala) on the onset of puberty has been suggested, but contradictory results need further clarification (Critchlow and Elwers Bar-Sela, 1967). Moreover, as puberty occurs at a precise stage of somatic development and is preceded, at least in man, by a spurt of growth, general metabolic changes must be taken into consideration in a complete explanation of the onset of puberty.

II. ESTROUS CYCLE

A. Definition and Characteristics

The estrous cycle lasts from 16 to 21 days in domestic mammals (ewe: 16–17 days; cow, sow, goat: 20–21 days). The duration of estrus differs according to species and varies from one female to another within the same species. The same holds true in respect to the time of ovulation, which occurs 24–30 hours after the onset of estrus in most ewes and cows, and 35–45 hours in the sow (Table 2).

The length of estrus and the time of ovulation vary in relation to internal and external factors. In the ewe, the interval between the onset of estrus and the LH ovulatory surge (and therefore, the interval between estrus and ovulation) lengthens with increasing numbers of ovulations (Fig. 4–4). Sexual stimulation reduces the length of estrus (cow, ewe, sow) and decreases the variability of ovulation time (ewe). In the sow, there is a tendency for cycles to lengthen in the spring.

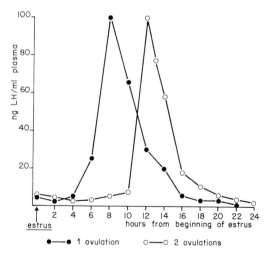

Fig. 4–4. Timing of LH surge in ewe according to number of ovulations (Ile-de-France breed). (*From Thimonier and Pelletier, 1971. Ann. Biol. Anim. Bioch. Biophys. 11, 559.*)

Table 4-2. Estrus Cycle. Estrus and Ovulation in Farm Animals

	Length of Estrous Cycle	Duration of Estrus	Time of Ovulation
Ewe	16–17 days	24–36 hours	24–30 hours from beginning of estrus
Goat	21 days (Also short cycles)	32–40 hours	30–36 hours from beginning of estrus
Sow	19–20 days	48–72 hours	35–45 hours from beginning of estrus
Cow	21–22 days	18–19 hours	10–11 hours after end of estrus
Mare	19–25 days	4–8 days	1–2 days before end of estrus

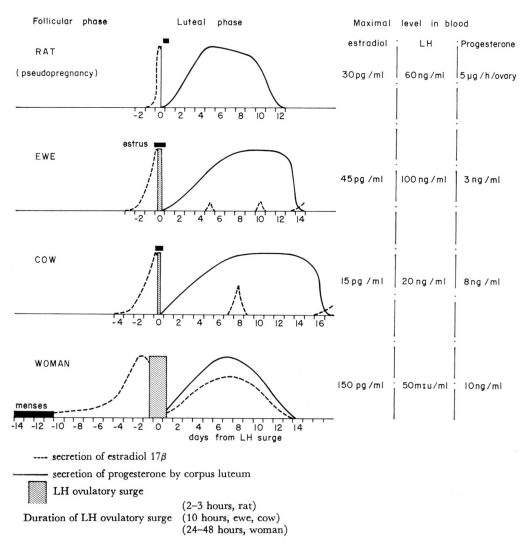

---- secretion of estradiol 17β

——— secretion of progesterone by corpus luteum

▨ LH ovulatory surge

Duration of LH ovulatory surge   (2–3 hours, rat)
(10 hours, ewe, cow)
(24–48 hours, woman)

FIG. 4–5. Comparison of estrous and menstrual cycles. During follicular phase, one or more graafian follicles actively secrete estradiol 17β. This surge of estradiol induces LH surge and ovulation. Note the different mode of corpus luteum regression in woman as compared with that of ewe and sow.

In the mare, 18% of estrus are abnormally long ($\geq$ 10 days) in March, whereas such cycles are not observed during the full breeding season at the beginning of summer.

The estrous cycle is divided into two very unequal phases: (a) a luteal phase which extends from the formation of the corpus luteum after ovulation until its regression at the end of the cycle; and (b) a phase of rapid follicular development which begins when the corpus luteum regresses and ends with estrus and the ovulation of one or more graafian follicles. The luteal phase lasts 14–15 days in the ewe and 16–17 days in the cow and sow. The follicular phase is short: 2–3 days in the ewe and goat; 3–5 days in cow and sow (Fig. 4–5).

B. Types of Sexual Cycles in Mammals

The estrous cycle in farm mammals differs from that of small rodents by the absence of a true luteal phase in the latter: the corpus luteum forms after ovulation but when there is no vaginal or cervical stimulation it does not become functional. On the other hand, if stimulations occur after mating with a vasectomized male, the corpus luteum secretes progesterone during a length of time comparable to that of the corpus luteum of domestic ruminants. This phenomenon in rodents is usually called pseudopregnancy. In fact, it is an estrous cycle with a true luteal phase which is very similar to the estrous cycle of domestic mammals.

The estrous cycle of domestic mammals differs primarily from the menstrual cycle of primates in the follicular phase which lasts about two weeks in nonhuman primates and women; the length of the luteal phase (about two weeks) in primates is similar to that of domestic mammals (Fig. 4–5).

In domestic mammals, rapid follicular growth leading to ovulation after fast regression of the corpus luteum at the end of the luteal phase, explains why estrus and ovulation generally occur 48–72 hours after withdrawal of synthetic progestagen (used to synchronize estrus). In women, suppression of the contraceptive steroid for five days permits menstruation but not ovulation.

In spite of these fundamental differences in the length of the follicular phase of the cycle in primates, or of the luteal phase in some rodents, many physiologic mechanisms are comparable, and part of our knowledge on the sexual cycle has been gained from studies done on domestic mammals.

Follicular growth is regulated by a basal level of FSH and LH. When estrogens secreted by these follicles reach a certain plasmatic level, they facilitate or stimulate the preovulatory gonadotropic surge in all the mammals studied (Fig. 4–6). The formation of the corpus luteum is also a result of hypertrophy of the granulosa cells and penetration of the theca internal cells into the granulosa cells and their distribution throughout the corpus luteum, accompanying conjunctive cells and blood vessels. In the woman, theca cells remain in islets and are responsible for the secretion of estrogens excreted at the same time as progesterone by the human corpus luteum.

C. Regulation of Cyclic Corpus Luteum Life

In order to develop and secrete progestins, the corpus luteum usually needs several hormones forming the luteotrophic complex (Table 4–3). The rapidity with which progesterone secretion rises after ovulation, and the level reached during the luteal phase in the ewe are proportional to the number of corpora lutea. When the progesterone level is known, it is possible to predict the number of spontaneous ovulations or ovulations induced by gonadotropin treatment.

The regression of the corpus luteum, leading to the completion of the estrous cycle and the appearance of a new estrus, is caused by a uterine luteolytic factor which appears to be prostaglandin F2α in the ewe, cow and sow, as well as in many other mammals such as small rodents. In sheep and cattle prostaglandin F2α mimics the luteolytic effect of the uterine factor. As the uterine luteolytic factor, prostaglandin can pass selectively through the wall of uterine vein and ovarian artery when they run side by side. Finally, a high level of prostaglandin has been detected during luteal regression in the venous uterine blood of sheep (Fig. 4–6). There is real prospect for the use of prostaglandins for estrus synchronization in domestic mammals by inducing corpus luteum regression at will.

In primates including man, the regression of the corpus luteum at the end of the menstrual cycle is not under the control of a uterine factor since hysterectomy does not lengthen the luteal phase as it does in domestic ruminants.

The secretion of prostaglandin is un-

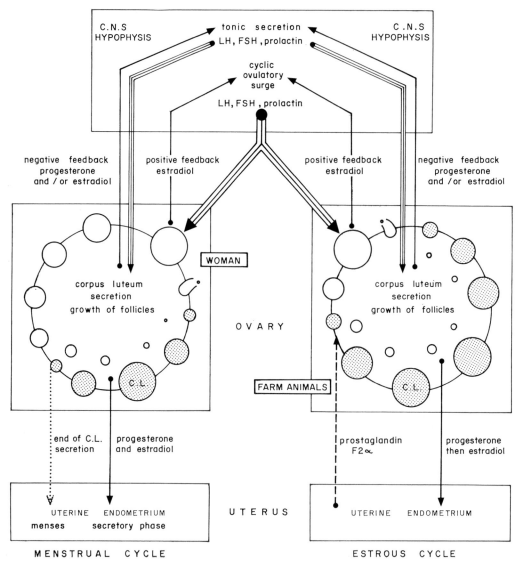

FIG. 4–6. Interrelationship between ovary and hypothalamo-hypophysis and between ovary and uterus.

Table 4-3. Luteotrophic Factors in Some Mammals

	Maintenance of Normal Corpus Luteum	Partial Maintenance of Corpus Luteum	Transitory Stimulation of Progesterone Secretion In Vivo or In Vitro	No Effect
Rat	Prolactin		LH	
Rabbit	Estradiol 17β*			
Sheep	Prolactin + LH	Prolactin	LH	FSH
Pig		LH		
Cow			LH, or better : LH + prolactin	
Man	LH			

* In normal rabbit, estradiol secretion is supported by LH.

der the influence of estrogen (McCracken et al., 1972). During the luteal phase in sheep, there are two transitory rises in plasma estrogen which cause rises of prostaglandin F2α in the uterine vein and slight temporary regressions of progesterone secretion. At about day 14 of the cycle, prostaglandin surges are greater and more frequent, and progesterone secretion regresses sharply; at that time one or two follicles complete growth. There is also a follicular development in the cow at about days 7–8 of the cycle accompanied by a temporary rise in estrogen. These transitory estrogen surges during the luteal phase of domestic ruminants result from abortive follicular growth, while in the woman estrogens are secreted by the corpus luteum.

D. Changes in Genital Tract During Estrous Cycle

The vagina, cervix, uterus and oviducts are steroid target organs, and their activity varies during the cycle. In rodents, the vaginal epithelium cycle is closely related to the ovarian cycle. In domestic mammals, variations of cellular types are not sufficiently clear-cut and the real time of ovulation cannot be determined. On the other hand, maintenance of the vaginal epithelium of a luteal type during the presumed period of estrus may be employed as an early pregnancy test.

This method seems to be utilizable for the sow.

The volume of genital secretions increases at estrus under the influence of estrogens, and the physicochemical changes which occur must be taken into account because they facilitate the transport, survival and capacitation of spermatozoa and permit fertilization and cleavage (see Chapers 6 and 7).

It has been demonstrated that physicochemical characteristics of cervical mucus during estrus act favorably on sperm survival and facilitate motile sperms to migrate from the vagina to cervical folds, then to the uterus.

Sperm density in the oviduct is very low, and as in vitro experiments have shown that high dilution is detrimental to bull and ram sperms, an equally unfavorable situation seems to exist in the oviduct. However, the potassium content of oviductal fluid is remarkably high, and it is known that increase in potassium in semen diluents prevents the detrimental effect of dilution. Moreover, bicarbonates, which stimulate sperm metabolism in vitro, are higher at the time of estrus than at any other period of the cycle. Finally, glycin, which has a sparing effect on glucose utilization by bull sperm in vitro, is the most abundant amino acid of the oviductal fluid in the two species studied—ewe and rabbit (Thibault, 1972).

E. Resumption of Sexual Activity After Parturition

In mares, there is an estrous cycle which may be fertile in the one to three weeks following parturition. However, parturition in the cow, ewe and goat is not followed by immediate resumption of estrous cycles (Casida et al., 1968; Hunter, 1968). The sow exhibits estrus within 48 hours following parturition, but there is no ovulation. The high plasma estrogen rise after parturition (Shearer et al., 1972) may explain this estrous behavior (see Chapter 10). Postpartum anestrus is longer when the female suckles. In the cow, the postpartum anestrus varies in length from three to seven weeks, depending on the breed. Cyclic ovarian activity reappears before estrous behavior reoccurs. Resumption of ovarian cycle with silent ovulation can occur as early as two to three weeks after parturition, and the number of these silent ovarian cycles is higher when cows suckle.

The same observations have been made in sheep by experimentally inducing pregnancy during anestrus so that lambing occurs at the onset of the breeding season, when seasonal anestrus and lactation anestrus do not overlap. Under these conditions, the weaned ewes usually return to estrus after about one month, while suckling ewes only present the first estrus two to three weeks later. Moreover, the majority of ewes have silent estrous cycles after parturition (Mauleon and Dauzier, 1965). In sheep and cattle, duration of postpartum anestrus, which varies with the breed, seems to be constant for the same female during successive pregnancies.

During the first estrus, fertility is low, particularly when the female suckles. Maximal fertility in the cow occurs 60 to 90 days after calving (Fig. 4–7). In Préalpes sheep, when estrus is induced during seasonal anestrus by combined progestagen + PMSG treatment, fertility during induced estrus increases when the delay after lambing is longer: 17%, 30% and 56% when the treatment began 17, 20 or 25 days after parturition, respectively (Thimonier et al., 1968).

III. SEASONALITY OF SEXUAL ACTIVITY

Seasonal variation of sexual activity occurs in most mammals. Several studies have also been devoted to an analysis of the effect of climatic factors on reproduction (Thibault et al., 1966). In domestic mammals, seasonality of sexual activity is very inconspicuous in cattle and swine, and very clear-cut in horses, sheep and goats. Several experimental studies have been done on the role of climatic factors on the breeding season of sheep.

A. Breeding Season

SHEEP

There are important breed differences in the duration of the sexual season. Préalpes and Mérino sheep are long-season breeders, whereas Blackface and Southdown are short-season breeders. The duration of the sexual season in

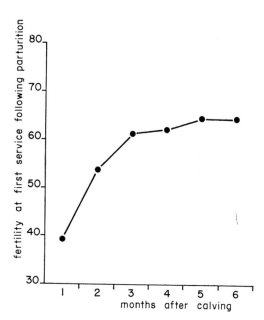

FIG. 4–7. Fertility of dairy cattle at first service following parturition. (*From Casida et al., 1968. Wisc. Expt. Sta. Research Bull. No. 270*).

these breeds is 260, 200, 139 and 120 days, respectively. Barbary ewes may exhibit two sexual seasons, one in October through January, and another in April through June. Finally, there are ewes living in the tropics which have no seasonal anestrus (Indian breed, Bikaneri). A long sexual season is a genetic dominant character. All Mérino crosses exhibit a long sexual season like the Mérinos. A cross of Dorset Horn and the Persian ewe has produced a breed— the Dorper—which has only a one-month anestrus (Fig. 4–8).

Silent ovulatory cycles always occur at the beginning and end of the sexual season. These ovarian cycles continue during the anestrous period in a variable number of ewes (Fig. 4–9). Sexual behavior and gonadotropin secretion are steroid-dependent. Since ovulation can occur without estrus, it seems that the sensitivity to steroids of nervous centers regulating sexual behavior is higher than the sensitivity of those involved in ovarian function. The frequency of silent cycles rises temporarily in spring. If a ram is present, behavioral estrus appears, thus permitting a second sexual season to occur naturally in the year in some breeds (Ile-de-France, Préalpes).

Although rams generally mate year-round, testes weight, testosterone, pituitary LH and hypothalamus LH–RH levels are minimal from January to May (Figs. 4–10, 4–19). They rise beginning

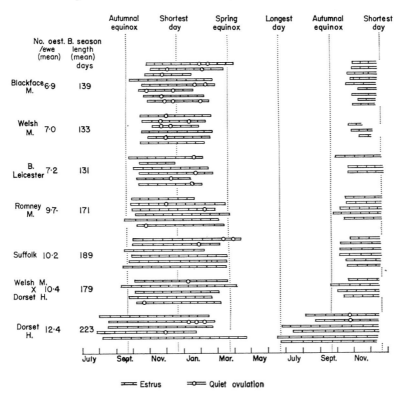

Fig. 4–8. Breed differences in the duration of the sexual season and nonsexual season in adult ewes in Great Britain. Some breeds such as the Dorset Horn had a prolonged sexual season whereas those such as the Welsh Mountain had a very restricted sexual season. In nearly all cases, the sexual season was within the period from the autumnal equinox and the spring equinox and the middle of the season corresponded rather closely to the shortest day of the year, or December 21. This illustrates the close relationship between the sexual season and length of day. Note that some estrous cycles double or triple the usual length occurred due to quiet ovulations or to the failure to detect heat in the nonpregnant females observed. (After E. S. E. Hafez, 1952. J. Agric. Sci. 42, 305.)

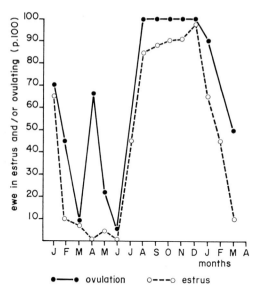

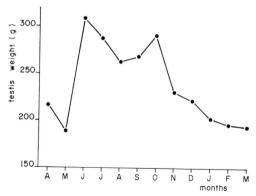

FIG. 4–9. Seasonal variation of estrous behavior and ovulation in the Ile-de-France ewe (47° N). (*From Thimonier and Mauleon, 1969. Ann. Biol. Anim. Bioch. Biophys. 9, 233*).

FIG. 4–10. Seasonal variation of testis weight in Ile-de-France ram (testis weight is adjusted to body weight) (47° N). (*From Pelletier, 1971. Ph.D. Thesis, University of Paris.*)

in May, reaching a maximum in July. Gonadotropic activity therefore resumes with the lengthening days. Seasonal decrease of plasma testosterone in November coincides with the shortening days (Fig. 4–12).

GOAT

In the goat, there is a well-defined sexual season in temperate climates. In the billy goat, the plasma testosterone

level is low from January to August (2 ng/ml) and rises suddenly in August to a maximum of 20 ng/ml, then drops slowly until December (Fig. 4–12). The ovaries in the Alpine goat are slightly active during February–March, quiescent in April–July, and activity is abruptly resumed in all goats in September (Fig.

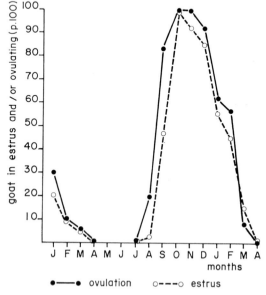

FIG. 4–11. Seasonal variation of estrous behavior and ovulation in goat of Alpine breed (47° N). (*From Cognie, 1973. World Anim. Prod., in press.*)

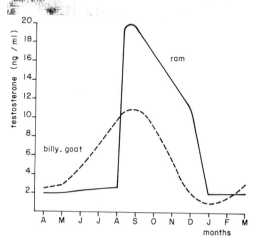

FIG. 4–12. Seasonal variations of testosterone plasma levels in Ile-de-France ram and Alpine billy goat. (*Ram: from Attal, 1970. Ph.D. Thesis, University of Paris; billy goat: from Saumande and Rouger, 1972. Compt. Rend. Acad. Sci. Paris D, 274, 89.*)

4–11). Quiet ovulations are less frequent than in ewes. As in sheep, in tropical climates, sexual activity in Creole goats never stops completely.

CATTLE

The seasonal sexual cycle is not clear-cut, and local breeding conditions may mask its expression. Seasonal variation in fertility in temperate climates can

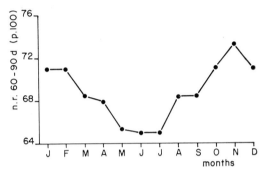

FIG. 4–13. Seasonal variation of fertility in cattle from 320,000 artificial inseminations over seven years (Montbéliard breed in French Jura, 47° N). The nonreturn rate is lower in spring and higher in autumn. (*From Courot et al., 1968. Ann. Biol. Anim. Bioch. Biophys. 8, 209.*)

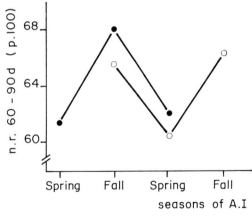

● —— ● semen collected in spring
○ —— ○ semen collected in fall

FIG. 4–14. Seasonal variation of fertility in cattle: relative influence of male and female. (Sperm has been collected either in spring or in fall, frozen and used for insemination fall and spring.) (*From Courot et al., 1968. Ann. Biol. Anim. Bioch. Biophys. 8, 209.*)

only be defined by studying the same herds over a period of several years (Fig. 4–13). Minimal fertility occurs in June and maximal fertility occurs in November. The aspect of the curve indicates that it may be directly linked to seasonal variation of the photoperiod rather than to that of temperature or feeding, which vary from one year to another. This variation in fertility depends on the female, as is shown by nonreturn rates after insemination in spring and fall with deep-frozen semen collected in fall or spring (Fig. 4–14). Slight variations in fertility have been observed in sows. Du Mesnil du Buisson and Signoret observed that farrowing rate is lower after artificial insemination in summer (52%) than in other months (62%) and the litter size is smaller (9.92 vs. 10.46).

B. Attainment of Sexual Maturity Related to Season

Since fertility is often narrowly limited during some months of the year, the age at puberty depends on season of birth. Romanoff ewes born in January attain puberty eight months later, whereas those born in April reach puberty at six months old (during full breeding season of adults in both cases). Similarly, Holstein heifers born in the spring attain sexual maturity at 12 months, and those born in the fall at 16 months of age.

C. Environmental Factors in Seasonal Variation

1. PHOTOPERIODICITY (DAYLIGHT RATIO)

There are only a few experiments on the action mechanisms of photoperiodicity in domestic mammals; most of the results have been obtained in sheep. However, it is possible to speculate that physiologic processes found by photoperiodic manipulation in this species are also involved in many light-sensitive vertebrates. The most conclusive experiment was done using two photoperiodic cycles per year.

Under such environments, the animals experience two breeding periods (Fig. 4–15). If the ewe is allowed to mate regularly, lambing occurs every $6\frac{1}{2}$ months since gestation (150 days) is proceeded by lactational anestrus (45 days) (Fig. 4–16).

Similar photoperiodic manipulation is also effective in the ram. With two photoperiodic cycles per year, there are two periods of decreasing spermatogenic activity coinciding with the two periods of increasing day length (Fig. 4–17).

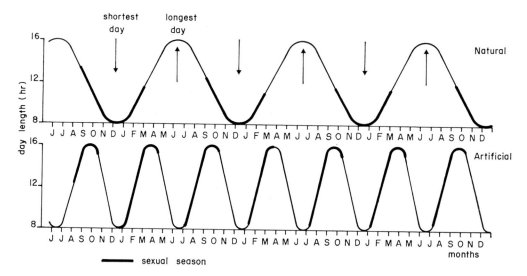

Fig. 4–15. Periods of sexual activity in Limousine ewes. *Top:* Under natural photoperiodicity, estrus normally occurs during decreasing daylight period. *Bottom:* Under six-month photoperiodic cycles, estrus occurs during the increasing daylight period. (*From Mauleon and Rougeot, 1962. Ann. Biol. Anim. Bioch. Biophys. 2, 209.*)

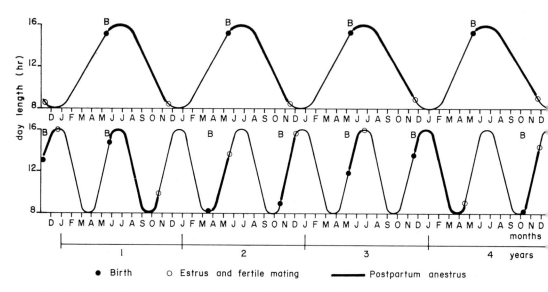

Fig. 4–16. Lambing and estrus in wild sheep (*Ovis ammon musimon*). *Top:* Under normal annual photoperiodicity. *Bottom:* Under six-month photoperiodic cycles. ((*From Rougeot, 1969. Ann. Biol. Anim. Bioch. Biophys. 9, 441.*)

The relative significance of endogenous and photoperiodic factors is shown when ewes are placed either under 12 hours of daylight every day, or under constant illumination for many years. Under such environments, a breeding season is maintained for one or two years, then estrus is more randomly distributed during the year.

Variations in gonadotropic and hypothalamic hormones: LH and FSH pituitary contents and plasma levels are higher in ewes and rams under short days than under long days (Figs. 4–18, 4–19, 4–20). Unexpectedly, LH–RH activity in rams under long days is higher than during short days (Fig. 4–19). In the ewe, LH–RH activity is also higher during anestrus (Jackson et al., 1971). The decrease of spermatogenic activity in the ram and anestrus in the ewe are not due to a deficient level of hypothalamic LH-RH, but to a lower release of this factor, causing a decrease in the secretion and release of pituitary gonadotropins.

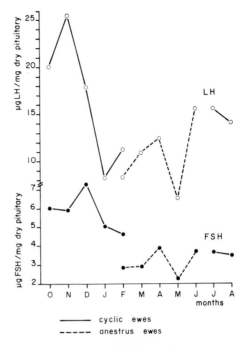

Fig. 4–18. LH and FSH pituitary contents of Ile-de-France ewes at day 12 of the estrous cycle or during full anestrus. (*Thimonier and Mauleon, 1969. Ann. Biol. Anim. Bioch. Biophys., 9, 233.*)

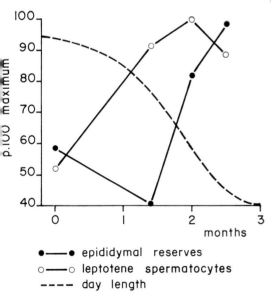

Fig. 4–17. Spermatogenesis in ram under day length decreasing from 16 hours to 8 hours per day. (Epididymal reserves and number of leptotene spermatocytes are given as a percentage of the maximal values observed.) (*From Ortavant, 1961. Proc. 4th Internat. Cong. Anim. Reprod. Artif. Insem., The Hague, 2, 236.*)

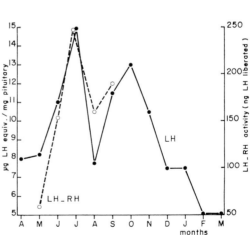

Fig. 4–19. Seasonal variation of pituitary LH and hypothalamic LH-RH (L-RF) in Ile-de-France ram. (*Pelletier and Ortavant, 1967, Colloque C.N.R.S. No. 172; Pelletier, 1971. Ph.D. Thesis, University of Paris.*)

Effects of photoperiod on the central nervous system: LH plasma level and LH pituitary content change with photoperiodic manipulation both in normal and castrated rams, although levels are always higher in castrated animals (Fig. 4–20). Thus, photoperiodicity directly regulates gonadotropic activity, whatever

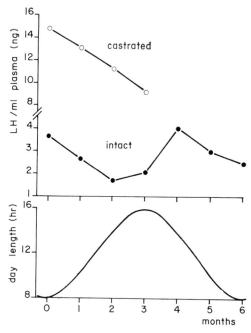

FIG. 4–20. LH plasma level in relation to length of daylight period under six-month photoperiodic rhythm in intact and castrated rams. (*Pelletier, 1971. Ph.D. Thesis, University of Paris.*)

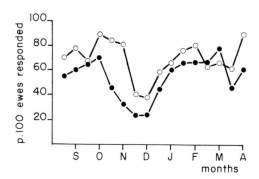

FIG. 4–21. Seasonal variation of steroid sensitivity in castrated ewes. Border Leicester × Mérino castrated ewes received 6 i.m. injections of 20 mg of progesterone every other day, followed by an estradiol benzoate injection (Australia). (*Gibson and Robinson, 1971. J. Reprod. Fert. 24, 9.*)

Table 4-4. LH Pituitary Content of Intact or Castrated Rams Under Short or Long Daylight and Injected With Testosterone Propionate*

| | | Average Pituitary Content (μg/mg) | |
		Short Daylight	Long Daylight
Intact	O	51.8	19.3
Intact	TP	16.3	1.9
Castrated	O	29.0	21.5
Castrated	TP	26.0	4.2

* 100 mg of testosterone propionate every other day for one month.
From Pelletier, 1971. Ph.D. Thesis, University of Paris.

the negative steroid feedback is. Moreover, the second effect of photoperiodicity is the modification of the sensitivity of the brain steroid receptors. If rams are injected with testosterone, LH plasma level is reduced to varying degrees, according to day lengths (Table 4–4), showing that negative steroid feedback is more sensitive under long days. In castrated ewe, estrus cannot be induced by the normal sequence of progesterone and estrogen with the same efficiency during anestrous season (Fig. 4–21). This double daylight regulation of the hypothalamo-hypophyseal gonadotropic function is probably a general physiologic law.

2. TEMPERATURE AND REPRODUCTION

Seasonal variation of temperature plays a major role in the regulation of sexual functioning in lower vertebrates, and particularly in reptiles. In mammals, when environmental temperatures remain within the limits in which thermoregulatory mechanisms are able to maintain body temperature or testis temperature in a narrow range, an effect of seasonal temperature variation on fertility is seldom reported (see Hafez, 1968).

An unfavorable interaction between photoperiodism and temperature was found in the boar. Sperm output, sperm motility and farrowing rate are severely depressed when boars are submitted to summer temperatures (35° C) under long days (16 hours) (Table 4–5).

Table 4-5. Combined Effect of Long Daylight and High Environmental Temperature on Sperm Production and Fertilizing Ability of the Boar*

Temperature	Light	Number of Spermatozoa $\times 10^9$	Farrowing Percentage
15° C	10L	68	57
	16L	48	51
35° C	10L	60	49
	16L	47	29

* From Mazzari, du Mesnil du Buisson and Ortavant, 1968, Proc. 6th Internat. Cong. Anim. Reprod. Artif. Insem., Vol. I, p. 305.

IV. SENESCENCE

Although there are few indications on the mechanisms of sexual aging in farm animals, these seem very similar to those described in such well-studied animals as rat, mouse and man.

As in other mammals, there is only a slight decrease in ovulation and fertilization rates in aged females, but embryonic mortality, stillbirth and postpartum losses increase due to rapid aging of the uterus and insufficient milk production.

There is a wide variation in the reproductive life of the pig which may range from 1 to 26 litters (Pomeroy, 1960). Maximal litter size occurs during the third, fourth and fifth pregnancies (Legault, 1969). On the contrary, in sheep most of the females are fertile up to 10 years. Maximal frequency of twin pregnancies occurs during the sixth or seventh year (Turner and Dolling, 1965). In cattle, reduction of fertility begins at about seven to eight years (Ahmed and Tantawy, 1959), although some cows are fertile for as long as 30 years. Thus, in sheep the average duration of good fertility is longer than that of the sow and probably the cow.

There is very little information on the male sheep. As in man, there is a decrease of testosterone secretion in bull with age after a maximum reached at about five years (Hooker, 1944), but fertility seems unaffected. The non-return rate does not differ significantly when cows of the same area are insem-

inated with sperm from 10–16 year-old bulls (63.4%) or younger animals (64.4%) (Jondet, personal communication).

REFERENCES

Ahmed, I. A. and Tantawy, A. C. (1959). Breeding efficiency of Egyptian cows and buffaloes. *Emp. J. Exp. Agr.* 27, 17–26.

Casida, L. E. (1968). Studies on the postpartum cow. *Wisc. Expt. Sta. Research Bull.* No. 270.

Courot, M. (1967). Endocrine control of the supporting and germ cells of the impuberal testis. *J. Reprod. Fert.* Suppl. 2, 89–101.

Courot, M. (1971). *Etablissement de la Spermatogenèse Chez L'agneau. Etude Expérimentale de son Contrôle Gonadotrope. Importance des Cellules de la Lignée Sertolienne.* Faculty Science Thesis, University of Paris.

Critchlow, V. and Elwers Bar-Sela, M. (1967). "Control of the Onset of Puberty." Chap. 20 in *Neuroendocrinology.* Vol. II. L. Martini and W. F. Ganong, New York, Academic Press.

Foster, D. L., Cruz, T. A. C., Jackson, G. L., Cook, B. and Nalbandov, A. V. (1972). Regulation of luteinizing hormone in the fetal and neonatal lamb. III. Release of LH by the pituitary *in vivo* in response to crude ovine hypothalamic extract or purified porcine gonadotrophin releasing factor. *Endocrinology* 90, 673–683.

Hafez, E. S. E. (1968). "Environmental Effect on Animal Productivity." Chap. 6 in *Adaptation of Domestic Animals.* E. S. E. Hafez, Philadelphia, Lea & Febiger.

Hooker, C. W. (1944). The postnatal history and function of the interstitial cells of the testis of the bull. *Am. J. Anat.* 74, 1–37.

Hunter, G. L. (1968). Increasing frequency of pregnancy in sheep. *Anim. Breed. Abstr.* 36, 347–378, 533–553.

Jackson, G. L., Roche, J. F., Foster, D. L., and Dziuk, P. J. (1971). Luteinizing hormone releasing activity in the hypothalamus of anestrous and cyclic ewes. *Biol. Reprod.* 5, 5–12.

Legault, C. (1969). Etude statistique et génétique des performances d'élevage des truies de la race Large White. I. Effets du troupeau, de la période semestrielle, du numéro de portée et du mois de naissance. *Ann. Génét. Sél. Anim.* 1, 281–298.

Liefer, R. W., Foster, D. L. and Dziuk, P. J. (1972). Levels of LH in the sera and pituitaries of female lambs following ovariectomy and administration of estrogen. *Endocrinology* 90, 981–985.

McCracken, J. A., Carlson, J. C., Glew, M. E., Goding, J. R., Baird, D. T., Green, K. and Samuelsson, B. (1972). Prostaglandin F2α identified as a luteolytic hormone in sheep. *Nature* (New Biol.) 238, 129–134.

Mauleon, P. (1973). Apparition et évolution de la prophase méiotique dans l'ovaire d'embryon de brebis placé dans des diverses conditions expérimentales. *Ann. Biol. Anim. Bioch. Biophys.* Suppl. 12, 89–102.

Mauleon, P. and Dauzier, L. (1965). Variations de durée de l'anoestrus de lactation chez les brebis de race Ile-de-France. *Ann. Biol. Anim. Bioch. Biophys.* 5, 131–143.

Odell, W. D., Swerdloff, R. S., Abraham, G. G., Jacobs, H. S. and Walsh, P. C. (1970. "Pituitary-Gonadal Interrelations." In *Control of Gonadal Steroid Secretion.* Pfizer Medical Monographs 6. D. T. Baird and J. A. Strong, Edinburgh, University Press.

Pomeroy, R. W. (1960). Infertility and neonatal mortality in the sow. I. Lifetime performance and reasons for disposal of sows. *J. Agric. Sci.* 54, 1–17.

Shearer, I. J., Purvis, K., Jenkin, G. and Haynes, N. B. (1972). Peripheral plasma progesterone and oestradiol 17β levels before and after puberty in gilts. *J. Reprod. Fert.* 30, 347–360.

Thibault, C. (1972). Physiology and physiopathology of the Fallopian tube. *Int. J. Fert.* 17, 1–13.

Thibault, C., Courot, M., Martinet, L., Mauleon, P., du Mesnil du Buisson, F., Ortavant, R., Pelletier, J., and Signoret, J. P. (1966). Regulation of breeding season and estrous cycles by light and external stimuli in some mammals. *J. Anim. Sci.* Suppl., 25, 119–142.

Thimonier, J., Mauleon, P., Cognie, Y. and Ortavant, R. (1968). Déclenchement de l'oestrus et obtention de la gestation pendant l'anoestrus postpartum chez les brebis à l'aide d'éponges vaginales imprégnées d'acétate de fluorogestone. *Ann. Zootech.* 17, 237–273.

Turner, H. N. and Dolling, C. H. S. (1965). Vital statistics for an experimental flock of Mérino sheep. II. The influence of age on reproductive performance. *Aust. J. Agric. Res.* 16, 699–712.

Wurtman, R. J., Axelrod, J. and Kelly, D. E. (1968). *The Pineal.* New York, Academic Press.

Chapter 5

Mammalian Semen

I. G. WHITE

Semen consists essentially of *sperm* suspended in a liquid or semigelatinous medium known as the *seminal plasma*. The sperm are produced in the testes but the seminal plasma is the mixed secretion of the accessory sex glands e.g. the epididymis, vesicular glands and prostate. Both sperm production by the testes and the secretory activity of the male accessory glands are controlled by hormones carried to them in the blood stream (see Chapter 3). The testes are regulated by pituitary FSH and LH and, in turn, produce testosterone which controls the development and secretion of the accessory glands.

The anatomy (see Chapter 1) and hence the relative contribution of the accessory glands to the seminal plasma, varies greatly with the species and it is not surprising, therefore, to find considerable differences in both the volume and composition of the semen. Bull and ram semen have a small volume and a high

Table 5-1. **Chemical Composition of Whole Semen.* Average Values (mg/100 cm³ of Semen Unless Otherwise Indicated) Are Given with the Range in Parentheses**

Constituent or Property	Bull		Ram		Boar		Stallion	
pH	6.9	(6.4–7.8)	6.9	(5.9–7.3)	7.5	(7.3–7.8)	7.4	(7.2–7.8)
Water, gm/100 cm³	90	(87–95)	85		95	(94–98)	98	
Sodium	230	(140–280)	190	(120–250)	650	(290–850)	70	
Potassium	140	(80–210)	90	(50–140)	240	(80–380)	60	
Calcium	44	(35–60)	11	(6–15)	5	(2–6)	20	
Magnesium	9	(7–12)	8	(2–13)	11	(5–14)	3	
Chloride	180	(110–290)†	86		330	(260–430)	270	(90–450)
Fructose	530	(150–900)	250		13	(3–50)	2	(0–6)
Sorbitol		(10–140)	72	(26–120)	12	(6–18)	40	(20–60)
Citric acid	720	(340–1150)	140	(110–260)	130	(30–330)	26	(8–53)
Inositol	35	(25–46)†	12	(7–14)	530	(380–630)	30	(20–47)
Glycerylphosphorylcholine (GPC)	350	(100–500)†	1650	(1100–2100)		(110–240)		(40–100)
Ergothioneine	Nil		Nil			(6–23)		(40–110)
Protein, gm/100 cm³	6.8		5.0		3.7		1.0	
Plasmalogen		(30–90)	380					

(Adapted from White, 1958. *Anim. Breed Abstr.*, 26, 109)

* Volume and sperm concentration are given in Chapter 22.
† Analysis of seminal plasma rather than whole semen.

sperm density; stallion and boar semen is much more voluminous with a lower sperm density (Table 5–1).

I. FORMATION OF SPERM

Sperm are formed in the testes (Mc-Kerns, 1969; Johnson, Gomes and Van-Demark, 1970), by a process known as *spermatogenesis* but undergo further maturation in the epididymis where they are stored until ejaculation takes place.

The sperm-producing capacity of the testes is predetermined by heredity and, during the life of the animal, controlled by the anterior pituitary and other factors acting either indirectly via the gland, or directly on the testes themselves.

Spermatogenesis commences at *puberty*, i.e., when the animal is sexually mature. Puberty is not reached suddenly

but gradually; the testes fully descend from the abdomen and both the seminiferous tubules and the interstitial cells become active. At birth the tubules have no lumen and only two types of cell are present—spermatogonia and indifferent cells (Fig. 5–1). During puberty the tubules acquire lumina and the germinal epithelium changes from a simple to a complex state characteristic of the sexually mature male (Fig. 5–1). Spermatogenesis will normally continue throughout life until senility sets in when progressive atrophy of the tubules occurs and only a few are capable of producing sperm.

Wild species breed only at certain seasons in the year and the cycle is usually dependent on changes in the day length. In such animals the testes regress completely during the non-breeding season

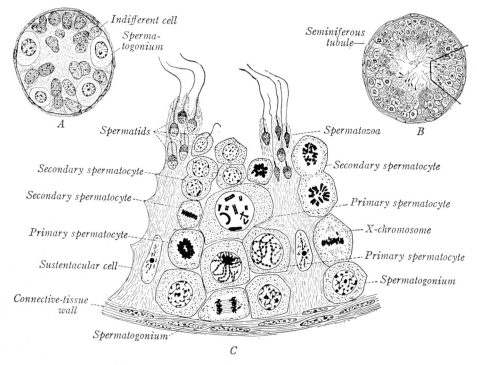

FIG. 5–1. Drawing of a transverse section through a seminiferous tubule of a mammal. *A*, Newborn (× 400); *B*, (× 115); *C*, detail of area outlined in *B* (× 900). Note that the spermatogonia lie against the basement membrane of the connective tissue wall and that the primary and secondary spermatocytes, spermatids and sperm—in that order—form layers extending to the lumen of the tubule. The cytoplasm of the large sustentacular or Sertoli cells is in intimate contact with all the other cells. (*From Arey, 1954. Developmental Anatomy. Philadelphia, courtesy of W. B. Saunders Co.*)

and the germinal epithelium returns to the state of the young, sexually immature male. In farm animals seasonal breeding is not so clearly defined, at least so far as the male is concerned. The testes usually remain in the scrotum after puberty and do not undergo marked cyclic changes.

Among the larger domestic species, the ram shows the most extreme example of seasonal breeding but even here there are great breed differences. The change in day length, acting on the testes via the pituitary, is probably the most important factor in controlling seasonal variation in spermatogenesis in the ram of European breeds. However, high environmental temperatures, quite independently of light, will also depress sperm production. In most mammals the testes are maintained in the scrotum at a temperature several degrees below that of the body by contraction or relaxation of the cremaster muscle (see Chapter 1). This lower temperature is essential for spermatogenesis in these species and degeneration of the seminiferous tubules occurs if the testes do not descend (a condition known as *cryptorchidism*) or if the descended testes are kept warm by insulation.

A. Spermatogenesis

Sperm are formed within the seminiferous tubules from spermatogonia or sperm mother cells which lie on the basement membrane (Fig. 5–2). The process is complex and involves cell division and differentiation, during which the number of chromosomes is halved and both nuclear and cytoplasmic components of the cell are extensively reorganized (Bishop and Walton, 1960a; Roosen-Runge, 1962; Leblond, Steinberger and Roosen-Runge, 1963; Ortavant, Courot and Hochereau, 1969).

1. FORMATION OF SPERM FROM
    SPERMATOGONIA IN THE TESTIS

Spermatogenesis in the ram is summarized in Figure 5–2; the process which can be divided into four phases, is probably similar in the bull but may differ slightly in the boar and some mammals. The developing sperm cells progressively migrate from the basement membrane to the lumen of the seminiferous tubule. During this time, however, they remain in contact with the Sertoli or sustentacular (supporting) cell cytoplasm which probably nourishes them.

Phase I. (15–17 days' duration) *Mitotic division* of spermatogonia, each into a dormant spermatogonium, to ensure continuity of spermatogonia, and an active one which divides four times giving more spermatogonia and finally 16 primary spermatocytes.

Phase II. (about 15 days' duration) *Meiotic division* of the primary spermatocytes during which the number of chromosomes is halved (Meiosis I).

Phase III. (a few hours) Division of the secondary spermatocytes into spermatids (Meiosis II).

Phase IV. (about 15 days' duration) Metamorphosis of the spermatids into sperm without further division. This involves radical alteration in cellular form during which most of the cytoplasm—including ribonucleic acid, water and glycogen—is lost. The spermatid is a fairly large rounded cell whereas the sperm is a compact elongated motile cell consisting essentially of a head and tail. The *Golgi apparatus* of the spermatid gives rise to the anterior cap or *acrosome* of the sperm and the *mitochondria* of the cytoplasm congregate in the tail which grows out from the *centriole* of the spermatid (Fig. 5–3). The sperm, 64 from one original spermatogonium, are finally released from the Sertoli cytoplasm and pass into the lumen of the seminiferous tubule.

After about seven days the dormant spermatogonia commence to divide in a similar manner so that the process repeats itself indefinitely. Phases I, II and III are often grouped together under the name of *spermatocytogenesis* and Phase IV is called *spermiogenesis*.

Hormonal Requirements for Spermatogenesis. Hypophysectomy in the rat, dog and

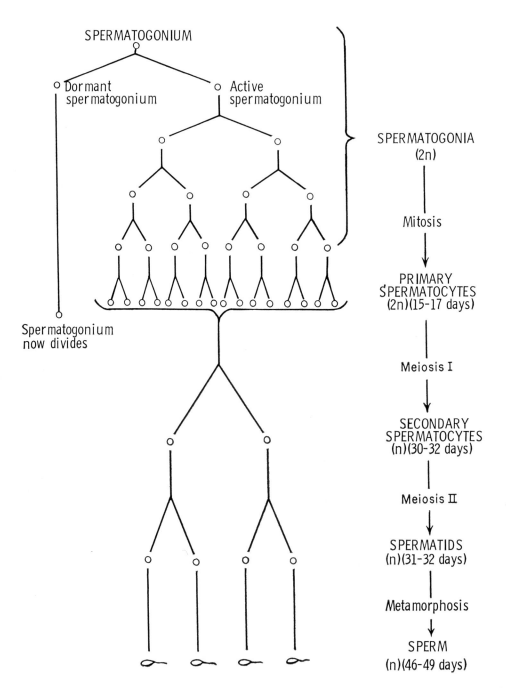

SPERMATOGONIUM

Dormant spermatogonium

Active spermatogonium

SPERMATOGONIA (2n)

Mitosis

Spermatogonium now divides

PRIMARY SPERMATOCYTES (2n)(15-17 days)

Meiosis I

SECONDARY SPERMATOCYTES (n)(30-32 days)

Meiosis II

SPERMATIDS (n)(31-32 days)

Metamorphosis

SPERM (n)(46-49 days)

FIG. 5–2. Diagram of spermatogenesis in the ram. The sequence of events is probably similar in the bull but may differ slightly in other mammals including the boar and stallion. The chromosome number and the time from the formation of the original spermatogonium are given in parentheses. All the primary spermatocytes divided in the manner indicated. (*Adapted from Ortavant, 1959. In Reproduction in Domestic Animals. Vol. 2. Cole and Cupps* [eds.], *New York, courtesy of Academic Press.*)

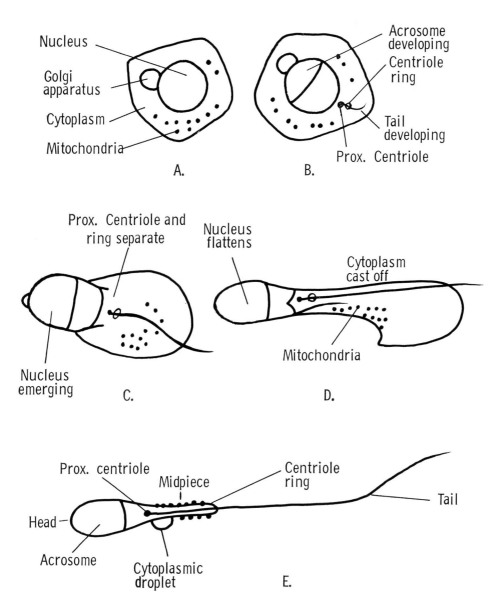

Fɪɢ. 5–3. Drawing illustrating the metamorphosis of a spermatid into a sperm (spermiogenesis) in the ram. The process is similar in other mammals including the bull, boar and stallion. *A*, The acrosome arising from the Golgi apparatus; *B*, the tail developing from the centriole which separates into a proximal and distal (ring) structure; *C*, the nucleus emerging from the cytoplasm and flattening to form the head; *D*, the casting off of the cytoplasm leaving only a droplet and the mitochondria that migrate to the midpiece; *E*, final stage. (*Adapted from Clermont, original and Ortavant, 1959. In Reproduction in Domestic Animals, Vol. 2. Cole and Cupps [eds.], New York, courtesy of Academic Press.*)

sheep causes a decrease in the weight of the testes and degeneration of the seminiferous epithelium which can be restored by giving gonadotrophin. There can be no doubt, therefore, that one or both of the gonadotrophins are concerned either directly or indirectly with the production of sperm by the seminiferous epithelium. In the rat, both FSH and ICSH stimulate spermatogenesis after hypophysectomy but the effect of the latter may well be due to androgen release from the interstitial cells, since the injection of high doses of testosterone has been shown to partially maintain spermatogenesis provided the general epithelium has not completely atrophied.

An attempt has recently been made to analyze the site of action of hormones during spermatogenesis using in vitro cultures of rat testicular tissue. It would appear that the formation of type A spermatogonia may require a hormone, possibly testosterone. Development can then occur through to the pachytene primary stage without hormones. The division of the pachytene spermatocyte to form a secondary spermatocyte does, however, require testosterone and the final stages of the metamorphosis of spermatid to sperm require FSH.

2. THE CYCLE AND WAVE OF THE SEMINIFEROUS EPITHELIUM

Examination of transverse sections of seminiferous tubules shows that the cells are arranged in concentric layers, the cells of any particular layer being at the same stage of development. Thus one finds, proceeding from the basement membrane to the lumen, layers of spermatogonia, spermatocytes and spermatids (Fig. 5–1). Up to five types of spermatogonia may be found, corresponding to the five generations of spermatogonia in Figure 5–2. Four successive types of primary spermatocytes can be seen during their long meiotic division into secondary spermatocytes and several types of developing spermatids are also easily distinguishable (Fig. 5–2).

If any one area of the seminiferous epithelium is considered, it is seen to consist of synchronously developing cells and their parallel evolution into sperm proceeds according to an ever repeating pattern. Thus spermatids at a given stage in their development are always found associated with the same types of spermatocytes and spermatogonia. As a consequence certain *cellular associations* arise which keep repeating themselves indefinitely in an orderly and regular manner. This succession of cellular associations would be seen by an observer if he could watch one section of the living tubule over a long period.

A complete series of these cellular associations, which follow one another in any given area of the seminiferous epithelium, is known as the *cycle of the seminiferous epithelium*. The duration of the cycle is the time required for the appearance of a complete series of cellular associations (i.e., the time between two successive appearances of the same cellular association) in one given area of the tubule. It is the same throughout the testis and is about 10 days; approximately five cycles are required for the complete evolution of a spermatogonium into a free sperm in the ram. The cycle of the seminiferous epithelium is a time concept and should not be confused with the distribution of the various cellular associations along the tubules which also follow in regular succession. The wave of the seminiferous epithelium does not, however, rigidly reproduce in space the constantly repeating cyclic changes occurring in time in any area of the seminiferous epithelium.

Estimation of Duration of Spermatogenesis. Radioactive phosphorus (P^{32}) is taken up from the blood stream by the testis and becomes incorporated into the deoxyribonucleic acid (DNA) of the young primary spermatocyte, but after this no further exchange of phosphorus occurs. It is possible, therefore, to estimate the duration of spermatogenesis by injecting P^{32} into animals and determining the time required for the labeled sperm to appear in the lumen of the seminiferous tubules. This takes about 30 days in the ram and it can be calculated that the total time for spermatogenesis is about 49 days. In the bull, labeled sperm apparently take longer (about 40 days) to be liberated into the lumen of the tubules.

II. MATURATION AND TRANSPORT OF SPERM

Sperm pass quickly from the seminiferous tubules into the rete testis and via the vasa efferentia into the head of the epididymis. They are infertile on first leaving the testis and undergo maturation in the epididymis which is also a reservoir for storage of sperm.

A. Rete Testis Fluid

The rete testis fluid that sweeps sperm out of the seminiferous tubules and into the epididymis is believed to be actively secreted by the Sertoli cells (Setchell, Scott, Voglmayr and Waites 1969; Setchell, 1970; White, 1973).

Recently it has proved possible to cannulate the rete testis and to obtain fluid from living animals over long periods of time. The composition of rete testis fluid is different from that of the blood and the lymph draining the testes and there must be a selectively permeable blood-testis barrier in or around the seminiferous tubules.

The fluid has a relatively low sperm concentration (10^8 per ml in ram and bull) and differs from blood plasma in the concentration of inorganic constituents. Perhaps the most striking difference between rete testis fluid and blood plasma is, however, in the glucose and inositol content. Rete testis fluid normally contains practically no glucose but about 100 times the concentration of inositol. Glucose is not, therefore, available to sperm as a nutrient during the two or three hours that it takes them to pass from the seminiferous tubules to the head of the epididymis and their ability to utilize inositol seems very limited.

Some lactic acid would, however, be available to them as an energy source if the oxygen tension is sufficiently high for its oxidation.

As might be expected from the existence of a blood-testis barrier, the protein concentration of rete testis fluid is less than that of blood plasma. This is also true of most of the free amino acids, except glycine, alanine, and glutamic and aspartic acid which appear to be synthesized from glucose within the seminiferous tubules.

Testosterone and dehydroepiandrosterone are found in the rete testis fluid of the ram and bull and the seminiferous tubules are clearly exposed to testosterone which plays a part in the maintenance of spermatogenesis. As previously indicated, the development and maintenance of the accessory glands and their secretions is dependent upon testosterone which reaches them from the testes predominantly in the blood stream.

B. The Epididymis

The epididymis is responsible for the transport, concentration, maturation and storage of sperm (Bishop, 1961; Risley, 1963; Martan, 1969; Waites and Setchell, 1969; Orgebin-Crist, 1969; White, 1973). The tubules also have both an absorptive and secretory function.

Transport of Sperm. Sperm are swept from the testis and through the efferent ducts by pressure of rete fluid in the testes, aided by cilia movement. Their passage through the duct itself is most probably at least partly dependent on peristaltic movements of the muscle wall. In mammals, this takes 5 to 25 days depending on the species (seven to nine days in the bull). Increased frequency of ejaculation probably has only a slight effect on the rate.

Maturation of Sperm. During their passage through the epididymis, sperm mature or ripen. Although the acrosome may undergo modification in some species, the most conspicuous change is migration of the remnant of cytoplasm known as the *cytoplasmic droplet*. This is attached near the neck of the sperm on leaving the testis but moves back so that the droplet is located further down the midpiece when sperm reach the tail of the epididymis. By the time semen is ejaculated most of these droplets have become completely detached from the sperm but the mechanism is obscure.

Other features of the ripening process in sperm are changes in membrane characteristics and an increase in specific gravity and fertilizing capacity. In some species at least, the sperm also increase their potential for movement. Decreases in DNA, protein, phospholipid and negative charge have also been reported along with changes in metabolic pattern and an increased susceptibility to cold shock. Studies in which rabbit sperm were retained in the caput or corpus epididymis by ligatures suggest that the morphologic and motility changes may take place without normal passage into the cauda. A sojourn in the cauda, however, appears essential for the attainment of full fertility.

Absorption and Secretion. Much of the fluid produced by the testis (99% in bull and boar) is absorbed in the head of the epididymis, and although some fluid may be added in the body, the concentration of sperm in the tail becomes very high (4×10^9/ml in the bull). Some sodium chloride is also absorbed in the head of the epididymis; glutamic acid and inositol are not absorbed to the same extent as water and appear in even higher concentrations in the epididymal plasma than in testicular fluids.

At least three substances are secreted by the epididymal tissue into the lumen— glycerylphosphorylcholine (GPC), sialic acid and carnitine; all are androgen-dependent, i.e., their concentration is reduced by castration and can be at least partially restored by injecting androgen. The GPC concentration rises sharply in the body of the epididymis where blood flow is high and reaches

enormous concentration (3% for the boar) in the fluid of the tail. GPC is not utilized by sperm in the epididymis but there is an enzyme in the female reproductive tract of some species which can break it down, thus providing a possible substrate for sperm after mating. Sialic acid forms part of the glycoprotein of fluid in the epididymis and carnitine is an important cofactor in fatty acid and coenzyme metabolism.

Storage of Sperm. Sperm are stored in the tail of the epididymis, where the concentration is high and the lumen of the duct wider. Conditions in the tail of the epididymis are very conducive to the survival of spermatozoa which can, for instance, remain fertile in the ligated epididymis of the bull for up to 60 days. This is usually attributed to the belief that sperm are immotile, or nearly so, in the epididymis and that their metabolism is in a quiescent state. The substrate for the basal metabolism of sperm in the epididymis is not known with any certainty; it may, however, be their own endogenous phospholipid.

After long periods of sexual rest, sperm may degenerate and be reabsorbed in the tail of the epididymis. It would seem, however, that when animals are not allowed to mate, sufficient sperm are normally lost in the urine or by masturbation to account for much of the continuous production of sperm by the testes. In evaluating semen it is good practice to examine a number of ejaculates to avoid the possibility of the first samples containing degenerating sperm.

C. Other Accessory Organs and Ejaculation

Vas Deferens and Ampulla. The vas deferens (ductus deferens) transports sperm from the tail of the epididymis to the urethra by peristaltic movements which may occur during courtship and precoital stimulation. The final segments of the vas deferens (i.e., the ampullae) are highly developed in the stallion and contribute ergothioneine—a sulfur-containing nitrogenous base—to the ejaculate.

Vesicular Glands or Seminal Vesicles. In the bull, boar and stallion, fructose, sorbitol, citric acid and inositol are produced by the vesicular glands, although to a different degree. The vesicular secretion of the bull contributes about half of the ejaculate and is often quite yellow due to its riboflavin content.

Prostate and Cowper's or Bulbourethral Glands. The relative contribution of the prostate to the volume of the ejaculate is small in the larger domestic species e.g. bull, stallion, ram and boar. However, the prostate of the dog is particularly well-developed and the bulk of the normal dog ejaculate is prostatic secretion.

In the boar, the paired Cowper's or bulbourethral glands are filled with a viscid, rubberlike, white secretion which is essential for the coagulation of the ejaculated semen. The "dribblings" from the prepuce of the bull before mounting are bulbourethral secretion and their function is to flush the urethra free of urine.

Before ejaculation, sperm from the ampullae of the bull and ram are mixed with accessory secretions in the pelvic part of the urethra and entry of semen into the bladder is prevented by engorgement of the colliculus seminalis.

Ram and bull semen is normally ejaculated very rapidly and there is complete mixing of the seminal components (Walton, 1960). On the other hand, the stallion (< 1 minute), boar (2–10 minutes) and dog (5–20 minutes) emit their semen over a longer period in fractions corresponding to the secretions of the different parts of the reproductive tract.

Boar and stallion semen can be readily collected in *pre-sperm, sperm-rich* and *post-sperm fractions.* The post-sperm fraction is usually gelatinous; however, gelatinous material may also sometimes be present in the pre-sperm fraction. The watery *postcoital penis drip* or tail end sample discharged by the stallion on dismounting from the mare is part of the post-sperm fraction. The ram, bull and dog ejaculate into the vagina and the stallion and boar into the uterus.

III. SEMINAL PLASMA

The chemical and physical properties of semen (Mann, 1964, 1967 and 1969; White and MacLeod, 1963) are largely determined by the seminal plasma which constitutes its bulk, particularly in the boar and stallion (Table 5–1). The primary function of the seminal plasma is to act as a vehicle for conveying sperm from the male to the female reproductive tract. It is well-adapted for this role; in most species it constitutes a buffered medium containing either a source of energy directly available to sperm (fructose and sorbitol) or one that can be unlocked on mixing with the female secretions (GPC). The function of the other unusual constituents of seminal plasma and the significance of the great variation in the volume produced by different species are not known.

Seminal plasma has a pH of about 7.0 and an osmotic pressure similar to blood (i.e., equivalent to 0.9% sodium chloride).

Sodium and potassium are the predominant cations in mammalian semen which contains lower concentrations of calcium and magnesium. The concentration of potassium is greater in the sperm than in the seminal plasma, while the reverse is true for sodium and these gradients seem to be maintained by active transport of the ions. Some of these cations, particularly potassium, influence the viability of the sperm. Semen also contains citrate and bicarbonate buffers but they may not maintain a neutral pH in face of the large amount of lactic acid which can be formed by ram and bull sperm from the fructose present in the seminal plasma (see Section V).

Seminal plasma is of great biochemical interest as it contains some unusual organic compounds (Fig. 5–4) (for example, *fructose, citric acid, sorbitol, inositol, glyceryl-phosphorylcholine* and *ergothioneine*) which are not found elsewhere in the body in such high concentrations. These substances are produced by the various accessory glands in response to testosterone from the testes and their estimation in ejaculated semen, or directly in the glands, can be used as an index of accessory gland function. Thus, after the animal's castration, each constituent disappears from the seminal plasma but reappears after testosterone is injected.

Fructose is the sugar found in bull, ram and boar semen and is derived from blood glucose. Stallion semen contains only traces of fructose and none is found in the semen of some species, such as the dog. The fructose concentration of ram and bull semen is particularly high and in these species it is an important nutrient for the sperm which break it down. Sorbitol, a sugar alcohol related to fructose, is also present in semen and can be oxidized to fructose by ram and bull sperm and thus also serves as a nutrient.

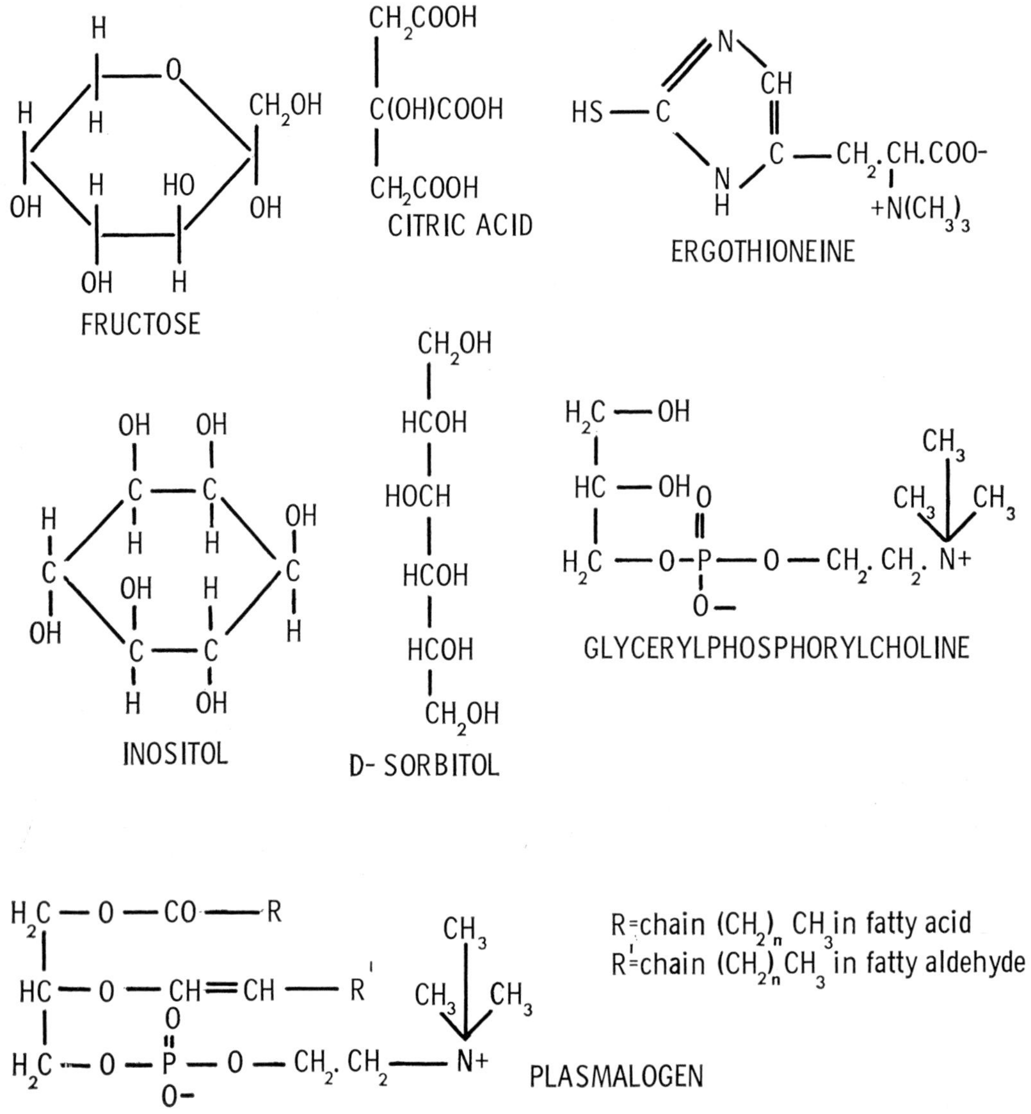

FIG: 5–4: Formulae of some semen constituents.

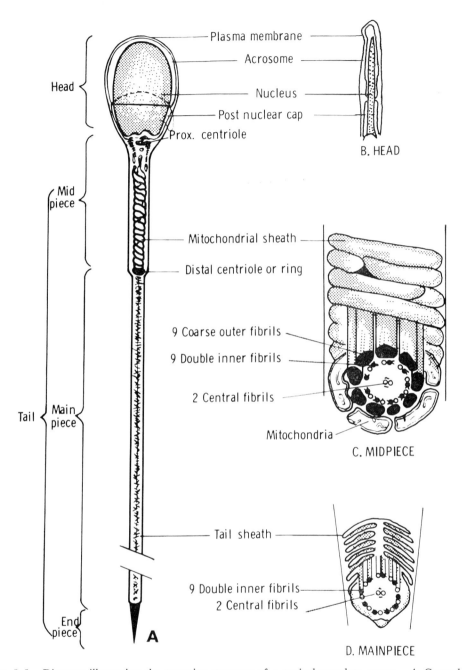

Plasma membrane

Acrosome

Nucleus

Post nuclear cap

Prox. centriole

Head

B. HEAD

Mid piece

Mitochondrial sheath

Distal centriole or ring

9 Coarse outer fibrils

9 Double inner fibrils

2 Central fibrils

Mitochondria

C. MIDPIECE

Tail

Main piece

Tail sheath

9 Double inner fibrils

2 Central fibrils

End piece

A

D. MAINPIECE

Fig. 5–5. Diagram illustrating the tentative structure of a typical ungulate sperm. *A*, General view (× 2700 approx.). *B*, Longitudinal section through the head in a plane at right angles to the paper (× 2700 approx.). *C*, Section through the midpiece showing the fibrils and surrounding mitochondrial sheath (× 30,000 approx.). *D*, Section through the mainpiece showing the fibrils and surrounding tail sheath (× 30,000 approx.). (*Adapted from Wu, 1966. Microstructure of mammalian spermatozoa. A. I. Digest, 14, No. 6, p. 7.*)

Bull semen contains the highest concentration of citric acid; however, it is not utilized by sperm and is of no importance to them as an energy source. Inositol is one of the major constituents of boar semen but, like citric acid, is not metabolized by sperm. The nitrogenous base, GPC, occurs in high concentrations in the seminal plasma of all the larger animals and is produced chiefly by the epididymis. Sperm are incapable of attacking GPC as such, but an enzyme in the secretions of the female genital tract can break it down into simpler units that sperm can utilize. GPC may, therefore, act as a source of energy for sperm in the female tract.

Ergothioneine, a sulfur-containing nitrogenous base, occurs in appreciable quantities only in stallion and boar semen. It is formed chiefly in the ampullae of the stallion, which are particularly well-developed, and the vesicular glands of the boar.

Mucoproteins, peptides, free amino acids, lipids, fatty acids, vitamins and a variety of enzymes may also be present in the seminal plasma of some species.

IV. STRUCTURE AND MOTILITY OF SPERM

The sperm is a highly specialized and condensed cell, which does not grow or divide (Bishop and Walton, 1960a; Hancock, 1966; Ortavant, Courot and Hochereau, 1969). It consists essentially of a head, containing the paternal heredity material, and a tail which provides a means of locomotion. It plays no part in the physiology of the animal that produces it and is solely concerned with fertilizing an ovum and thus producing new individuals of the kind from which it arose. Sperm lack the large cytoplasm characteristic of most cells; the volume of a bull sperm, for instance, is only about one twenty-thousandth of an ovum to which is it is equivalent in hereditary significance. On the other hand, sperm are produced in much greater numbers and a good bull ejaculate contains about 10,000 million sperm which is sufficient to inseminate 1000 cows.

The normal sperm of farm mammals are 50–60μ long and are similar in appearance and size (Fig. 5–5, Plates 6 and 7). Some abnormally shaped sperm are, however, frequently encountered in all species.

The surface of the sperm is covered by a membrane of lipoprotein. When the cell dies the permeability of the membrane increases, particularly in the head region, and this provides the basis for staining techniques which distinguish living from dead sperm. The stain most commonly used is eosin or Congo red against a background of nigrosin. During senescence or aging there is also a passage of material out of sperm and their heads show a tendency to stick to each other or to a glass surface.

PLATE 6. Photomicrographs and electronmicrographs of sperm.

A, Bull sperm in ultraviolet light showing the strong absorptions of radiation by the head ($\times$ 2000 approx.) (*Hancock, 1952. J. Exp. Biol. 29, 445.*)

B, Boar sperm (Giemsa stain; $\times$ 2500). (*Hancock, 1947. J. Roy. Micr. Soc. 76, 84.*)

C, Ram sperm from the head of the epididymis (Eosin-nigrosin stain; $\times$ 1000 approx.). Note the cytoplasmic droplet in the proximal region of the midpiece. (*Voglmayr, original.*)

D, Ram sperm from the tail of the epididymis (Eosin-nigrosin stain; $\times$ 1000 approx.). Note the cytoplasmic droplet has moved back to the distal region of the midpiece and is absent from some sperm.

E, Stereoscan electronmicrograph of a bull sperm showing the paddle-shaped head. The acrosome has been removed by cold shocking ($\times$ 12,500 approx.). (*Dott, 1969. J. Reprod. Fert. 18, 133.*) The insert shows a sperm from the same material photographed under the phase contrast microscope ($\times$ 2000 approx.)

F, G, Electronmicrographs showing a cross and longitudinal section through a cytoplasmic droplet in the neck region of a ram sperm obtained directly from the testis via a catheter ($\times$ 30,000 approx.). (*Voglmayr, Setchell and White, 1971. J. Reprod. Fert. 24, 71.*)

PLATE 6

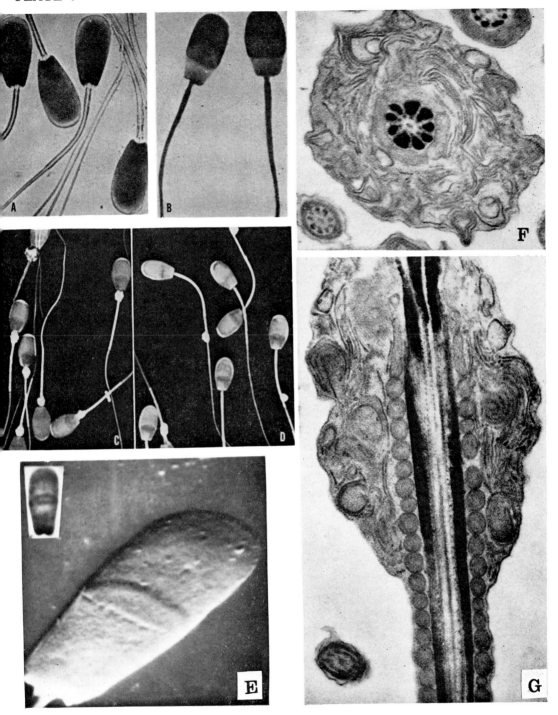

Legend on facing page.

PLATE 7

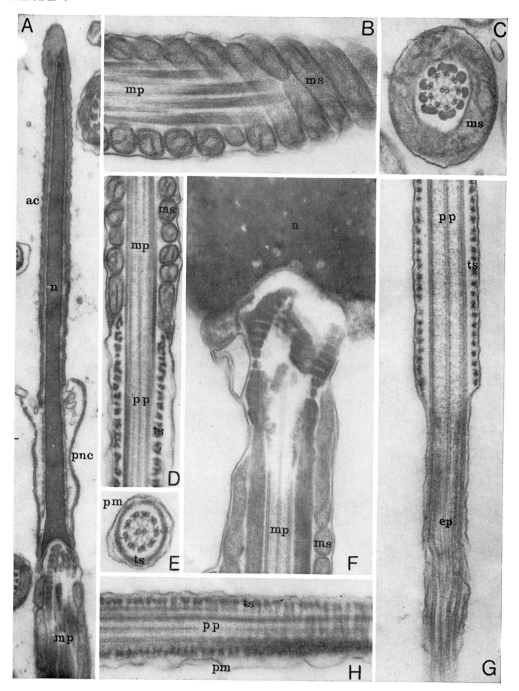

Legend on facing page.

A. The Head

In farm mammals the head is a flattened ovoid structure (approximately $8 \times 4 \times 1 \mu$) made up chiefly of a nucleus covered anteriorly by the *acrosome* and posteriorly by the *postnuclear cap*. The head of the sperm of some animals has quite an exotic shape; the rat, for instance, has a hook-shaped head and a well-defined pronged structure—the *perforatorium*—lies at the apex beneath the acrosome.

The nucleus is composed of deoxyribonucleic acid (DNA) conjugated with protein. The genetic information carried by the sperm is in some way "coded" and stored in the DNA molecule, which is made up basically of many *nucleotides*.

In mammals the hereditary properties of the sperm nucleus include the determination of the sex of the embryo. As a result of the reduction division which occurs during spermatogenesis, sperm contain only half the amount of DNA present in the somatic cells of the same species, and two kinds of sperm are formed: those that carry the X chromosome produce female embryos and those that carry the Y chromosome produce males. Attempts have been made to separate the two kinds of sperm by electrophoresis, sedimentation and centrifugation with the aim of controlling the *sex ratio*. The production of either female or male offspring at will, upon artificial insemination with the appropriate sperm, would be a tremendous advantage in the dairying industry, for example, but so far little success has been achieved.

The acrosome forms a caplike structure over the anterior of the nucleus. It arises from the Golgi apparatus of the spermatid as it differentiates into a sperm and apparently has some vital role in fertilization, since bull and boar sperm with deformities of the acrosome are sterile. The acrosome consists of protein-bound polysaccharide, composed of fucose, mannose, galactose and hexosamine—which is responsible for the periodic acid Schiff (PAS) staining reaction of sperm. It also contains lysomal enzymes, some of which may be involved in fertilization.

B. The Tail

The long (40–50 μ) thin tail is differentiated into three parts, *midpiece, mainpiece* and *endpiece*, and arises from the spermatid centriole during spermatogenesis. It propels the sperm by waves which are generated in the implantation region and pass distally along the tail like a whiplash.

The anterior end of the midpiece connecting the head is known as the implantation region. Separation of the head and tail may occur here, e.g. in

---

PLATE 7. Electronmicrographs of ram sperm. (*Cleland, original.*)

A, Section edge-on through head. The head is mostly nucleus (*n*) surrounded by the acrosome (*ac*) and the postnuclear cap (*pnc*).

B, Longitudinal section through the midpiece. Note the mitochondrial sheath (*ms*) wound around the fibrils of the midpiece (*mp*).

C, Transverse section through the midpiece. Note the mitochondrial sheath (*ms*) surrounding the nine outer, the nine double inner and two central fibrils.

D, Longitudinal section though the junction of the midpiece (*mp*) and the mainpiece (*pp*) where the mitochondrial sheath (*ms*) is replaced by the tail sheath (*ts*).

E, Transverse section through the mainpiece. Here the nine double inner and two central fibrils are surrounded by the tail sheath (*ts*). The nine outer fibrils are absent.

F, Longitudinal section through the junction of the head (flat side on) and midpiece (*mp*) showing the complex implantation region.

G, The terminal region of the mainpiece (*pp*) tapering into the thinner endpiece (*ep*). The endpiece, unlike the mainpiece, is not surrounded by the tail sheath.

H, Longitudinal section through the mainpiece (*pp*) showing the tail sheath (*ts*) surrounding the fibrils. The plasma membrane (*pm*) is easily seen in the tail.

bulls with a specific hereditary defect, in animals with fever, or in animals when heat is applied to the testes. The electron microscope shows that the implantation region has a complex structure and contains the *proximal centriole* which in bovine sperm appears as a cylindrical structure.

An *axial core* consisting of two central fibrils surrounded by a concentric ring of nine double fibrils runs from the region of implantation through to the end of the tail, a pattern common to cilia and flagella generally (Fawcett, 1961).

The midpiece (10–15 μ) is the thickened region of the tail between the head and mainpiece and may be regarded as an important power house for supplying energy to the sperm. Here the central axial core of 11 fibrils is surrounded by an additional outer ring of nine coarser fibrils. Individual mitochondria are wrapped spirally around these outer fibrils to form the mitochondrial sheath, which contains the enzymes concerned in the oxidative metabolism of the sperm. The midpiece is rich in the phospholipids, lecithin and plasmalogen. The latter contains a fatty aldehyde and a fatty acid linked to glycerol as well as phosphoric acid and choline (Fig. 5–1). The fatty acid can be oxidized and may represent an endogenous store of energy for sperm activity.

The mainpiece (about 30 μ) is the longest part of the tail and provides most of the propellant machinery. The coarse nine fibrils of the outer ring diminish in thickness and finally disappear leaving only the 11 inner fibrils in the axial core for much of the length of the mainpiece. Through the main-

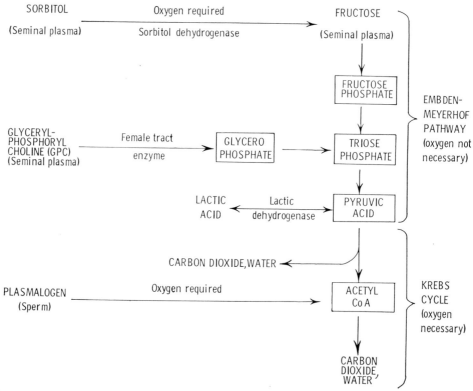

Fig. 5–6. A diagram of the probable routes by which sperm metabolize substances in semen. The origin of the four substrates—fructose, sorbitol, glycerylphosphorylcholine and plasmalogen—is shown in parentheses. The end products are lactic acid, carbon dioxide and water. Intermediates which do not accumulate are enclosed by boxes. The concentration of the seminal plasma substrates (see Table 5–1) and hence their relative importance in sperm metabolism varies from one species to another.

piece the fibrils are surrounded by a fibrous tail sheath. This probably consists of branching and anastomosing semicircular strands or "ribs" held together by their attachment to two bands that run lengthwise along opposite sides of the tail.

In the short terminal portion of the tail (3μ) or endpiece, the filament is not surrounded by a sheath and the outer nine fibrils are absent.

C. Motility of Sperm

The most striking feature of sperm is their motility (Bishop and Walton, 1960b; Bishop, 1962; Nelson, 1967) which makes them admirably suited for physiologic studies and also provides a simple means of evaluating semen for artificial insemination. Such microscopic observations are, however, subjective and much effort has been directed toward obtaining more critical methods for measuring sperm movements. The best known device is the *impedance bridge* which measures the rate of change of the electrical resistance of a sperm suspension. The impedance change frequency (I.C.F.), as it is called, is correlated with the activity of dense sperm suspensions.

The sperm tail contains all the apparatus necessary for motility and tails that have become separated from heads can be fully motile. The nine larger outer fibrils of the tail are thought to be the main contractile elements and to be capable of propagating localized contractions along their length. The smaller inner fibers may be specialized for the rapid conduction of impulses, arising rhythmically at the neck and coordinating the localized contractions in the outer fibers.

Waves of sperm swimming in the same direction are a characteristic feature of undiluted ram and bull semen when viewed under the microscope. The speed of the sperm varies with the medium and temperature but is of the order of 100 μ per second at 37° C. The motility of sperm probably plays an im-portant part in the final encounter of the gametes and, in general, there is a fairly good correlation between motility and fertilizing capacity.

V. METABOLISM OF SPERM

This has been the subject of a number of recent reviews (Salisbury and Lodge, 1962; White and MacLeod, 1963; Mann, 1964, 1967 and 1969). There are at least four substances in semen which can be utilized either directly or indirectly by the sperm as energy sources for the maintenance of motility. These are fructose, sorbitol, GPC and phospholipid (see Section III and Table 5-1); the first three are constituents of the seminal plasma, but phospholipid is present in the sperm itself. All four can be utilized by sperm in the presence of oxygen, which would normally be available in most parts of the female tract, provided in the case of GPC it has first come into contact with an enzyme in the female tract secretions; none are broken down by the seminal plasma.

The *oxygen uptake* or *respiration* of most semen will reflect the overall oxidation of these substances and is between 5 and 20 μl per hundred million sperm/hr. at 37° C. Lactic acid, which accumulates in semen as the result of the metabolism of the sperm, is derived from the plasma constituents. In addition to these four physiologic substrates sperm can metabolize a large number of related substances that do not occur in semen or are only present in low concentrations, such as pyruvic and acetic acids.

The rate of oxygen uptake and utilization of fructose by spermatozoa is greatest in medium of pH 6-8 isotonic with semen. In general, small amounts of potassium, phosphate and carbon dioxide in the medium stimulate metabolism but high concentrations inhibit it.

A. Metabolism of Fructose and Sorbitol

The metabolism of fructose by sperm proceeds via the *Embden-Meyerhof pathway* which is common to most animal tissues. Fructose phosphates, triose phosphates and pyruvic acid are involved as transient intermediates leading to lactic acid, which tends to accumulate although in the presence of oxygen it is further oxidized to carbon dioxide and water. The oxidative phase of metabolism takes place over the *Krebs tricarboxylic acid cycle* as in other tissues.

If the sperm are deprived of oxygen, for instance by letting semen stand in long narrow tubes, oxidation of the lactic acid cannot take place and the fructose is quantitatively converted into the acid. It is convenient to measure the fructose breakdown (or *fructolysis*) under these essentially anaerobic conditions, and at 37 °C the rate for bull and ram sperm is 1.5–2 mg/10^9 motile cells/hour but is much lower for the boar (0.2–1.0). In the absence of oxygen, sperm depend solely upon the breakdown of fructose to lactic acid as a source of energy. Ram and bull sperm survive quite well under such conditions provided fructose is present. Boar sperm, on the other hand, becomes immotile even when adequate fructose is available; this is presumably due to their reduced ability to metabolize fructose to lactic acid, which cannot be further utilized without oxygen.

Although fructose is the sugar normally found in semen, two other hexoses — glucose and mannose — are similarly metabolized by sperm when added to semen. The sugar-alcohol, sorbitol, which occurs in seminal plasma, is oxidized to fructose by ram and bull sperm, but not by boar sperm. The enzyme sorbitol dehydrogenase is responsible for the oxidation which only takes place in the presence of oxygen. The fructose so formed will be metabolized in the same way as the fructose initially present in the seminal plasma, and the sorbitol can, therefore, act as a nutrient for the sperm.

There is now evidence that some of the pyruvic acid formed from fructose by bull and ram spermatozoa may be converted by dismutation to lactic acid, acetic acid and carbon dioxide. Some fructose may also be incorporated into sperm lipid.

B. Metabolism of Glycerol and Glycerylphosphorylcholine (GPC)

In view of the widespread use of glycerol in diluents for the deep freezing of semen, it is of interest that ram and bull sperm can utilize it under aerobic conditions. Glycerol presumably enters the glycolytic cycle at the triose phosphate stage, a step which requires oxygen, and is converted into lactic acid which can be further oxidized.

Sperm are incapable of attacking GPC as such, but there is an enzyme in the secretions of the female duct system that splits off choline thus liberating phosphoglycerol which, like glycerol, is utilized by sperm. GPC may therefore constitute an additional source of energy for sperm after semen has been ejaculated into the female duct system.

C. Metabolism of Phospholipid

In addition to fructose, sorbitol and GPC, which are available from the sperm themselves carry a reserve nutrient. This endogenous source of energy, which is probably brought into play when others are exhausted, may be plasmalogen (Fig. 5–4). One fatty acid residue can be detached from the plasmalogen molecule and oxidized as two carbon atom fragments (i.e., "active acetate") via the Krebs tricarboxylic acid cycle giving carbon dioxide and water.

D. Relation Between Metabolism, Motility and Fertility

Both the rate of fructolysis and respiration are correlated with the motility of sperm. In general, however, the correlation between metabolic rate and fertility is not sufficiently high to give such tests an advantage over the much simpler microscopic examination of semen (see Chapter 22).

Adenosine triphosphate (ATP) and *ATPase*, which play an important part in muscle contraction, are present in sperm and provide a link between the energy yielding reactions and motility. The breakdown of ATP probably furnishes the energy necessary for the contraction of the sperm fibrils in much the same way as it does in muscle fibrils; the loss of ATP is made good in sperm by the energy yielding reactions of fructolysis and respiration.

VI. FACTORS AFFECTING THE SURVIVAL OF SPERM IN VITRO

The physical and chemical properties of the diluent, the degree of dilution, and factors such as temperature and light are important in handling and storing semen for artificial insemination (Parkes, 1960; Mann, 1964).

A. Composition of Media

In general, sperm are most active and survive for the longest period at a pH of about 7.0. There is a fairly rapid falloff in motility on either side of the optimal pH, but at least partial motility is observed between 5 and 10. Although sperm are fairly rapidly immobilized by acid conditions, motility can be restored in some species if the pH is brought back to neutral promptly. Bull and ram sperm produce large amounts of lactic acid from seminal fructose, and it is necessary to have a buffer such as phosphate, citrate or bicarbonate in the medium. The presence of a sugar, such as fructose or glucose which would nourish the sperm, might be expected to be beneficial in a diluent.

Sperm remain motile for the longest period in media having about the same tonicity as semen or blood. In general, they are less readily affected by hypertonic than by hypotonic conditions in the range of 50 to 150% of normal tonicity.

Potassium is necessary for the normal functioning of the sperm and it would seem advisable to include some potassium and possibly also magnesium in semen diluents. On the other hand, calcium and high concentrations of phosphate and potassium depress motility and should be avoided. Copper and iron are toxic to sperm, but the danger of contaminating semen with heavy metals in artificial insemination practice is not great if glass vessels and distilled water are used.

Low concentrations of bicarbonate stimulate the metabolism of sperm, but the motility and metabolism of bull sperm can be reversibly suppressed by high concentrations of carbon dioxide and bull semen can be preserved in this way for several days at room temperature without great loss of fertility. A carbonated media has also proved successful for storing boar semen.

The short (blue) wavelengths of visible light can damage sperm, particularly under aerobic conditions, and semen should not be unnecessarily exposed to light.

B. Effect of Dilution

Moderate dilution of semen, particularly in a buffered, isotonic medium containing a sugar such as fructose, is not harmful to sperm motility and may stimulate activity and increase their life span. It is even possible to revive the activity of senescent samples in this way.

Excessive dilution (greater than 1 in 1000) even in an optimal medium, however, depresses motility. The depression of motility at low cell concentration would seem to be due to both dilution of macromolecules in the seminal plasma and loss of substances from the sperm. The precise nature of these substances is not known, but a variety of high molecular substances (e.g., proteins and starch) have a protective action on diluted ram and bull sperm and presumably act by preventing the escape of intracellular constituents. In storing semen it would seem advisable to avoid excessive dilution and to add a protective agent to the diluent. The protective agent widely used in artificial insemination practice is egg yolk; the active constituent is probably a lipoprotein.

The fertility of bull semen falls off when it is diluted to less than 8 million sperm per ml and inseminated in 1 ml doses. It is not yet known if this is a true dilution effect, as described above, or is due to the reduced number of sperm inseminated. Although the motility of ram sperm is less affected than the bull by high dilution, fertility may fall off much more strikingly and some investigators report that a dilution of only 1 in 4 in egg yolk-citrate can cause a 50% decrease in conception rate.

The seminal plasma of the boar and stallion is much less favorable for survival of the sperm which, in fact, live longer if they are centrifuged down and concentrated in an artificial medium.

C. Effect of Temperature

A 10° C increase above ambient temperature will more than double the metabolic rate of sperm and there is a corresponding decrease in life span. Semen quality should be evaluated at a constant temperature and 37° C is usually chosen for mammalian species. At body temperature the life of the sperm in vitro is only a few hours, because of the exhaustion of substrate, the fall in pH of semen due to accumulation of lactic acid, senescent changes in the sperm, and the growth of bacteria. Above 50° C sperm suffer an irreversible loss of motility within five minutes.

Cold Shock. When ejaculated ram, bull, boar and stallion sperm are quickly cooled to about freezing point they suffer an irreversible loss of viability called *cold shock.*

The most obvious sign of cold shock is loss of motility which is not regained on warming the semen. There is also an increase in the proportion of cells staining with dyes like eosin or Congo red. Changes in cell permeability occur with leakage of potassium and proteins (e.g., cyto-

chrome) and damage to the acrosome. There is also a decrease in the rate of fructose breakdown by the sperm, and a decrease in oxygen uptake and a fall in ATP, which can no longer be resynthesized and used to supply energy for motility.

Cold shock can be avoided by cooling bull, ram and stallion semen slowly in the critical region from 15–0° C. When the sperm are cooled in this way there is a reduction in motility and metabolism, but on rewarming full activity is restored. Cold shock can also be prevented in these species by adding egg yolk to semen; the active principle is most probably phospholipid bound to protein. Ejaculated boar sperm are particularly susceptible to cold shock and should not be cooled below 15° C. The sperm of farm animals suffer severe damage if they are cooled much below 0° C unless glycerol is included in the medium.

Low Temperature Storage. At temperatures near freezing point it is possible to keep sperm for several hours or days, particularly if antibiotics, which are innocuous to mammalian sperm in bacteriostatic quantities, are added to the diluent. A slow cooling technique (to 2–5° C for the bull, ram and stallion and to 15° C for the boar), usually using an egg yolk or milk diluent, provides the basis for routine storage of liquid semen for artificial insemination.

If glycerol is added to the medium it is possible to deep freeze bull sperm successfully at −79° C (dry ice alcohol) or −196° C (liquid nitrogen) in ampules, plastic straws or as pellets. In this way, it is possible to preserve bull sperm indefinitely, although for maximum fertility such semen should not be stored for more than one year. The fertility of deep-frozen ram semen has proved poor despite satisfactory recovery of motility; however recent results using pelleted semen gives some promise of success. Goat and stallion semen has also been successfully deep frozen, boar semen does not freeze well and dog sperm are rendered infertile.

VII. CAPACITATION AND METABOLISM OF SPERM IN FEMALE REPRODUCTIVE TRACT

Normally, mating takes place early in heat and ovulation towards the end or after this period, depending on the species. In view of their rapid ascent of the female reproductive tract, therefore, sperm usually reach the site of fertilization several hours before the ova. A sojourn in the uterus or oviducts is, in fact, a prerequisite for penetration of ova and the process is known as the capacitation of the sperm (Adams and Williams, 1967; Bedford, 1970; Srivastava and Williams, 1971).

Although it is almost 20 years since Austin and Chang first discovered the need for the capacitation of rabbit sperm, little is known about the process and it remains one of the most challenging problems in reproductive physiology. As well as the rabbit, there is now evidence of the need for capacitation in the rat, sheep, ferret, hamster and mouse and it is possible to capacitate the sperm of the last two species outside the body.

Decapacitation. Capacitation seems to involve the removal, probably by enzymic action, of macromolecular material located on the surface of the sperm. Re-exposure of the sperm to seminal plasma leads to loss of capacitation and the process may well involve the restoration of the surface layer or "decapacitation factor" which normally coats the sperm as they pass through the male reproductive tract.

The decapacitation factor (DF) is present in the seminal plasma of the bull, boar, stallion, man and monkey as well as the rabbit, but is absent from dog semen. It occurs in the epididymal fluid of the rabbit but must also be produced in the male reproductive tract below the vas deferens since DF is present in seminal plasma from vasectomized males. The exact nature of this factor is still unknown (Williams, Abney, Chernoff, Dukelow and Pinsker, 1967). The active substance would seem to have a molecular weight as low as 500 although it must be associated with a macromolecule in rabbit semen as it can be spun out of the plasma at 100,000 g.

Sperm Enzymes and Inhibitors. Hyaluronidase is present in the acrosome of sperm and no doubt plays a part in the passage of sperm through the cumulus. More recently it has been demonstrated that the acrosome of rabbit and ram sperm also contain a corona penetrating enzyme (CPE) and a trypsinlike enzyme (TLE) that seems essential for penetration of the zona pellucida.

DF inhibits CPE and it is possible that capacitation may involve removal of this inhibitor and thus activation of the enzymes clearing a passage for sperm through the corona radiata. An inhibitor of TLE is also present in seminal plasma; the enzyme-enzyme inhibitor relationship appears analogous to the CPE-DF relationship and part of capacitation may also involve removal of the inhibitor from TLE.

Sperm Metabolism in Female Genital Tract. The oxygen uptake and glycolytic activity of rabbit sperm seems to increase after incubation in the uterus of the doe and there is some activation of pentose shunt activity (White, 1972). It is not clear, however, if these metabolic changes are a prerequisite for capacitation.

There is no shortage of possible substrates for sperm in the female genital tract (Section V). Apart from the three constituents that are potentially available from the seminal plasma (fructose, sorbitol, GPC), uterine and oviduct fluid contains glucose and lactic acid. Furthermore, when all exogenous sources of energy are exhausted, the sperm can utilize its own phospholipid reserves.

REFERENCES

Adams, C. E. and Williams, W. L. (1967). Capacitation. *Biol. Reprod. 10*, 177–186.

Bedford, J. M. (1970). Sperm capacitation and fertilization in mammals. *Biol. Reprod. 2*, 128–157.

Bishop, D. W. (1961). "Biology of Spermatozoa." In *Sex and Internal Secretions*. Vol. 2. W. C. Young (ed.), London, Balliere, Tindall & Cox.

Bishop, D. W. (1962). Sperm motility. *Physiol. Rev.* 42, 1–59.

Bishop, M. W. and Walton, A. (1960*a*). "Spermatogenesis and the Structure of Mammalian Spermatozoa." In *Marshall's Physiology of Reproduction.* A. S. Parkes (ed.), London, Longmans.

Bishop, M. W. and Walton, A. (1960*b*). "Metabolism and Motility of Mammalian Spermatozoa." In *Marshall's Physiology of Reproduction.* Vol. I, part 1. A. S. Parkes (ed.), London, Longmans.

Fawcett, D. S. (1961). "Cilia and Flagella." In *The Cell.* Vol. 2. J. Brachet and A. E. Mirsky (eds.), New York, Academic Press.

Hancock, J. L. (1966). "The Ultrastructure of Mammalian Spermatozoa." In *Advances in Reproductive Physiology.* Vol. 1. A. McLaren (ed.), London, Logos.

Johnson, A. D., Gomes, W. R. and VanDemark, N. L. (1970). *The Testis.* Vol. 1. New York, Academic Press.

Leblond, C. P., Steinberger, E. and Roosen-Runge, E. C. (1963). "Spermatogenesis." In *Mechanisms Concerned with Conception.* C. G. Hartman (ed.), Oxford, Pergamon.

McKerns, K. W. (1969). *The Gonads.* Part 3. Amsterdam, North-Holland.

Mann, T. (1964). *The Biochemistry of Semen and of the Male Reproductive Tract.* London, Methuen.

Mann, T. (1967) "Sperm Metabolism." In *Fertilization.* C. B. Metz and A. Monroy (eds.), New York, Academic Press.

Mann, T. (1969). "Physiology of Semen and of the Male Reproductive Tract." In *Reproduction in Domestic Animals.* 2nd Ed. H. H. Cole and P. T. Cupps (eds.), New York, Academic Press, pp. 277–312.

Martan, J. (1969) Epididymal histochemistry and physiology. *Biol. Reprod.* Suppl. *1*, 134–154.

Nelson, L. (1967) "Sperm Motility." In *Fertilization.* Vol. 1. C. B. Metz and A. Monroy (eds.), New York, Academic Press.

Orgebin-Crist, M. C. (1969). Studies on the function of the epididymis. *Biol. Reprod.* Suppl. *1*, 155–175.

Ortavant, R., Courot, M. and Hochereau, M. T. (1969). "Spermatogenesis and Morphology of the Spermatozoon." In *Reproduction in Domestic Animals.* 2nd Ed. H. H. Cole and P. T. Cupps (eds.), New York, Academic Press.

Parkes, A. S. (1960). "The Biology of Spermatozoa and Artificial Insemination." In *Marshall's Physiology of Reproduction.* Vol. 1, part 1. A. S. Parkes (ed.), London, Longmans.

Risley, P. L. (1963). "Physiology of the Male Accessory Organs." In *Mechanisms Concerned with Conception.* C. G. Hartman (ed.), Oxford, Pergamon.

Roosen-Runge, E. C. (1962). The process of spermatogenesis in mammals. *Biol. Rev. 37*, 343–377.

Salisbury, G. W. and Lodge, J. R. (1962). Metabolism of spermatozoa. *Advances in Enzymology 24*, 35–104.

Setchell, B. P. (1970). "Testicular Blood Supply, Lymphatic Drainage and Secretion of Fluids." In *The Testis.* A. D. Johnson, W. R. Gomes and N. L. VanDemark (eds.), New York, Academic Press.

Setchell, B. P., Scott, T. W., Voglmayr, J. K. and Waites, G. M. (1969). Characteristics of testicular spermatozoa and the fluid which transports them into the epididymis. *Biol. Reprod.* Suppl. *1*, 40–66.

Srivastava, P. N. and Williams, W. L. (1971). "Sperm Capacitation and Decapacitation Factor." In *Control of Human Fertility. Nobel Symposium 15.* E. Diczfalusy and U. Borell (eds.), New York, Wiley.

Waites, G. M. and Setchell, B. P. (1969). "Physiology of the Testis, Epididymis and Scrotum." In *Advances in Reproductive Physiology.* Vol. 4. A. McLaren (ed.), London, Logos, pp. 1–63.

Walton, A. (1960). "Copulation and Natural Insemination." In *Marshall's Physiology of Reproduction.* Vol. 1, part 1. A. S. Parkes (ed.), London, Longmans, pp. 130–160.

White, I. G. (1972). Biochemistry and semen and interactions in the female reproductive tract. *Search 122*, 22–30.

White, I. G. (1973). Biochemical aspects of spermatozoa and their environment in the male reproductive tract. *J. Reprod. and Fert.* Suppl. *18*, 225–235.

White, I. G. and MacLeod, J. (1963). "Composition and Physiology of Semen." In *Mechanisms Concerned with Conception.* C. G. Hartman (ed.), Oxford, Pergamon.

Williams, W. L., Abney, T. O., Chernoff, H. N., Dukelow, W. R. and Pinsker, M. C. (1967). Biochemistry and physiology of decapacitation factor. *J. Reprod. and Fert.* Suppl. *2*, 11–23.

Chapter 6

Gamete Transport

E. S. E. HAFEZ

While the mammalian female sheds one or two ova (or 10–15 ova in the case of swine) each estrous cycle, the male discharges massive numbers of spermatozoa at each copulation. Since the survival time of ova and spermatozoa is relatively short (20–48 hours), fertilization depends primarily on the synchronous transport of the gametes in the female reproductive tract. Gamete transport is the result of the inherent motility of the female tract as modified by central nervous system reflexes and hormonal activity. There are pharmacologically active substances in the semen which stimulate and modulate the motility of the female reproductive tract. The oviductal cilia and fluids, the cervix, uterotubal junction and ampullary-isth-mic junction may play a role which is as yet undetermined.

This chapter deals with the physiologic mechanisms involved in the transport, survival, capacitation and loss of spermatozoa, and the transport and development of eggs in the oviduct.

I. SPERM TRANSPORT IN THE FEMALE TRACT

There are species differences in the site at which the penis deposits semen in the female reproductive tract during copulation (Table 6–1). In cattle and sheep, the small volume of semen is ejaculated into the cranial end of the vagina and on to the cervix. In horses and swine the voluminous ejaculate is

Table 6-1. Species Differences in the Site of Ejaculation and Semen Characteristics in Several Mammals

Site of Ejaculation		Semen Characteristics	Species
Vagina	Incipient plug	Slight coagulation of ejaculate	Man Rabbit
	Incipient plug	Instant coagulation of ejaculate	Monkey
	Little accessory fluid	Semen with high sperm concentration	Cattle Sheep
Uterus	Voluminous	Distension of cervix	Horse
	Voluminous	Retention of penis during copulation	Dog Pig
	Vaginal plug	Spasmodic contraction of vagina	Rodents

deposited through the relaxed cervical canal into the uterus.

A. Patterns of Sperm Transport

The cervix, uterus and oviduct are lined with different types of nonciliated secretory cells and ciliated cells. In general, secretory cells have a dome-shaped surface covered with numerous microvilli and their cytoplasm contains numerous secretory granules (Fig. 6–1). The percentage of ciliated cells in the epithelium, which varies in different parts of the reproductive tract, is maximal in the fimbriae and oviductal ampullae and minimal in the uterus and cervix (Hafez, 1972). The ciliated cells are covered with kinocilia which beat rhythmically toward the vagina.

Immediately after deposition of semen into the vagina, spermatozoa penetrate the micelles of the cervical mucus where some are quickly transported through the cervical canal. This phase takes minutes and may be facilitated by increased contractile activity of the myometrium and mesosalpinx during courtship and/or coitus. Rapid transport gets the potentially fertilizing sperm out of the vagina where sperm survival is poor, and into the uterine horns where survival is better.

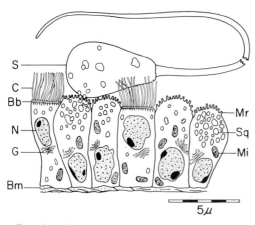

Fig. 6–1. The size of spermatozoa (*S*) in relation to the nonciliated secretory cells and ciliated cells of the cervical epithelium. *Bb*, basal bodies of cilia; *Bm*, basal membrane of epithelium; *C*, cilia; *G*, Golgi apparatus; *Mi*, mitochondria; *Mr*, Microvilli of nonciliated cells; *N*, nucleus; *Sg*, Secretory granules.

Thus spermatozoa have the opportunity for capacitation in the most favorable environment, the uterus and/or the oviduct. Meanwhile, the constituents of the seminal plasma may play a role in stimulating or modulating uterine function.

The cervical crypts and probably the uterotubal junction constitute the major "selective reservoirs" for spermatozoa. The orientation of micelles of cervical mucus leads to their point of secretory origin: the cervical epithelium. These orientation lines tend to guide a majority of motile spermatozoa to the mucosa of the cervical crypts (Fig. 6–2). Fewer numbers of spermatozoa are found in the mucus of the cervical canal, the endometrial glands and the endometrial fluid, but very few are found in the oviductal fluid. Concentration gradients of spermatozoa are established within a short time interval after copulation (Fig. 6–3). After adequate sperm reservoirs are established within the reproductive tract, the spermatozoa may be released sequentially to the site of fertilization.

B. Sperm Transport in the Cervix

The endocervical mucosa is an intricate system of clefts or grooves grouped together. Several functions have been ascribed to the cervix and its secretion: (*a*) receptivity to sperm penetration at or near ovulation and inhibition of migration at other phases of the cycle; (*b*) sperm reservoir; (*c*) protection of spermatozoa from the hostile environment of the vagina and from being phagocytized; (*d*) provision of spermatozoa energy requirements; (*e*) filtration of defective and immotile spermatozoa; and (*f*) possible capacitation of spermatozoa (Moghissi, 1972; Hafez, 1973). However, the cervix is considered an anatomic barrier well-adapted to prevent excessive numbers of spermatozoa from reaching the uterus. In sheep, for example, the normal ejaculate contains about 3 billion sperm, but less than a million pass the cervix.

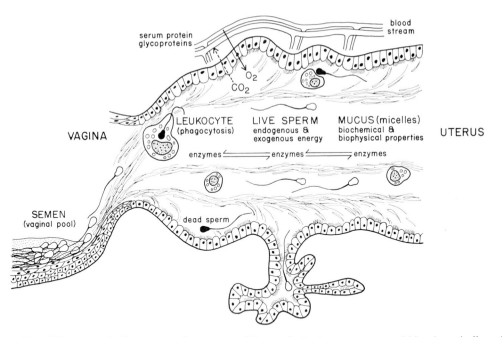

FIG. 6-2. Diagrammatic illustration of transport of live and dead spermatozoa within the micelles of cervical mucus in relation to cilia beat toward the vagina.

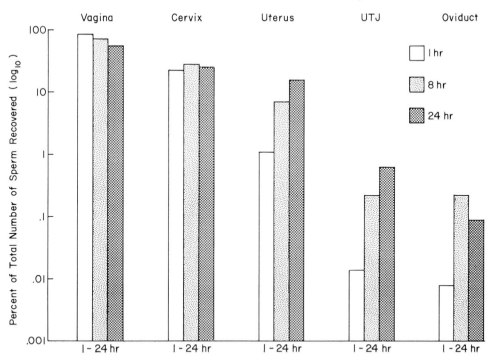

FIG. 6-3. Sperm distribution in the reproductive tract of cattle at different hours after insemination as a percentage of the total number of sperm recovered (logarithmic scale). The numbers of spermatozoa in the vagina and cervix decreases progressively following insemination. Sperm numbers in the uterotubal junction and the oviducts reach a maximum eight hours after insemination. (*Dabrowolski and Hafez, 1970. J. Anim. Sci. 31, 940.*)

Despite this filtering process the number of sperm in each ejaculate is greatly in excess of the minimum required for optimal fertility. With cattle artificially inseminated, high conception rates are obtained with 10 million sperm, i.e. .002 to .001 of the number in a normal bull ejaculate. With excessive dilution of semen, fertility declines and it is not clear if this is due to a reduction in the number of spermatozoa or to the dilution of protective substances in the seminal plasma.

Ejaculated spermatozoa rapidly penetrate the watery cervical mucus during estrus aided principally by sperm motility, as well as the micro- and macrorheologic properties of the mucus. In the rabbit, sufficient numbers of vaginal spermatozoa for normal fertility enter the cervix within five minutes after coitus (Bedford, 1970) and enter the oviduct within 30–40 minutes after coitus. Early passage of spermatozoa into the cervix is much enhanced by a second successive coital stimulus: destruction of all vaginal spermatozoa only two minutes after double mating caused a high fertilization rate of 90% (Bedford, 1971).

Little is known about the effect on spermatozoa transport of various steroids and related compounds administered orally or vaginally to synchronize estrus in farm animals. Synthetic progestogens alone or in combination with estrogen inhibit the secretion of watery cervical mucus to varying degrees and may prevent sperm migration in the micelles of cervical mucus. The cervical mucus becomes scanty, highly viscid and cellular, and its ferning pattern disappears and the spinnbarkheit is reduced.

C. Sperm Transport in the Uterus

Little is known about the pattern of sperm migration into the uterine lumen. Some spermatozoa invade the endometrial glands, whereas phagocytosis of living and probably dead spermatozoa by leukocytes occurs in the uterine lumen. The contractile activity of the vagina and myometrium plays a major role in the transport of spermatozoa into and through the uterus. Uterine contractions increase during estrus and are greatly augmented on copulation, due most probably to reflex stimulation of the posterior pituitary gland and subsequent secretion of oxytocin.

The intrauterine device (IUD) reverses the direction of the uterine contraction wave so that the majority of the waves are propagated toward the cervix, rather than toward the oviduct as in control ewes. This may explain in part the possible contraceptive mechanism of the IUD.

D. Sperm Transport in the Oviduct

Although massive numbers of spermatozoa are ejaculated, only a few hundred reach the site of fertilization in the oviduct (Fig. 6–4). Even so, the chances of a sperm encountering an ovum are high and it has been estimated that this occurs every two minutes in the rabbit during the first four hours after ovulation. The pattern and rate of sperm transport through the oviduct are controlled by several mechanisms, such as peristalsis and antiperistalsis of oviductal musculature, complex contractions of the oviductal mucosal folds and of the mesosalpinx, fluid movements created by ciliary action, and possibly the opening and closing of the uterotubal junction. The relative importance of these mechanisms in sperm transport through the oviduct is unknown.

In ruminants some spermatozoa reach the site of fertilization, i.e. the ampullae of the oviducts, in less than 15 minutes. This is too short a time to be accounted for solely by sperm motility and must be largely accomplished by contraction of the uterus and oviduct. Dead sperm and even inert particles, such as india ink, are also quickly transported up the reproductive tract and demonstrate the small contribution made by sperm motility. Oviductal contractions alter the configuration of the oviductal compartments

PLATE 8

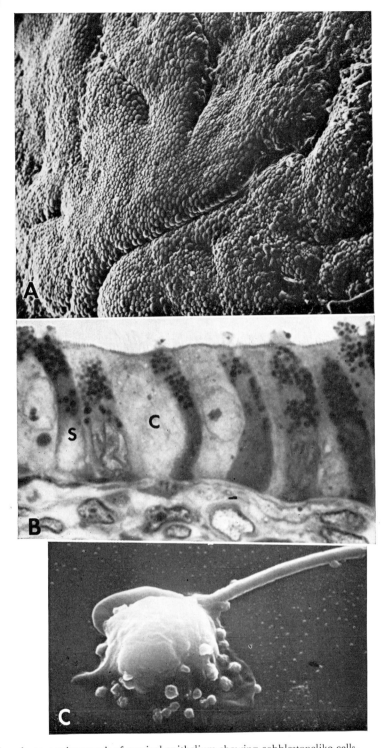

A, Scanning electron micrograph of cervical epithelium showing cobblestonelike cells.
B, Cervical epithelium showing secretory cells (*S*) with secretory granules, and ciliated cells (*C*).
C, Scanning electron micrograph of phagocytosis of the mammalian spermatozoa by a leukocyte.
(*Hafez and Kanagawa, 1973. Fertil. Steril. 25, 776.*)

PLATE 9

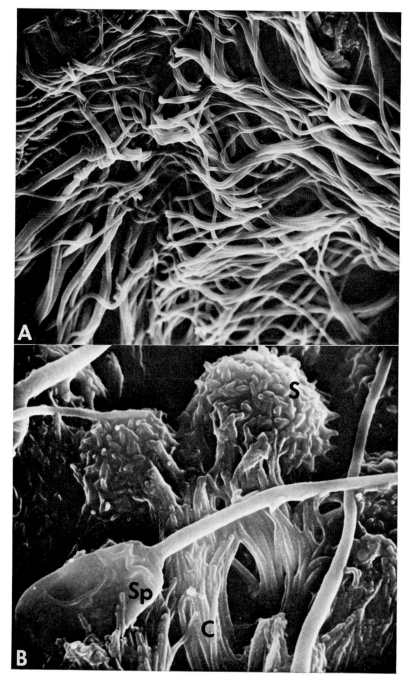

Scanning electron micrographs.

A, Spermatozoa on the cervical epithelium one hour postcoitum. Note the arrangement of sperm tail in a parallel formation. (*Hafez, 1973. J. Reprod. Med.*)

B, Cervical epithelium showing nonciliated secretory cells (*S*) with microvilli, ciliated cells (*C*) with kinocilia. Note the relative size of the spermatozoa (*Sp*) and the kinocilia.

momentarily, so that fluids and sperm may be transported toward the fimbriae from one compartment to the next. In the oviduct of the pigeon and painted tortoise there are two systems of kinocilia, one beats toward the ovary and the other toward the cloaca (Parker, 1931). These two ciliary systems are capable of moving particles in opposite directions. The isthmus normally limits passage of spermatozoa to the ampulla and prevents polyspermy (Thibault, 1972). In the pig, fertilization took place in the absence of the isthmus, which was re-sectioned, followed by end-to-end anastomosis of the ampulla and of the lower part of the isthmus proximal to the uterotubal junction (Hunter and Leglise,

1971). However, an increased number of spermatozoa was found in the zona pellucida of the eggs, together with a remarkable increase in polyspermic fertilization.

The role of the uterotubal junction in sperm transport varies with the species (Hafez and Black, 1969). In the horse and pig, a large volume of semen is ejaculated directly into the uterus. In the pig, most of the ejaculate disappears from the uterus within two hours, leaving a high concentration of spermatozoa at the uterotubal junction. This sperm reservoir persists for 24 hours and then disappears within the next 48 hours. These reservoirs may provide a continuous flow of sperm to the ampullae.

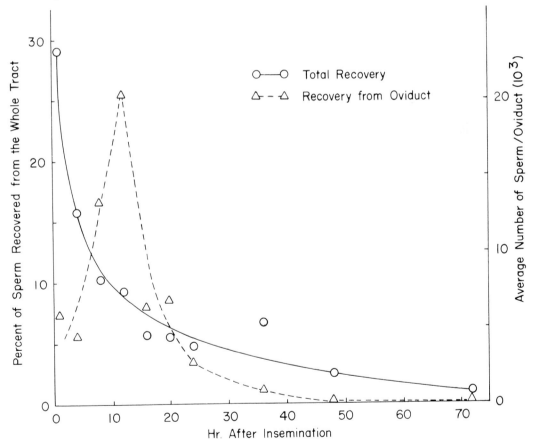

Fig. 6–4. Total sperm recovered from the rabbit female tract in relation to the number of sperm recovered from the oviduct at different intervals after insemination. Note that the number of spermatozoa recovered from the oviducts increased with time following insemination, although the total number of sperm recovered from the whole tract was decreasing. (*El-Banna and Hafez, 1970. Fertil. Steril. 21, 534.*)

In primates the volume of the ejaculate is relatively small. Semen is deposited in the vagina, and spermatozoa are found in high concentration in the uterotubal junction. In these species, the cervix and uterotubal junction appear to act as simple mechanical barriers which maintain graded concentrations of sperm throughout the reproductive tract.

E. Physiologic Mechanisms

Sperm transport in the female reproductive tract is influenced by several factors: (*a*) the mechanics of copulation; (*b*) the biophysical aspects of luminal fluids; (*c*) the intrinsic and spontaneous motility of the female reproductive tract; (*d*) the release of hormones (such as oxytocin) by the female during mating, which stimulate the motility of the female tract; and (*e*) the presence of uterine stimulatory substances, such as prostaglandins and spasmogens, in the semen (Fig. 6–5). The sequence of events of sperm transport is summarized in Table 6–2.

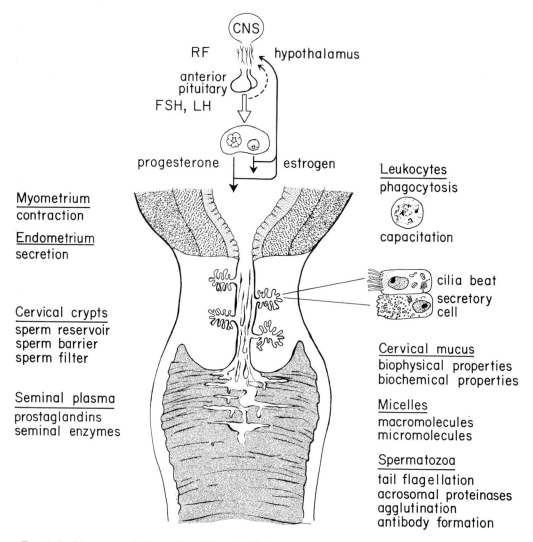

Fig. 6–5. Diagrammatic illustration of the physiologic, biophysic, anatomic and biochemical mechanisms involved in sperm transport in the cervix.

Table 6-2. Summary of Sequence of Major Physiologic Phenomena Associated with Sperm Transport in the Male and Female Reproductive Tract

Site	Physiologic Phenomena	Mechanisms Involved
Male reproductive tract	1. Sperm stored in cauda epididymis undergo maturation	Neuromuscular
	2. At ejaculation, sperm released from epididymis are mixed with male accessory secretions	Metabolic
Vagina	3. Semen deposited in several ejaculatory pulsations	Copulatory motor activities
	4. Semen mixed with vaginal and cervical secretions	
Cervix	5. Sperm migrate through micelles of cervical mucus	Biophysical
	6. Abnormal sperm filtered (gross selection of sperm) through cervical canal	Biochemical
	7. Cervical crypts establish "sperm reservoir" or rid excessive sperm causing massive reduction in sperm number	Mechanical (kinocilia of epithelium)
Uterus	8. Sperm separated from seminal plasma and transported to oviduct	Myometrial contraction
	9. Surface plasma of sperm removed	Agglutination of sperm
	10. Metabolic changes and capacitation of sperm	Phagocytosis of sperm by leukocytes
	11. Acrosomal proteinase (trypsin-enzyme) inactivated by trypsin inhibitors from seminal plasma	Enzymatic
Uterotubal junction	12. Quantitative selection of sperm	Mechanical
Isthmus	13. Sperm numbers reduced	
Ampullary-isthmic junction	14. Control of egg transport in oviduct	Neural
	15. Sperm plasma membrane changes (acrosome reaction), sperm capacitation	Biochemical
Ampulla	16. Sperm motility increases in oviductal fluid to be able to penetrate corona radiata and zona pellucida	Mechanical Metabolic Enzymatic
	17. Reduction division of gametes completed	
	18. Acrosomal proteinases released	Biophysical
	19. Selection at egg surface (receptors?) by sperm	
Fimbriae	20. Excessive sperm lost into peritoneal cavity	Sperm motility

1. SPERM MOTILITY

Normal spermatozoa have an oval-shaped head which is flattened in one dimension. Spermatozoa must be morphologically normal in order to migrate between the micelles of cervical mucus. Sperm migration through the cervical mucus is facilitated by the morphology of the sperm head which resembles a hydroplane. In order to maintain vigorous directional motility, the spermatozoon must overcome the resistance of the surrounding viscid medium (such as cervical mucus and endometrial fluid), the resistance from the proximity of other spermatozoa, and the stiffness of its flagellum. Flagellation of the tail enables the sperm cell to penetrate the cervical mucus.

2. ENDOCRINE AND NEURAL CONTROL

Ovarian hormones affect the structure, ultrastructure and secretory activity of the cervical, uterine and oviductal epithelium; the contractile activity of the utero-oviductal musculature; and the quantitative and qualitative characteristics of cervical mucus, and uterine and oviductal secretions. Changes are noted in protein content, enzyme activity, electrolytes, surface tension and conductivity in these fluids. Increasing amounts of endogenous estrogen during the preovulatory phase of the cycle or the administration of synthetic estrogens produce copious amounts of thin, watery, alkaline secretions. Endogenous progesterone during the luteal phase of the cycle or in pregnancy produces scanty, viscid, cellular cervical mucus with low spinnbarkheit and absence of ferning (Fig. 6–6).

The seminal plasma contains physiologically significant amounts of prostaglandins, particularly in the ram. Prostaglandins, unsaturated C_{20} acids secreted by the seminal vesicles, may play a part in the transport of sperm through the female reproductive tract. Seminal prostaglandins, introduced with the ejaculate, are readily absorbed into the female reproductive tract and exert their effect on uterine and oviductal musculature.

During copulation the posterior pituitary is stimulated and oxytocin is released, which in turn stimulates the utero-oviductal musculature. Oxytocin has been detected in the peripheral blood following stimulation of the vulva in sheep (Roberts and Share, 1968). The central nervous system in the female must mediate and integrate the large number of stimuli associated with mating, such as pheromones; visual, auditory and olfactory stimuli; mounting, intromission, coitus and ejaculation. Psychologic stress and psychosomatic factors may inhibit sperm transport. The number of spermatozoa reaching the ampullae of the tube is depressed in ewes showing signs of discomfort or fright (Thibault and Wintenberger-Torres, 1967). It is possible that this depression is mediated through the release of adrenalin which in turn reduces the contractility of myometrium in response to oxytocin.

3. BIOCHEMICAL MECHANISMS

During their transport to the site of fertilization spermatozoa are significantly diluted with luminal secretions from the female reproductive tract and are susceptible to changes in the pH of luminal fluids. Acidity and excessive alkalinity (8.5) of mucus immobilize spermatozoa, whereas alkaline mucus enhances their motility. Follicular, oviductal, peritoneal and amniotic fluids increase the activity and speed of propulsion of the spermatozoa.

Glycolytic and metabolic enzymes in the sperm tail and respiratory enzymes in the mitochondria are required for the biochemical reactions of the Embden-Meyerhof pathway, tricarboxylic acid cycle, fatty acid oxidation and electron transport system. The cervical mucus secreted at the time of ovulation provides an environment suited to the maintenance of metabolic activity of spermatozoa. This mucus undergoes biochemical changes, such as a decrease in

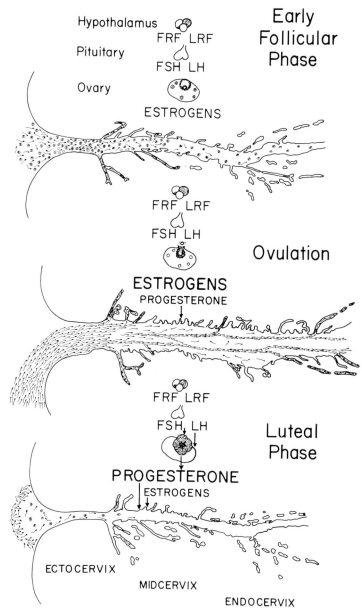

Fig. 6–6. Diagrammatic illustration of sperm transport through the cervical canal at three different stages of the estrous cycle. If copulation takes place at the time of estrus and ovulation, many spermatozoa aggregate along strands of cervical mucus: some of these spermatozoa migrate directly to the uterine lumen, whereas others move along mucous strands originating in the cervical crypts where sperm may form reservoirs If artificial insemination (A.I.) takes place during the early follicular phase, sperm transport is largely inhibited and cervical mucus contains large numbers of polymorphonuclear leukocytes mixed with the sperm. If copulation or A.I. takes place during the luteal phase, sperm progression is almost completely inhibited as a result of unfavorable biophysical characteristics of cervical mucus. (*Jaszczak and Hafez, 1973. Amer. J. Obst. Gynec. 115, 1070.*)

albumin, alkaline phosphatase, peptidase, antitrypsin, esterase and sialic acid, and an increase in mucins and NaCl.

Transport of spermatozoa into the uterus may influence capacitation in that the sperm are separated from an excess of the "decapacitation factor" and from other enzyme inhibitors in the seminal plasma (Srivastava and Williams, 1971).

Both the seminal plasma and spermatozoa contain chymotrypsinlike and trypsinlike enzymes, a fibrinolytic enzyme and peptidases (Fig. 6–5). "Acrosomal proteinase" (previously known as "acrozonase" or "acrosin") is similar but not identical to pancreatic trypsin, localized in the sperm acrosome.

4. Immunologic Mechanisms

Motility, agglutination and migration of spermatozoa in the cervical mucus are influenced by immunologic factors. Cervical mucus contains immunoglobulins A (IgA), G (IgG) and secretory IgA as shown by immunohistologic studies. Small numbers of IgG and IgA staining lymphoid cells are noted in the cervical stroma.

Semen and spermatozoa contain various antigenic systems such as species-specific and sperm-specific antigens, blood group, spermatozoa coating and seminal plasma antigens. These antigenic systems elicit various responses in the female reproductive tract. The exposure of cervical epithelium to sperm antigens elicits local immune response, which is probably unrelated to systemic rate of antibody production. These immune antibodies in the cervical secretion react with their specific antigens, thus causing agglutination or immobilization of the spermatozoa during their transport to the uterine cavity. Viscid cervical mucus seems to provide a protective barrier to foreign and potentially antigenic semen.

Hereditary factors and quality of spermatozoa may affect spermatozoa transport. There is evidence of species-specificity in sperm transport in the cervix. Sheep spermatozoa migrate

through the sheep's oviduct more quickly than those of goat in their species (Hancock and McGovern, 1968).

II. SURVIVAL OF SPERMATOZOA

Survival of spermatozoa in the female reproductive tract is an important factor in fertility. Much information is known about the duration of sperm motility, but little is known about the duration of fertilizing capacity which is lost long before motility. The fertile life of spermatozoa in the female is usually short. Spermatozoa can remain viable in the uterus for months in hibernating mammals such as bats, and for years in some insects and reptiles. In the reproductive tract of the cow, ewe and sow, sperm remain fertile for one to two days. In the mare viable sperm have been recovered after six days, which is significant in view of the long estrous period and the possibility that mating may occur several days before ovulation.

III. CAPACITATION

In most species spermatozoa are not capable of fertilization immediately upon entering the female reproductive tract and must spend at least two to four hours in the uterus or oviduct. Capacitation involves such morphologic, physiologic and biochemical changes which occur to the sperm to be capable of penetrating through the cumulus oophorus, corona radiata and zona pellucida of the egg.

Normally copulation takes place during early estrus, and ovulation toward the end or after this period, depending on the species. In view of their rapid transport in the female reproductive tract, spermatozoa reach the site of fertilization several hours before the eggs. During this time spermatozoa undergo capacitation.

Capacitation seems to involve sequential time release of a series of hydrolytic enzymes, allowing the sperm to digest a passage through the cellular and macromolecular investments of the egg. During the course of studies on fertilization in vitro many agents have been proposed to participate in or influence capacitation—calcium ions, fertilizins, viruses, β-amylase, β-glucuronidase, serum and steroids. Changes in the cell membrane during capacitation probably occur near the acrosome. Theoretically these changes might involve the form and disposition of helical protein molecules associated with the peptide-lipid framework of the membrane (Austin, 1969). Increase in permeability of the sperm membrane is associated with an increase in the number of aqueous channels crossing the membrane. Capacitation seems to be under endocrine control since estrogens stimulate capacitation and progesterone inhibits it.

The seminal plasma contains several compounds including proteinase inhibitor(s) and a decapacitation factor (DF). Spermatozoa upon exposure to these compounds in the epididymis or during ejaculation become infertile. It seems likely that removal of certain compounds from the spermatozoon results in the activation of the acrosomal enzymes so that the spermatozoa can penetrate the various layers of the ovum. DF, a nondialyzable constituent of seminal plasma, reverses the process of capacitation, rendering spermatozoa incapable of fertilization. Williams (1972) believes that sperm leave the vicinity of a high concentration of excess DF by migrating through the cervix into the uterus. There DF, attached to membranes and the corona-penetrating enzyme, is destroyed or removed, possibly by digestion with acrosin, a trypsin- and plasminlike acrosomal enzyme. Other acrosomal changes result in the release of hyaluronidase and removal of seminal plasma trypsin and acrosin inhibitors from the acrosomal membrane, thus expediting the passage of spermatozoa through the cumulus and the zona.

IV. LOSS OF SPERMATOZOA

Although millions of spermatozoa are deposited into the reproductive tract of the female, few ever reach the egg at the site of fertilization. Most spermatozoa perish at the selective barriers: the uterine cervix, uterotubal junction and oviductal isthmus. Residual sperm degenerate and are removed within a few days as a result of phagocytosis by leukocytes which can pass through the wall of the female tract and into the lumen. A continual loss also occurs in the vaginal and peritoneal cavities.

Loss Through Phagocytosis. The introduction of semen in the uterine cavity initiates a leukocytic response: the appearance of polymorphonuclear leukocytes. Electron microscopic studies demonstrated that most leukocytes in the stroma and lumen of the endometrium are mature heterophils, characterized by primary and secondary granules, extensive nuclear lobation, condensation of nuclear chromatin and by small dense particles of glycogen distributed throughout the cytoplasm. Active phagocytosis is not noted until 8–16 hours after sperm deposition, and maximal leukocytic response is slower in the oviduct (12–16 hours) than in the uterus (6–8 hours) (Howe, 1967).

The phagocytic removal of spermatozoa from the female genital tract and the intensity of leukocytic response to various stimuli depend on the endocrine state of the female. Phagocytosis is less efficient in the progesterone-dominated phase (Haynes, 1967).

The biologic relationship between leukocytes and spermatozoa with respect to capacitation and/or sperm survival is not known. In the bovine cervix, the majority of leukocytes occur in the central mass of the mucin, a fact indicating that most of them have invaded the cervix from the uterus (Mattner, 1966). Most viable spermatozoa, lodging in the cervical crypts, escape the leukocytes so that an adequate population of them survives.

Spermatozoa can be inactivated in the uterus by disconnection at the junction of the head and midpiece. Sperm cell breakage in the uterus could be increased by endogenous or exogenous estrogen or by the presence of an intrauterine contraceptive device, and decreased by endogenous or exogenous progesterone (Conley and Hawk, 1970).

Loss in the Vagina. Damaged spermatozoa are carried back passively through the ectocervix with the help of ciliated cells beating toward the vagina. Such spermatozoa, advancing only a short distance into the cervical mucus core, do not reach the cervical crypts and decrease in number within a few hours after copulation. Since spermatozoa, which become immotile elsewhere, are not so rapidly eliminated the ratio of immotile spermatozoa is higher in the cervix than in other segments of the female reproductive tract. Large amounts of cervical mucus are produced and numerous spermatozoa are expelled with the mucus through the vulva in cattle.

Loss in the Peritoneal Cavity. Spermatozoa reaching the fimbriae are released into the peritoneal cavity. In rabbits, when one uterine horn was ligated above the cervix prior to mating, 10% of the animals had implantations in both horns (Rowlands, 1957), indicating sperm migrated through the peritoneal cavity from the unligated side to effect fertilization and ovum development on the ligated side of the female reproductive tract. Several hours after intraperitoneal insemination 4000–70,000 sperm are present in the oviduct and cervix of the rabbit and occasionally in the uterus.

V. RECEPTION OF EGGS

The viscid mass of cumulus oophorus containing oocytes and corona cells adheres to the stigma and remains attached to it unless it is removed by the action of the kinocilia of the fimbriae. The fimbriae at the time of ovulation are widely separated and engorged with blood, and are brought into close contact with the surface of the ovary by the muscular activity of the mesotubarium. The

ovary is moved slowly to and fro and around its longitudinal axis by contractions of the ligamentum ovarii proprium. This chain of reactions is controlled by anatomic and hormonal mechanisms.

Anatomic Mechanisms. The ovary is located inside the ovarian bursa to which the ampulla of the oviduct and part of the fimbriae are attached. The ovary can move readily from this location to the surface of the fimbriae, which is positioned at the open portion of the ovarian bursa. This movement is controlled by both the *ligamentum ovarii proprium* and the mesovarium which hold the ovary and oviduct in position.

The fimbriae and the infundibulum are basically composed of erectile structure, which is rich in vascular and muscular tissues. During estrus the fimbriae undergo distension from the increased blood flow; furthermore, the margins of the fimbriae become edematous and translucent.

Hormonal Mechanisms. The contractile activities of the fimbriae, oviduct and ligaments are partly coordinated by hormonal mechanisms involving the estrogen-progesterone ratio. Egg reception is most efficient about the time of estrus but occurs to some degree throughout the cycle. In some species there are also neurohormonal mechanisms which stimulate the contractile activity of the fimbriae at the time of copulation; however, these are not as yet understood.

VI. TRANSPORT OF EGGS

The oviduct is a highly specialized organ with two more or less distinct anatomic parts—the ampulla and the isthmus. Histologic examination of these two parts suggests mechanisms by which the egg is transported from the infundibulum to the uterus. The transport appears to depend largely on the normal function of the cilia, smooth muscle activity and functions of the ampullary-isthmic junction and the uterotubal junction. The interrelationships of these physiologic activities are controlled by the ovarian hormone's balance after ovulation and can thus be greatly altered by ovariectomy or administration of steroid hormones.

After its relatively quick transport through the ampulla, the ovum is arrested at the ampullary-isthmic junction for two days (Fig. 6–7). This junction is a physiologic rather than anatomic sphincter, the nature of which and the mechanisms which regulate its function are unknown. The junction is biochemically characterized by high acid phosphatase activity compared with the isthmus or ampulla, but during ova passage a "surge" of activity of this enzyme occurs uniformly throughout the oviduct. Acid phosphatase may play a role in ova denudation and removal of denuded cumulus and corona cell debris.

The physiologic closure of the distal portion of the isthmus may be due to stimulation of noradrenergic neurons, since infusion of noradrenaline or stimulation of the hypogastric nerve produces isthmic constriction. The role of prostaglandins in the control of egg transport is unknown: F prostaglandins (PG F_2) stimulate and E prostaglandins (PG E_2) inhibit contractions of the isthmus. When ova are transported prematurely from the oviduct to the uterus they do not implant. Similarly, acceleration or retardation of ova passage through the oviduct causes their rapid degeneration.

Table 6-3. Transport Time of Ova in the Oviduct of Farm Animals Compared with Some Other Mammals

Species	Transport Time of Ova in Oviduct (Hours)
Cattle	90
Sheep	72
Horse	98
Pig	50
Cat	148
Dog	168
Monkey, rhesus	96
Opossum	24
Woman	48–72

The length of time taken for the oviductal transport of ovum in farm mammals ranges from 50 to 98 hours (Table 6–3). As a result of delay at the ampullary-isthmic junction, the ova enter the isthmus slowly, remain in the distal portion for a while and then pass out of it rapidly. Apparently the uterotubal junction exerts a regulatory influence on ova entry into the uterus.

Differences in the rate of egg transport in the ampulla and the isthmus are due to differences in their histologic characteristics, such as thickness of musculature, number of cilia, size of lumen, distensibility, innervation and concentration of adrenergic nerve endings and characteristics of the secretory epithelium. Several physiologic mechanisms have been suggested to explain blocking of eggs at the ampullary-isthmic junction, such as tem-

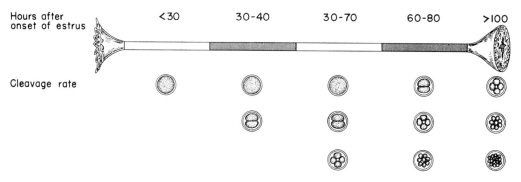

FIG. 6–7. Rate of transport and cleavage of eggs in swine. Eggs pass through the first half of the oviduct very rapidly and they remain in the third quarter which contains the ampullary-isthmic junction until 60–75 hours after onset of estrus. The eggs enter the uterus between 66 and 90 hours after onset of estrus. (*Data from Oxenreider and Day, 1965. J. Anim. Sci. 24, 413.*)

porary inactivity of the epithelial cilia, localized edema of the isthmic region, oviductal motility toward the ovary, constriction or inactivity of the isthmic circular muscles and tubal locking by specific sphincteric muscle(s) in the isthmus. El-Banna and Hafez (1970) measured planimetrically the lumen size of all segments of the oviduct in cattle and found that the ampullary-isthmic junction was the smallest at all reproductive stages, although it enlarged two-fold in the third to fourth days following estrus.

A. Role of Contraction of Oviductal Musculature

Several techniques were used to study oviductal contractions in vivo and in vitro, such as the simultaneous recordings of extracellular electric activity and intraluminal pressure. The pacemaker region is mainly localized to a small area in the isthmus at the ampullary-isthmic junction. The contractions spread in both the uterine and ovarian direction from the pacemaker area.

Segmental contractions start in different segments of the oviduct, run for a short distance toward the uterus, then disappear. These contractions are more powerful and more regular in the isthmus than in the ampulla. Since the inner circular fibers contract first, the oviduct stretches at the beginning of the contraction; it then shortens, owing to contraction of the outer fibers, before it relaxes again. When the contraction has subsided, the previously contracted segment becomes tensely engorged with blood, and this may cause that segment of oviduct to become double its normal thickness (Koester, 1970).

The frequency and amplitude of spontaneous contractions vary with the phases of estrous cycle. In sheep, strong antiperistaltic contractions of the isthmus, lasting 2 to 2.5 days after ovulation, prevent the peristaltic contractions from forcing the ova through the isthmus. Antiperistaltic contractions then weaken, and the eggs are transported through the whole isthmus into the uterus in a short period of time under the influence of peristaltic contractions (Wintenberger-Torres, 1961). Unlike intestinal peristalsis, oviductal peristalsis instead of pushing the ovum forward actually tends to delay its progression slightly. Further, instead of playing a role in ovum transport, oviductal contractions cause a thorough mixing of oviductal contents, promote fertilization and help to denude ova.

B. Role of Oviductal Secretions and Cilia

During midcycle, epithelial height and secretory activity reach their maximum. After ovulation, the secretory material is evacuated from the secretory cells, and epithelial height decreases. These variations become progressively more pronounced from the isthmus to the infundibulum. The structure and number of the secretory granules depend on the ovarian function. Dense homogeneous granules are most common in estrus, whereas less dense, nonhomogeneous granules are characteristic of the postovulatory phase. Several types of granules $(1-5 \ \mu$ in diameter) are found in the secretory cells of the oviduct.

The major flow of oviductal fluid is toward the infundibulum into the peritoneal cavity. Since the cilia beat toward the uterotubal junction, a small proportion of the flow of oviductal fluid in proximity to the ciliary surface may be toward the uterus. It is possible that on days when the uterotubal junction is open the fluid current reverses its direction thereby enabling the ova to travel with it through the last portion of the isthmus and the uterotubal junction into the uterus. The velocity of the fluid current depends on the cross sectional diameter of the oviduct; in the narrower isthmus the flow is faster than in the wider ampulla (Fig. 6–8).

The kinocilia beat rapidly toward the uterus, moving the egg against the flow of oviductal secretion and causing constant rotation of the egg. The fluid flows so slowly in the ampulla that it is overcome by the power of the cilia. At the ampullary-isthmic junction the oviductal lumen narrows and there is a sharp increase in the rate of flow. At the same time there is a decline in the number of ciliated cells compared with the secretory cells (Fig. 6–8). Thus the power of the cilia driving the egg toward the uterus is less effective against the countercurrent of the secretion (Koester, 1970).

C. Endocrine Mechanisms

It is critical that fertilized eggs reach the uterus at an appropriate progestational stage of the estrous cycle. Both estrogen and progesterone must be present at appropriate levels to induce the normal egg transport. Ovarian hormones affect the structure, ultrastructure and secretory activity of the oviductal epithelium, the contractile activity of the oviductal musculature, the quantitative and qualitative characteristics of the oviductal secretions, the actions of the ampullary-isthmic and uterotubal junctions, and the pattern and rate of gamete transport. As progesterone activity begins to increase toward the end of the third day of the egg's passage through the oviduct, the rate of cilia beating increases and the secretion rate in the isthmus declines.

The physiologic mechanisms which control egg transport in the oviduct will be altered by exogenous synthetic and natural estrogens and progestins, certain synthetic nonsteroids, gonadotropins, prostaglandins and other adrenergic drugs influencing the contraction of smooth musculature. Such compounds may accelerate the rate of egg transport to the uterus or cause tube-locking of the eggs at the ampullary-isthmic junction. Both alterations result in embryonic mortality. The effects of estrogen and progesterone on egg transport vary with their proportions and the timing of their administration in relation to the endocrine status.

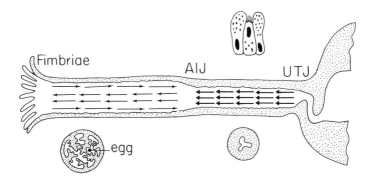

Day 1 (estrogen)	Cilia beat moves egg to AIJ against flow of secretion	Counter-current of secretion more effective than cilia
	Weak flow of secretion	Less ciliated cells
Day 3 (progesterone)	More cilia beat (+20%)	Less secretion in isthmus

FIG. 6–8. Diagram to show the rate of estrogen and progesterone on the flow of oviductal secretions in the ampulla and isthmus (shown by arrows toward the fimbriae), the flow of oviductal cilia (shown by arrows toward the uterotubal junction [UTJ]), and egg transport in the oviduct. AIJ = ampullary-isthmic junction.

VII. FERTILIZABLE LIFE AND AGING OF EGGS

The fertilizable life of the egg is the maximum period during which it remains capable of fertilization and normal development. In most species the egg is capable of being fertilized for some 12–24 hours (Table 6–3). It rapidly loses its fertilizability upon reaching the isthmus and is completely nonfertilizable after reaching the uterus.

The egg may be fertilized near the end of its fertilizable life as a result of delayed breeding. Such eggs may or may not implant and, if so, produce mostly nonviable embryos. Guinea pigs show a high percentage of abnormal pregnancies and decrease in litter size as the age of the egg increases prior to fertilization (Fig. 6–9). Fertilization of aged eggs in swine is associated with polyspermy and hence abnormal embryonic development. In single-bearing animals, aging of the egg may cause abortion, embryonic resorption or abnormal development of the embryo. Similar abnormalities may result from aged sperm.

In general, fertilization of aged gametes involves one of the following possibilities:

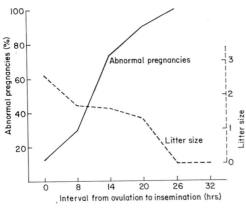

Aged egg + freshly ejaculated sperm
Aged egg + aged sperm
Freshly ovulated egg + aged sperm
Nonviable embryos resulting from any of the above combinations may cause low conception rates in certain herds and flocks. There is good evidence that fertilization with aged sperm increases subsequent embryonic mortality in swine and poultry. At present there is insufficient evidence in farm mammals concerning the relative deleterious results of gamete aging on fertilization, implantation, and prenatal development. It is possible that some of the congenital abnormalities in postnatal life are a consequence of aged gametes.

If the egg is not fertilized, it will fragment into several cytoplasmic segments of unequal size, and in some cases it may even resemble a fertilized egg. All unfertilized eggs eventually disappear through complete disintegration or phagocytosis in the uterus.

VIII. TRANSUTERINE MIGRATION AND LOSS OF EGGS

Transuterine migration of the egg through the common body of the uterus is quite common in ungulates. For example, when one of the ovaries is removed from a sow, approximately half of the embryos develop in each uterine horn, irrespective of which ovary was removed. There is also a tendency in the normal sow for the number of embryos to be equalized between the two horns. Transuterine migration is more common in swine and horses than in cattle and sheep. Nonetheless, cattle and sheep, which have double ovulations from one ovary, usually have one embryo in each uterine horn. The physiologic mechanisms which govern the movement of eggs, both within individual horns and between horns, is unknown.

Transperitoneal migration of eggs can be accomplished by suitable experimental conditions: e.g., removal of one ovary leaving the fimbriae and oviduct intact and ligation of the other oviduct. In

Fig. 6–9. The effect of aging of ova (delayed insemination) on percentage of abnormal pregnancies and litter size in guinea pigs. Note that the ova were fertilized and implanted when the animals were inseminated at 26 hours after ovulation, yet the embryos did not continue development. (*Data from Blandau and Young, 1939. Am. J. Anat. 64, 303.*)

this case, the remaining oviduct has the ability to pick up the ova released by the contralateral ovary and a normal pregnancy may follow. Transperitoneal migration may be aided by currents and surface tension of the peritoneal fluid.

The egg may never reach the infundibulum due to many causes. For example, eggs entrapped in the ruptured follicles can be in the developing corpus luteum. It may also be lost into the peritoneal cavity; such an egg usually degenerates but in rare cases may result in ectopic pregnancy (pregnancy located outside the uterus). Egg loss in the peritoneal cavity may be caused by the immobilization of the oviduct as a result of faulty rectal palpation of the ovaries, postpartum or postabortum infections, endometritis or nonspecific abdominal infections.

IX. EMBRYONIC DEVELOPMENT IN OVIDUCT

The oviduct seems to take an active part in maintaining eggs and preparing them for fertilization and segmentation. The oviductal fluid is rich in substrates and cofactors involved in ovum development, such as pyruvate and bicarbonate, free amino acids, oxygen, CO_2, carbohydrates, perhaps lipids, nucleosides, steroids and other compounds. Apparently these substances are contributed by the cells of the oviductal mucosa to the luminal fluid milieu.

Oviductal epithelium is most active near the developing embryo. In the rabbit, epithelial uptake of radioactive sulfate increases in the region containing the egg. Using an immunofluorescence technique, Glass and McClure showed that two- to four-cell mouse eggs take up proteins similar to those in the mother's blood serum.

In vitro studies indicate that oviductal fluid also supports blastulation. In an artificial medium, two-celled mouse eggs did not develop, except for some late two-cell embryos that continued cleavage after the addition of lactate. Early two-cell embryos did cleave when cultured in extirpated oviducts, growing best in oviducts from late estrous and metestrous mice. Endocrine factors thus seem important in the early development of embryos in the oviduct.

The biochemical composition of the uterine fluid is very different from that of the oviductal fluid. This would indicate that eggs at early cleavage stages require specific substances provided by the oviduct for their development. The premature entry of the morulae into the uterus, therefore, will cause their degeneration. After a certain time, the blastocysts need to enter the uterus for final development and implantation.

REFERENCES

Austin, C. R. (1969). "Sperm Capacitation—Biological Significance in Various Species." In *Advances in the Biosciences, 4: Schering Symposium on Mechanism Involved in Conception*. Berlin, Pergamon Press-Vieweg, pp. 5–11.

Bedford, J. M. (1970). "The Saga of Mammalian Sperm From Ejaculation to Syngamy." In *Mammalian Reproduction*. H. Gibian and E. J. Plotz (eds.). Berlin, Springer-Verlag, pp. 124–188.

Bedford, J. M. (1971). The rate of sperm passage into the cervix after coitus in the rabbit. *J. Reprod. Fert. 25*, 211–218.

Conley, H. H. and Hawk, H. W. (1970). Intensification by intrauterine devices of sperm loss from sheep uterus. *Biol. Reprod. 2*, 401–407.

El-Banna, A. A. and Hafez, E. S. E. (1970). Sperm transport and distribution in rabbit and cattle female tract. *Fertil. Steril. 21*, 534–540.

Hafez, E. S. E. (1972). Scanning electron microscopy of rabbit and monkey female reproductive tract. *J. Reprod. Fert. 30*, 293–296.

Hafez, E. S. E. (1973). "The Comparative Anatomy of the Mammalian Cervix." In *The Biology of the Cervix*. R. J. Blandau and K. S. Moghissi (eds.). Chicago, The University of Chicago Press, pp. 23–56.

Hafez, E. S. E. and Black, D. L. (1969). "The Mammalian Uterotubal Junction." In *The Mammalian Oviduct Comparative Biology and Methodology*. E. S. E. Hafez and R. J. Blandau (eds.). Chicago, The University of Chicago Press, pp. 85–126.

Hancock, J. L. and McGovern, P. T. (1968). The transport of sheep and goat spermatozoa in the ewe. *J. Reprod. Fert. 15*, 283–287.

Haynes, N. B. (1967). "The Influence of the Uterine Environment on the Phagocytosis of Spermatozoa." In *Reproduction in the Female Mammal*. G. E. Lamming and E. C. Amoroso (eds.). London, Butterworths.

Howe, R. G. (1967). Leukocytic response to spermatozoa in ligated segments of the rabbit vagina, uterus and oviduct. *J. Reprod. Fert. 13,* 563–566.

Hunter, R. H. F. and Leglise, P. C. (1971). Polyspermic fertilization following tubal surgery in pigs, with particular reference to the role of the isthmus. *J. Reprod. Fert. 24,* 233–246.

Koester, H. L. (1970). "Ovum Transport." In *Mammalian Reproduction.* H. Gibian and E. J. Plotz (eds.). Heidelberg, Springer-Verlag, p. 189.

Mattner, P. E. (1966). Formation and retention of the spermatozoan reservoir in the cervix of the ruminant. *Nature* (Lond.) *212,* 1479–1480.

Moghissi, K. S. (1972). The function of the cervix in fertility. *Fertil. Steril. 23,* 295–306.

Parker, G. H. (1931). The passage of sperms and eggs through the oviducts in terrestrial vertebrates. *Phil. Trans. Roy R. Soc (B) 219,* 381.

Roberts, J. S. and Share, L. (1968). Oxytocin in plasma of pregnant, lactating and cycling ewes during vaginal stimulation. *Endocrinology 83,* 272–278.

Rowlands, S. W. (1957). Insemination of the guinea pig by intraperitoneal injection. *J. Endocr. 16,* 98–106.

Srivastava, P. N. and Williams, W. L. (1971). "Sperm Capacitation and Decapacitation Factor." In *Control of Human Fertility.* E. Diczfalusy and U. Borell (eds.). New York, Wiley, p. 73.

Thibault, C. (1972). Physiology and physiopathology of the fallopian tube. *Int. J. Fert. 17,* 1–8.

Thibault, C. and Wintenberger-Torres, S. (1967). Oxytocin and sperm transport in the ewe. *Int. J. Fert. 12,* 410–415.

Williams, W. I. (1972). "Biochemistry of Capacitation of Spermatozoa." In *Biology of Mammalian Fertilization and Implantation.* Chap. 2. K. S. Moghissi and E. S. E. Hafez (eds.). Springfield, Charles C Thomas, pp. 19–53.

Winterberger-Torres, S. (1961). Movements des trompes et progression des oeufs chez la brebis. *Ann. Biol. Anim. Bioch. Biophys. 1,* 121, *1* (2), 121–133.

Chapter 7

Fertilization, Cleavage and Implantation

ANNE MCLAREN

I. FERTILIZATION

The entire process of sexual reproduction is centered around the act of fertilization: yet fertilization is not itself a reproductive process. On the contrary, it consists essentially of the fusion of two cells, the male and female gametes, to form one single cell, the zygote. Fertilization is a dual process:

(*a*) In its *embryologic* aspect, it involves activation of the ovum by the sperm. Without the stimulus of fertilization, the ovum does not normally begin to cleave, and no embryologic development occurs. In some animals experimental treatments are known which mimic this aspect of fertilization, inducing development in the unfertilized ovum.

(*b*) In its *genetic* aspect, fertilization involves the introduction of hereditary material from the sire into the ovum. By this means it is possible for beneficial characters arising far apart in time and space eventually to become combined in a single individual. The importance of this process for natural and artificial selection can hardly be overestimated. According to current genetic belief, the essential hereditary material is the chromosomal DNA in the sperm nucleus: fusion of male and female nucleus in the process of *syngamy* is therefore often thought of as the central process of fertilization. Although attempts have been made to inject foreign DNA into the ovum experimentally, this aspect of fertilization has not yet been mimicked in the laboratory.

In fertilization, two cells combine to form one, the first cell of the new individual: yet the number of chromosomes remains constant in every generation. This is because the two gametes each contain only half the number of chromosomes characteristic of the species.

Throughout this chapter, much use will be made of information derived from the study of mice, rats and rabbits, since comparatively little direct work has so far been done on fertilization and early development in the larger and more expensive farm animals. Indeed, the most detailed and exact information which we possess about fertilization relates to the sea urchin (Monroy, 1965); but how far one is justified in generalizing from sea urchins to mammals is unknown, and this work will therefore not be referred to here.

A. Description of Fertilization Process

1. THE OVUM: ITS POSITION AND STATE

In most mammals, fertilization begins after the first polar body has been extruded, so that the sperm penetrate the ovum while the second reduction division is in progress (see Chapter 5 for description of meiosis). In the horse and dog, however, sperm may enter the ova before the second reduction division has begun.

6

143

Table 7-1. Estimates of the Fertile Life of Sperm and Ova, and of
the Tempo of Embryonic Development

Species	Fertile Life* (Hours) of:		Days After Ovulation† for:				
	Sperm	Ovum	2-Cell	8-Cell	Into Uterus	Blastocyst	Birth
Cattle	30–48	8–12	1	3	3–3½	7–8	275–290
Horse	72–120	6–8	1	3	4–5	6	335–345
Man	28–48	6–24	1½	2½	2–3	4	252–274
Mouse	10–12	6–15	1	2½	3	3½	19–20
Rabbit	30–36	6–8	1	2½	3	4	30–32
Rat	12–14	8–12	1½	3	3½	4	20–22
Sheep	30–48	16–24	1	2½	3	6–7	145–155
Swine	24–48	8–10	14–16 hr.	2	1¾–2	5–6	112–115

* "Fertile life" is a relative concept, since fertility declines progressively over a period of hours. For sperm, only the period in the female genital tract is included. The life of ova is timed from ovulation. For both, longevity probably depends on a variety of factors, including the hormonal state of the female. It is therefore impossible to give precise figures.

† These estimates are only approximate, since developmental rate is subject to considerable variation both among individuals and among breeds. In addition, accurate information on the time of ovulation is lacking in several species.

The site of fertilization in all farm and most other mammals is the lower portion of the ampulla of the oviduct. When it enters the ampulla, the ovum in its mucoprotein coat (the *zona pellucida*) is still surrounded by a cluster of granulosa cells which were shed with it from the ovarian follicle, and which at this stage are often called *cumulus cells*. In swine, the cumulus cells surrounding the several ova join together in the fimbriae of the oviduct to form a single cluster, the "egg plug." This normally disintegrates soon after ovulation, but Spalding and associates report that, when ovulation is induced, the egg plug may persist until after fertilization. In farm animals other than swine, cumulus cells are absent from those ova which have been examined a few hours after ovulation (e.g. within 9 to 14 hours of ovulation in the cow). They are probably shed before fertilization begins, since unpenetrated ova recovered from the oviduct are usually found to be almost or quite denuded.

In most species, ova that are not fertilized degenerate within a few days, but in the horse they remain in the fallopian tube for several months.

2. The Sperm: The Encounter with the Ovum

The entry of sperm into the female genital tract and their transport to the site of fertilization have been dealt with previously (see Chapter 6). Here we wish to emphasize three points only: (*a*) Although the total number of sperm in an ejaculate is measured in hundreds or thousands of millions, the number travelling as far as the ampulla is relatively small, probably not much more than 1000 in any mammal. (*b*) Some sperm reach the site of fertilization very quickly, within about 15 minutes of mating. (*c*) In the rabbit, mouse, hamster, sheep, ferret and cat, and probably in the cow and rhesus monkey, the sperm have to undergo some change or set of changes (called *capacitation*) before they can activate the ova.

In those species where capacitation is necessary, the ova arrive in the ampulla a sufficient length of time after the sperm to ensure that full capacitation has occurred. The fertile life of ova is short (Table 7–1), usually less than 24 hours. In all mammals except bats, the fertile life of sperm is also fairly short, though successful fertilization five days after insemination has been reported in the mare. Possibly sperm may lose their ability to induce viable embryos before they lose their ability to fertilize. The relatively brief fertile life of both sperm and ovum renders timing a matter of the utmost importance in mating and artificial insemination. For instance in the cow, which normally ovulates about 14 hours after the end of heat, the conception rate from inseminations made at the time of ovulation is very low, and the best time for insemination is from 6 to 24 hours before ovulation (Fig. 7–1).

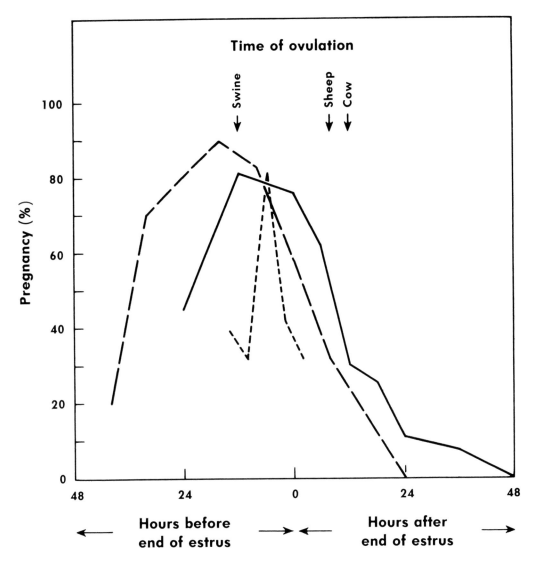

Fig. 7–1. Relationship between time of insemination and fertility in the cow, sheep and swine. ———— Cow; — — — — Swine; - - - - - Sheep. (*After Trimberger and Davis, 1943, Neb. Agric. Expt. Sta. Res. Bull. No. 129, for cow. Dziuk, 1970, J. Reprod. Fert. 22, 277, for sheep. Willemse and Boender, 1967, Tijdschr. Diergeneesk, 92, 18, for swine.*)

Fertilization rate in most mammals is remarkably high. In one strain of rabbits, only 1.4% of ova failed to be fertilized, though the overall prenatal loss amounted to 30%. The scanty data on fertilization losses in farm animals have been reviewed by Hancock (1962). What information we possess on the size of the ampulla of the tube, the number of sperm present, the rate at which they swim, and the surface area of the ovum, suggests that some factor other than chance is probably operating to ensure that ovum and sperm meet. Yet no such factor has so far been demonstrated experimentally. The mass of cumulus cells might be thought to facilitate contact by trapping sperm in the neighborhood of the ovum but, at least in the rabbit, it has been shown that sperm make contact just as readily with ova from which the cumulus mass has been experimentally removed. In the mouse, there is evidence that the cumulus plays a role in sperm capacitation.

It has long been assumed that fertilization is an entirely random process—i.e. that there is an equal chance of any sperm fertilizing any ovum. Although direct evidence is hard to come by, this is probably not entirely true. Experiments with mixed inseminations demonstrate that the sperm from different males may differ in their fertilizing capacity. Bateman (1960) has shown that true selective fertilization may in certain circumstances occur in the mouse. When sperm of a particular type were presented with a choice of ova, they united more frequently with one type than with another.

3. Entry of Sperm into Ovum

To enter the ovum, the sperm (see Piko, 1969) has first to penetrate (a) the cumulus mass, if this is still present;

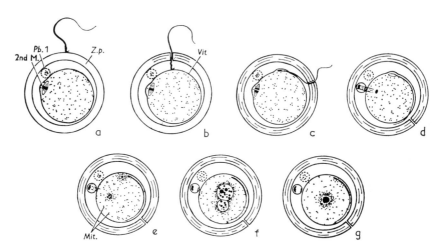

Fig. 7–2. Diagram illustrating the processes occurring during fertilization in the rat.

(a) The sperm in contact with the zona pellucida (Z.p.). The first polar body (Pb.1) has been extruded; the nucleus of the ovum is undergoing the second meiotic division (2nd M).

(b) The sperm has penetrated the zona pellucida, and is now attached to the vitellus (Vit). This evokes the zona reaction, which is indicated by shading as it passes round the zona pellucida.

(c) The sperm head is taken into the vitellus and lies just below the surface which is becoming elevated above it. The zona has rotated relative to the vitellus.

(d) The sperm is now almost entirely within the vitellus. The head is swollen (see Plate 10B). The vitellus has decreased in volume, and the second polar body has been extruded.

(e) Male and female pronuclei develop. Mitochondria (Mit.) gather around the pronuclei.

(f) The pronuclei are fully developed and contain numerous nucleoli. The male pronucleus is larger than the female.

(g) Fertilization is complete. The pronuclei have disappeared and been replaced by chromosome groups, which have united in the prophase of the first cleavage division.

(After Austin and Bishop, 1957. Fertilization in mammals. Biol. Rev. 32, 296.)

PLATE 10

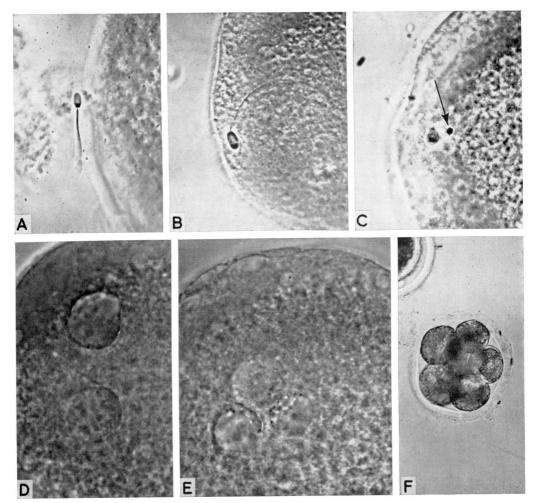

A, B, C, Different regions of a single pig ovum, eight hours postcoitum ($\times$ 450). A shows a nonfertilizing sperm in the zona pellucida; B shows the penetrating sperm which has entered the vitellus (note its swollen head); C shows the nuclear apparatus of the ovum, with the recently formed nucleus of the 2nd polar body to the left, and the presumptive female pronucleus (arrow) to the right.

D, Male and female pronuclei in a normal pig ovum during fertilization ($\times$ 500 approx.).

E, Pig ovum six hours postcoitum, with three conjugating pronuclei. The chromatin clumps on the nuclear membranes of the two smaller pronuclei suggest that these are both female ($\times$ 460).

F, Unfertilized ovum recovered from a sow's uterus 72 hours after mating to a vasectomized boar. The ovum has fragmented in a manner strikingly reminiscent of normal cleavage ($\times$ 185).

(A, B, C, E from Hancock, 1961. Fertilization in the pig. J. Reprod. Fert. 2, 307; D from Hancock, 1958. The examination of pig ova. Vet. Rec. 70, 1200; F from Hancock, 1958. The fertility of natural and of artificial matings in the pig. Studies on Fertility 9, 146.)

PLATE 11

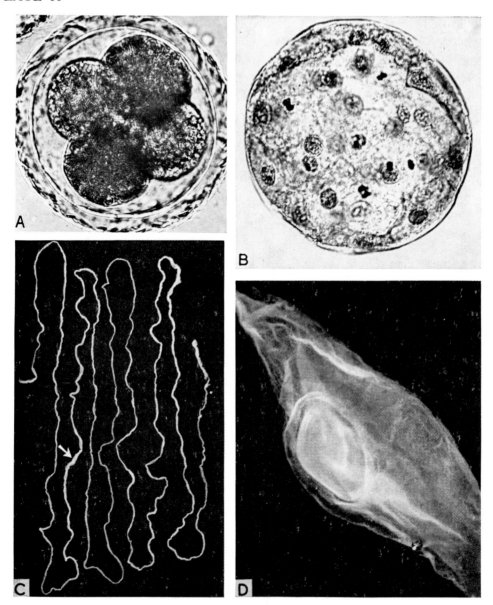

A, Four-celled ovum recovered from a sow's oviduct 48 hours after artificial insemination. The vitellus stains black because of the dense concentration of yolk platelets ($\times$ 300).

B, Late morula of pig, 144 hours postcoitum. Several mitotic figures can be seen ($\times$ 330).

C, Pig blastocyst 13 days after mating at the beginning of estrus. The total length of this blastocyst was 157 cm. The position of the embryonic disc is indicated by an arrow.

D, Enlarged view of a pig embryonic disc 14 days after mating.

(*A* from Hancock, 1958. *Studies on Fertility 9, 146; B from Hancock, 1961. J. Reprod. Fertil. 2, 307; C, D from Perry and Rowlands, 1962. J. Reprod. Fertil. 4, 175.*)

(*b*) the zona pellucida; and (*c*) the vitelline membrane.

The sperm makes its way through the cumulus mass by virtue of its own motility, dissolving a tunnel through the hyaluronic acid matrix as it goes. Sperm have been shown to contain the enzyme *hyaluronidase*, and this may help, at least in rodents, to dissolve the cumulus mass. The final disintegration of the cumulus mass is a separate process, which is not necessarily the result of fertilization and may in fact precede it in farm animals.

The next obstacle to sperm entry is the zona pellucida (Fig. 7–2*a*). There is some evidence for the existence of a mechanism to ensure that the sperm remains attached at this stage. The ovum is said to produce a substance (*fertilizin*) which reacts with the sperm and specifically agglutinates it. Although well established in sea urchins, such a phenomenon has not yet been demonstrated unequivocally in mammals. At any rate the agglutination process cannot permanently immobilize the sperm, since it continues to swim through the zona pellucida, leaving (at least in rodents) a narrow tunnel behind it. At this stage the acrosome, loosened during capacitation, is finally lost, exposing the *perforatorium*. Probably the action of a proteolytic enzyme associated with the perforatorium facilitates passage through the zona. An extract of acrosomes isolated from ram, bull or rabbit sperm has been reported effective in dissolving the zona, as well as dispersing the corona radiata of rabbit ova.

The last stage in the penetration of the ovum involves the attachment of the sperm head to the surface of the vitellus (Fig. 7–2*b*). This period, which lasts for up to 30 minutes in those rodents where it has been measured, is a vital one since it is at this time that *activation* occurs. Stimulated by the proximity of the sperm, the ovum awakes from its dormancy and development begins. The sperm head, and in some species the tail as well, then enters the vitellus. A projection on the surface of the vitellus marks for some hours the point of sperm entry (Fig. 7–2*c*).

A combination of transmission and scanning electron microscopy has been used to study the actual process of penetration of the sperm into the vitellus. In the hamster, studies by Yanagimachi and Noda (Fig. 7–3) suggest that the microvilli on the surface of the ovum actively participate in sperm-ovum association. The microvilli grasp the sperm head; the plasma membranes of sperm and ovum then rupture and fuse with one another to form a continuous cell membrane over the ovum and outer surface of the sperm. As a result, the sperm

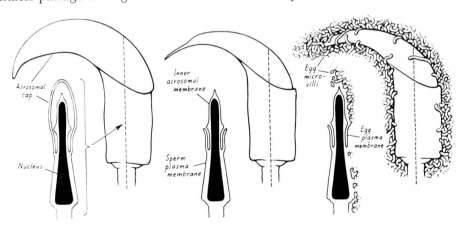

Fig. 7–3. Changes in hamster sperm before and during fertilization. *A*, intact epididymal sperm; *B*, sperm after capacitation; *C*, initial stage of sperm entry into the vitellus. (*After Yanagimachi and Noda, 1972. Scanning electron microscopy of golden hamster spermatozoa before and during fertilization. Experientia 28, 69.*)

comes to lie inside the vitellus, leaving its own plasma membrane incorporated into the vitelline membrane.

Experiments on various combinations of rodent species (mouse, rat, hamster, guinea pig) suggest that the major block to the penetration of ova by foreign sperm resides in the zona pellucida. Capacitated sperm of one species are unable to get through the zona of another species but may penetrate the vitellus if the zona has first been removed.

4. PRONUCLEUS FORMATION

One striking result of activation in some species (e.g. pig, cow) is that the vitellus shrinks in volume, expelling fluid into the perivitelline space. At the same time the sperm head in the vitellus swells and acquires the consistency of a gel, losing its characteristic shape (Plate 10*B*, Fig. 7–2*d*). The perforatorium and the tail drop off. Within the sperm nucleus, a number of nucleoli appear and subsequently coalesce, and a nuclear membrane develops around its periphery. The final structure, which resembles the nucleus of a somatic cell much more closely than it does a sperm nucleus, is termed the male pronucleus (Fig. 7–2*e*).

Little is known of the subsequent fate of sperm constituents other than the nucleus. In some species the numerous mitochondria of the sperm midpiece all go to one of the two daughter cells at the first cleavage division, in others they are released into the cytoplasm of the ovum and are distributed equally.

In most species the second polar body is extruded from the ovum soon after sperm entry, and formation of the *female pronucleus* then begins (Plate 10*C*). This resembles the male pronucleus in the appearance of nucleoli and the formation of a nuclear membrane. The two pronuclei develop synchronously, increasing in volume during the course of several hours to an extent that was estimated at 20-fold for the rat. In some species, including the pig, the two pronuclei are of roughly similar size (Plate 10*D*); in others the male pronucleus is the larger (Fig. 7–2*f*). Pronucleus formation has been studied in detail by phase-contrast microscopy in living rat eggs by Austin (1951). This has not yet been done in farm animals because the texture of the cytoplasm in the living ovum makes the pronuclei almost impossible to see. In farm animals the pronuclei are smaller than in the rat, and the nucleoli are few and small; but in other respects the formation of pronuclei in the rat is probably representative of mammals in general.

5. SYNGAMY

At some stage during their maximum development, the male and female pronuclei come into contact. After a time they begin to shrink, and at the same time to coalesce. The nucleoli and nuclear membrane disappear, and the pronuclei can no longer be seen. In those mammals in which it has been measured, the total life span of the pronuclei extends over a period of 10–15 hours. As the time of the first cleavage draws near, two chromosome groups become visible. These are the maternal and paternal chromosomes respectively. They unite to form a single group, which represents the prophase of the first cleavage mitosis (Fig. 7–2*g*). Fertilization is now complete.

As in any other mitotic division, each chromosome splits longitudinally, and the halves separate to opposite ends of the cleavage spindle. The fertilized ovum undergoes its first cleavage, to produce a two-celled embryo. Each daughter cell now contains the normal diploid number of chromosomes characteristic of the species, half having been derived from the ovum and half from the sperm.

The duration of fertilization, i.e. the total time interval from the penetration of sperm to the metaphase of the first cleavage division, has been estimated as 12 hours in the rabbit, 12–14 hours in the pig, 16–21 hours in the sheep, 20–24 hours in the cow and approximately 36 hours in man.

6. Zona Reaction and Vitelline Block

If ova are classified according to the number of sperm they contain, those penetrated by one sperm only are found to be commoner than would be expected by chance. Often ova are observed with several sperm clustering around the outside of the zona pellucida, but only a single one within. It is therefore inferred that the zona pellucida can undergo some change after the passage of the first sperm which renders it less easy to penetrate subsequently. This change is termed the *zona reaction*. It is further inferred that the reaction consists of a propagated change in the zona, set off when the first sperm makes contact with the surface of the vitellus, and mediated by some substance passing out from the vitellus to the zona. Possibly this substance may be liberated by the *cortical granules*, particles 0.1–0.5μ in diameter which have been seen in the ova of the golden hamster, rat, mouse, guinea pig, rabbit, coypu and swine and which disappear after the first sperm has entered the ovum.

Extra sperm which succeed in passing through the zona pellucida into the perivitelline space are called *supplementary sperm*. In some species (sheep, dog, hamster) the zona reaction is relatively quick and effective, and supplementary sperm are found rarely if at all. In other species (mouse, rat) they are more common. In the pig, extra sperm enter the zona pellucida but do not normally succeed in passing right through it (e.g. see Hunter, 1972). The rabbit shows no zona reaction, and up to 200 supplementary sperm have been observed in the perivitelline space of the fertilized ovum.

The other defense mechanism against entry of more than one sperm is shown by the vitellus itself, and is termed the *vitelline block* or the *block to polyspermy*. The fertilizing sperm is actively engulfed by the vitellus; but subsequently the vitelline surface becomes unresponsive to contact, and no further sperm are engulfed. Sperm which have been damaged by x irradiation may make contact with the vitelline surface without activating the ovum. In this case the contact also fails to induce the vitelline block. In the golden hamster ovum, the vitelline block is completed two to three hours after sperm penetration.

Extra sperm which succeed in entering the vitellus, in spite of both the zona reaction and the vitelline block, are called *supernumerary sperm*, and the ovum is said to show *polyspermy*. The effectiveness of the vitelline block varies from species to species. When polyspermic ova are found, but supplementary sperm are seldom or never seen (e.g. sheep, dog, hamster), the vitelline block must either be absent or must be delayed until after the zona reaction is in operation. On the other hand in a species such as the rabbit, with many supplementary sperm in the perivitelline space but a very low incidence of polyspermy, there must exist a rapid and efficient vitelline block.

In mice the incidence of polyspermy depends upon the strain of the female; on the other hand the proportion of ova containing supplementary sperm is related to the strain of the male and does not depend on the female.

B. Polyspermy

The existence of the zona reaction and the vitelline block would lead one to expect that polyspermy in mammals was a disadvantageous state. This is indeed so. All the processes of fertilization, including the factors which regulate the number of sperm reaching the ampulla, are coordinated to ensure that, while every ovum is fertilized, polyspermy is kept to a minimum. The incidence of polyspermic ova in most mammalian species is normally only one to two percent. In birds, on the other hand, polyspermy is usual.

The incidence of polyspermy may be increased experimentally, either by increasing the number of sperm in the ampulla, or by lowering the barriers which prevent extra sperm from entering the ovum. Conditions which reduce the zona reaction also tend to reduce the vitelline block. These include aging of the ovum or heating of the ovum or of the whole animal. Delayed copulation, since it leads to fertilization of aged ova, is an effective method of increasing the incidence of polyspermy in pigs, rats and rabbits. A high incidence of polyspermy has also been observed in hamster ova fertilized in vitro.

When polyspermy does occur, the one or more supernumerary sperm often form pronuclei in the normal way, though in such a case all the pronuclei are reduced in size. Ova with three or occasionally four pronuclei have been observed in many species, including the cow, sheep and pig (Plate 10*E*). Sometimes it is possible to prove that two of the three pronuclei are male in origin and hence that polyspermy has occurred; sometimes, however, the extra pronucleus will be female in origin (*digyny*), having arisen through failure of polar body formation at one or other reduction division. Digyny is rare in cows and ewes, even after delayed fertilization; but in sows Thibault has found that if copulation takes place more than 36 hours after the beginning of estrus, more than 20% of the ova are digynic. At syngamy the three pronuclei, whatever their origin, give rise to three chromosome groups which then unite. In this way a *triploid* embryo is formed, with three sets of chromosomes in every cell instead of the normal two sets. Triploid rat and rabbit embryos can survive to midgestation, and three children and one cat with three sets of chromosomes in most of their cells have been reported; but the great majority of triploid embryos die at a very early stage of development. In the sow, triploid embryos do not survive implantation. The disadvantage of polyspermy to the organism is thus that it leads to triploidy, which is a lethal condition.

C. Parthenogenesis, Gynogenesis, Androgenesis

In *parthenogenesis* the ovum is activated by some means other than by a sperm.

In *gynogenesis* the ovum is activated by a sperm which takes no further part in fertilization. In *androgenesis* the ovum is activated by a sperm in the normal manner, but the nucleus of the ovum takes no part in fertilization. In all three phenomena only a single pronucleus is formed, female in origin in the first two, and male in the third. Gynogenesis and androgenesis rarely if ever occur spontaneously in mammals but may be induced experimentally, for instance by irradiating either the sperm or ova. The embryos that result are *haploid*, with only a single set of chromosomes, and do not survive beyond the early stages of development.

Spontaneous cleavage of unfertilized ova has been reported in a large number of mammalian species. In some of these cases, degenerative fragmentation may have been mistaken for parthenogenetic development. True activation has been achieved in vitro by chemical or temperature treatment, or by culture of unfertilized ova in appropriate conditions (rabbits, mice), and in vivo by cooling, anesthesia, oxygen lack or electrical treatment (rabbits, mice, rats, sheep). Development may proceed to blastocyst stage or even (in mice) to midway through gestation, but a report that induced parthenogenesis in rabbits led to the birth of live young has not been confirmed. According to the nuclear mechanism involved, parthenogenetic embryos may be either haploid or diploid. In the second case, they would be difficult to distinguish from normal young.

Gynogenesis, androgenesis, parthenogenesis and other abnormalities of fertilization are discussed at length by Beatty (1957).

D. Fertilization in Vitro

Experiments on the mechanism of fertilization in mammals require a reliable procedure for achieving fertilization of mammalian ova outside the body of the female. Numerous attempts to obtain fertilization in vitro before the phenomenon of sperm capacitation had been discovered were largely vitiated by the use of uncapacitated sperm. Cleavage of the supposedly fertilized ovum in culture may have been due to parthenogenetic activation; while ova treated with sperm in vitro and shortly afterward transferred to the fallopian tubes of a recipient female may have had sperm attached to their surfaces which only achieved fertilization after transfer.

These difficulties were first overcome in the work done on rabbits by Dauzier and his associates. Rabbit ova were removed at body temperature, to lessen the risk of parthenogenetic activation. They were mixed with sperm which had been previously capacitated in the genital tract of female rabbits. Subsequent microscopical examination showed that the ova had been penetrated by sperm and were undergoing pronucleus formation, and apparently normal cleavage. Chang, using the same technique, returned the embryos to recipient female rabbits and obtained live young. The percentage of embryos which developed successfully was rather low.

Fertilization in vitro has been achieved not only in the rabbit, but also in the mouse, Syrian hamster, Chinese hamster, guinea pig, cat and man. Only in the rabbit and mouse has in vitro fertilization been followed by cleavage and birth of live young. Experience on in vitro fertilization in farm animals is so far not encouraging.

E. Sex Determination

Every cell in the mammalian body except the gametes contains a pair of *sex chromosomes*. In females the two members of the pair resemble one another and are known as *X chromosomes*; in males the sex chromosomes differ, one being an X chromosome, the other smaller, known as a *Y chromosome*. The sex chromosome constitutions of females and males are therefore referred to as XX and XY respectively. The gametes, being haploid, contain only a single sex chromosome: an X chromosome in the female (*homogametic sex*) and either an X or a Y chromosome in the male (*heterogametic sex*). Faulty sharing out (*nondisjunction*) of sex chromosomes, either to the gametes (Fig. 7–4) or, after fertilization, to the products of early cleavage, occasionally gives rise to individuals whose cells contain only a single X chromosome (XO), or an extra X or Y chromosome (XXX, XXY or XYY). XXY individuals are male and XO individuals female, proving that maleness must be determined in the first instance by factors on the Y chromosome.

In normal fertilization, the embryo develops as a female or as a male according to whether the ovum (carrying an X chromosome) is fertilized by a sperm carrying an X or Y chromosome. If the two types of sperm are present in equal numbers, the ratio of males to females at conception (the *primary sex ratio*) should be equal to one. Attempts to control the sex ratio in farm animals usually depend on treating the semen in such a way as might be expected to alter the proportions of X-bearing and Y-bearing sperm (Chapter 3).

		OVA		
		Normal	Nondisjunctive	
		X	XX	O
SPERM Normal	X	XX (= normal female)	XXX	XO
	Y	XY (= normal male)	XXY	YO
SPERM Nondisjunctive	XY	XXY		
	XX	XXX		
	YY	XYY		
	O	XO		

FIG. 7–4. Diagram showing how normal and abnormal sex-chromosome constitutions can arise at fertilization. An *O* sperm or ovum is one which carries neither an *X* nor a *Y* chromosome. Nondisjunctive gametes arise through faulty sharing out (nondisjunction) of the sex chromosomes. *YO* individuals are probably not viable; *XXX* individuals, in man, are abnormal females.

II. CLEAVAGE

After syngamy is completed, there ensues a period of several days during which the fertilized ovum, zygote or *embryo* as it should be called once development has started, leads a free-living existence in the fallopian tubes and uterus of the mother. During the latter part of this period, the embryo may be nourished by uterine secretions; but not until implantation has taken place does the embryo derive any nourishment from the maternal blood stream.

At the beginning of the free-living period the ovum is a single cell, of relatively enormous volume compared with other cells of the body and therefore with a very large ratio of cytoplasm to nuclear material. Reserve nutrients are stored in the cytoplasm in the form of yolk (*deutoplasm*). This single cell divides and redivides many times without any accompanying increase in volume of cytoplasm, though some increase in volume occurs through the uptake of water. The total amount of cellular material in the embryo actually decreases by about 20% in the cow, and by as much as 40% in the sheep. The total protein content of mouse embryos has been shown by Brinster to decrease by 25%. The process of cellular division without growth is called *cleavage*. It continues until implantation, by which time cell size has been reduced more or less to the size characteristic of the species. During the early stages of cleavage, up to the appearance of the blastocoele, the embryonic cells are often known as *blastomeres*.

A. Normal Course of Cleavage

The unfertilized ovum already possesses some polarity and an axis of symmetry. The nucleus lies in the *animal pole* where the cytoplasm is usually dense and rich in ribonucleoproteins and mitochondria. In the opposite half (*vegetal pole*) the cytoplasm is more vacuolated and contains fewer mitochondria. In species (e.g. guinea pig) where the ovum contains fat globules, it is in the vegetal pole that they chiefly accumulate.

The plane of the first cleavage division is not related to the plane of symmetry of the ovum. It usually passes through the area where the male and female pronuclei were situated at the beginning of syngamy, passing from the animal to the vegetal pole. The second cleavage divisions occur at right angles to the first, the third more or less at right angles to the second. But the divisions are not perfectly synchronized, so that three-celled and five-, six-, seven-celled stages may be found. Nor are the divisions always equal since the cells containing the more vacuolated cytoplasm tend initially to be larger than those at the animal pole. (This tendency is less marked in the cow than in other farm animals.) All the divisions are mitotic, and consequently each cell of the embryo, from the fertilized ovum onwards, contains the diploid number of chromosomes ($2n$). Considerable amounts of DNA are synthesized during cleavage.

By the 16- to 32-celled stage, the cells are crowded together into a compact group within the zona pellucida. The embryo is now known as a *morula* (Plate 11*B*). Fluid begins to collect in the intercellular spaces, and an inner cavity or *blastocoele* appears. Once this has begun to expand, the embryo is known as a *blastocyst*. A single peripheral layer of large flattened cells, the *trophoblast* layer, surrounds a knob of smaller cells which lie to one side of the central cavity. This knob, the *inner cell mass*, will give rise to the adult organism, while the cells of the trophoblast form the placenta and embryonic membranes.

Electron microscope investigations have shown that the membranes of trophoblast cells are closely apposed and interdigitated, and are linked at intervals by the "tight junctions" characteristic of highly integrated tissues. In the rat, Enders and Schlafke have observed these tight junctions as early as the 8- to 16-celled stage around the periphery of the embryo.

In swine and horses, where the ovum is very rich in yolk, the surplus is eliminated (*deuto-plasmolysis*) into the perivitelline space during cleavage and later is found in the blastocoele. In the horse, this yolk is extruded asymmetrically on the side of the ovum which is farthest from the nucleus.

1. Fate of Cleavage Products

In mammals, cleavage is believed to be of the *indeterminate* type. This implies that, until quite a late stage of development, it is impossible to tell which particular organs of the body are going to be formed by which cells. It also implies that all the cells of the early embryo retain all their original potentialities, in the sense that any cell would be capable of giving rise to an entire new embryo if the environmental conditions were suitable.

How indeterminate the mammalian embryo is at different stages we do not yet know (McLaren, 1969). Normal embryonic development can follow destruction of one cell of a two-celled mouse embryo, or seven cells of an eight-celled rabbit embryo. Identical twinning has been achieved experimentally in mice by separating the two blastomeres at the two-celled stage, allowing each to develop independently in culture, then transferring them to the uterus of a recipient female. Embryos derived in this way from a reduced number of blastomeres remain smaller than normal throughout the first part of gestation but are of normal size by the time they are born.

Further evidence that the development of the mouse embryo before implantation is capable of a remarkable degree of regulation comes from experiments on experimental aggregation of embryos (Mintz, 1971). Techniques for removal of the zona pellucida and aggregation of cleavage stages were developed independently by Tarkowski and by Mintz. The double embryo develops into a single blastocyst containing twice the normal number of cells, and development after

transfer to a recipient female proceeds normally, giving rise to a chimeric individual containing two genetically distinct cell populations. Experimental chimeras of this sort have so far been obtained only in mice: they have yielded valuable information on cell lineages in development, phenotypic interactions between cells of different genotype, and sexual differentiation.

The technique of embryo aggregation not only demonstrates the lability of early embryonic development but has also been used to investigate the extent to which the fate of cleavage products depends on their spatial arrangement within the embryo. Hillman, Sherman and Graham (1972) combined mouse embryos and parts of embryos during cleavage and traced the developmental fate of their cells either by prelabeling some cells with a radioactive isotope or by combining cells which synthesized different electrophoretic variants of an enzyme. They found that each blastomere of a four-celled stage could contribute to both trophoblast and inner cell mass. When blastomeres of a four-celled embryo were placed on the outside of other four-celled embryos, their descendants were usually to be found in the trophoblast at the blastocyst stage and in the trophoblast and yolk sac on the 10th day of pregnancy. Conversely, four- to eight-celled embryos which were completely surrounded by other blastomeres did not contribute to the trophoblast.

These experiments confirm, at least for the mouse, the view first put forward by Tarkowski and Wroblewska (1967), that differentiation into inner cell mass and trophoblast is not determined by any fixed segregation of preexisting heterogeneity in the cytoplasm of the fertilized ovum, but is environmental in origin, depending on whether or not the blastomere is on the outside of the embryo. On the other hand the "inside-outside" explanation cannot apply in the same form to species like the goat and swine, where the inner cell mass develops not from the inside cells of the morula but from a group of larger, less actively dividing cells at one pole. A group of small, rapidly dividing cells at the other pole subsequently spreads out to form an outer layer to the embryo which in turn gives rise to the trophoblast.

The properties of the inner cell mass and trophoblast have been examined in rodents. Dalcq observed that proteins, nucleoproteins and alkaline phosphatase are concentrated in the inner cell mass of the rat embryo, while the outer trophoblast layer produces most of the mucopolysaccharides and acid phosphatase. The microsurgical experiments of Gardner and Johnson (1972) on mouse blastocysts established that isolated trophoblastic vesicles, lacking an inner cell mass, would pump fluid and were capable of inducing an implantation reaction in the uterus but would not stick together. Isolated inner cell masses would aggregate but would neither pump fluid nor induce an implantation reaction. Trophoblast proliferation after implantation depends on an inductive influence coming from the inner cell mass.

2. Cleavage Rates

Approximate estimates of the time taken by the embryos of various mammals to reach certain stages of development are given in Table 7–1. The rate of cleavage up to the blastocyst stage tends to be faster in species where the total gestation period is short. Little is known about the influence of the maternal environment, though Wintenberger-Torres has shown that in sheep the rate of cleavage of blastocysts after seven days is significantly increased either by superovulation, when there is an abnormally high number of corpora lutea, or during treatment with exogenous progesterone.

3. Activation of the Embryonic Genome

At the beginning of cleavage, the embryonic genome appears to be inactive genetically, so that the control of

early development must depend on maternal products transmitted through the cytoplasm of the ovum. The first embryonic genes to be activated are those concerned with the protein-synthesizing apparatus. In the mouse, both ribosomal and transfer RNA begin to be synthesized at the four-celled stage and ribosomal RNA is transferred to the cytoplasm at the eight-celled stage. From the morula stage onward, RNA synthesis increases rapidly, and synthesis of proteins encoded by the embryonic genome begins at the late morula or blastocyst stage. In the rabbit, on the other hand, although transfer RNA is again synthesized early in development, no ribosomal RNA is synthesized until blastocyst formation, by which time the embryo contains several hundred cells. Presumably the larger rabbit ovum contains sufficient storage products to support development at least until blastocyst formation. No information is yet available on the stage at which the embryonic genome is activated in any farm animal.

4. The Blastocyst

The metabolic activity of ova, whether judged by oxygen uptake or carbon dioxide output in vitro, increases little during the early cleavage divisions but rises sharply between the morula and blastocyst stage in both mouse and rabbit. In the rabbit, this is partly due to a shift from the hexose monophosphate oxidation pathway to the more efficient Embden-Meyerhof pathway and the tricarboxylic acid (TCA) cycle. The changing metabolic requirements of early embryos are discussed by Biggers (1971).

As the embryo swells up with fluid, the blastocoele enlarges. The extent of this enlargement varies from species to species. In farm mammals it is considerable, the blastocyst becoming within a few days a thin-walled fluid-filled sac, more or less filling the uterine lumen.

The loss of the zona pellucida has been extensively studied in rats and mice. In normal pregnancy, the zona is dissolved at the onset of implantation, probably by a proteolytic enzyme emanating from the estrogen-sensitized uterus. During lactation, on the other hand, or when females are ovariectomized early in pregnancy so that the uterus is deficient in estrogen, or when blastocysts are cultured in vitro, loss of the zona is delayed. Under these conditions the blastocyst "hatches" from the zona, probably aided by rhythmic contractions and expansions such as have been observed in culture, and no lysis of the zona occurs. In the hamster, loss of the zona is again by lysis and is strongly progesterone-dependent. The zona pellucida of the guinea pig becomes penetrated at the time of implantation by slender pseudopodial processes from the abembryonic trophoblast cells of the blastocyst but is not lost until after attachment has been effected. In farm mammals, the zona is lost before attachment, but the mechanism of loss is not known.

The biochemistry of the fluid in the blastocoele has been studied in the rabbit before and during implantation. Use of radioactive tracers makes it possible to determine the passage of substances into and out of the blastocyst. Before implantation the blastocoelic fluid is very rich in potassium and bicarbonate, which appear to be actively drawn into the blastocyst from the uterine fluid. As implantation proceeds, potassium and bicarbonate fall to the levels found in the maternal serum. At the same time protein and glucose, previously present in small amounts only, increase up to maternal serum levels. Phosphorus and chlorides also increase greatly in concentration. Water-soluble vitamins (thiamine, riboflavin, B_{12}, nicotinic acid) are present in small amounts in the blastocyst fluid before implantation.

It is clear that the blastocyst cannot be considered as a passive vesicle into which substances enter by simple diffusion. It possesses a high degree of metabolic selectivity, actively controlling the rate of entry of substances from the surrounding fluid. The role of uterine secretions in embryonic nutrition is discussed later.

B. Twinning

Two distinct types of twins are known: *monozygotic (identical) twins* and *dizygotic (fraternal) twins*.

Dizygotic twins originate from a double ovulation in a normally monotocous species. Two ova may be shed in the same estrous period, and fertilized by two different sperm. The resulting young do not resemble one another genetically any more closely than do ordinary brothers and sisters. In fact a pair of dizygotic twin lambs of opposite sex differ on average more from one another in birth weight than do male and female lambs born in twin pairs of like sex. This is known as the *enhancement effect*, and is probably due to some form of competition between embryos in the uterus. Multiple ovulation may be induced, and hence the frequency of dizygotic twins increased, by the injection of pituitary or chorionic gonadotropins. This technique is already finding practical application in sheep breeding.

Monozygotic twins, on the other hand, originate from a single fertilized ovum. In theory they could occur in all species, including species such as the pig which normally bear several young at a time, but so far they have been recognized in only a few species, notably man and cattle. Even here they are relatively rare: up to five percent (depending on breed) of all births in cattle will be twin births, but only about one in every thousand births will be monozygotic twins. There is no evidence that any twin births in sheep or swine are monozygous.

Since monozygotic twins represent, as it were, two halves of a single individual, they resemble one another very closely indeed in all genetically determined characteristics. For instance, they are always of the same sex. Because of their identical hereditary equipment, monozygotic twin pairs of cattle are particularly valuable experimental material for studying the effect of varied environmental conditions on such characters as milk yield in dairy breeds, and weight gain and conformation in beef breeds.

Many cases of monozygotic twinning probably originate fairly late in development, after implantation. A single blastocyst implants, and the single inner cell mass then differentiates two primitive streaks, giving rise to two separate individuals. Such twins have a common chorion, and also sometimes a common amnion. Alternatively, the inner cell mass in a single blastocyst may duplicate before implantation. This condition has been reported in both sheep and pigs. The resulting embryos will have separate amnions and placentas. It is not impossible that some monozygotic twins may originate still earlier, as a result of blastomeres separating inside the zona pellucida.

In swine, the pregnant uterus normally contains several embryos. There is rarely fusion between either the allantois or the chorion of neighboring embryos. In sheep, where more than one embryo occurs frequently but not always, there is fusion of the chorion but not of the allantois; while in cattle, where multiple pregnancies are found relatively seldom, there is usually fusion of both chorion and allantois, with consequent anastomosis of blood vessels between neighboring embryos. This means that in cattle the majority even of dizygotic twins have a common blood circulation, while in other farm mammals the proportion is much lower. Where the twin partners are of opposite sex, the common blood circulation leads to the development of free-martins (Chapter 21).

This fusion of the embryonic blood vessels is responsible for an unexpected difficulty in distinguishing monozygotic from dizygotic twins in cattle. In general, monozygotic twins are recognized by their striking similarity in all traits that depend mainly or only on heredity. Sex, coat color and pattern, nose prints, serum globulin type and presence or absence of horns are particularly useful traits for this purpose in cattle. But the genetically determined blood groups have to be used with caution in classifying cattle twins, because the common embryonic circulation results in a mixture of the blood-forming cells, so that each dizygotic twin shows not only his own blood group but also that of the other twin. If detected, this mosaicism is of course itself evidence of dizygosity. It has been estimated that, if all diagnostic tests are used, the error in diagnosing monozygosity in cattle twins can be reduced to about one percent.

C. Interspecific Hybrids

Crosses between related species have been reported from many different mammalian groups (Gray, 1972). Those of economic importance are summarized in Appendix I. Note that species crosses, even when viable, tend to be partly or totally sterile (see Chapter 19). As J. B. S. Haldane first pointed out, fertility is most affected in the heterogametic (XY) sex, i.e., in mammals the male.

Failure of hybridization may occur in a number of ways. Some examples are given in Appendix I. Mating preferences can be overcome by artificial insemination. The foreign sperm may fail to reach the site of fertilization in normal numbers, as observed by Braden for ewes inseminated with goat semen. If they reach the site of fertilization, they may fail to penetrate the egg: this has been reported, for instance, after artificial insemination of rats with bull, mouse, guinea pig or rabbit sperm, and may be due to some specificity residing in the zona pellucida. Where fertilization occurs, the hybrid embryo often cleaves normally, only to fail at the blastocyst stage. Possibly development up to the blastocyst stage is more or less independent of the embryo's nuclear apparatus, and it is only when a normal nucleo-cytoplasmic ratio has been restored as a result of cleavage that deficiencies or chromosomal imbalances in the embryonic nucleus make themselves apparent in development.

Differences between the results of reciprocal crosses have often been reported. Thus, in ewes inseminated with goat semen, Warwick and Berry found no evidence of fertilization, while Hancock found a small proportion of eggs cleaving; no embryos succeeded in implanting. In goats inseminated with ram semen, on the other hand, the hybrid embryo survives until about halfway through pregnancy. Buttle and Hancock have confirmed the hybrid constitution of these embryos by chromosome counts. The successful gestation of sheep-goat hybrids by both ewes and nanny goats has been claimed by Bratanov and Dikov (1962) after experimental treatment of the females. The procedures used were designed to improve the chances of fertilization by foreign sperm and to reduce the maternal response believed to be responsible for rejection of the hybrid embryo.

III. IMPLANTATION

The embryo is said to be implanted or attached when it becomes fixed in position, and physical contact with the maternal organism is established.

The term implantation seems most appropriate for those species where the embryo becomes buried in the wall of the uterus. In many rodents, for example, the blastocyst comes to lie in a pocket (*crypt*) of the uterine wall, forming an intimate association with the maternal tissues; while in other species, including man, the blastocyst implants by passing through the uterine epithelium and is thus entirely cut off from the uterine cavity. In farm mammals, on the other hand, the embryo remains in the uterine cavity, and whatever attachment it forms with the wall of the uterus prior to the formation of the placenta is of an extremely loose nature. Movement of the blastocyst within the uterus becomes increasingly restricted as it expands; in the sheep, a mucous substance sticking the blastocysts to the uterine wall has been reported.

The loose and gradual nature of the attachment process in farm mammals has led to considerable controversy as to when implantation actually begins. Estimates range from the 10th to the 22nd day postcoitum for the sheep, and from the 11th to the 40th day for the cow.

A. The Embryo

1. Spacing

In polytocous species, the blastocysts become distributed down the length of the uterine horn as a result of the muscular churning movements of the uterine wall. In many rodents, the two uterine horns are largely or entirely separate, but this is not so in farm mammals. In swine, for example, blastocysts can pass freely between the two horns, and Dhindsa and his associates timed the distribution of embryos entering the uterus from one side only. The proportion of the total uterus which was occupied increased from 13% on the 6th day of gestation to 86% on the 12th day.

The presence of large blastocysts such as those of the rabbit may modify the muscular movements of the uterus so as to result in some regularity of spacing of embryos during implantation. By seven days postcoitum, rabbit blastocysts are spaced along the length of the uterine horn more regularly than random. In the sow also, the distribution of embryos between the two horns is much more even than would be expected by chance, and once elongation has been achieved there is surprisingly little overlap between adjacent blastocysts.

Smaller blastocysts, e.g. those of rats and mice, show little regularity of spacing. There is no evidence that an implanting blastocyst exerts any inhibitory influence on the implantation of another blastocyst close by. When the muscular movements which normally promote mixing of the uterine contents are inhibited, mouse blastocysts will successfully implant in close proximity to one another. After implantation embryos often become more uniformly spaced in the uterus, owing to differential growth of the uterine wall later in pregnancy.

**Table 7-2. Orientation of Blastocyst and Fetal Membranes
Relative to the Uterus**

	Rodents	Carnivores	Cattle, Sheep, Swine	Monkey, Man
Embryonic disc	Meso	Anti	Anti	Anti
First trophoblast attachment	Anti	Central	Central	Anti
First allantoic attachment	Meso	Anti	Meso	Anti

Meso = mesometrial; anti = antimesometrial. (Modified from Mossman, 1971. Chapter 3 in *Biology of the Blastocyst*. R. J. Blandau (ed.), Chicago, University of Chicago Press.)

The apparent absence of any inhibitory effect of an implanting blastocyst upon its neighbors implies that the upper limit to the number of embryos which can implant in a single uterus may be very high, at any rate in normally polytocous species. Implantation rate is therefore unlikely to be a limiting factor in any attempts to increase litter size artificially in swine (for instance by inducing superovulation with gonadotropic hormones). Such attempts are more likely to founder at a later stage of pregnancy, owing to the inadequacy of the vascular supply or to the mechanical effects of crowding on the embryos; while the young which are born may well be undersized and difficult to rear.

2. Orientation

In any given species, the position in which the blastocyst implants is usually fixed (Table 7–2). In swine, for example, the embryonic disc is always situated on the antimesometrial side of the uterine horn. On the other hand the embryos at the end of pregnancy are equally likely to be facing up or down the uterine horn.

If the mesometrial-antimesometrial axis of the rat uterus is reversed surgically, blastocysts still implant on the antimesometrial edge. The position of implantation is therefore determined by the relationship between blastocyst and uterine wall, rather than by the action on the blastocyst of any external factor such as gravity. That it is the uterus itself which plays the chief determining role was

shown by Wilson (1960). Using pseudopregnant female mice as recipients, he showed that pieces of muscle or tumor put into the uterus at the appropriate time would become "implanted" on the antimesometrial side just as do blastocysts in a normal pregnancy.

3. Gastrulation

Gastrulation is a stage of embryonic development occurring, though in various different guises, in all vertebrates. It succeeds formation of the blastocyst and precedes organ formation. Essentially, gastrulation consists of movements of cells or groups of cells in such a way as (*a*) to convert the embryo from a two-layered into a three-layered structure; and (*b*) to bring the future organ-forming regions into their definitive positions in the embryo.

In mammals, gastrulation involves the cells of the embryonic disc only. From the embryonic disc three types of tissue differentiate, *endoderm, mesoderm* and *ectoderm* (see prenatal development in Chapter 8). From these, all the fetal tissues develop, and also the embryonic membranes which connect the embryo and fetus to the mother. Cells migrate or split off from the inner cell mass of the embryonic disc to give rise to a layer of endoderm which spreads round the interior of the blastocyst, forming the *bilaminar omphalopleure*. The notochord and mesoderm are formed by invagination of cells in the *primitive streak* region of the embryonic disc.

4. MORPHOGENESIS AND ORGANOGENESIS

During gastrulation, the embryonic disc thickens considerably. The primitive streak formed in the midline defines the antero-posterior axis of the embryo. As the mesoderm spreads outwards from the primitive streak between the endoderm and ectoderm, it splits into two layers, separated by the *coelom*. The notochord develops from the anterior end of the primitive streak (*Hensen's node*).

Dorsal to the notochord, the ectoderm thickens to form a *neural plate*. After a few days, *neural folds* grow up and fuse to form a *neural tube*, the forerunner of the brain and spinal cord. Meanwhile *somites*, paired condensations of dorsal mesoderm, have appeared on either side of the notochord. As the embryo elongates, additional paired somites continue to develop, so that somite number can be used as an accurate index of the age of the early embryo. At about the seven-somite stage the anterior somites differentiate into three parts, destined to form the skeletal muscles, the skeleton and the connective tissue, respectively. Shortly after this, the primordia of the ears and eyes are apparent on the head of the embryo, and the early heart develops. The gut is formed from pockets in the endoderm and mesoderm.

B. The Uterus

1. PREIMPLANTATIONAL CHANGES

While the embryo is undergoing cleavage and blastocyst formation, the uterus too is undergoing changes, preparing the way for implantation.

During this *progestational* period, there is a decrease in the muscular activity and tonicity of the uterus, which may help to retain the blastocysts in the uterine lumen. At the same time, an increased blood supply to the uterine epithelium develops. In some species the increased vascularity is greatest along the side of the uterus at which implantation takes place. The amino acid and protein content of the uterine fluid changes markedly at the time of implantation. These changes have been followed in detail only in the rabbit. The concentration of most amino acids is much higher in rabbit uterine fluid at implantation than in blood serum. Glycine, alanine, taurine and glutamic acid are particularly abundant and their concentration is progesterone-dependent. In cows also, the concentration of free amino acids in uterine fluid is high and has been reported to undergo cyclic variation; it has not been studied during pregnancy. Certain protein fractions, detectable by electrophoresis, appear in rabbit uterine fluid only at the time of implantation. The first such fraction to be observed was variously termed β_1, *U-globulin*, *blastokinin* or *uteroglobin*; in the last few years several other fractions have been isolated but in no case has their biologic function yet been established.

Associated with the increased blood supply to the uterus, changes occur in the secretory activity of glandular and surface epithelium of the endometrium. High molecular weight compounds (proteins, carbohydrates, mucopolysaccharides) are broken down, and low molecular weight derivatives, along with glycogen and fats, accumulate. This material, along with cellular debris and extravasated leukocytes in the uterine lumen, forms the *histotrophe* (uterine milk) which provides nourishment for the embryo in the early period of uterine life, before the chorioallantoic placenta is established. There is evidence that histotrophe plays an important role in embryonic nutrition from about 80 hours postcoitum in the rabbit, and from the nine-day blastocyst stage onward in the sheep. In farm animals, where the placenta is of the epitheliochorial or syndesmochorial type, the association between fetal and maternal blood is not very close. Histotrophic nutrition is therefore important not only in the early stages of uterine life, but throughout gestation.

The epithelial lining of the uterus undergoes striking histologic changes during the estrous cycle. Proliferation of the epithelium during the progestational phase provides increased uterine surface and also increased glandular activity. In the absence of fertilization the epithelium regresses.

The hormonal basis of implantation varies so widely among those species which have been studied that generalization is difficult. Progesterone plays the leading part in determining the preimplantational changes in the uterus, just as at an earlier stage in the cycle estrogen predominated. The balance between estrogen and progesterone is probably more important than the absolute levels of either alone. In rats, estrogen priming seems to be required before a progestational endometrium can be established; estrogen is also necessary on the fourth day of pregnancy to sensitize the endometrium for implantation or deciduoma induction. In almost all mammals, ovariectomy immediately after ovulation prevents both implantation and the uterine developments associated with it.

2. Relative Roles of Embryo and Uterus

Both the embryo and the uterus undergo a characteristic sequence of changes during the course of implantation. These changes will be described for the mouse, for which most information is available, but similar principles probably apply more widely.

As the time of implantation approaches, the metabolic activity of the embryo increases. In the environment of an estrogen-sensitized uterus, lysis of the zona pellucida occurs, the trophoblast cells enlarge and their nuclear DNA content increases (*trophoblast giant cell transformation*). The trophoblast invades through the uterine epithelium and into the underlying stroma, ingesting dead epithelial cells by phagocytosis. Under conditions where the uterus is not estrogen-sensitized (e.g. during lactation), zona lysis and trophoblast enlargement and outgrowth do not occur, implantation does not take place, and the embryo enters a period of metabolic dormancy. Evidence suggests that an inhibitory substance is present in the uterus under these circumstances and is only removed under the influence of estrogen (McLaren, 1973). Outside the uterus, e.g. in vitro or in ectopic sites, trophoblast outgrowth occurs irrespective of the presence of hormones.

If embryos are present in the estrogensensitized uterus, a decidual reaction is stimulated locally around each embryo. The earliest indication of decidualization is a local increase in capillary permeability in the uterus, leading to edema. This precedes trophoblast outgrowth and does not require loss of the zona pellucida. It seems likely that the increase in capillary permeability is triggered by some chemical stimulus emanating from the blastocyst, perhaps a product of its heightened metabolism (McLaren, 1969a).

In farm animals, the uterus does not undergo a decidual reaction, but the requirement for precise synchronization (see next section) suggests that an equally complex interaction occurs between embryo and uterus, with each partner providing stimuli essential for the further development of the other. In rodents, the first stimulus from the embryo to the uterus appears to coincide with the onset of implantation, but in farm animals the presence of embryos must be signalled well before implantation begins so as to prevent the regression of the corpora lutea. Perhaps the very striking expansion of the trophoblast serves a dual role, not only ingesting "uterine milk" for the nutrition of the embryo, but also synthesizing some hormonelike substance which inhibits the uterus from producing a luteolysin.

3. Relation Between Embryo and Maternal Environment

The importance of the tubal environment in embryonic development has been underlined by the findings of Wintenberger-Torres. If the passage of sheep embryos through the fallopian tubes to the uterus is hastened by hormone treatment, development is slowed down and cleavage rate does not return to normal for nine days. If, on the contrary, the embryos are retained in the tubes by a ligature, they continue to develop normally for a week; for the next two days they develop more slowly but will recover if transferred to the uterus; while if they remain for more than 10 days in the tubes, they will undergo no further development even if transferred to the uterus.

In species other than the sheep, the role of the tubal environment has been little studied. Data on the role of the uterine environment, on the other hand, has been obtained by experiments on embryo transfer (see Chapter 6). The critical importance of *synchronization* between the stages of development of the embryo and of the uterus has been established in several rodents. Embryo transfers to the uterus give a high success rate if donor and recipient female are at the same stage of development, or better still (since transfer may slightly retard embryonic development) if the donor is somewhat more advanced in development than the recipient. But if the embryo is "younger" than the recipient uterus, so that it is not ready to implant at the moment when the uterine endo-metrium is ready for implantation, the success rate is usually very low. Once the receptive phase of the uterus is past, the uterine environment becomes extremely toxic and the embryos rapidly die.

Synchronization has also been found necessary for the normal development of sheep embryos transferred between the 2nd and 11th day after estrus (Rowson and Moor, 1966). When estrus in donors and recipients was exactly synchronized, 75% of all recipients became pregnant; but when they were three days or more out of phase, the pregnancy rate fell to 8% (Fig. 7–5). This is the more surprising, since from the seventh day onward, the level of progesterone in the ovarian venous blood of the ewe remains relatively constant. In cattle, Rowson and his colleagues have found that the synchronization requirements for successful egg transfer are more acute than in the sheep, since a variation of ±1 day gives a fall in the percentage of animals which became pregnant (Fig. 7–5). Synchronization is also required in swine.

In farm mammals, the period of implantation accounts for a significant proportion of all reproductive losses. De-

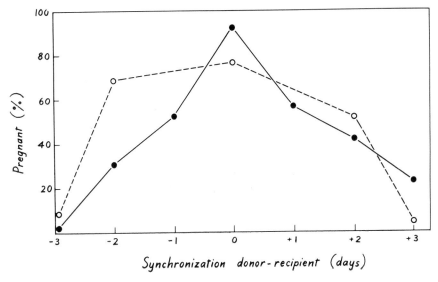

FIG. 7–5. The effect of the degree of synchronization of donor and recipient on pregnancy rate after egg transfer in sheep (- - - -) and cattle (———). (*Redrawn from Rowson, Lawson, Moor and Baker, 1972. Egg transfer in the cow: Synchronization requirements. J. Reprod. Fert. 28, 427.*)

layed development of the embryos, their delayed passage into the uterus or precocious development of the endometrium will all lead to failure of synchronization between blastocyst and endometrium at the critical time, and hence may be responsible for failures of implantation. The more precise the synchronization requirements (e.g., cow), the more likely is it that reproductive losses will be due to this cause.

C. Course of Implantation

It has been stressed earlier that implantation is a gradual process in all farm mammals, and that the early attachment of the trophoblast to the endometrium is of an extremely loose nature. We shall now briefly consider the morphologic processes concerned in implantation in farm mammals. The differentiation of the embryonic membranes, by which the embryo is connected to the mother, and the subsequent formation of the chorioallantoic placenta, will be described in Chapter 8.

Swine. The initial period of attachment lasts from about the 12th to the 24th day after fertilization. By about the seventh day, the zona pellucida surrounding the blastocyst has been shed, so that the trophoblastic cells are in direct contact with the uterine epithelium. The trophoblast now starts to proliferate rapidly, which leads to a folding of the trophoblast wall, presumably because the accumulation of fluid in the blastocyst cavity does not keep pace with the expansion of the wall. Endoderm appears and the blastocyst changes in the course of a few days from a small spherical vesicle to an exceedingly elongated threadlike tube, sometimes attaining a length of several feet (Plate 11). The embryonic disc occupies a short enlarged section in the central part of this tube. At this time the wall of the uterus is deeply folded, and the outer layer (*chorion*) of the elongated blastocyst is apposed to the uterine epithelium, following the course of the folds. Throughout this period the nutri-

tion of the embryo depends on the absorption of histotrophe or "uterine milk." The great length of the pig blastocyst provides a large absorptive surface for this purpose. The course of organogenesis in the pig embryo has been well-described by Rugh (1964).

Sheep. The early development of the sheep blastocyst is similar to that of the pig. Some degree of attachment has been reported as early as the 10th day, but blastocysts can still be shaken out of the uterus 16 to 17 days after mating. Elongation is less extreme, and does not start until the 11th or 12th day, but the blastocyst may still attain a length of 20 cm by the third week (Chang and Rowson, 1965). The process of implantation is different from that of the pig, since the sheep uterus contains permanent "attachment organs," the *caruncles*. By about the 18th day, the chorion has expanded until it fills the uterine lumen, bringing the trophoblast into close contact with the uterine epithelium over the caruncles. Trophoblastic cells invade and eventually destroy this epithelium, and later, all other uterine epithelium with which they are in contact. This destructive process lays the foundation for a more intimate relationship between chorion and maternal tissue than is formed in the pig or horse. By about four to five weeks, the process of implantation is completed.

Cattle. The course of implantation in the cow is essentially similar to that in the sheep but starts later. The zona pellucida is shed at about eight days, in the early blastocyst stage, and a few days later the blastocyst begins to elongate. Gastrulation is complete at 13 days. By the 33rd day, the chorion has formed a fragile attachment with two to four of the cotyledons surrounding the fetus; within a few days maternal and fetal tissues have become so intimately interdigitated that the embryo is being nourished by these cotyledons. Growth of the cotyledons is probably stimulated by progesterone.

Horse. For nearly two months the horse blastocyst, although pressed up against the uterine epithelium by the pressure of its own fluids, is not attached in any way. It attains a diameter of at least 5 cm and elongates only very slightly. During the third week, the blastocyst acquires an albumen coat, 3 to 4 mm in thickness. At the end of the third week, groups of columnar cells, the *trophoblastic discs*, can be seen on the trophoblastic wall of the chorion. Some of the cells have processes which are thought to be phagocytic. The discs may help in attachment but are more likely to be concerned with the ingestion of "uterine milk." By the 10th week, *villi* (projections) from the chorion are penetrating into the mucosal folds of the uterine wall, and by the 14th week attachment is complete.

REFERENCES

Austin, C. R. (1951). The formation, growth and conjugation of the pronuclei in the rat egg. *J. Roy. Micr. Soc. 71*, 295–306.

Bateman, N. (1960). Selective fertilization at the T-locus of the mouse. *Genet. Res. Camb. 1*, 226–238.

Beatty, R. A. (1957). *Parthenogenesis and Polyploidy in Mammalian Development.* Cambridge, Cambridge University Press, p. 132.

Biggers, J. D. (1971). "Metabolism of Mouse Embryos." In *Experiments on Mammalian Eggs and Embryos.* J. S. Perry (ed.). *J. Reprod. Fert.* Suppl. *14*, 41–54.

Chang, M. C. and Rowson, L. E. A. (1965). Fertilization and early development of Dorset Horn sheep in the spring and summer. *Anat. Rec. 152*, 303–316.

Gardner, R. L. and Johnson, M. H. (1972). An investigation of inner cell mass and trophoblast tissues following their isolation from the mouse blastocyst. *J. Embryol. Exp. Morph. 28*, 279–312.

Gray, A. P. (1972). *Mammalian Hybrids.* 2nd ed. CAB Tech. Commun. No. 10 (revised), England, Commonwealth Agricultural Bureaux.

Hancock, J. L. (1962). Fertilization in farm animals. *Anim. Breed Abstr. 30*, 285–310.

Hillman, N., Sherman, M. I. and Graham, C. (1972). The effect of spatial arrangement on cell determination during mouse development. *J. Embryol. Exp. Morph. 28*, 263–278.

Hunter, R. H. F. (1972). Fertilization in the pig: Sequence of nuclear and cytoplasmic events. *J. Reprod. Fert. 29*, 395–406.

McLaren, A. (1969). "Recent Studies on Developmental Regulation in Vertebrates." In *Handbook of Molecular Cytology.* A. Lima-de-Faria (ed.), Amsterdam, London: North-Holland, pp. 639–655.

McLaren, A. (1969a). Stimulus and response during early pregnancy in the mouse. *Nature 221*, 739–741.

McLaren, A. (1973). "Blastocyst Activation." In *Regulation of Mammalian Reproduction.* Bethesda, National Institutes of Health. S. J. Segal, R. Crozier, P. Corfman, P. Condliffe (eds.), pp. 32–328.

Mintz, B. (1971). "Allophenic Mice of Multi-embryo Origin." In *Methods in Mammalian Embryology.* J. C. Daniels, Jr. (ed.), San Francisco, Freeman & Co., pp. 186–214.

Monroy, A. (1965). *Chemistry and Physiology of Fertilization.* New York, Holt, Rinehart & Winston., 150 pp.

Piko, L. (1969). "Gamete Structure and Sperm Entry in Mammals." In *Fertilization.* Vol. 2. C. B. Metz and A. Monroy (eds.), New York, Academic Press, pp. 325–403.

Rowson, L. E. A. and Moor, R. M. (1966). Embryo transfer in the sheep: The significance of synchronizing oestrus in the donor and recipient animals. *J. Reprod. Fert. 11*, 207–212.

Rugh, R. (1964). *Vertebrate Embryology.* New York, Harcourt, Brace and World, 600 pp.

Tarkowski, A. K. and Wroblewska, J. (1967). Development of blastomeres of mouse eggs isolated at the 4- and 8-cell stage. *J. Embryol. Exp. Morph. 18*, 155–180.

Wilson, I. B. (1960). Implantation of tissue transplants in the uteri of pseudopregnant mice. *Nature* (Lond.) *185*, 553–554.

Chapter 8

Gestation, Prenatal Physiology and Parturition

E. S. E. HAFEZ AND M. R. JAINUDEEN

During the course of mammalian evolution, there have been marked anatomic, endocrinologic and physiologic changes. Among the more obvious are: economy in the production of gametes, reduction in the size of the egg, internal fertilization, development of the corpus luteum as a temporary endocrine organ, and development of the placenta as a nutritive, excretory, endocrine and protective organ. The main effect was to insure the continuation of the species.

This chapter deals with the intricacy of gestation: duration, formation and functions of the placenta, hormonal requirement, fetal nutrition, prenatal physiology and parturition.

I. GESTATION

A. Length of Gestation

Gestation extends from fertilization to birth. Length of gestation is calculated as the interval from fertile service to parturition (Table 8-1). The duration of gestation is genetically determined although it can be modified by maternal, fetal and environmental factors.

Maternal Factors. The age of the dam influences the duration of pregnancy in different species. A two-day extension from the normal occurs in the eight-year-old ewe. Heifers which conceive at a relatively young age carry their calves for

Table 8-1. **Differences in Gestation Periods of Farm Mammals**

Animal	Average (Range)
Cattle (dairy breeds)	
Ayrshire	278
Brown Swiss	290 (270–306)
Dairy Shorthorn	282
Friesian	276 (240–333)
Guernsey	284
Holstein-Friesian	279 (262–359)
Jersey	279 (270–285)
Swedish-Friesian	282 (260–300)
Zebu (Brahman)	285
Cattle (beef breeds)	
Aberdeen-Angus	279
Hereford	285 (243–316)
Beef Shorthorn	283 (273–294)
Sheep	148 (140–159)
Swine	
Domestic	114 (102–128)
Wild Pig	(124–140)
Horse	
Arabian	337 (301–371)
Belgian	335 (304–354)
Clydesdale	334
Morgan	344 (316–363)
Percheron	(321–345)
Shire	340
Thoroughbred	338 (301–349)

(From the literature.)

166

a slightly shorter period than those which conceive at an older age.

Fetal Factors. An inverse relation between the duration of gestation and litter size is well-documented in several polytocous species except in the pig. Multiple fetuses in monotocous species also have shorter gestation periods. Twin calves are carried three to six days less than single calves.

The sex of the fetus may also determine gestation length. Male calves and foals are carried one to two days longer than females. The size presumably affects gestation length by hastening the time of parturition initiation. The duration of pregnancy may be influenced by the endocrine functions of the fetus. Circumstantial evidence points to fetal pituitary and adrenal hypofunction in cows and ewes which undergo prolonged gestations.

Genetic Factors. The small variations in pregnancy duration that exist within breeds (Table 8–1) may be due to genetic, seasonal or locality effects. The genotype of the fetus is recognized as playing a part in the duration of pregnancy in cows. The extreme expression is known in genetically conditioned prolonged gestation among dairy cows which carry a fetus homozygous for an autosomal recessive gene.

The influence of equine fetal genotype on gestation length is noted in the hybrids between the horse and the donkey. For example, the duration of pregnancy of a mare carrying a foal to a stallion is 340 days, whereas to a jack is 355 days. Serum gonadotropin levels are 10 times higher in mares carrying foals to a stallion than those bred to a jack (Clegg et al., 1962). Thus this influence may be mediated through a hormonal mechanism or merely reflects the influence of fetal size. A sex-linked gene in Arabian mares or fetuses affects the duration of their pregnancy (Rollins et al., 1956). Differences in gestation between mutton breeds of sheep with different wool types are due to genetic factors.

Physical Environment. The longer gestation of mares following winter breeding has been attributed to delayed implantation, different seasonal feeding conditions or to seasonal fluctuations in the production of ovarian hormones. There is some evidence that high temperature may prolong pregnancy in rodents, but no information is available for domestic animals. In both mares and ewes, plane of nutrition influences pregnancy length.

II. MATERNAL PHYSIOLOGY IN PREGNANCY

A. Reproductive Organ Changes

Vulvar and Vaginal Changes. Edema and vascularity are the major reactions of the vulva to pregnancy, the edema increasing with the progress of pregnancy. These vulval changes, noted more in cattle than in horses, occur around the fifth and seventh months of gestation in heifers and cows respectively. The vaginal mucosa is pale and dry during most of gestation but becomes edematous and pliable toward the end of pregnancy.

The Cervix. During pregnancy endocervical crypts increase in number and produce a very viscid mucus, serving to seal the cervical canal by the so-called "mucus plug of pregnancy." Prior to parturition this seal breaks down and is discharged in strings. During pregnancy the external os of the cervix remains tightly closed.

Uterine Changes. As pregnancy progresses, the uterus undergoes gradual enlargement to permit expansion of the fetus, but its muscular walls remain quiescent to prevent premature expulsion. Three phases can be identified in the adaptation of the uterus to accommodate the products of conception—proliferation, growth and stretching—the duration of each varies with the species.

Endometrial proliferation occurs before blastocyst attachment and is characterized by a marked preparatory progestational sensitization of the endometrium. Characteristic changes of the endometrium initiated by hormones, mainly progesterone, are increased vascularity, growth and coiling of the uterine glands, and leucocyte infiltration.

Uterine growth commences after implantation. Uterine growth comprises muscular hypertrophy, an extensive increase in connective ground tissue substance and an increase in fibrillar elements and collagen content. Modification of the ground substance is significant both in uterine adaptation to the conceptus and in the process leading to involution. The structural changes which take place in the gravid uterus are reversible but are differentially restored after parturition.

During the period of uterine stretching, uterine growth diminishes while its contents are growing at an accelerating rate.

Ovarian Changes. Ovarian changes commence with the transformation of the graafian follicle to a corpus luteum. During a nonfertile estrous cycle, a luteolytic mechanism causes the corpus luteum to regress; however, if an embryo is present, the corpus luteum persists as the corpus luteum of pregnancy (corpus luteum verum) and subsequent estrous cycles are suspended. During early pregnancy this suspension may not be complete as considerable follicular development occurs in the ovaries and some may even reach preovulatory size, however, these follicles eventually become atretic.

The corpus luteum of pregnancy in the cow persists at a maximal size throughout pregnancy. It is golden brown in color and does not protrude above the surface of the ovary.

In the mare four distinct stages are recognized: (*a*) During the first 40 days of pregnancy, a single corpus luteum verum is present. (*b*) Between the 40th and 150th day of pregnancy marked ovarian activity occurs. As many as 10 to 15 follicles (over 1 cm in diameter) undergo luteinization to form the accessory corpora lutea. The presence of ova in the oviducts suggests that some of these follicles even ovulate. Usually each ovary contains three to five accessory corpora lutea. (*c*) From the fifth to seventh month, these corpora lutea and the large follicles regress completely but the mare does not show signs of estrus because the placenta secretes progesterone until the end of gestation. (*d*) From the seventh month onward only vestiges of the corpora lutea and small follicles are present. During the last two weeks of gestation follicular activity is observed in preparation for the postpartum estrus ("foaling heat").

Utero-ovarian Relationship. Normal regression of the corpus luteum (C.L.) at the end of the estrous cycle is retarded when implantation occurs; the lytic effect of the uterus is apparently overcome by the embryo. *Uterine luteolytic factor* (ULF) is probably inactivated during later pregnancy by placental or embryonic secretion. The period of pregnancy when ULF activity is maximal seems to vary with the species. Most information about embryonic influence on ovarian function has been obtained by transplanting eggs or embryos to different parts of the uterus.

If no embryo is present in the ovine uterus by day 12 of the cycle, C. L. regression begins. In sheep with intact uteri, corpora lutea were maintained by embryos transferred to either adjacent or opposite horns. Embryos transferred to one isolated uterine horn maintained luteal function in the ipsilateral ovary in 80% of cases and had no effect on C. L. in contralateral ovaries. However, if the horn ipsilateral to the C. L. were removed, embryos transferred to the contralateral horn regularly maintained C.L. life span during pregnancy. When recipient ewes had corpora lutea in both ovaries, transfer of embryos to one isolated horn resulted in C. L. maintenance only in the ovary adjacent to the gravid horn. These results suggest the existence of a local unilateral relationship between embryo and C. L. in sheep (Moor and Rowson, 1966).

Pregnancy does not occur in pigs when one horn is surgically isolated (Anderson et al., 1966). A pregnancy continues, however, when most of the nongravid horn is removed, even though corpora lutea regress on the ovary adjacent to the surgery. These and other findings suggest that at or just prior to day 12 of the cycle the embryo or uterus of the pig begins to emit some sort of signal leading to C. L. maintenance during gestation.

Pelvic Ligaments and Pubic Symphysis. Relaxation of the pelvic ligaments gradually occurs during the course of pregnancy but is more noted with approaching parturition. Relaxation of the pelvic ligaments is more marked in the cow and ewe than in the mare and is related to high levels of estrogens in late pregnancy and to the action of relaxin. The caudal part of the sacro-sciatic ligament, which is cordlike in the nonpregnant cow, becomes more relaxed and flaccid at term.

The pubic symphysis of the young female undergoes sufficient demineralization to permit some separation at parturition. In the guinea pig and mouse, prolonged treatment with estrogen, with or without progesterone, induces relaxation of the pubic symphysis due to bone resorption, connective tissue proliferation and water retention. Relaxation is more rapid with relaxin due to breakdown of the collagenous fibers into thin strands with a depolymerization of the ground substance.

B. Hormones of Pregnancy

A balance of certain hormones is necessary for pregnancy. The corpus luteum persists throughout pregnancy in all farm species except the horse. The ovary is essential for maintenance of pregnancy in the cow, goat and sow (Table 8–2) but because the placenta produces progesterone the mare and the ewe can dispense with it. Hypophysectomy before implantation terminates pregnancy and after the 44th day of gestation causes abortion in the goat but not the ewe.

Early in pregnancy neither the fetus nor the placenta alone has the necessary enzyme systems for the synthesis of steroids from cholesterol. Later, the increased production of estrogens and progesterone results from the mother, placenta and fetus functioning as a unit—"feto-placental unit." Such a unit is involved in estrogen production in late

Table 8-2. Effect of Ovariectomy and Hypophysectomy on the Maintenance of Gestation in Farm Animals and Other Mammals

Animal Species	Length of Gestation (Days)	Ovariectomy		Hypophysectomy	
		First Half	Second Half	First Half	Second Half
Cow	282	—	±	n.d.	n.d.
Ewe	148	—	+	—	+
She-goat	148	—	—	—	—
Sow	113	—	—	—	—
Mare	350	—	+	n.d.	n.d.
Woman	280	+	+	+	+
Monkey	165	+	+	+	+
Rat	22	—	±	—	+
Rabbit	29	—	—	—	—
Bitch	61	—	n.d.	—	±

+ = Fetuses survive; — = abortion; ± = some fetuses survive; n.d. = not determined.
(Adapted from the literature and Heap, 1972. In Austin, C. R. and Short R. V. (eds.). *Reproduction in Mammals.* Cambridge, Cambridge University Press.)

Table 8-3. Estrogen and Related Compounds in the Urine During Pregnancy

Female	Estrone	17β- Estradiol	17α- Estradiol
Cow	+	—	+
Ewe	—	—	+
Goat	—	—	+
Mare*	+	+	+
Sow	+	—	—

* Mare's urine also contains equilin, equilenin, 17α- and 17β- dihydroequilenin.
(From the literature.)

gestation in the goat (Thorburn et al., 1972).

Species differences are observed in urinary excretion of estrogen (Table 8–3). In the mare, estrogen excreted from the 90th day of pregnancy reaches a peak between the 8th and 10th month. In the pig, elevated urinary levels of estrone occur during the fourth and fifth weeks after breeding (Lunas, 1962) and during the last six weeks prior to parturition (Edgerton and Erb, 1972). In the cow, maximum excretion of 17α-estradiol and to a lesser degree estrone occurs at nine months of gestation (Hunter et al, 1970).

Blood progesterone remains constant throughout pregnancy in the ewe and cow or attains a high level early in pregnancy in the sow. In the mare, plasma progesterone levels are high during the first 10 weeks, but after the corpora lutea have regressed it cannot be detected even though the placenta produces adequate quantities to maintain pregnancy (Short, 1960). Pregnanediol, the urinary metabolite of progesterone in the mare, has not been detected in other farm species.

In sheep, the conceptus prolongs the life of the corpus luteum by neutralizing the ULF. Up to about day 50 of pregnancy the conceptus is unable to produce sufficient luteotropins to maintain pregnancy, since hypophysectomy before that time causes abortion (Fig. 8–1). In women, on the other hand, the formation of the corpus luteum of pregnancy probably depends upon the secretion of *human chorionic gonadotrophin* (HCG) by the trophoblast within a few days of implantation. In the mare, *pregnant mare serum gonadotropins* (PMSG) appear on day 40 of pregnancy, reach a peak on day 70–80, then decline afterward (Fig. 8–2).

C. Physical Changes

Usually animals gain weight during pregnancy due to growth of the conceptus as well as to increases in maternal body weight. In young females, nutrient retention due to growth may mask actual weight increase due to pregnancy and would necessarily be continued in weight gains during pregnancy. Litter size and maternal weight gain in swine appear to be independent of each other.

Considerable alteration in the distribution of water occurs in pregnancy. Some of this is mechanical and is related to the increase in venous pressure of the enlarging uterus. Edema extending from the udder to the umbilicus is frequently observed during late gestation in cows and mares.

D. Maternal Adjustments

Pregnancy demands rather considerable adjustments in the physiology of the maternal organism, namely changes in (*a*) blood volume and composition; (*b*) cardiovascular dynamics; (*c*) respiratory functions; (*d*) renal function; (*e*) alimentary function; and (*f*) metabolism of water, electrolytes, minerals and vitamins.

Blood Volume and Composition. As pregnancy advances blood volume increases in ewes (Metcalfe and Parer, 1966) and cows (Reynolds, 1953) due to an increase in plasma volume. But unlike in human pregnancy this increase in plasma volume is not associated with a decrease in hemoglobin concentrations in the blood. The phenomenon of "physiologic anemia of pregnancy" observed in man does not occur in farm animals.

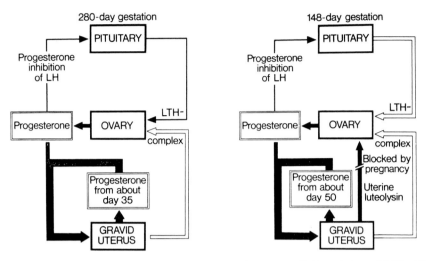

FIG. 8–1. The hormonal maintenance of gestation in the woman (*left*) and sheep (*right*). (*Heap, 1972. In Reproduction in Mammals, Book 3. Austin and Short [eds.], Cambridge, Cambridge University Press.*)

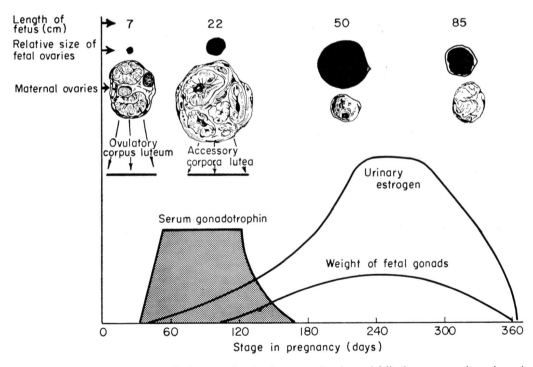

FIG. 8–2. Sequence of events during gestation in the mare: the rise and fall of serum gonadotropin and urinary estrogen in pregnant mares, and the effects of these hormones on the fetal gonads. Note that the corpus luteum verum degenerates and that accessory corpora lutea are formed when the serum gonadotropin level rises. Fetal ovaries (*upper row*) grow in response first to gonadotropin and later to estrogen and become larger than the atrophic maternal ovaries. (*Redrawn from Nalbandov's Reproductive Physiology, 1964. Cole et al., 1933, Anat. Rec. 56, 275, 1955, Brit. Med. Bull. 11.*)

Cardiovascular Dynamics. The pregnant uterus depends on its circulation to perform various functions. In sheep cardiac output increases during pregnancy and provides the gravid uterus with additional blood supply. The uterine blood flow, which amounts to two percent of the cardiac output in non-pregant ewes, increases to 20% at term pregnancy but does not keep pace with the growth of the fetus which extracts increasing amounts of oxygen from the maternal blood allotted to it. The quantity of blood in the uterus tends to increase in proportion to its contents and is related more to fetal than placental weight (Barron, 1970).

The blood pressure of the ewe tends to fall during late gestation. An increasing cardiac output with a decrease in blood pressure indicates a decrease in peripheral resistance. Probably the uterus contains an area of low vascular resistance similar in respects to an arteriovenous fistula.

III. PLACENTA

The following discussion deals with the morphologic and physiologic aspects of the placenta and its role in the intrauterine development of the fetus.

A. Morphology

1. FORMATION

In its broadest sense the placenta is defined as a union of fetal and maternal tissues for physiologic exchange. During early gestation the morphogenesis of the placenta is closely related to those extraembryonic membranes which are differentiated into *amnion, allantois, chorion* and *yolk sac* (Fig. 8–3). The amnion surrounds the fetus. The chorion, the outermost membrane, is in contact with the endometrium. The allantois, located between the amnion and chorion, is continuous with the anterior extremity of the bladder by way of the *urachus*, which passes through the umbilical cord. The

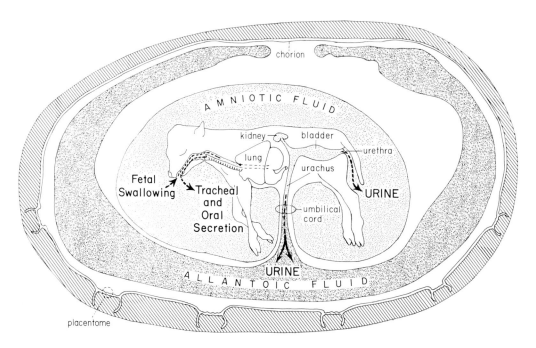

FIG. 8–3. Diagram of a chorionic sac typical of the cow or ewe late in the fetal period. Urinary bladder opens to the amniotic cavity by way of the urethra and allantoic cavity by way of the urachus. Note the relation of fetal urine and tracheal fluid to amniotic fluid. (*Adapted from Harvey, 1959. In Reproduction in Domestic Animals. Cole and Cupps [eds.], New York, courtesy of Academic Press.*)

inner layer of the allantois is fused to the amnion; the outer layer is fused to the chorion, forming the *chorioallantois*. By this fusion the fetal vessels in the allantois come into close apposition to the umbilical arteries and veins located in the connective tissue between the allantois and chorion. These vessels are important for exchange between the fetus and placenta.

2. Gross Shape

A feature of the chorioallantoic placenta is the highly increased area at the feto-maternal junction either by formation of chorionic villi protruding into uterine crypts or by the formation of chorionic labyrinths. The shape of the placenta is determined by the distribution of villi over the chorionic surface. In the pig and horse villi develop over the entire surface (*diffuse placenta*). In sheep, goat and cow villi arise in tufts which are distributed over the chorion (*cotyledonary placenta*). These fetal cotyledons fuse with maternal *caruncles* or projections of the uterine mucosa to form functional units called *placentomes*. Maternal caruncles are convex in the cow and concave in the ewe (Fig. 8–4). The number of placentomes ranges from 90 to 100 in the ewe and from 70 to 120 in the cow. In cattle placentomes start to form immediately around the fetus and progress toward the distal limit of the chorioallantois in the nongravid horn (about 12 to 13 weeks). During pregnancy they also enlarge to several times their original diameter. Those situated about the middle of the gravid horn develop to a larger size than those at the extremities. During this growth, they change from flat, plaquelike bodies to rounded, pedunculated, mushroomlike structures which, except for an area around the pedicle, are completely engulfed by the chorioallantois.

Normally the chorioallantois extends into the nongravid horn; its caruncles hypertrophy, but the degree of development is usually less than in the gravid horn. When there is a lack of placentomes due to uterine disease, primitive placental structures (adventitious placenta, accessory placentomes) simulating a diffuse placenta develop.

3. Microscopic Structure

Grosser's classification of placentas (Table 8–4) is based on the number of

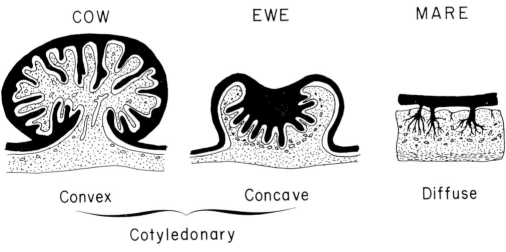

COW   EWE   MARE

Convex   Concave   Diffuse

Cotyledonary

Fig. 8–4. Epitheliochorial placenta of cow (*left*), ewe (*middle*) and mare (*right*). Villi from chorioallantois (*black*) invade crypts in maternal uterine epithelium (*stippled*). The apposition of maternal and fetal tissues is diffuse (mare) or localized as placentomes (cow and ewe). Each placentome is composed of fetal cotyledon and maternal caruncle. (*Adapted from Mossman, 1937. Contributions to Embryology, Carnegie Inst., 26, 129. Courtesy of the Carnegie Institution of Washington, Washington, D.C.*)

**Table 8-4. Classification of Mammalian Placenta Based on Tissue Layers
Separating Maternal and Fetal Blood**

| Placental Microscopic Structure | | | | |
Type	Maternal Tissue	Fetal Tissue	Gross Form	Species
Epitheliochorial			Diffuse	Pig, horse
			Cotyledonary or multiplex	Sheep, goat, cow*
Endothelio-chorial			Zonary	Dog, cat
Hemochorial			Discoid	Man, mouse, rat, guinea pig, rabbit
Hemoendothelial			Discoid	Rabbit

* A syncytium is present between trophoblast and maternal connective tissue as observed under the electron microscope. The placentas of sheep, goats and cows are therefore classified as epitheliochorial rather than syndesmochorial (Bjorkman, 1965, *J. Anat.* 99, p. 283).

(Adapted from Amoroso, 1952. In Parkes, A. S. (ed.) *Marshall's Physiology of Reproduction.* New York, Longman.)

tissue layers separating the fetal and maternal circulations (Plates 12 and 13). They are named in such a way as to identify the maternal and fetal tissues actually in contact in the order maternal→ fetal.

In the horse and pig, the maternal epithelium and chorion lie in apposition (*epitheliochorial placenta*). Electron microscopic studies revealed that the junction between fetal and maternal tissues in the epitheliochorial placenta of the pig is the interdigitation of microvilli from the trophoblast and the uterine epithelium (Plate 13). An analogous arrangement between the trophoblast and cryptal lining occurs in the placentomes of cattle, sheep and goat indicating the maternal origin of the cryptal lining. Thus the ruminant placenta is classified as epitheliochorial rather than *syndesmochorial* (Bjorkman, 1965). The allantochorion

PLATE 12

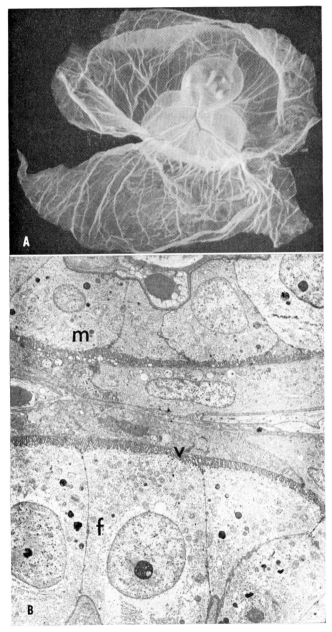

A, Equine conceptus at 7 weeks. Note that the chorion is oval rather than cylindrical as in the cow, and the chorion is more distended with fluid which facilitates early pregnancy diagnosis. Note the amnion in relation to yolk sac and chorioallantois. (*Arthur, 1964. Wright's Veterinary Obstetrics, London, Bailliere, Tindall & Cox.*)

B, Electronmicrograph of the placenta of the mare in late gestation. Fetal (*f*) and maternal (*m*) epithelia with zone of microvilli (*v*). Fetal and maternal epithelia consist of discrete cells resting on distinct basement membranes. ✕ 4800. (*Photographs by Prof. Nils Bjorkman.*)

PLATE 13

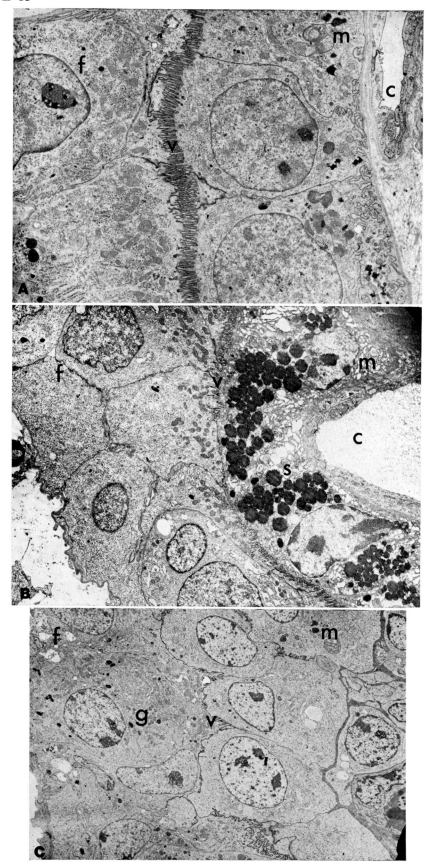

Legend on facing page:

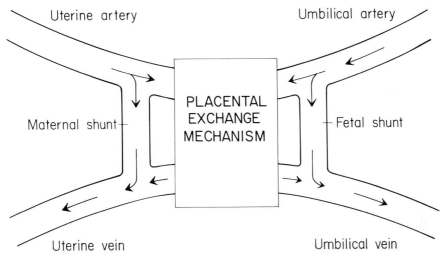

FIG. 8–5. A scheme illustrating the presence of shunts in maternal and fetal circulation through which blood flows without participating in gas exchange.

of the epitheliochorial placenta is folded (pig) or is villous (horse, cow, sheep and goat), and the endometrial components fit closely to the fetal part by corresponding folds or crypts. The epithelial lining of the folds or crypts is cellular (pig, horse, cow) or is syncytial (sheep). The trophoblast forms a single layer of cuboidal and binucleate giant cells (ruminants).

In the dog uterine vessels impinge directly on the chorion (*endotheliochorial placenta*). In man, maternal blood bathes the chorionic villi (*hemochorial placenta*). In the rabbit placenta, near term, only the fetal endothelium separates maternal from fetal blood (*hemoendothelial placenta*). Estimates of the total tissue thickness separating maternal and fetal in different placental types range from 1–100 μ.

Chorionic Villi. The formation of chorionic villi is an important step in the development of the placenta. The chorionic villi consist of vascular mesenchymal cones surrounded by cuboidal trophoblastic and giant binucleate cells. These either penetrate directly into the endometrium or simply interdigitate with vascular foldings of the endometrial surface (farm animals); their function is to bring the fetal (allantoic) vessels into proximity with the maternal blood vessels.

The control of villus formation is obscure. The earliest stage in the formation of the chorionic villi (27th day in the ewe) is marked by the appearance of a series of parallel ridges opposite each uterine caruncle. Separate prominences

---

Electronmicrographs of the feto-maternal junction in the epitheliochorial placentas of swine, sheep and cattle.

A, Swine, late pregnancy. Fetal (f) and maternal (m) epithelia, microvilli (v) and maternal capillary (c). × 4250.

B, Sheep, 30 days pregnancy. The feto-maternal junction is characterized by interdigitating microvilli (v); capillaries (c) are seen on fetal (f) and maternal (m) sides. The caruncular epithelium contains fat droplets (*black*). Note syncytial (s) nature of the uterine epithelium. × 3060.

C, Cattle, 61 days pregnancy. Note the presence of microvillous interdigitation (v) between fetal (f) and maternal (m) epithelia. The cryptal lining is cellular. A binucleate giant cell (g) is observed in the fetal epithelium. × 2500. (*Photographs A, B by Prof. Nils Bjorkman and C by Dr. Vibeke Wang Sorenson.*)

appear on the summits of these ridges and by continued growth give rise to the primary villi of the intervening ridge segments. Each primary villus, by terminal budding, forms three to six generations of branches and sub-branches. During early pregnancy, the villi are slender; thereafter, they become plump due to the enormous increase in mesenchymal tissues during late pregnancy. The maximum length of the villi is attained during late gestation; thereafter, they shorten somewhat.

Uteroplacental Circulation. Vascularization of the placenta marks the onset of hemotropic nutrition. The uterine arteries and veins show a characteristic coiling pattern, running through the maternal septum of the caruncles and crypts. On the other hand, branches of the umbilical arteries and veins penetrate the villus without coiling. The capillaries connect the axial arterial branch at one end and the axial venous tributary at the other and penetrate into a multitude of minute tufts. Not all the blood entering the placenta participates in gas exchange across the placenta; this is due to the

presence of shunts in maternal uterine and fetal umbilical circulations (Fig. 8–5). However in late gestation in sheep, most of the uterine and umbilical flows perfuse the placentomes and only a small part is shunted to nonplacental structures (Fig. 8–6).

The rate of uterine blood flow increases during pregnancy and is related to weight of the fetus (Barron, 1970). On the fetal side the rate of umbilical flow increases with advancing pregnancy due to a decrease in umbilical vascular resistance earlier in pregnancy (90–115 days of gestation) and to an increased arterial blood pressure later (Dawes, 1962). In sheep, progesterone released by the placenta may play a role in the chronic regulation of the uteroplacental circulation after the 80th day of gestation (Barron, 1970).

Of all these factors discussed, the increases in uterine and fetal umbilical blood flows which occur with advancing gestation, exert the greatest influence on the transfer of oxygen across the placenta.

B. Physiology

The development of the placenta is associated with viviparity. The placenta functions as a multi-organ performing many functions and substituting for the fetal gastrointestinal tract, lung, kidney, liver and endocrine glands. In addition, the placenta separates the maternal and fetal organism, thus ensuring the separate development of the fetus.

1. Placental Exchange

The blood of fetus and dam never come into direct contact. Yet, the two circulations are close enough at the junction of chorion and endometrium so that oxygen and nutrients can pass from the maternal blood to the fetal blood, and waste products in the opposite direction.

The original Grosser concept considered the placental barrier a semipermeable membrane and implied that a reduction in the number of layers in the placental barrier was equivalent to in-

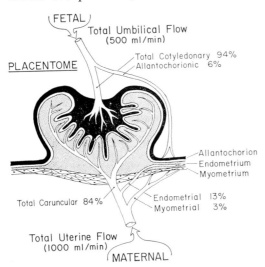

Fig. 8–6. Distribution of uterine and umbilical blood flows in a pregnant sheep at term. Percentage of blood flows is given in parenthesis. Note that in late gestation only a small part of blood is shunted away from placenta. (*Adapted from Makowski et al., 1968.) Circulation Res. 23, 623.*)

Table 8-5. Classification of the Transfer of Substances Across the Placenta

Group	Physiologic Role	Substances	Exchange Mechanism
I	Maintenance of biochemical homeostasis or protection against sudden fetal death	Electrolytes, water and respiratory gases	Rapid diffusion
II	Fetal nutrition	Amino acid, sugars and most water-soluble vitamins	Predominantly by active transport systems
III	Modification of fetal growth or the maintenance of pregnancy	Hormones	Slow diffusion
IV	Immunologic or toxic importance	Drugs and anesthetics	Rapid diffusion
		Plasma proteins, antibodies and whole cells	Pinocytosis or leakage through pores in placental membrane

(Adapted from Page, 1957. *Amer. J. Obstet. Gynec. 74*, 705.)

creased placental efficiency. This concept regarded the placenta as an inert barrier with pores of a size that prevent the transfer of large molecules but permit the transfer of small molecules. Electron microscopic studies show that the number of layers within the placental barrier is of lesser significance. Instead each component of the vital placental membrane participates in one or more metabolic functions (Wynn, 1968).

The placental membrane controls the transfer of a wide range of substances by several processes. *Simple diffusion*, the movement of molecules from an area of high concentration to an area of low concentration, was formerly thought to be of primary importance. Most molecules of physiologic importance are transferred by some *active transport*, thus they can be "pumped" against a concentration gradient allowing the embryo to accumulate higher concentrations of nutrients than exist in the maternal blood. *Phagocytosis* (engulfing food) and *pinocytosis* (engulfing water) are also important mechanisms of mediating physiologic exchange in the placenta. Page (1957) has suggested that placental exchange be discussed in terms of four groups of substances (Table 8–5).

The energy needed to drive these processes is derived from energy-yielding metabolic reactions. The placenta contains enzymes for glycolysis, the citric acid cycle and the pentose phosphate pathway. Placental oxygen consumption is comparable to such organs as the liver or kidney.

Respiration (O_2 and CO_2). The most complete account of the gas exchange in the pregnant uterus has been provided by

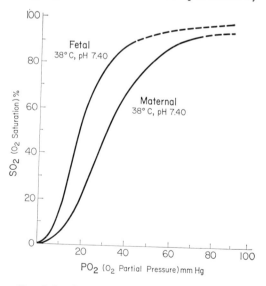

FIG. 8–7. Oxyhemoglobin dissociation curves for fetal and maternal goat blood. At a given oxygen tension fetal blood contains more oxygen than maternal blood. (*Redrawn from Metcalfe et al., 1967. Physio. Rev., 47*, 782.)

Metcalfe et al. (1967). There are many similarities between the gas exchange across the placenta and the lungs. The major difference, however, is that in the placenta it is a fluid-to-fluid system whereas in the lung it is a gas-to-fluid system. The process of transfer of oxygen from maternal to fetal blood involves its dissociation from the maternal blood, its diffusion through the placental membrane and finally its combination with fetal hemoglobin. The umbilical arteries carry unoxygenated blood from the fetus to the placenta, while the umbilical veins carry oxygenated blood in the reverse direction.

The placental transfer of oxygen is a function of several factors. These include the characteristics of maternal and fetal oxyhemoglobin dissociation curves; the spatial relation of the maternal and fetal exchange vessels; and the maternal and fetal blood flow rates.

Oxygen is chemically bound to hemoglobin within the red cell. The relationship between oxygen and the rate of combination with hemoglobin can be demonstrated in the form of the oxygen dissociation curve. For this purpose blood samples are equilibrated with different oxygen tensions (pO_2) at standard pH and temperature and then the oxygen contents (SO_2) in these samples are determined. In this way, curves can be plotted for maternal and fetal blood (Fig. 8–7). The fetal dissociation curve

CONCURRENT SYSTEM

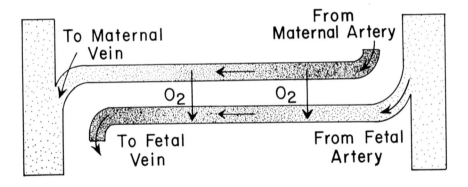

COUNTERCURRENT SYSTEM

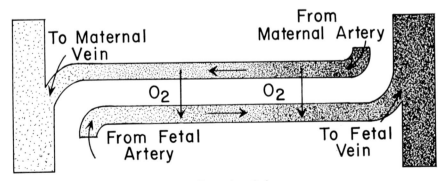

Fɪɢ. 8–8. Legend on facing page.

MULTIVILLOUS STREAM SYSTEM

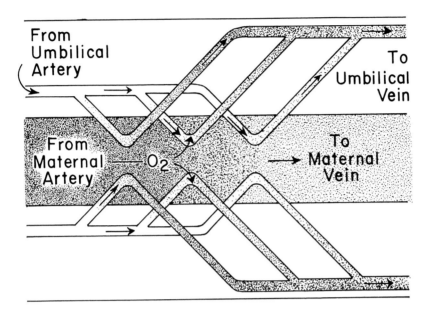

POOL SYSTEM

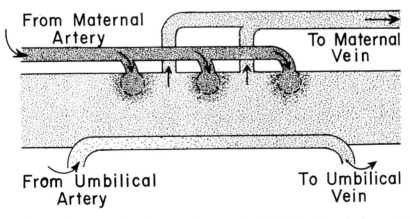

Fig. 8–8. Schematic presentation of pattern of maternal and fetal blood flow during gas exchange. Oxygenation of blood is proportional to intensity of shading. Note that the efficiency of oxygen exchange varies with the particular system and is greatest in the countercurrent system. (*Adapted from Metcalfe et al., 1967. Physio. Rev. 47, 782.*)

for oxygen lies to the left of the maternal curve in all species studied so far except the cat. This means that at any given oxygen tension the fetal blood will contain more oxygen than will the maternal blood. It was believed that fetal hemoglobin had a greater affinity for oxygen than maternal hemoglobin, but its specific physiologic role is obscure (Barron, 1970). It is also not certain if the fetal oxygen affinity changes during the latter part of gestation.

Four basic systems have been suggested for gas exchange in the placenta (Fig. 8–8). The efficiency of oxygen exchange varies with the particular system. It is greatest in the countercurrent system and least in the concurrent system. The efficiency of the multivillous system is intermediate between the above mentioned systems. In the pool system, gas exchange is less than in the multivillous system but is comparable to a concurrent system. It is difficult to ascertain which of these systems is primarily involved in a particular species and probably some species may contain more than one system. Gas exchange in the sheep placenta contains a mixture of countercurrent and concurrent flow systems.

Carbon dioxide diffuses freely from the fetal to the maternal circulation as in the blood of nonpregnant adults (Fig. 8–9). The transfer of CO_2 from the fetus to the mother is facilitated by certain physiologic mechanisms. The maternal and fetal dissociation curves for CO_2 show that fetal blood has a lower affinity for CO_2 than maternal blood during placental oxygen transfer. This favors the diffusion of CO_2 from fetal to maternal blood. The presence of significant fetal-maternal pCO_2 gradients suggests that factors other than simple diffusion may be involved in the placental transfer of CO_2. As in the case for O_2 transfer, fetal and maternal shunts, and placental metabolism may be significant. Furthermore, the low concentration of fetal carbonic anhydrase which catalyzes the interconversion of H_2CO_3 to CO_2 (Fig. 8–9) may limit the elimination of CO_2

during the passage of fetal blood past the region of gas exchange and contribute to the CO_2 tension gradient between fetal and maternal blood.

Nutrition. The placenta provides for the transport of sugars, amino acids, vitamins and minerals to the fetus as substrates for fetal growth; it serves as a storage organ for glycogen and certain other substances such as iron.

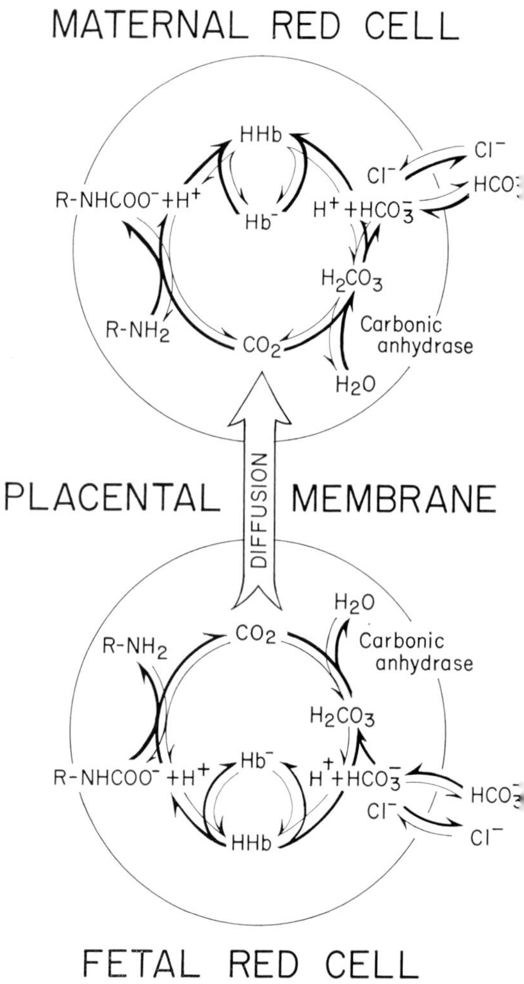

FIG. 8–9. Diagrammatic illustration of the processes involved in the transfer of CO_2 from fetal to maternal blood across the placenta. CO_2 diffuses as physically dissolved gas and not as bicarbonate ion. Note the presence of carbonic anhydrase in fetal blood which catalyzes the interconversion of H_2CO_3 to CO_2. (*Redrawn from Metcalfe et al., 1967. Physio. Rev. 47, 782.*)

(a) Water and electrolytes. The placental membranes are freely permeable to water and electrolytes. Of the 78 gm of substance that a 115-day sheep fetus gains daily, 60 gm is water. One puzzling problem of placental physiology arises from the fact that water moves from the mother to the fetus against (1) an osmotic gradient; and (2) a lower plasma protein concentration. The osmolarity of the fetal and maternal bloods and their respective concentrations of Na^+ Cl^- and K^+ are identical.

(b) Minerals. Iron tends to be higher in the fetus than the mother and is stored in the liver, spleen and bone marrow. The pig fetus obtains its iron from the uterine lumen by way of the areolae. Copper readily traverses the bovine placenta. Copper and iron tend to accumulate in the liver. Manganese is found in the calf fetal liver but apparently does not accumulate. Calcium and phosphorus enter fetal blood against a concentration gradient (cf. Newland et al., 1960).

(c) Carbohydrates, fat and proteins. In ruminants, the fetal blood sugar level is higher than in the mother. Peculiarly, fructose comprises about 70–80% of the sugar in fetal blood, while glucose is predominant in maternal blood. In sheep the rate of transfer of C^{14}-labeled glucose from mother to fetus is about 10 mg/minute and varies with the level of glucose. The placenta converts a certain fraction into fructose, which is then transferred to the fetus independently. C^{14}-labeled fructose injected into the fetus crosses the placenta in a very small amount.

The lipid content of maternal blood is higher than in the fetus of sheep and pigs fed a high fat diet (bacon). The placenta is not permeable to fat as such, but constituents, namely fatty acids and glycerol, pass freely.

Proteins as such are not transferred. Amino acids cross readily against a concentration gradient; a high fetal-maternal amino acid ratio exists in sheep (2–3:1). A concentration of amino acids by the placenta occurs during transport. The accumulation occurs on the maternal side and the transfer from trophoblast to the fetal circulation takes place by simple diffusion down a concentration gradient. Immunoglobulins are transmitted in man and some animals but not in farm animals. This can be explained by structural differences in the various placental types.

(d) Vitamins and hormones. The fat-soluble vitamins (A, D and E) seem to be impeded by the placenta, thus at birth the concentration of these is lower than normal (the mother). The role of water-soluble vitamins in fetal nutrition has not been studied since they are formed in the rumen or are present in high concentration in normal feeds.

The placenta is probably permeable to all hormones and certainly to insulin, steroids and gonadotropins. Steroid hormones pass through the placenta in sufficient quantities to cause enlargement of the responsive fetal organs. For example, in the pregnant mare the peak of gonadotropin secretion is followed by a peak of estrogen secretion (both from the placenta). These hormones pass through the placenta to the fetus resulting in marked enlargement of the fetal gonads (Fig. 8–2). A direct demonstration of estrogen transfer was done by injecting (i.v.) C^{14}-labeled estradiol into a female guinea pig. The large amounts of radioactivity appearing in the fetal plasma coincided with the rapid disappearance from the maternal blood.

2. Placental Enzymes

The placenta contains a complex enzyme system which has been classified into five major groups: oxidoreductases, transferases, hydrolases, lyases and isomerases (Hafez, 1964). In the sheep placenta (which produces large amounts of fructose from glucose) sorbitol is an intermediate in fructose synthesis. Two enzymes are involved in this reaction: aldose reductase and sorbitol dehydrogenase. A hydroxysteroid dehydrogenase specific for 17α-estradiol is found in sheep and cattle placenta.

The localization of enzymes in the placenta varies with the type of placentation. In the epitheliochorial placenta of pigs, succinic dehydrogenase is localized in the whole trophoblastic epithelium and particularly in that of the areolae and of the bottom of the chorial fossae. In the placenta of cows and sheep, the enzymes are localized in the trophoblastic epithelium of the intercaruncular chorion and that of the bases of the villi.

Alkaline phosphatase is widely distributed in all types of placenta but, ironically, in maximal concentration in those species producing no fructose, and in minimal concentration in the fructogenic species. This enzyme is found histologically in sheep's maternal placenta and absent in the chorionic villi and other fetal parts.

3. IMMUNOLOGIC ASPECTS

The placenta may be considered as a homograft since it differs genetically from the host. Although it is intimately united with the maternal tissue, it is not rejected until parturition, a period far in excess of the time taken to elicit a homograft reaction; theoretically it should be rejected within two to three weeks. The fact that the placenta does not follow such immunologic concepts has resulted in much speculation and experimentation.

Several hypotheses have been suggested to explain this phenomenon: (a) the hormonal changes of pregnancy may induce a transient state of tolerance; (b) fetal tissues produce large quantities of histamine that may prevent vascular ischemia associated with graft rejection; (c) the syncytium is haploid and contains only maternal chromosomes; (d) the uterus is an immunologically privileged site; or (e) the trophoblast does not contain or is unable to express tissue antigen. Of all these hypotheses, the last appears to be the true explanation.

IV. PRENATAL PHYSIOLOGY

The fetus is the end result of a series of orderly differentiational processes that transform a single-celled zygote into a replica of the species. During the early cleavage of a fertilized egg, the cell size progressively decreases with little change in shape. During later embryonic development the cell size does not significantly change as the cell number increases.

A. Embryology

During early differentiation, cells at one pole of the blastocyst, the germ-disc,

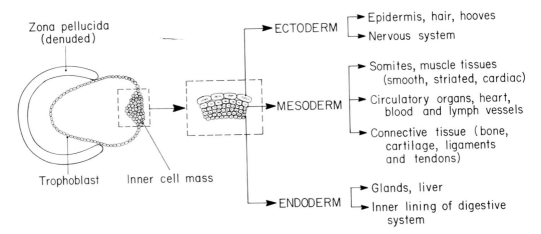

FIG. 8–10. The derivation of various body organs by progressive differentiation and divergent specialization. The origin of all fetal organs can be traced back to the primary germ layers which originate from the inner cell mass.

give rise to three separate layers of cells (Fig. 8–10). The innermost layer, the *endoderm*, forms the lining of the gut, its glands and the bladder. The outermost layer, the *ectoderm*, forms an elongated ridge in the central axis of the germ-disc quite early in development. This elongated ridge, *neural ectoderm*, subsequently gives rise to all adrenal medulla, brain, spinal cord and all other derivatives of the nervous system, i.e. optic vesicles, posterior pituitary and ganglia. The ectodermal cells lateral to the neural ectoderm give rise to the anterior pituitary, skin and all its derivatives, i.e. mammary glands and other skin glands, nails, hair, hooves and lens of the eye. The third germ layer, the *mesoderm*, between the ectoderm and endoderm, gives rise to connective tissues, vascular systems, bones and muscle as well as the adrenal cortex. The *primary sex cells* may be derived from either the mesoderm or the ectoderm. At present there is equal evidence for each.

The body segments or *somites*, which develop from the outer layer (somatic layer) of mesoderm, differentiate into three regions which will form different parts of the fetal body. The first region develops into the vertebrae which encase the neural tube. The second region, the upper part near the neural tube, forms the skeletal muscles. The third region, which is the lower part of the somites, forms the connective tissues of the skin. Differentiation of the somite regions starts on the 19th day after ovulation in cattle, the number increases rapidly to 25 on the 23rd day, 40 on the 26th day and 55 on the 32nd day (Greenstein and Foley, 1958).

Organogenesis. The first formation of most organs and body parts takes place from the second to the sixth week of gestation in cattle. During this period the digestive tract, lungs, liver, and pancreas all develop from the primitive gut. The beginnings of muscular, skeletal, nervous and urogenital systems are established. On the 21st day the heart starts to beat and blood circulation is initiated.

Growth Curve. Useful designations for describing growth are: absolute and relative. *Absolute growth* is the change in volume, crown-rump length or weight of the fetus per unit time. *Relative growth* is the absolute growth per initial dimension of the interval measured. The absolute growth of the fetus is not linear but increases exponentially up to parturition, reaching a maximum during late gestation, whereas relative growth declines about midgestation. In cattle over one-half of the increase in fetal weight occurs during the last two months of gestation. At term the weight of the fetus makes up approximately 60% of the total weight of the conceptus.

The organs of the fetus grow at differential rates, resulting in continuous conformational changes of the organs. The growth pattern of the fetus follows a definite order. For example at birth, the head, limbs and forequarters are relatively more developed than the muscles. Skeletal formation and growth is fairly uniform; however, some dimensions increase more rapidly than others and result in changing body proportions.

Age Determination of Embryo and Fetus. The main criteria used in determining the age of embryos and fetuses have been the time of copulation and ovulation, or the

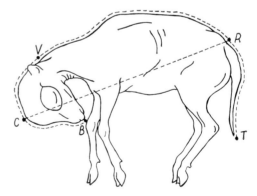

Fig. 8–11. Diagram to illustrate measurements used for estimation of age and growth rate of mammalian fetuses. (*From Harvey, 1959. In Reproduction in Domestic Animals. Cole and Cupps [eds.], New York, courtesy of Academic Press.*)
BCVRT, total length; C, crown-rump; CVR, curved crown-rump; VR, vertebral column length; and VRT, vertebral column and tail length.

Table 8-6. Some Outstanding Horizons in Development of Embryos and Fetuses of Farm Mammals

Developmental Horizons	Cow (Days)	Ewe (Days)	Sow (Days)
Morula	4–7	3–4	3.5
Blastula	7–12	4–10	4.75
Differentiation of germ layers	14	10–14	7–8
Elongation of chorionic vesicle	16	13–14	9
Primitive streak formation	18	14	9–12
Open neural tube	20	15–21	13
Somite differentiation (first)	20	17 (9 somites)	14 (3–4 somites)
Fusion of chorioamniotic folds	18	17	16
Chorion elongates in nonpregnant horn	20	14	—
Heart beat apparent	21–22	20	16
Closed neural tube	22–23	21–28	16 (11 somites)
Allantois prominent (anchor-shaped)	23	21–28	16–17
Forelimb bud visible	25	28–35	17–18
Hindlimb bud visible	27–28	28–35	17–19
Differentiation of digits	30–45	35–42	28+
Nostril and eyes differentiated	30–45	42–49	21–28
Cotyledons first appear on chorion	30	—	—
Allantois replaces exocoelom of pregnant horn	32	21–28	—
First attachment (implantation)	33–	21–30(?)	24–
Allantois replaces all of exocoelom	36–37		25–28
Eyelids close	60	49–56	
Hair follicles first appear	90	42–49	28
Horn pits apparent	100	77–84	
Tooth eruption	110	98–105	(160 mm pig)
Hair around eyes and muzzle	150	98–105	—
Hair covering body	230	119–126	—
Birth	280	147–155	112

(for the cow: Adapted from Salisbury and VanDemark, 1961. *Physiology of Reproduction and Artificial Insemination of Cattle.* San Francisco, Freeman & Co. for the ewe: Cloette, 1939. *Onderstep. J. Vet. Sci. Anim. Ind. 13,* 417. for the sow: Patten, 1948. *Embryology of the Pig.* Philadelphia, The Blakiston Co.)

weight and crown-rump length of the fetus. A reliable measurement taken from the tip of the nostril to the tip of the tail over the back in a sagittal plane (Fig. 8–11) has been used to estimate fetal age in sheep. The length of the foot or head has been used in cattle fetuses. All these methods are subject to variation since the precise time of ovulation cannot be determined as well as the dependence of fetal weight and length on breed and strain, maternal age, litter size and season of birth.

An ideal method of determining the age of embryos and fetuses would be to use the differentiation and development of embryonic and fetal structures, so-called *"developmental horizons"* as a guide; but this information is incomplete for farm mammals (Table 8–6).

B. Fetal Nutrition and Metabolism

The supply of nutrients in prenatal life is achieved in four stages. In the first, the cleaved egg obtains its nutrition from its own deutoplasm, which in farm animals is only temporary. In the second, the blastocyst absorbs fluids and nutrients from the uterine luminal fluid. As the blastocyst increases in size it can no longer absorb sufficient "uterine milk" to supply nutritive material adequately by diffusion. In the third, histotropic nutrition is aided during implantation by the vitelline circulation in the yolk sac and the trophoblastic cells. Fluid, unaltered fats and tissue debris (endometrial fragments) may be engulfed at this time by phagocytosis. After the formation of the placenta, absorption of nutrients from the maternal circulation occurs across the placental membrane as already discussed.

The fetus may be regarded as a parasite living within the mother and is assumed to have priority in the event of insufficient maternal nutrition so that its development can proceed unimpaired. However, it is known that the fetus is most likely to suffer from maternal undernutrition when it occurs toward the end of gestation.

The fetus needs carbohydrates, proteins, vitamins and minerals for maintenance, for differentiation, and for subsequent development and growth. The fetus synthesizes all its proteins from the amino acids of its mother which cross the placenta against a concentration gradient. Throughout gestation retention of calcium, phosphorus and iron increases relative to fetal body weight as well as absolutely. In addition, the fetus has the unique ability to deplete maternal skeletal stores of calcium if feeds are very low in calcium. Iron is used for hemoglobin synthesis but little is known about its distribution and metabolism. Glucose is probably the main source of energy for the fetus. Heat produced in fetal metabolism is dissipated by the placenta, amniotic fluid and various maternal tissues surrounding the fetus.

C. Factors Affecting Prenatal Development

Prenatal development is influenced by several factors: namely, heredity, size, parity and nutrition of the mother, duration of pregnancy, litter size, position of the fetus within the uterine horn, competition between litter mates, relative embryo and endometrium preimplantation development, placental size and ambient temperature.

1. HEREDITY

The maternal contribution to variability in fetal size is greater than the paternal contribution. It has been esti-

mated that 50–75% of the variability in birth weight is due to maternal factors. Species, breed and strain differences in fetal size are partly due to differences in the rate of cellular division. Holstein fetuses at birth weigh about 35% more than Jersey calves and about 15% more than the average dairy calf.

2. SIZE AND AGE OF DAM

The size of the dam has been positively correlated with prenatal growth—the larger the faster. In reciprocal crosses between the large Shire and the small Shetland breeds of horses, the crossbred foal from the Shire mare at birth was three times larger than the crossbred foal from the Shetland mare. In postnatal life the difference between the reciprocal crosses diminished, but after four years the weight of the foal from the Shire mare was still $1\frac{1}{2}$ times that of the foal from the Shetland mare. This difference was maintained throughout adult life. Similar size differences occur in reciprocal horse and donkey matings. The horse gives birth to a much larger mule than the hinny born to the small donkey.

A small cow bred to a large bull, provided she is normally developed for her age, has a small maternal environment and thus should restrict the fetal size sufficiently to allow parturition. On the other hand, if the mother is the larger parent, she will exercise a beneficial influence on the offspring's birth size.

Reciprocal crosses between the large Border Leicester and the small Welsh breeds of sheep resulted in slight maternal effects on the lamb size (Hunter, 1956). Comparison of these crossbreds with correspondingly purebred lambs showed that the small ram had less influence on the size of the lambs from large dams than the large ram had on the reverse cross.

The maternal effect on fetal size may be more pronounced in horses and cattle than in sheep and laboratory mammals due to the relatively long gestation, in which the maternal tissue competes with fetal growth over a longer period and thus effectively controls the size of the fetus.

A comparison of maternal and genetic factors involved in size was made by transferring embryos reciprocally between ewes of the large Lincoln breed and of the small Welsh Mountain breed (Dickinson et al., 1962). The genotype of the lamb was the most important factor influencing weight and cannon length at birth, although maternal-fetal interactions were present. As the uterine environment became smaller, the genotypic differences were still distinct but much less in magnitude. It was concluded that the ewe was able to respond to progressively greater demands imposed by the genetic potential for larger lambs but in accordance with the law of diminishing return.

3. Caloric Intake

During the first two-thirds of gestation, growth of the fetus is apparently independent of both the caloric intake of the dam and the litter size. In the last trimester of pregnancy, there are marked differential changes in fetal weights, which reflect variations in genetic factors, litter size, nutritional status and health of the dam. Maternal nutrition exerts an important influence on fetal growth, notably in sheep. Undernutrition of the ewe during the latter part of gestation leads to the production of stunted lambs even though a normal level had been present earlier. Conversely a reversed type of feeding program results in normal-sized lambs.

One of the biochemical results of poor maternal nutrition during late gestation is the reduction of glycogen in fetal muscle and liver. Normally this supply is built up during late gestation and is used as a source of energy immediately after birth.

4. Litter Size

In polytocous species, increased litter size reduces the rate of individual prenatal growth because of the competition among the fetuses in utero. This relationship is determined partly by variation in placental function and duration of pregnancy. In large litters of pigs one piglet may weigh only one-half or even one-third as much as its litter mate. In monotocous species, twin fetuses are generally smaller than single fetuses.

The influence of litter size on fetal development (or growth) is manifested through a *local effect* and a *general effect* (Fig. 8–12). The local effect refers to the effects of other fetuses within the same horn. The general effect refers to the influence of any fetuses in the uterus, independent of their horn distribution. Thus an increase in the number of fetuses within the same horn could have both a local and a general effect, whereas an increase in the opposite horn could only have a general effect. The degree of local effect to general effect varies with the species. In small litters, only a general effect is apparent. A local effect becomes manifested only when the litter size becomes large (above the physiologic level), presumably because the number of fetuses per horn becomes larger.

5. Placental Size

Placental retarded growth can be a result of (*a*) placental size; (*b*) conditions affecting the nutrient content of the maternal blood or its supply to the placenta;

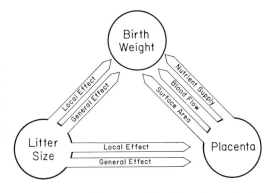

Fig. 8–12. Diagrammatic illustration of the factors and mechanisms affecting prenatal development and birth weight. The "local" effect denotes the influence of other conceptuses within the same uterine horn; the "general effect" denotes the influence of other conceptuses within both uterine horns.

(c) poor development, damage to or specific abnormalities of the placental membrane, thus affecting transport across it; or (d) disorders of the fetal placental circulation.

Prior to the period of maximum embryonic growth the placenta is growing at its peak rate. The placental size can be limited by various maternal processes, which subsequently may retard fetal growth. This indirect maternal effect may account for the close relationship between placental size and fetal weight. Thus the neonate piglets with stunted growth (runts) often observed in large litters may be due to the very small placenta that they developed. Similarly young born from dams with excessive internal fat which prevents full expansion of the uterus and placenta are often small.

The importance of maternal blood supply was shown by inducing partial ischemia of one uterine horn during the latter part of pregnancy in the rat. Such experiments resulted in fetal stunting of all degrees of severity. A form of fetal vascular abnormality which may be important in affecting prenatal growth is the *intrauterine transfusion syndrome*. This is a shunting of blood from one monozygotic twin to the other via placental anastomoses. In such cases, the donor twin is nearly always the smaller.

6. AMBIENT TEMPERATURE

High ambient temperature during pregnancy affects fetal size in some species. Exposure of pregnant ewes to heat stress reduces fetal growth, the degree of reduction being proportional to the length of exposure. This dwarfing is a specific effect of temperature and not due to reduced feed intake during pregnancy. Analysis of some of the bones, organs and endocrine glands of stunted lambs show the effect to be distinct from the undernutritional syndrome. Heat-induced dwarfs are well-proportioned miniatures, whereas underfed ewes have long-legged, thin lambs.

Animals which are undersized at birth are physiologically premature and are subject to neonatal mortality owing to both their poor heat regulating mechanisms and their inability to withstand stress in a new environment.

D. Fetal Circulation

Course of the Circulation. In goats and sheep oxygenated blood from the placenta flows into the fetus through the umbilical veins and passes rapidly through the liver into the caudal vena cava; from there it flows through the foramen ovale into the left atrium soon to appear in the aorta and arteries of the head. Venous

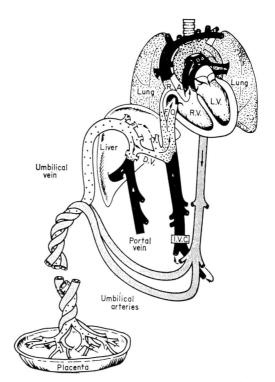

Fig. 8–13. Schematic representation of fetal circulation in the lamb. *DV*, ductus venosus; *DA*, ductus arteriosus; *FO*, foramen ovale. The decrease in the oxygen concentration in blood is illustrated by intensity of shading. Note that oxygenation of fetal blood occurs in the placenta rather than in the lungs and that blood in the umbilical vein is more oxygenated than in the umbilical arteries. The ductus venosus, foramen ovale and ductus arteriosus act as shunts directing the oxygenated blood away from the liver, right ventricle and functionless lungs respectively. (*Modification of Dawes by Rhodes, 1968. Reproductive Physiology for Medical Students. London, Churchill.*)

blood from the lower extremities and the head passes predominantly into the right atrium, ventricle and then into the pulmonary artery (Fig. 8–13).

Fetal circulation is essentially similar to that of the adult except that oxygenation of blood occurs in the placenta rather than in the lungs. In addition, it is provided with several shunts or bypasses to direct oxygenated blood to the tissues. To avoid metabolism by the liver, a major portion of the blood in the umbilical vein is shunted through the *ductus venosus* in the liver into the caudal vena cava. The *crista dividens*, a structure projecting from the border of the *foramen ovale*, separates the caudal vena cava flow into two separate streams before the atria are reached; the stream from the ductus venosus is guided largely into the left atrium, thereby directing oxygenated blood to the head and developing the left ventricle in the neonatal period. And finally, the ductus arteriosus shunts most of the pulmonary arterial blood flow into the aorta away from the functionless lungs. Blood leaving the aorta above the ductus is distributed to the descending aorta. The two umbilical arteries are long, highly contractile with thick muscular layers, originating from the caudal end of the descending aorta. They carry blood to the placenta.

Blood Volume. In fetal lambs blood volume increases during gestation (Assali et al., 1968) with a linear relationship between blood volume and fetal body weight. A certain pool of blood also exists in the fetal placenta; this pool was 48% of the blood volume at 130 days of gestation and 26% at near term.

Regional Distribution of Blood. The proportion of blood (34–91%) passing through the ductus venosus increases with increasing umbilical blood flow. The shunting of blood through the foramen ovale depends on whether the fetus is hypoxic or acidemic. Total cardiac output and output per kilogram body weight is higher in the latter stages of gestation than at near term. Shunting

of the blood through the foramen ovale and ductus arteriosus allows approximately 55% of the total cardiac output of the fetus to return directly to the placenta (Fig. 8–14), 10% through the lungs and 35% to perfuse the body tissues (Rudolph and Heymann, 1967).

Blood Pressure. Striking differences are noted in fetal and adult blood pressures. The higher blood pressure in the right side of the fetal heart than in the left side keeps the foramen ovale patent. Likewise the pressure in the right ventricle is greater than in the left ventricle. Also the higher pressure in the pulmonary artery than in the aorta, probably due to the high vascular resistance existing in the fetal lungs, causes blood to flow from the pulmonary artery into the aorta via the ductus arteriosus.

The mean aortic pressure of the fetal lamb rises gradually during the latter stages of pregnancy to reach approximately 65 mm Hg; the biggest drop in pressure in the fetal circulation occurs in

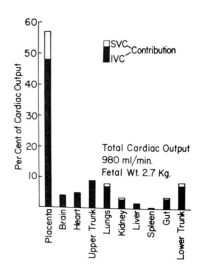

FIG. 8–14. Distribution of blood flow as a percentage of cardiac output to placenta and other fetal tissues. *SVC*, superior vena cava; *IVC*, inferior vena cava. Note that shunting of blood through the foramen ovale and ductus arteriosus allows approximately 55% of the total cardiac output to return to the placenta. (*From Rudolf, 1970. In Fetal Growth and Development. Waisman and Kerr, New York, McGraw-Hill.*)

the fetal placenta (Fig. 8–15). It follows that the umbilical circulation (from the origin of the umbilical arteries to the entry of the umbilical vein into the liver) offers 85% and the liver and ductus venosus together only 15% of the total vascular resistance. This means that changes in the vascular resistance of the umbilical circulation is one of the most important factors that determines umbilical flow. In lambs, between the 90th and 115th days of their 147-day gestation period, a doubling of umbilical blood flow is achieved by a decrease in umbilical resistance. As pregnancy proceeds, a further doubling of umbilical blood flow is brought about by the rising arterial blood pressure of the maturing fetus (Dawes, 1962).

Fetal Heart Rate. Fetal heart rate differs in various species as well as at differ-

ent stages of gestation within each species. In general heart rates are higher in the fetus than in the adult. The fetal heart rate ranges from 170 to 220 per minute in sheep and from 120 to 140 per minute in cattle. The reason for such a fast heart rate is not clear. It may provide the fetus with a higher cardiac output needed to meet its metabolic activities.

E. Amniotic and Allantoic Fluids

The necessity for a liquid medium for life is very marked in embryogenesis. The embryo is immediately surrounded by the amniotic fluid contained within the amniotic membrane. Surrounding this membrane is the allantoic fluid contained within the allantoic membrane. Unlike the human conceptus, the embryos of farm animals have well-developed allantoic vesicles.

Origin. The fluid compartments comprising the fetal environment are closely associated with the renal mechanisms of the developing fetus (Fig. 8–3). In the fetal lamb, urine formed by the mesonephros passes into the allantoic cavity through the urachus up to about 90 days of gestation. Thereafter urine passes in increasing quantities into the amniotic sac due to occlusion of the urachus and patency of the urethra. Thus fetal urine forms a major source of amniotic fluid in the latter part of pregnancy in sheep. In other species sources besides the fetal kidney may influence the amount and composition of amniotic fluid: (*a*) secretions from fetal salivary glands, buccal mucosa, lungs and trachea; and (*b*) dynamic interchange between maternal-fetal-amniotic fluid compartments. In the pig the initial accumulation of allantoic fluid is the result of the secretory activity of the allantoic membrane; later in gestation, however, the fetal urine provides most of the allantoic fluid.

In cattle the character of the fluids gives some indication as to source(s) of the fetal fluids. During early pregnancy the fetal fluids resemble urine. Probably the fetal urine enters these two cavities

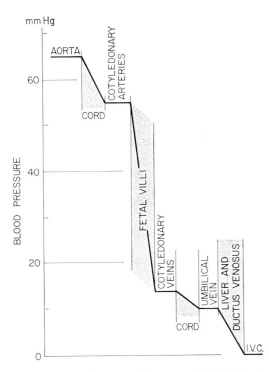

FIG. 8–15. Diagram to illustrate the fall in blood pressure from the aorta through the umbilical circulation and liver to the inferior vena cava (*IVC*) in mature lamb. The biggest drop in pressure occurs in the placenta. (*Redrawn from Dawes, 1962. Amer. J. Obst. Gynec. 84, 1634.*)

by way of the urachus (allantoic) and the urethra (amniotic). At mid-pregnancy, the amniotic fluid changes from a watery to a mucilaginous fluid. This is probably because the sphincter of the bladder begins to function and prevents further flow of urine through the urethra into the amniotic cavity.

Volume. The relative volumes of fluid in the amniotic and allantoic cavities show much fluctuation during pregnancy (Fig. 8–16). These variations probably reflect the contributions of the fetal and maternal compartments. Fetal fluids increase throughout gestation in all species but in the pig they tend to decline at term. The amniotic fluid in the ewe reaches a maximum during mid-pregnancy, falling thereafter; in the mare it equals the volume of the allantoic fluid during the latter stages of pregnancy. Similarly the allantoic fluid increases during the course of gestation and especially in the cow a considerable increase occurs a few weeks prior to calving. The volume of allantoic fluid is relatively higher than amniotic fluid during pregnancy, the exception being the ewe at midgestation.

The mechanisms that control the volume of the fluids in these two cavities are as yet unknown. Removal must occur at approximately the same rate as its formation or it would lead to a gradual or sudden increase in volume of these

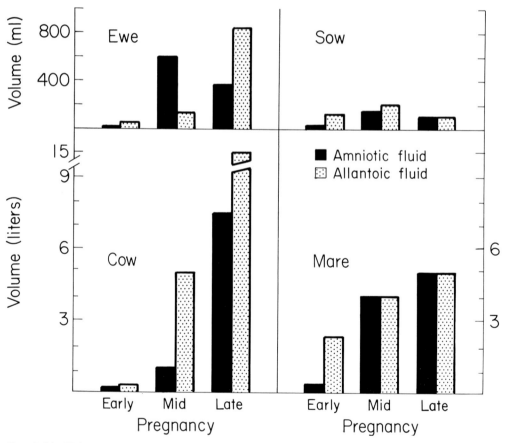

FIG. 8–16. Volume of amniotic and allantoic fluids at three stages of gestation. Note that the volume of fetal fluids increases throughout gestation in ewe, cow and mare, whereas in the sow it decreases. In the cow allantoic fluid increases considerably during the last few weeks of gestation. (*Data from Arthur, 1969. J. Reprod. Fert. Suppl. 9, 45.*)

fluids. The hypothesis originally proposed by Harvey that the fetus removes fluid by swallowing has found recent support. Intrauterine respiratory movements may be another route for removal of amniotic fluid, but it is highly unlikely that these occur under normal conditions (Fig. 8–3).

The volume of amniotic and allantoic fluid in sheep is under hormonal control (Alexander and Williams, 1968). In pregnant ovariectomized ewes progesterone injections (7 mg/day) cause an excessive accumulation of allantoic fluid which could be inhibited by estrogens (5 mg/day). Apparently these large volumes of allantoic fluid are related to the absence of luteal rather than other ovarian tissue and interestingly were associated with high concentrations of Na^+, K^+ and low concentrations of proteins and urea. This suggests that the ability of the fetus to retain Na^+ is apparently affected by progesterone and estrogen.

Functions. One important function of the amniotic fluid is to provide a watery medium in which the embryo can develop free from distortions that would arise from being pressed against rigid surrounding structures. Moreover it prevents adhesions of embryonic skin to the amniotic membrane. It may aid the initial steps of implantation when the expanding chorionic sac is brought into close apposition with the endometrium. During late gestation the lubricant property of amniotic fluid facilitates the expulsion of the fetus. Allantoic fluid, composed of hypotonic urine, maintains the osmotic pressure of the fetal plasma and prevents fluid loss to maternal circulation. In the pig the chorioallantoic membrane possesses secretory properties and is capable of actively removing sodium from the allantoic cavity thereby maintaining the allantoic fluid hypotonic relative to bladder or serum. To what extent the amniotic and allantoic fluids help retain or detoxify fetal excretions and secretions has not been ascertained.

The pressure of the fetal fluids upon their membranes aids the dilation of the cervix at parturition.

Composition. Amniotic and allantoic fluids are slightly alkaline and contain protein, fat, glucose, fructose and inorganic salts. Urea and creatinine are also present, supporting the hypothesis that they have a urinary origin. In ruminants the inner lining of the amnion, particularly near the umbilicus, contains numerous raised, discrete, round foci called "amnionic pustules," which disappear late in gestation. They are rich in glycogen and have no known function. *Hippomanes* are smooth, discoid, rubberlike dark brown masses and float in the allantoic fluid. These are probably aggregations of fetal hair and meconium, having no functional significance.

F. Adaptations to Extrauterine Life

A complex series of structural and physiologic changes adjusts the fetus for extrauterine life. During fetal life, the cardiovascular system is modified to bypass the unexpanded lungs. Thus when the placenta functions as the respiratory organ the lungs are in parallel with the systemic circulation; the foramen ovale allows blood to pass from the right auricle to the left, and the ductus arteriosus shunts blood from the pulmonary artery to the aorta. With the cessation of the umbilical circulation and the commencement of lung ventilation at birth, the flow in the ductus arteriosus is reversed due mainly to a drop in pulmonary arterial pressure and blood then flows from left to right. This results in a distinctive murmur heard over the left thoracic wall in the neonate. This flow ceases promptly after the first few breaths due to an increase in blood oxygen tension. Once this vessel is totally obliterated the ventricles and the lungs begin to function in series. The closure of the ductus arteriosus is one of the most important adjustments in the fetal circulation for extrauterine life. The whole volume of blood now traverses the lungs during each circuit of the body. The rapid decline in blood pressure in the right auricle due to the interruption of the umbilical flow and the increasing left auricular pressure causes closure of the foramen ovale within a few hours in the foal and toward the end of the first week of life in the lamb.

Upon birth the neonate must make thermoregulatory adjustments to fluctuating environmental conditions, contrasting to the relatively constant temperature and nutrient supply present in utero during pregnancy. The efficiency of such adjustments depends primarily on the degree of physiologic immaturity of the species at birth, glycogen reserves and presence of brown adipose tissue. Swine and sheep are particularly susceptible to low ambient temperatures; the rectal temperature of lambs falls 2° to 3° C, while that of piglets declines 2° to 5° C in the first hour after birth.

Newborn animals are also not well adapted to withstand high temperatures early in life, lambs and calves being especially susceptible. For example, lambs between two and seven days of age cannot survive longer than about two hours at 38° C, or more than three hours of solar radiation exposure.

After birth the continuous supply of glucose and other nutrients obtained from the mother is severed. To prepare for the new environment, the fetus accumulates large amounts of glycogen in the liver and both the skeletal and cardiac muscles during the later part of gestation. These carbohydrate reserves are utilized rapidly after birth, the rate of decline occurring much faster at lower environmental temperatures. Thus any undue stress puts the neonate under a severe nutritional requirement that must be overcome for it to survive.

The offspring is born without a supply of maternal antibodies or immunoglobulins. The persistence of the uterine epithelium in the epitheliochorial placenta of farm animals, as previously noted, probably makes the placenta impermeable to maternal antibodies. Immediately after birth, immunoglobulins are transferred to the newborn via colostrum during a period of 24–36 hours.

V. PARTURITION

Parturition or labor is defined as the physiologic process by which the pregnant uterus delivers the fetus and placenta from the maternal organism.

A. The Pregnant Uterus at Term

Before normal labor begins, the fetus usually assumes a position in the uterus characteristic of the species. It is presented at the time of parturition in a position which poses the least amount of difficulty in passing through the pelvic girdle. In monotocous species, the fetus is on its back during intrauterine life.

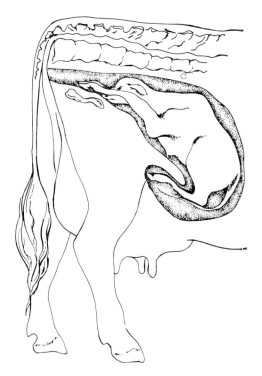

Fig. 8–17. Position of the calf in the uterus after it has been oriented for normal delivery. (*Redrawn from Salisbury and VanDemark, 1961. Physiology of Reproduction and Artificial Insemination of Cattle. San Francisco, Freeman & Co.*)

Prior to labor it rotates to an upright position with its nose and forelegs directed toward the posterior end of the dam. The *anterior presentation* is the most common in ruminants; the front feet of the fetus emerge first with the nose between them; the head is extended and the dorsum of the fetus is in contact with the sacrum of the dam (Fig. 8–17). This presentation coordinates the natural curvature of the birth canal and the curvature of the fetus. Posterior presentation or entry of the fetus with hind feet first, hocks up, occurs less frequently in cattle.

In horses, the fetus is carried to a larger extent in the body of the uterus, rather than in one uterine horn as in cattle. The foal is nonetheless presented essentially the same as that of a calf. In swine, delivery of the individual fetuses from the two uterine horns proceeds in an orderly fashion beginning at the cervical end. Fetuses are presented either anteriorly or posteriorly with equal facility.

B. Initiation of Parturition

The onset of parturition is regulated by a complex interaction of endocrine, neural and mechanical factors. Many theories have been proposed to explain the initiation of parturition: (*a*) progesterone blocks myometrial activity in pregnancy and its withdrawal induces parturition; (*b*) a rise in estrogen secretion; (*c*) a release of pharmacologically active substances, e.g. oxytocin, prostaglandins, catecholamines etc.; (*d*) neural mechanisms; and (*e*) fetal hypothalamic-pituitary-adrenal axis. Whether any one of these plays a unique or a primary role remains to be determined. Moreover it is not clear if the stimulus that triggers parturition is of maternal or fetal origin.

1. Mechanical Factors

Distension of the uterus during rapid fetal growth may cause increased sensitivity of the uterine musculature to both estrogen and oxytocin. Twins are born earlier than singles, suggesting that the mass of the conceptus or optimal uterine stretch may play a part in the initiation of uterine contractions and subsequent parturition. But in view of the gradual nature of these changes, it is unlikely that increase in uterine volume alone can cause the initiation of parturition.

2. Hormonal Factors

Oxytocin. Oxytocin released from the maternal posterior pituitary plays an important role in the initiation of parturition. Failure to inhibit parturition by hypophysectomy does little to disprove the role of oxytocin since in these instances the hypothalamus might have produced oxytocin. Oxytocin levels show little change during the initial stages of labor but rise to a peak during expulsion of the fetus and decrease thereafter. The control mechanisms for oxytocin are unknown. The uterine sensitivity to oxytocin varies among different species. The pregnant ovine uterus when not in labor does not respond to even larger doses of oxytocin; it becomes sensitive, however, when the uterus is in labor. In the goat the reflex release of oxytocin following vaginal stimulation is inhibited by progesterone and increased by estrogens. The possibility exists that the fall in progesterone or the rise in estrogens, a feature of the final stages of pregnancy, could cause a release of oxytocin from the posterior pituitary which in turn initiates parturition. It is also possible that as labor proceeds and cervical dilation occurs, oxytocin is reflexly secreted by the posterior pituitary in response to stimuli originating in the cervix and vagina.

The presence of an enzyme oxytocinase in the blood of primates during pregnancy led to the hypothesis that this enzyme protected the uterus during pregnancy by inactivating the endogenous oxytocin, and at term this inhibition is removed allowing oxytocin to initiate labor. Experimental proof for this concept is lacking since the activity of this enzyme changes little during labor. No abrupt increase in oxytocin levels is observed during labor in women; but the human fetus releases oxytocin during labor which reaches maximal levels at the time of delivery (Chard, 1972).

Progesterone. In most mammals progesterone performs an essential function in maintaining pregnancy. The concept has been advanced that progesterone blocks myometrial contractility, and that withdrawal of this block may constitute an important stimulus to the onset of parturition. It has also been argued that the ratio of uterine volume and pro-

gesterone concentration (V/P) triggers parturition. In farm animals the concentration of progesterone in maternal plasma falls just before parturition, but in pregnant women the highest levels are found at the time of parturition. Thus it remains unclear whether progesterone is the main trigger to parturition.

Estrogens. The levels of estrogen gradually increase during pregnancy reaching a peak at (sheep, goat) or just before (cow) the onset of parturition (Fig.

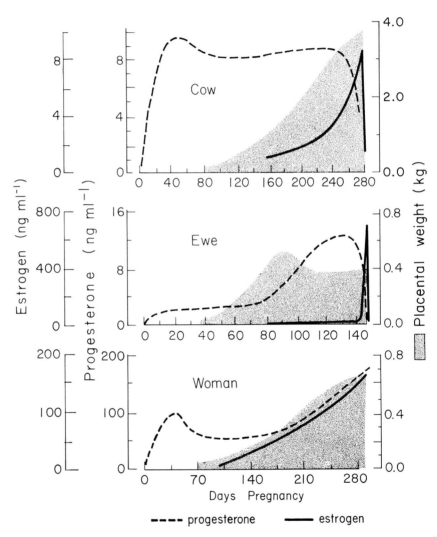

---- progesterone   —— estrogen

Fig. 8–18. Circulating hormonal levels and placental weights during pregnancy. In the cow and ewe the rise in estrogen coincides with a decline in progesterone. In women estrogen and progesterone continue to rise at parturition. (*Redrawn from Bedford, 1972. J. Reprod. Fert. Suppl. 16, 1; placental weight of cow from Swett et al., USDA Tech. Bull. 964.*)

8–18). Estrogen increases the spontaneous contractility of the myometrium. Thus it may play a role either by overcoming the progesterone block or by a direct, stimulatory effect on myometrial contractility.

3. PHARMACOLOGICALLY ACTIVE SUBSTANCES

Prostaglandins. The compounds collectively known as prostaglandins of the F group (PGF) may be of importance in the process of parturition in sheep and goat (Liggins et al., 1972; Thorburn et al., 1972). In these species a major increase in the concentration of PGF in the uterine vein blood is closely associated with the onset of labor. It is not certain whether the PGF is produced in the placenta or in the myometrium. The stimulus for PGF synthesis may be cortisol secreted by the fetal adrenals or the extremely high levels of unconjugated estrogens in maternal plasma just prior to and during parturition. It is assumed that PGF plays an active role in stimulating myometrial activity during labor.

Neurohumoral Mediators. Endogenously released catecholamines and acetylcholine may play a role in labor.

4. FETAL FACTORS

The fetal hypothalamic-pituitary-adrenal axis initiates parturition in sheep and goats (Liggins et al., 1972; Thorburn et al., 1972). Very high levels of fetal corticosteroids occur just prior to the onset of parturition, and induction of parturition by the administration of adrenocorticotropic hormone (ACTH) or dexamethasone to fetal lambs suggests that fetal corticosteroids may be involved in the onset of labor.

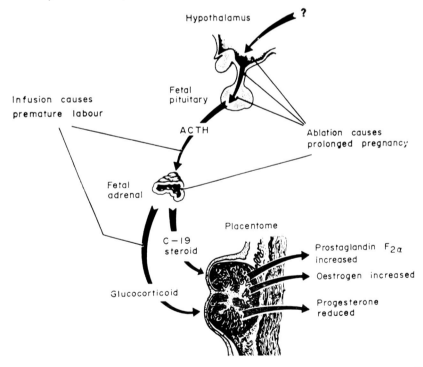

FIG. 8–19. Diagrammatic illustration of fetal mechanisms regulating parturition in sheep. Fetal hypothalamic-pituitary-adrenal axis initiates parturition by stimulating the production of prostaglandin $F_{2\alpha}$ and unconjugated estrogens in the placenta. Following a fall in progesterone prior to parturition induced by prostaglandin, the unconjugated estrogens initiate spontaneous uterine contractility. Oxytocin secreted by the maternal posterior pituitary alone or in combination with prostaglandin causes a rapid delivery of the fetus. (*From Liggins et al. 1972. J. Reprod. Fert. Suppl. 16, 85.*)

Table 8-7. Time (Hours) Required for Various Stages of Parturition in Farm Mammals

Animal		Preparatory Stage	Expulsion of Fetus(es)	Expulsion of Placenta(s)	Labor Should be Completed (Hours)	Involution of Uterus (Days)
Cow	Range	0.5–24	0.5–3 to 4	0.5–8	8	45
	Average	2–6	0.5–1	4–5		
	Trouble if		In multiparous			
	more than	6–12	2–3	12		
Ewe	Range	0.5–24	0.5–2	0.5–8	1–2	30
	Average	2–6	–	–		
	Trouble if					
	more than	6–12	2–3	12		
Sow	Range	2–12	1–4	1–4	1 hr/young	28
	Trouble if					
	more than	6–12	6–12	–		
Mare	Range	1–4	10–30 min.	12	1–3	13–25
	Average	–	–	0.5–3		
	Trouble if					
	more than	4	20–30 min.	12		

(Adapted from *Robert's Veterinary Obstetrics and Genital Diseases*, 1956. Published by the author, Ithaca, N.Y.)

5. INTEGRATED MECHANISM

During pregnancy progesterone inhibits myometrial contractility, and the enlarging fetus distends the uterine musculature. A few days prior to parturition the fetus secretes large quantities of corticosteroids which enter the maternal circulation and stimulate the production of prostaglandins and also increase the concentration of unconjugated estrogens in maternal plasma (Fig. 8–19). The prostaglandins ($PGF_2\alpha$) inhibit the secretion of progesterone by the placenta (sheep) or corpus luteum regression (goat) leading to a fall in progesterone levels prior to parturition. Following a decrease in maternal progesterone concentration, spontaneous uterine activity is initiated by increasing levels of unconjugated estrogens. In response to stimuli originating in the cervix and vagina, oxytocin is reflexly secreted by the maternal posterior pituitary. Oxytocin acting alone or in combination with prostaglandins causes a rapid delivery of the fetus.

C. Stages of Labor

Parturition, the normal conclusion of a pregnancy, may be divided into three stages: (a) preparatory; (b) expulsion of the fetus; and (c) expulsion of the placenta. Although the physical events are described as they occur, the mechanisms underlying the cessation of tolerance for the fetus and the placenta by the mother are not clearly understood. The time for expulsion of fetus is the shortest of the three stages and for farm mammals the mare has the quickest (Table 8–7). The following description primarily refers to these stages in the cow.

1. FIRST STAGE: PREPARATORY STAGE

Prior to the preparatory stage the uterus is relatively quiescent and the available energy reserves are building up to very high levels. Toward the end of gestation actomyosin, the contractile protein of muscle, also begins to increase in quantity and improve in quality. Thus the uterus has developed the necessary proteins and energy supply required to expel the conceptus.

Once the initiating stimuli are received uterine contractions are more effective during labor than abdominal contractions. They are in fact responsible for 90% of the expulsive force and are directly proportional to fetal resistance. Gillette and Holm (1963) electronically recorded these two types of contractions in cows near and during calving by means of microballoons embedded in the uterine mucosa and connected to external strain gauges. A transitional period of irregular, brief, noncoordinated contractions (prepartum period) was transformed into regular, long, coordinated, propagated contractions at parturition. The preparatory stage is characterized by dilatation of the cervix and rhythmic contractions (peristalsis and segmentation) of the longitudinal and circular muscles of the uterus. These contractions force the fetal fluids and membranes against the relaxed cervix, causing it to dilate.

In monotocous species the contractions start at the apex of the cornua, while the caudal part remains quiescent. In polytocous mammals contractions begin just cranial to the fetus nearest the cervix, the remaining part of the uterus remains quiescent. These uterine contractions are the result of extrinsic autonomic neural reflex mechanisms and characteristic automatic contractility of smooth muscle. The neural reflex may be enhanced by fetal motility, while the intrinsic mechanism is enhanced by hormones, particularly oxytocin.

At the end of this stage, the cervix expands, allowing the uterus and vagina to become a continuous canal. The fetus and the chorioallantois are forced into the pelvic inlet where the chorioallantois ruptures, resulting in allantoic fluid flowing from the vulva. The first stage is followed shortly by the second stage.

2. Second Stage: Expulsion of Fetus

The distended amnion along with the head and part of the extremities are forced into the pelvic inlet. This initiates reflex and voluntary contractions of the diaphragm and abdominal muscles. The fetal extremities within the amnion now appear through the vulva. The passage of the fetus through the cervix into the vagina along with rupture of one or both water bags initiates reflex contractions which force the fetus through the birth canal.

In ruminants, the fetus is expelled while still attached to the fetal membranes. The caruncles continue to supply oxygen from the mother even if the expulsion is prolonged. The last fetal cotyledons are not detached from the maternal caruncles until after the young is born, thus ensuring an oxygen supply until the young is able to breathe independently. In swine and horses which have diffuse placentas, most of the placental connections are broken soon after the onset of the first stage of parturition. It is imperative that the second stage be fairly rapid or the fetus will suffocate.

3. Third Stage: Expulsion of Placenta

The expulsion of the fetal membranes is an active process associated with uterine contraction. Peristaltic contractions originating at the apex of the uterine horn causes the inversion of the chorioallantois, which facilitates its expulsion. Loosening of the chorionic villi from the crypts of the caruncles may be the result of much of the blood being removed from the villi and maternal caruncles by the strong uterine contractions occurring during expulsion of the fetus. Such contractions can squeeze some 20% of the total fetal blood from the placenta into the lamb as it is being born.

Normally the placenta of the cow and ewe is delivered within 12 hours following parturition. Frequently in cases of abortion, dystocia, premature birth, and multiple pregnancy, the placenta is not loosened and its expulsion is delayed. Delayed expulsion (over three hours) of the placenta in the mare may cause metritis.

VI. PUERPERIUM

The puerperium extends from the time of expulsion of the placenta until the maternal organism returns to its normal nonpregnant state. Among the most important changes which occur during this period are regeneration of the endometrium, uterine involution and resumption of estrous cycles.

A. Regeneration of the Endometrium

Cow. *Lochia,* or the uterine discharge which normally occurs during the puerperium, is composed of mucus, blood, shreds of fetal membranes and caruncular tissue. For the first two or three days lochia is blood-stained and then becomes paler in color; between the 7th and 14th day it is mixed with an increased quantity of blood due to hemorrhage from caruncular tissue sloughing. The involution of the maternal caruncle involves degenerative vascular changes, peripheral ischemia, necrosis and sloughing (Fig. 8–20). The surface of the bovine caruncle, which is devoid of epithelium immediately after parturition, begins to regenerate by 12–14 days postpartum by proliferation from the surrounding tissue and is completely reestablished in most normal cows within 30 days postpartum (Wagner and Hansel, 1969).

Mare and Sow. During the first week postpartum the mare may discharge small quantities of lochia. At "foal heat" (11 days postpartum) the endometrium is highly disorganized and contains large numbers of leucocytes. By 13–25 days it is fully regenerated in normal foaling mares. Regeneration of the uterine epithelium in the sow begins at one week and is completed by the third week postpartum.

B. Involution of Uterus

Following expulsion of the fetus and placenta, the return of the uterus to its normal nonpregnant size is termed *uterine involution.* This process is associated with enzymatic breakdown of mucopolysaccharides, a rapid shrinkage of cytoplasm in the cells and at the end of the involution period, a grouping of the nuclei of the muscle cells.

Cow. For the first few days postpartum, uterine contractions increase to one contraction every three minutes. During the next three to four days they gradually diminish to one every 10–12 minutes. These contractions cause shortening of the elongated uterine muscle cells. Involution of the uterus and cervix can be detected by rectal palpation and is completed by 45 days postpartum. The nongravid horn regresses almost completely, whereas the pregnant horn and cervix remain larger than before, even after involution is completed. Such a

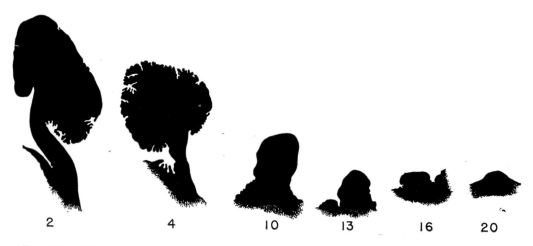

| 2 | 4 | 10 | 13 | 16 | 20 |

Fig. 8–20. Diagram showing postpartum regression of maternal caruncles in cow. Numbers indicate days after calving. Note the dissolution and sloughing of the caruncle between days 10 and 13, and also regression of a caruncle to normal size by day 20. (*Redrawn from Rasbech, 1950. Nord Vet. Med. 2, 265.*)

uterus in a heifer that has not calved indicates that she has conceived and subsequently aborted. Involution of the uterus is faster in cows suckling calves and in primiparous cows and is delayed after dystocia, twin births and retained placenta.

Mare, Ewe and Sow. Involution of the uterus in the mare is rapid though it may not be completely involuted by the onset of the foal heat. In the ewe, at least 24 days are needed for complete involution and another 10–12 days before ewes could conceive; as in the cow, retention of placenta delays uterine involution. In the pig, involution is completed within 28 days of farrowing.

C. Resumption of Estrous Cycles

Cow. The corpus luteum of the previous pregnancy regresses very rapidly. The interval from parturition to first estrus ranges from 30 to 72 days for dairy cows and 46 to 104 days for beef cows. The interval is prolonged when a calf is suckled, and by increasing the frequency of milking (four milkings *vs.* two milkings per day). The removal of the calf shortens this interval. During the postpartum period the first ovulation occurs earlier than the first observed estrus. Following the reestablishment of ovulatory cycles, a short first cycle occurs in the early postpartum period, especially in high-producing dairy cows. It is not clear whether or not this short estrous cycle is related to a deficiency of progesterone by the corpus luteum. Ovarian activity after calving occurs more often on the ovary on the side of the previously nongravid horn. This tendency decreases as the interval from parturition to ovulation increases.

Mares. Most mares exhibit a "foal heat" within 6 to 13 days postpartum. It is a routine practice to breed mares at the foal heat despite the lower conception rates and higher incidence of nonviable foals and abortions.

Sow. Corpora lutea of pregnancy regress rapidly following parturition. Anovulatory estrus occurs three to five days after farrowing. In the majority of animals, however, estrus and ovulation are generally inhibited throughout lactation, although some nursing sows may exhibit estrus after 40 days from farrowing. Sows which do not nurse their litters during the first week of farrowing show estrus and ovulation within two weeks. Removing the piglets or weaning them at any time induces estrus and ovulation to occur in three to five days.

REFERENCES

Alexander, G. and Williams, D. (1968). Hormonal control of amniotic and allantoic fluid volume in ovariectomized sheep. *J. Endocr. 41*, 477–485.

Anderson, L. L., Rathmacher, R. P. and Melampy, R. M. (1966). Uterus and unilateral regression of corpora lutea in the pig. *Amer. J. Physiol. 210*, 611–614.

Arthur, G. H. (1969). The fetal fluids of domestic animals. *J. Reprod. Fert.* Suppl. *9*, 45–52.

Assali, N. S., Bekey, G. A. and Morrison, L. W. (1968). "Fetal and Neonatal Circulation." In *Biology of Gestation II. Fetus and Neonate.* N. S. Assali (ed.). New York, Academic Press.

Barron, D. H. (1970). "The Environment in Which the Fetus Lives: Lesson Learned Since Barcroft." In *Prenatal Life.* H. C. Mack (ed.). Detroit, Wayne University Press.

Bjorkman, N. (1965). Fine structure of the ovine placentome. *J. Anat. 99*, 283–297.

Chard, T. (1972). The posterior pituitary in human and animal parturition. *J. Reprod. Fert.* Suppl. *16*, 121–128.

Clegg, M. T., Cole, H. H., Howard, C. B. and Pigon, H. (1962). The influence of genotype on equine gonadotropin. *J. Endocr. 25*, 245,

Dawes, G. S. (1962). The umbilical circulation. *Am. J. Obstet. Gynec. 84*, 1634–1648.

Dickinson, A. G., Hancock, J. L., Hovell, G. J. R., Taylor, St. C. S. and Weiner, G. (1962). The size of lambs at birth—A study involving egg transfer. *Anim. Prod. 4*, 64–79

Edgerton, L. A. and Erb, R. E. (1972). Metabolites of progesterone and estrogen in domestic sow urine. I. Effect of Pregnancy. *J. Anim. Sci. 32*, 515–524.

Gillette, D. D. and Holm, L. (1963). Prepartum to postpartum uterine and abdominal contractions in cows. *Amer. J. Physiol. 204*, 1115–1121.

Greenstein, J. S. and Foley, R. C. (1958). The early embryology of the cow with notes on comparable human development. *Int. J. Fert. 3*, 67–79.

Hafez, E. S. E. (1964). Uterine and placental enzymes. *Acta Endocrin. 46*, 217–229.

Hunter, G. L. (1956). The maternal influence on size in sheep. *J. Agric. Sci. 48*, 36–60.

Hunter, D. L., Erb, R. E., Randel, R. D., Garverick, H. A., Callahan, C. J. and Harrington, R. B. (1970). Reproductive steroids in the bovine. I. Relationships during late gestation. *J. Anim. Sci. 30*, 47–59.

Liggins, G. C., Grieves, S. A., Kendall, J. Z. and Knox, B. S. (1972). The physiological roles of progesterone, oestradiol-17β and prostaglandin F₂α in the control of ovine parturition. *J. Reprod. Fert.* Suppl. *16*, 85–103.

Lunas, T. (1962). Urinary estrogen levels in the sow during oestrous cycle and early pregnancy. *J. Reprod. Fert. 4*, 13–20.

Metcalfe, J., Bartels, H. and Moll, W. (1967). Gas exchange in the pregnant uterus. *Physio. Rev. 47*, 782–838.

Metcalfe, J. and Parer, J. T. (1966). Cardiovascular changes during pregnancy in ewes. *Amer. J. Physiol. 210*, 821–825.

Moor, R. M. and Rowson, L. E. A. (1966). Local maintenance of the corpus luteum in sheep with embryos transferred to various isolated portions of the uterus. *J. Reprod. Fert. 12*, 539–550.

Newland, H. W., Davis, G. K. and Wallace, H. D. (1960). Placental transfer of phosphorus in sows

maintained on high and low levels of dietary manganese. *Amer. J. Physiol. 198*, 745–748.

Reynolds, M. (1953). Measurements of bovine plasma and blood volume during pregnancy and lactation. *Amer. J. Physiol. 175*, 118–122.

Rollins, W. C., Laben, R. C. and Mead, S. W. (1956). Gestation length in an inbred Jersey herd. *J. Dairy Sci. 39*, 1578–1593.

Rudolf, A. M. and Heyman, M. A. (1967). The circulation of the fetus in utero: Methods for studying distribution of blood flow, cardiac output and organ blood flow. *Circulation Res. 21*, 163–184.

Short, R. V. (1960). Blood progesterone levels in relation to parturition. *J. Reprod. Fert. 1*, 61–70.

Thorburn, G. D., Nicol, D. H., Bassett, J. M., Shutt, D. A. and Cox, R. I. (1972). Parturition in the goat and sheep: Changes in corticosteroids, progesterone, oestrogens and prostaglandin F. *J. Reprod. Fert.* Suppl. *16*, 61–84.

Wagner, W. C. and Hansel, Wm. (1969). Reproductive physiology of the *post partum* cow. I. Clinical and histological findings. *J. Reprod. Fert. 18*, 493–500.

Wynn, R. M. (1968). "Morphology of the Placenta." In *Biology of Gestation* I. *The Maternal Organism.* N. S. Assali (ed.). New York, Academic Press.

Chapter 9

Lactation

A. T. COWIE AND H. L. BUTTLE

Lactation is the final phase of the reproductive cycle of mammals and the physiologic state of the mammary gland is linked to the reproductive state of the animal. The duration of lactation varies considerably in different species but in nearly all, the milk provides the only nourishment available to the young in the postnatal period. Lactation is thus an essential phase of reproduction and, with few exceptions, failure to lactate means failure to reproduce. The milk, and in particular the colostrum, is in ruminants, horse and pig the chief route whereby antibodies are transmitted from mother to offspring.

The cow, the goat, the sheep and the water buffalo have been domesticated and selectively bred by man so that they produce milk in quantities far in excess of the needs of their young and which is harvested by man for his own and his children's nurture.

In this chapter the cow and small ruminants will be used as the main models in discussing the physiology of lactation but brief reference will be made to other species when such reference becomes pertinent. The basic components of the functional mammary gland, i.e. the alveoli and ducts embedded in a stromal or supporting tissue, are common to the mammary glands of all mammals from the lowly monotremes which have retained the reptilian practice of laying eggs to the higher placental mammals. There are, however, species differences in the numbers and positions of the mammary glands and in their shape and detailed architecture.

I. ANATOMY OF THE UDDER

A. Gross Structure

The bovine udder consists of four mammary glands or "quarters." Each gland is a separate entity drained by its own duct system and with its own storage cistern and teat. Under no normal circumstances can milk secreted in one gland pass into an adjacent gland. Within the udder the four glands are in close apposition; on the ventral surface the right and left "halves" are demarcated by a distinct groove on the skin; internally they are separated by the double-layered medial suspensory ligament. The boundary between the glands of the same side cannot be distinguished but the glandular systems are separate as can be demonstrated by the injection of suitable dyes through the teat into the duct system.

The opening through the tip of the teat (the *teat* or *streak canal*) leads into the teat cistern or cavity within the teat (Fig. 9–1). Where the canal opens into the teat cistern there occurs a series of four to eight radiating folds in the mucosal lining of the sinus known as the

Fürstenberg rosette. Within the teat cistern there are numerous irregular annular and longitudinal folds in the mucosal lining. At its upper end the teat cistern communicates through a circular opening with the gland cistern. The size and shape of the gland cistern can vary considerably; it has a multilocular appearance because of the pockets formed by the openings of the large ducts. The cistern region merges into the more solid glandular substance, the presence of numerous smaller ducts giving sections of the glandular substance a spongelike appearance; more dorsally the glandular substance becomes dense and fleshy.

B. Microscopic Structure

In the mammary gland are two main types of tissue: first the parenchyma or glandular tissue and, secondly, the stroma or supporting tissue. In the lactating animal the parenchyma comprises the alveoli in which the milk is secreted and the duct system through which the milk flows to reach the cisterns. The alveoli are minute, saclike or pear-shaped structures the walls of which consist of a single layer of epithelial cells (Fig. 9–2). The alveoli occur in clusters or lobules in which there may be up to 200 alveoli, bounded by a thin fibrous septum. The majority of the alveoli within the lobule open individually into their respective terminal intralobular ducts, but groups of two or three may open together into a common duct and occasionally even large clusters may form a common opening into a terminal duct. The lobules of alveoli themselves form larger clusters or

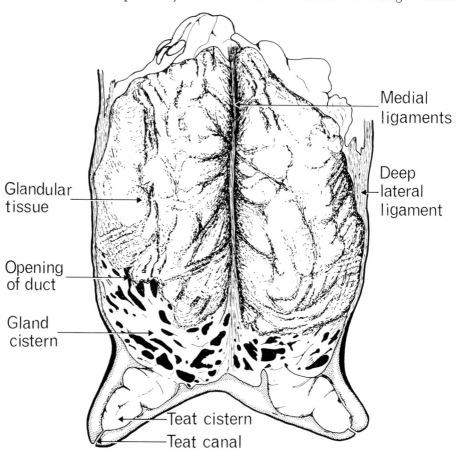

FIG. 9–1.   Section of bovine udder showing teat canal, teat cistern, gland cistern and the openings of large ducts.

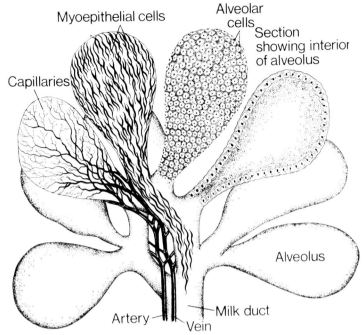

FIG. 9–2. Diagram of clusters of alveoli. (*From Cowie, 1972.* In *Reproduction in Mammals. Austin and Short [eds.], Cambridge, Cambridge University Press.*)

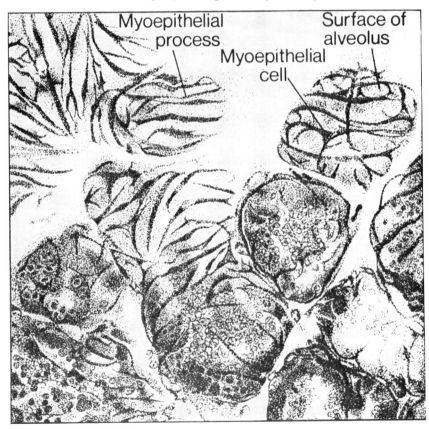

FIG. 9–3. Myoepithelial cells on the outer surface of alveoli. (*From Cowie, 1972.* In *Reproduction in Mammals. Austin and Short [eds.], Cambridge, Cambridge University Press.*)

205

lobes bounded by thicker septa. In the cow the mammary parenchyma is arranged in the form of a series of leaves lying more or less parallel to the surface of the gland. The connective tissue separating these lobes is connected to the ligaments forming the suspensory mechanism. Immediately overlying the alveoli are the myoepithelial cells. These are stellate cells with long contractile processes (Fig. 9–3) which contract in response to oxytocin in the blood and compress the alveoli thereby ejecting the milk into the duct system. Each alveolus is surrounded by a delicate stroma in which lies a fine capillary network. As the ducts become confluent and increase in diameter their epithelial lining changes from a single to a double layer of epithelial cells. On the outer surface of the ducts the myoepithelial cells are arranged in a longitudinal manner so that on contraction they cause the ducts to shorten and thereby increase in diameter so facilitating the flow of milk. A double-layered epithelium lines the gland and teat cisterns and is continuous with that of the ducts. Accessory glandular tissue in the form of small periductal lobules may occur in the wall of the teat cistern and even in the wall of the teat canal. At the junction of the teat cistern with the teat canal the two-layered epithelium of the cistern changes abruptly into a thick squamous epithelium similar to and continuous with that of the skin of the teat. There is no sharply defined sphincter muscle surrounding the teat canal as often alleged but the canal is kept closed by an ill-defined circular network of smooth muscle and elastic fibers (Fig. 9–4).

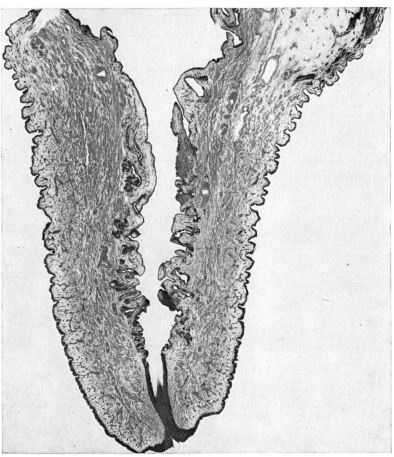

Fig. 9–4.   Section through the teat of a cow. Note the thick stratified epithelium lining the teat canal and the absence of any well-defined sphincter muscle around the teat canal. (*From Cowie and Tindal, 1971. In The Physiology of Lactation. London, Monographs of the Physiological Society, Arnold.*)

C. Vascular System

In high-yielding animals one volume of milk is produced from about 500 volumes of blood passing through the mammary gland. This ratio decreases to 1000:1 in low-yielding animals. Most of the blood is supplied through the large paired external pudendal (or external pudic) arteries, which leave the abdomen by way of the inguinal canal.

The main venous drainage of the udder is through paired external pudendal veins which pass through the inguinal canal. At the base of the udder these veins have anastomotic connections with the caudal superficial epigastric (subcutaneous abdominal or milk) veins and with the perineal vein, these connections forming the so-called venous circle at the base of the udder. In the heifer the caudal superficial epigastric vein drains *into* the venous circle as does the perineal vein, but in the pregnant or lactating animal the valves in the caudal superficial epigastric vein become incompetent and these paired veins then drain blood away from the venous circle (Linzell, 1961).

The lymphatic system carries tissue fluid and lymph from the connective tissue spaces in the interlobular and interalveolar areas to the lymph nodes and thence by way of the thoracic duct into the venous system. In the normal lactating animal the composition of the mammary lymph indicates that it is formed by diffusion of plasma from the capillaries with no back diffusion from the mammary alveoli or ducts. If, however, the udder becomes acutely distended with milk there may be a movement of protein, lactose and probably other milk constituents into the lymph. The rate of lymph flow greatly increases soon after parturition when copious milk secretion is being established.

D. Nervous System

The mammary nerves contain somatic sensory and sympathetic motor fibers; there is no evidence of a parasympathetic innervation. The sympathetic components supply the smooth muscle elements in the walls of the larger ducts and cisterns. Sensory nerve endings are present

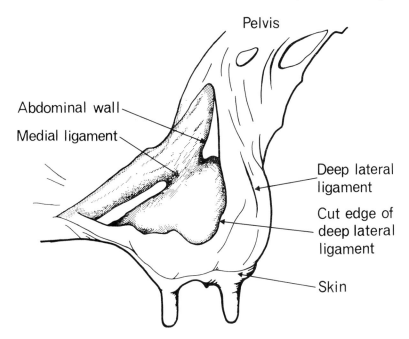

FIG. 9–5: Ligaments of the udder.

in the teat but the degree of sensory innervation of mammary tissue proper is still uncertain (Linzell, 1971). The main nerves are the 1st and 2nd lumbar, the inguinal and perineal nerves.

E. Suspensory Mechanisms

The udder of a high-yielding cow is a relatively large organ which, inclusive of blood and milk, may weigh over 40 kg; it therefore requires adequate support. The skin plays only a minor role in supporting and stabilizing the udder, and the suspensory mechanism proper consists of a series of strong ligaments and tendons by which the udder is attached and suspended both directly and indirectly to the bony pelvis (Fig. 9–5). Superficial and deep lateral ligaments arise from the subpelvic tendon; the superficial ligaments sweep downward and forward over the sides of the udder and then reflect off the udder to the inner aspect of the thigh, the deep lateral ligaments extend down over the sides of the udder and virtually envelop it. These deep ligaments join with the medial ligaments which pass down between the two udder-halves. The elastic medial ligaments arise from the strong tendons of the abdominal wall at a point located over the center of gravity of the udder. The deep lateral and medial ligaments thus form slings for the suspension of each udder-half. The connective tissue septa of the udder tissues are also connected to these ligaments so that the glandular tissue is supported in a series of layers thereby preventing the lower part of the glands from being compressed by the weight of the upper part.

F. Species Other Than Cow

The udder of the sheep and goat is made up of two mammary glands separated by the medial suspensory ligament. The mammary glands of the sow extend over the entire abdominal wall, the number ranging from four to nine pairs. Each nipple is traversed by two canals; each of these canals leads into a small dilatation of the duct or sinus, beyond which the duct ramifies into its own sector of lobulo-alveolar tissue, i.e. there are two separate sectors to each gland each of which is drained by its own duct system and each opening separately at the tip of the nipple. There are a pair of inguinal mammary glands in the mare, each nipple has two canals each leading to a sinus beyond which the duct ramifies in its own sector of lobulo-alveolar tissue.

II. MAMMARY GROWTH

A. Embryonic and Fetal Period

The glandular tissues of the mammary gland are derived in the embryo from ectoderm, the various stages of their morphogenesis being similar in all mammals. At an early age two parallel ridges of ectoderm appear on either side of the midline of the fetus—the *milk lines*; these lines then diminish in length and become interrupted to form a series of nodules of ectodermal cells, the number and positions depending on the species. These nodules sink into the dermis to form the *mammary buds*. Initially these buds are lenticular in shape but become spherical and then conical. There is then generally a distinct pause in their development after which the deep end of the bud, i.e. the apex of the cone, elongates to form a cordlike sprout—the *primary mammary cord*. The primary cord becomes canalized and the lumen so formed at the growing tip dilates to form a miniature gland cistern which becomes well-outlined when the fetus is four to five months old; also about this time canalization of the base of the primary cord has formed the rudiment of the teat cistern. Secondary cords grow out from the gland cistern representing the future ducts; later tertiary cords may appear. At birth the ducts are still restricted to a relatively small zone around the gland cistern. The stroma of the udder is now, however, well-developed; even as early as 13 weeks the stromal tissue has assumed the characteristic udder form.

In the goat and sheep, apart from the retention of only one pair of mammary buds, the pattern of fetal mammary development is similar to that described above for the cow. In the pig embryo the number of pairs of mammary buds corresponds to the number of paired mammary glands of the adult; two primary sprouts grow from each bud, these become canalized and secondary sprouts appear.

Factors Regulating the Growth of the Mammary Rudiments in the Fetus. Little information is so far available about the mechanisms regulating fetal mammary growth in ruminants and other farm animals. In rats and mice, however, it has been demonstrated that certain sex differences in mammary development, e.g. the absence of nipples in male rats and mice, are brought about by modifications in the growth pattern induced by androgens from the fetal testes in the male. Whether sex differences in growth of the mammary glands in fetal ruminants, such as the failure of the stroma to form an udder in the male fetus, are induced by sex hormones is not known.

B. Postnatal Mammary Growth

1. BEFORE PREGNANCY

At birth the mammary gland of the female calf has gland and teat cisterns that are essentially mature in form, further changes being largely increases in size; the mammary ducts are still short and confined to the region of the gland cistern; the gland stroma is well-organized, forming a large pad of fatty and connective tissues. For some time after birth there is minimal mammary growth. Slight extension of the ducts occurs although in some animals there may be a considerable increase in size of the stromal pad. However some two months in advance of the first estrus there begins a period of rapid parenchymal growth which lasts for some four months, declining in rate when the heifer is about one year old. Cyclic changes have been described in the duct system during the

estrous cycle; at estrus a secretion is present in the lumina of the smaller ducts and their epithelium is cuboidal whereas in the progestational phase of the cycle the ducts are empty and shrunken and their epithelium is columnar. These changes suggest that some cellular proliferation occurs at estrus but that later in the cycle some regression occurs.

2. DURING PREGNANCY

In the early months of pregnancy further extension of the duct system occurs but the intensity of this depends on the age of the heifer since in older animals considerable duct growth will have occurred before conception. There is further branching of the duct system, the small interlobular ducts are formed and the alveoli begin to appear; the stroma now contains numerous islands of parenchyma made up of collections of small ducts and alveoli. By the fourth to the fifth months the glandular lobules are well-formed. These lobules increase in size both through the formation of new alveoli, i.e. true growth or hyperplasia, and through hypertrophy or increase in volume of the existing alveolar cells and distension of the alveoli with the onset of secretory activity. Secretion containing fat globules is present within the alveoli during the fifth month. By the sixth month much of the stroma has been occupied by lobules which have so increased in size that they are now only separated from each other by thick bands of stromal tissue. During the last two months of pregnancy the alveoli become further distended with secretion rich in fat globules, and the stroma is represented only by the thin sheets of connective tissue that divide the parenchyma into lobes and lobules.

3. MAMMARY GROWTH IN SPECIES OTHER THAN COW

The general pattern of postnatal mammary growth in the goat and sheep is similar to that in the cow. In virgin goatlings there can be great individual

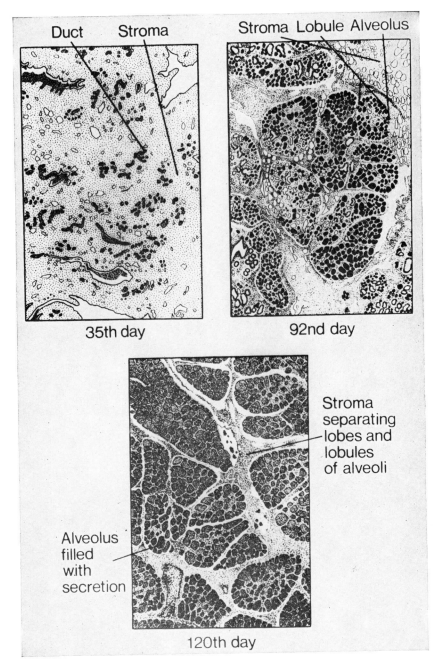

Duct   Stroma

Stroma Lobule Alveolus

35th day

92nd day

Stroma
separating
lobes and
lobules
of alveoli

Alveolus
filled
with
secretion

120th day

FIG. 9–6.   Sections of the mammary gland of the goat during pregnancy (gestation length 150 days). 35th day—Small collections of ducts scattered throughout the stroma. 92nd day—Lobules of alveoli forming into groups or lobes; secretion is present in many of the alveoli, 120th day—Lobules and lobes are now well-developed; the alveoli are full of secretion; stroma is reduced to thin bands of tissue. (*From Cowie, 1972. In Reproduction in Mammals. Austin and Short [eds.], Cambridge, Cambridge University Press.*)

variation in the degree of mammary development; generally the udder pad is compact, containing well-defined gland cisterns and limited duct systems but in some individuals the cisterns and large ducts become greatly distended with secretion thereby increasing the size of the udder; small areas of true alveoli adjacent to the cistern may be found in the mammary stroma of such animals. In the pregnant goat a very rapid extension of the lobulo-alveolar tissue sets in early in the second half of pregnancy, i.e. 80–100 days; there is also marked secretory activity at this period, the alveoli becoming distended with secretion rich in fat globules (Fig. 9–6).

In the pig at birth the mammary glands contain relatively few secondary ducts; by five months of age the duct system is well-developed though still mostly confined to the region of the gland cistern. Lobulo-alveolar development is present by day 45 of pregnancy and is well-developed by day 60. Traces of secretion appear in the alveoli but they do not become distended as in ruminants; even nine days before parturition, the alveoli are small, their secretion is still scanty and hyaline in appearance with no evidence of fat globules; the intralobular stroma is broad. Four days before parturition the alveoli have become wider and the intralobular stroma much thinner; fat globules first appear in the alveolar cells and in the secretion about two days before parturition (Cross et al., 1958).

4. HORMONAL CONTROL OF MAMMARY GROWTH

Analysis of the hormones required for full lobulo-alveolar growth has been carried out in hypophysectomized-ovariectomized-adrenalectomized rodents, the removal of the endocrine glands avoiding the difficulties of interpretation that occur when endogenous hormones are also present. In such triply-operated on rodents, the atrophied mammary duct system can be stimulated to full lobulo-alveolar growth equivalent to that of

late pregnancy by suitable injections of estrogen + progesterone + growth hormone + prolactin + adrenocortical hormones. The anterior pituitary hormones by themselves will induce some lobulo-alveolar growth whereas the steroids in the absence of the anterior pituitary hormones have little effect, the anterior pituitary hormones thus appear to be of major importance. In pregnancy in rodents it is most probable that a protein hormone secreted by the placenta also participates in the stimulation of mammary growth. There is evidence that the steroids act in part by sensitizing the mammary tissue to the growth stimulating effects of the pituitary and placental hormones. Estrogens, moreover, can in some species act indirectly by stimulating the release of anterior pituitary hormones. Although an extensive analysis of the hormones involved in mammogenesis has not been made in ruminants, a preliminary study in hypophysectomized-ovariectomized goatlings suggests that the general pattern may be similar to that described for rodents. The anterior pituitary hormones (and probably the recently discovered placental hormone) appear to be of major importance, and if these are absent the steroid hormones are ineffective.

In the intact nonpregnant ruminant the ovarian steroids can be used to induce mammary growth and milk secretion (Cowie and Tindal, 1971.) Estrogen alone will induce extensive lobulo-alveolar growth although the alveoli tend to be abnormally large resulting in a reduction in the normal area of secretory epithelium; treatment with estrogen-progesterone combinations induces a more normal alveolar structure. This effect may be associated with the ability of progesterone to inhibit the onset of secretory activity by the alveolar cells. Animals treated with estrogen alone tend to produce milk more quickly whereas those receiving estrogen + progesterone, while they come into milk production more slowly, eventually give higher milk yields. In the hormonal induction of lactation in ruminants it is important to recognize the role played by the milking stimulus itself. Indeed mammary growth and milk secretion may be induced in virgin ovariectomized goatlings merely by the regular application of the milking stimulus. When attempting to induce udder growth and lactation in ruminants by ovarian hormone therapy it would seem desirable to start regular milking at an early stage in the treatment.

Although considerable milk yields have been obtained from ruminants in which udder growth and milk secretion have been induced by ovarian hormone therapy, in very few animals has the degree of udder growth or subsequent milk yield been comparable to those occurring in normal pregnancy and lactation. It is now probable that in ruminants in normal pregnancy a prolactinlike hormone secreted by the placenta (Buttle et al., 1972) plays an important role in mammary growth. This hormone has recently been detected in the placenta of the cow, sheep, goat and deer. In the goat the levels of the placental hormone in the blood increase markedly early in the second half of pregnancy at the time when a phase of rapid lobulo-alveolar growth sets in. There can be little doubt but that the growth of the mammary gland in pregnancy is associated with the increased levels of the ovarian hormones in the blood There is, however, less evidence of increased levels of prolactin but as already noted the placental mammotrophic hormone may well play a major role in the lobulo-alveolar growth of the mammary gland.

III. LACTATION

A. Characteristics

The onset of copious lactation occurs around parturition; with some high-yielding cows, however, it may be necessary to begin the regular withdrawal of milk before parturition to relieve the pressure that develops in the udder. Initially the secretion is of colostrum which is high in fat, protein and immunoglobulins and low in lactose (Table 9–1).

Table 9-1. Components of Colostrum

Components	Cow	Ewe	Goat	Sow	Mare
	(grams/liter)				
Water	733	588	812	698	851
Lipid	51	177	82	72	24
Lactose	22	22	34	24	47
Protein	176	201	57	188	72
Ash	10	10	9	6	6

Data taken from Long, C. (ed.), 1961. *Biochemists' Handbook*, London, E. & F. N. Spon, Ltd.

Over the first four days or so of lactation (in cattle) the composition of the secretion changes from that of colostrum to that of normal milk. The yield in cattle, under present husbandry methods, increases steadily to reach a peak at eight or nine weeks and then steadily declines over the remainder of the lactation period. Lactation can be maintained for long periods at a reduced level as long as the animal remains nonpregnant, receives enough food, stays healthy and is milked regularly. Cows, however, are usually mated at the first or second estrous period after parturition (8–12 weeks), and if pregnancy ensues the animal will be dried off about the 28th week of pregnancy (some 40 weeks after the previous parturition) when the milk yield will be low and she will remain dry until the next parturition. This dry period is essential if high yields are to be obtained in the next lactation. The composition of the milk changes slightly during lactation. Once the changes from colostrum to milk have taken place there is then a small but steady decline in protein, fat and lactose (i.e., total solids) until the peak of lactation occurs; thereafter there is a very slight increase in the total solids until cessation of lactation. In other words, within an individual there is an inverse relationship between yield and composition. Factors which limit the increase in daily yield to a peak at eight to nine weeks, such as genetic capabilities, disease, accidents or poor husbandry, will adversely affect the daily yield for the remainder of that lactation.

In other domestic species the lactation characteristics vary according to the reproductive cycles of the animal. In sheep and goats lactation is seasonal because of the nature of the breeding cycle, but otherwise similar in characteristics to that of cattle. In pigs, litters obtain milk from their mothers once every hour throughout the day, and the daily yield increases to a peak at four weeks after parturition and then declines. The sow remains anestrous for the duration of lactation except for the sterile estrous period immediately postpartum; cyclic ovarian activity is resumed after the removal of the litter.

B. Histology and Cytology of Milk Secretion

The secretory changes occurring in the mammary gland both at the level of the light and electron microscope are essentially the same in all the farm animals and indeed in most species so far studied. In the recently milked gland the alveolar cells are tall, the alveolar walls are wrinkled, the alveolar lumina are cleftlike, interlobular septa are conspicuous and wide whereas in the gland fixed when full of milk the alveoli have stretched walls, their cells are flattened, the lumina are wide and the interlobular septa are inconspicuous and thin (Fig. 9–7).

The milk fat is believed to be synthesized by the rough endoplasmic reticulum, the fat droplets then move toward the apex of the cell, possibly aided by the microtubules and microfilaments within the cell, and eventually cause the apical cell membrane to protrude. As the process of protrusion continues the cell membrane envelops the droplet and constricts behind it so that a narrow neck is formed, the walls of the neck fuse and the fat droplet enveloped in membrane drops free into the lumen (Fig. 9–8). The extrusion of the droplet may be facilitated by the numerous vesicles of Golgi origin which come to underlie the globule and which progressively open to the surface, their membranes fusing with the apical cell membrane (Wooding, 1971). Not infrequently a portion of cytoplasm

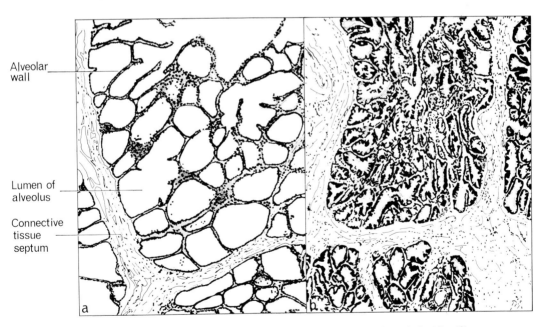

Alveolar wall

Lumen of alveolus

Connective tissue septum

Fig. 9–7.   (a) Part of a lobule from left half of a goat's udder fixed when distended with milk.
(b) Part of a lobule from the right half of udder of the same goat which was milked out before fixing. Note the contracted lobules with collapsed alveoli.

Table 9-2. Components of Milk

	Unit	Cow	Ewe	Goat	Sow	Mare
		(Values given per liter)				
Water	gm	873	837	866	788	890
Lipid	gm	37	53	41	96	16
Lactose	gm	48	46	47	46	61
Total proteins	gm	33	55	33	61	27
Caseins	gm	27.3				
Albumin	gm	3.0				
Globulin	gm	1.3				
Proteose	gm	1.4				
Calcium	mg	1250	1930	1300	2100	1020
Chloride	mg	1030	540	1590	—	—
Magnesium	mg	120	—	160	—	90
Phosphorus (total)	mg	960	995	1060	1500	630
Potassium	mg	1380	1900	1810	—	640
Sodium	mg	580	—	410	—	—
Sulphur	mg	300	310	160	800	323
Vitamin A	I.U.	1460	1460	1340	1760	—
Ascorbic acid	mg	16	40	14	110	118
Biotin	µg	35	—	63	14	—
Choline	mg	130	43	130	122	30
Folic acid	µg	2.3	2.2	2.7	3.9	1.3
Inositol	mg	130	—	210	—	—
Nicotinic acid	µg	850	3930	2730	8350	580
Pantothenic acid	mg	3.5	3.7	2.9	4.3	3.3
Pyridoxine	µg	480	—	70	200	—
Riboflavin	µg	1570	4360	1140	1450	400
Thiamine	µg	420	600	480	980	160
B$_{12}$	µg	5.6	1.4	0.2	1.05	0.02

Data taken from Long, C. (ed.) 1961. *Biochemists' Handbook*, London, E. & F. N. Spon, Ltd.

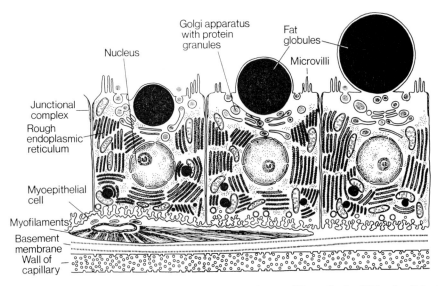

FIG. 9–8. Diagram of the ultrastructure of alveolar epithelium. (*From Cowie, 1972.* In *Reproduction in Mammals. Austin and Short [eds.], Cambridge, Cambridge University Press.*)

becomes entrapped within the enveloped fat droplet.

The milk protein appears as fine granules within the vesicles of the Golgi apparatus; these vesicles move to the apex of the cell, fuse with the cell membrane which then ruptures and allows the granules to escape into the lumen.

C. Formation of Milk Constituents

The concentrations of components in the milk of the cow, sheep, goat, pig and horse are given in Table 9–2. The major component of milk is water in which the solid constituents are dissolved or suspended. The solid constituents— the proteins, fats and lactose—are all formed within the one type of alveolar cell.

All components of milk are derived from the blood, either in the same chemical form or in different chemical forms as precursors. In general the major components, protein, fat and lactose are formed within the epithelial cell from precursors and the minor components such as vitamins, salts and immunoglobulins are secreted by selective passage of materials, without change of chemical form, across the epithelial cells into the lumen along with the water. During active secretion the mammary gland utilizes some energy sources (glucose and acetate) for its own tissue metabolism.

The cell content of milk is used as an index of the health of the gland, the cells are polymorphonuclear leukocytes and sloughed epithelial cells. The cells and their contents do not contribute much to the solid constituents of milk; in diseased states the cell content increases sharply above the normal level of 10^4–10^5 cells/ml milk for the majority of individual quarters.

1. Proteins

Casein forms the major part of the milk proteins and this protein has been classified into four subtypes: α-casein, $\varkappa$-casein, β-casein and γ-casein. The caseins are aggregated to form insoluble granules called micelles. The other major proteins in milk β-lactoglobulin and α-lactalbumin are present in soluble form. All the major milk proteins are formed from free amino acid precursors in the blood, with very little or no milk protein being formed from the breakdown of the plasma proteins. However, the minor protein constituents of milk (γ-casein, immunoglobulins and serum albumin) are apparently absorbed as such from the blood, but these only constitute some 5–10% of the total milk proteins. The essential amino acids are absorbed as such and utilized in the formation of milk proteins, whereas a certain amount of interconversion of the nonessential amino acids does occur, both from the essential amino acids and between the nonessential amino acids.

The amino acids absorbed by the epithelial cells are assembled into short chain peptides by both the free and the bound ribosomes of the cells, from whence the peptides migrate in soluble form to the Golgi apparatus of the cell. Further condensation of the peptides occurs within the Golgi apparatus to form the various insoluble granules of the caseins and also the soluble β-lactoglobulin. The Golgi vesicles containing the casein granules then migrate to the luminal surface of the cell.

2. Lactose

Lactose is the main carbohydrate occurring in milk and is also peculiar to milk, not occurring elsewhere in mammalian tissues. Lactose is formed by the condensation of one glucose molecule with one galactose molecule and the reaction is catalyzed by the two-protein enzyme, lactose synthetase. Glucose is absorbed as such from the blood supply and the majority of galactose is formed from glucose although there is also a minor conversion from acetate into galactose within the mammary gland cells. Hence the lactose in milk is formed almost entirely from glucose in the blood. The enzyme lactose synthetase consists of two proteins, known as the A protein which is a galactosyltransferase, and the B protein which is α-lactalbumin. The A protein is present in the epithelial cells within the Golgi apparatus, but α-lactalbumin is formed within the endoplasmic reticulum when it migrates to the Golgi apparatus, complexes with the A protein (which is presumably bound to or is part of the Golgi membranes) when the complex is able to act as lactose synthetase. The lactose content is largely responsible for the osmotic pressure exerted by milk, hence the lactose concentration will determine to a great extent the amount of water secreted or resorbed by the epithelial cells.

3. Fats

Almost all of the fat content of milk occurs in the form of triglycerides, which are elaborated within the granular endoplasmic reticulum of the epithelial cell into the form of fat droplets. It has been suggested that fat droplets form at sites in the cell where the granular endoplasmic reticulum is in close association with mitochondria. The fatty acids composing the tryglycerides are C_4-C_{18} fully saturated acids, with oleic acid being the only significant unsaturated fatty acid, but the proportions of fatty acids in milk can be substantially altered by the diet. The precursors of the C_4-C_{18} fatty acids in cow and goat milk appear to be obtained from the breakdown of triglycerides contained in the blood chylomicra and of lipoproteins, but acetate and hydroxybutyrate are also utilized to a considerable extent in the formation and interconversion of these C_4-C_{16} acids by the mammary gland cells. The saturated C_{18} acid (stearic acid) is obtained as such entirely from this residue in the blood trigylcerides, the oleic acid (unsaturated C_{18} acid) is obtained both from the same source and by conversion from stearic acid. In ruminants synthesis of fatty acids by the mammary gland cells from glucose does not occur to any extent. The glycerol component of the milk tryglycerides is obtained mainly by conversion from glucose and also from the glycerol of the blood triglycerides. Further information on the biochemistry of milk secretion will be found in the review by Schmidt (1971).

D. Milk Removal

1. Milk-Ejection Reflex

Milk removal, the second phase of lactation, is the process whereby the milk stored in the mammary gland is made available to the suckling young or, in the dairy cow, to the milking machine. In the cow during the interval between milkings a considerable volume of milk, up to half that secreted, passes into the larger ducts and cisterns whence it is readily available to the milker. The remainder of the milk, being the portion stored in the fine ducts and alveoli, cannot be obtained until it is ejected or expelled from these regions into the larger ducts and cisterns. The process by which this transfer of milk occurs is known as *milk ejection* or in farm parlance—the "let down" of milk. Milk ejection is brought about by the operation of a reflex; stimulation of the teat triggers the nerve receptors in the skin and nerve impulses ascend the spinal cord to reach the hypothalamus where they cause the release into the circulation of the hormone *oxytocin* from the posterior lobe of the pituitary gland. Oxytocin is carried in the blood to the mammary gland where it causes the myoepithelial cells to contract thereby expelling the milk from the alveoli, forcing it along the duct system toward the gland and teat cisterns and causing the internal pressure in the cisterns to rise. It will be noted that unlike the classic reflex in which both afferent and efferent arcs are nervous, the efferent arc in the milk-ejection reflex is hormonal and the reflex is thus known as a neuroendocrine reflex.

In the dairy cow the usual stimulus for triggering the reflex is the application of the teat cups. Like other reflexes the milk-ejection reflex can become conditioned so that the reflex occurs in response to visual or sound stimuli which the cow has come to associate with the act of milking, for example the appearance of the milker or sight or sound of the milking apparatus. For efficient milking the cow should be milked as soon as possible after milk ejection has occurred and it is thus important to maintain a regular routine in preparation for milking. Studies on the levels of oxytocin in the jugular blood reveal that further releases of oxytocin may occur during milking.

Dairy cattle have been selectively bred for high milk production and this selection has therefore favored cows in which milk ejection is readily induced. In the more primitive breeds of cattle the induction of milk ejection to ensure efficient milking can be troublesome and difficult and for over 4000 years primitive peoples have practiced various stratagems to facilitate milk ejection; for example, the cow may be milked in the presence of its calf or the calf may be allowed to suckle at one teat while the other teats are hand milked. An odder technique, referred to by Herodotus and one that is still used by primitive peoples in both Africa and Asia, is the induction of milk ejection by vaginal stimulation—usually blowing air into the vagina. It is now known that vaginal stimulation can cause the release of oxytocin so that this ancient practice has a sound physiologic basis.

As is true for other reflexes, the milk-ejection reflex can be inhibited under conditions of stress, hence the importance of a disturbance-free environment in the milking parlor. Inhibition is believed to be effected through a block of the release of oxytocin.

The importance of the reflex, however, is not the same in all species. In some breeds of goats and sheep complete milking can be achieved in the absence of the reflex although recent studies on the blood levels of oxytocin indicate that the reflex does commonly operate particularly in early lactation. Why the reflex should be less essential in the small ruminants than in the cow is not known but it has been suggested that slight differences in the duct architecture permit the milk to drain more readily from the alveoli and fine ducts. It is also possible that the myoepithelial cells may be more sensitive in these species and that they may contract in response to direct physical or mechanical stimuli as occurs in the rabbit in which a sharp tap on the skin overlying the lactating mammary gland will cause some contraction of the myoepithelial cells; this has been termed the "tap reflex." In the lactating sow the proper functioning of the milk-ejection reflex is vital to the maintenance of lactation and to the survival of the litter for unless milk ejection occurs virtually no milk can be removed from the mammary gland.

2. Ascending Paths of the Suckling (Milking) Stimulus

The pathways in the spinal cord carrying the impulses from the mammary gland have yet to be properly delineated (Cowie and Tindal, 1971). In the brain stem the paths concerned in oxytocin release have been traced out in some detail in the goat, guinea pig and rabbit and have been found to pass through the midbrain and diencephalon to the lateral hypothalamus and thence to the paraventricular nuclei from where the neurosecretory axons sweep down to the posterior pituitary gland. Studies on the pathways to the anterior pituitary are still in progress; in the rabbit a prolactin release path has been traced to the medial forebrain bundle in the lateral hypothalamus. There is, moreover, a prolactin release pathway from the orbitofrontal region of the neocortex. This latter path from a "higher" region of the brain may be involved in modulating anterior pituitary function and hence lactation in response to environmental changes.

3. The Mechanism of Suckling and Milking

It is a common belief that the young obtain milk by sucking the teat or nipple of the mother but this is not so; only the milking machine obtains milk in this manner. Cineradiographic studies have shown that the young ruminant strips the milk from the teat with its tongue in an action somewhat similar to that of the fingers in hand milking. The base of the teat is compressed between the tongue and hard palate, the milk trapped in the teat from the base toward its tip against the hard palate; the pressure on the base of the teat is then released to allow the teat cistern to refill with milk and the whole action is repeated. A negative pressure is certainly created within the mouth which undoubtedly aids in the removal of the milk but suction is not an essential component of the act of suckling.

Cineradiographic studies of the action of the milking machine have shown that once the teat is placed in the teat cup all the surface of the teat except the tip is in contact with the liner and that the tip of the teat is exposed to the constant vacuum in the milk liner throughout the whole period of milking. When the liner is in the expanded position the milk is sucked out of the teat, when the liner collapses below the teat the milk flow ceases because the pressure of the liner walls closes the teat canal. There is thus no squeezing of the milk from the teat in machine milking; the rhythmic collapse of the liner around and below the teat helps to maintain a reasonably normal blood circulation in the teat.

E. The Maintenance of Lactation

Once lactation has been established it can be maintained for long periods of time, especially in ruminants, as long as the conditions within the animal do not preclude lactation. The need for adequate food, maintenance of health and regular milking are self-evident factors concerned with adequate husbandry, but the hormonal factors necessary for the maintenance of lactation require some further discussion.

The evidence for the implication of hormones in the maintenance of lactation is based upon the cessation of lactation after removal of endocrine glands and the subsequent maintenance or restora-

tion of lactation by administering the hormone in question. The conditions required by mammary gland cells for their maintenance and continued secretion when in culture in vitro also give an indication of the requirements for maintenance of the cells in vivo. The removal of the pituitary gland results in the immediate and complete cessation of lactation and hence it is considered to be of prime importance in the maintenance of lactation. The removal of other endocrine glands such as the adrenals or the destruction of the islets of Langerhans will also result in a depression and eventual cessation of lactation, but as the specific effects of ablation of these glands on the mammary gland cannot be readily distinguished from the general metabolic effects (i.e. death of the animal) the hormones secreted by these glands may or may not have definitive effects on the maintenance of lactation. Removal of the thyroids or parathyroids will diminish milk secretion. In the hypophysectomized rabbit prolactin alone is capable of restoring milk secretion. In the goat and sheep—the only ruminants in which the restoration of lactation after hypophysectomy has been studied—prolactin, growth hormone, thyroid hormone and adrenal steroids are all necessary for full milk restoration (Fig. 9–9). Once lactation has been restored in the hypophysectomized goat it may be maintained at least for a time in the absence of prolactin provided growth hormone, thyroid hormone and adrenal steroids are given. Recent studies in normal lactating ruminants with 2-Br-α-ergcoryptine—an ergot alkaloid which inhibits the release of prolactin—have also shown that blood prolactin levels can be markedly depressed without affecting milk yields and it is thus possible that the main role of prolactin in ruminants is in the initiation of lactation rather than in the maintenance of milk secretion.

A protein hormone (placental lactogen) with lactogenic properties is also sedreted by the placenta. However, placental lactogens are not normally present during the majority of lactation and so cannot be considered of importance in maintaining lactation.

Additional evidence for the involvement of these pituitary gland hormones in the maintenance of lactation comes from the comparison of the concentrations of hormones found in blood during lactation with those found to occur while the animal is in other physiologic states. It has been shown that high levels of both prolactin and growth hormone occur throughout the day during lactation, with peaks of hormone activity being found after each suckling or milking episode, indicating that large releases of prolactin and of growth hormone occur as a reflex response to milking. The uptake by the mammary gland of goats of adrenal steroid hormones has been measured, larger amounts of cortisol being extracted from blood by the mammary gland during lactation than during pregnancy (Paterson and Linzell, 1971).

Various attempts to determine the hormones concerned in the control and maintenance of lactation have been made by injecting intact lactating ruminants with hormone preparations and measuring any change in milk yields. Thyroid hormone or growth hormone can boost declining milk yields whereas prolactin is usually ineffective; adrenal steroids tend to depress milk yields. Such studies in intact animals are, however, difficult to interpret since the injection of hormones may depress the secretion of the animal's own hormones.

F. Initiation of Lactation

At or around the time of parturition the mammary gland changes from actively growing tissue which is, according to the species, nonsecretory or is secreting only small amounts of colostrum, to one that has almost ceased to grow but is secreting large volumes of milk. The likely stimuli for these changes are the changes in blood hormone concentrations associated with parturition. During pregnancy high blood levels of progesterone,

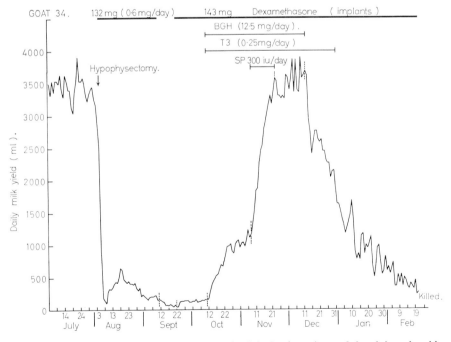

Fig. 9–9. Daily milk yields of a goat hypophysectomized during lactation and then injected and implanted with hormones to restore lactation. (*Cowie, 1969.* In *Lactogenesis. Reynolds and Folley* [eds.], *Philadelphia, University of Pennsylvania Press.*)

estradiol, adrenal steroids and placental lactogen occur whereas prolactin levels are variable but on the whole low. After parturition the levels of all these hormones change, estradiol and progesterone concentrations are low, the levels of adrenal steroids decrease somewhat and placental lactogen is absent but prolactin is present in high concentrations. Also during pregnancy, a corticoid-binding globulin is present in plasma in large quantities and this may be responsible for "inactivating" the high levels of adrenal steroids: after parturition this corticoid-binding protein disappears from the circulation thus "liberating" adrenal steroids for use by the mammary gland and other tissues.

The precise hormonal mechanisms involved in the initiation of copious milk secretion in ruminants have yet to be determined. In the rat the high levels of progesterone in the blood inhibit the onset of secretory changes in the mammary gland during pregnancy by directly preventing the alveolar epithelium from responding to the lactogenic effects of anterior pituitary and placental hormones. A change in the steroid metabolism of the ovary shortly before parturition results in a fall in blood progesterone and a removal of this block to the action of the lactogenic hormones. In ruminants other mechanisms are involved since secretory activity first occurs in the mammary alveoli about mid-pregnancy when progesterone levels in the blood are still high. In the ruminant, moreover, prepartum milking is possible although increases in yield will still occur soon after parturition.

G. Regression of the Mammary Gland

Regression of parenchymal tissue of the udder may be a rapid or slow process depending on the pertinent circumstances. It will occur rapidly if milk removal is suddenly stopped in the fully lactating animal but usually it is a gradual process, associated with the natural decline in milk yield, becoming somewhat acceler-ated when weaning finally occurs or milking ceases. In animals that are allowed to live out their natural lives there occurs another type of regression or senile involution of the gland when reproductive activities have waned.

The histologic and cytologic changes in the acute or rapid type of mammary regression have been extensively studied in laboratory animals and to a lesser extent in ruminants. The alveoli soon become distended, their walls stretched and the alveolar cells flattened, the capillaries become compressed and the blood supply is greatly reduced; within three to four days the alveolar cells break up, lysosomal hydrolytic enzymes are released and there is digestion of the cellular components and resorption of the secretory products which diffuse into the interstitial spaces and are carried away in the lymph. By day 5 the alveoli collapse and disappear and the gland is infiltrated by phagocytic cells. In ruminants macrophages play a major role in the removal of fat from the regressing gland. Eventually the lobular structure disappears and the stroma predominates, the parenchyma being reduced to a duct system.

The tissue and cellular changes associated with the "drying off" of the udder at the end of lactation have been little studied. In the course of normal lactation both the amount of parenchyma and its secretory activity decline despite the continued regular milking. The decline in yield is more marked if a new pregnancy supervenes since the resulting hormonal patterns of raised estrogen and progesterone levels will tend to depress secretory activity while maintaining or even stimulating parenchymal growth. In the lactating cow which is pregnant and which is "dried off" two months before parturition there is little evidence that any marked regression occurs in the lobulo-alveolar tissue. Although there is little known about the histologic and cellular changes, the customary dry period of two months is of considerable physiologic importance for if it be unduly curtailed there will be a serious depression of milk yield in the subsequent lactation. This deleterious response is due to some local effect, as yet undetermined, which apparently interferes with the normal renewal or regeneration of the alveolar cells.

The histologic picture of the mammary tissue in the nonpregnant cow that is being "dried off" at the end of lactation is somewhat different since the hormone levels are not such as to stimulate or even maintain mammary growth. The lobules decrease in size, the alveoli become collapsed and folded and there is a marked increase in stromal tissue. The changes progress until ultimately the lobules are reduced to a few branching ducts.

Further information on the physiology of lactation will be found in the review by Cowie and Tindal (1971).

REFERENCES

Buttle, H. L., Forsyth, I. A. and Knaggs, G. S. (1972). Plasma prolactin measured by radioimmunoassay and bioassay in pregnant and lactating goats and the occurrence of a placental lactogen. *J. Endocr.* *53*, 483–491.

Cowie, A. T. and Tindal, J. S. (1971). *The Physiology of Lactation.* London, Monographs of the Physiological Society, Arnold.

Cross, B. A., Goodwin, R. F. W. and Silver, I. A. (1958). A histological and functional study of the mammary gland in normal and agalactic sows. *J. Endocr.* *17*, 63–74.

Linzell, J. L. (1961). Recent advances in the physiology of the udder. *Vet. Ann.* *3*, 44–53.

Linzell, J. L. (1971). "Mammary Blood Vessels, Lymphatics and Nerves." In *Lactation.* I. R. Falconer (ed.), London, Butterworths, pp. 41–50.

Paterson, J. Y. F. and Linzell, J. L. (1971). The secretion of cortisol and its mammary uptake in the goat. *J. Endocr.* *50*, 493–499.

Schmidt, G. H. (1971). *Biology of Lactation.* San Francisco, Freeman & Co.

Wooding, F. B. P. (1971). The mechanism of secretion of the milk fat globule. *J. Cell Sci.* *9*, 805–821.

Chapter 10

Sexual and Maternal Behavior

G. Alexander, J. P. Signoret and E. S. E. Hafez

The behavior of animals plays an important role in reproduction, affecting both the success of mating and survival of the young. Behavioral patterns, associated with courtship and copulation, with birth, and with maternal care and sucking attempts of the newborn have a dramatic quality that has attracted students of mammalian behavior and has led to the development of an extensive literature that covers wild and domestic animals.

This chapter deals with the patterns in the domestic ungulates, patterns that have been muted by domestication and restricted or modified by conditions imposed in accordance with husbandry requirements. These requirements include confinement in paddocks, yards or indoor pens, segregation of sexes, controlled mating, Cesarean delivery, enforced weaning, imposed proximity with other individuals, and the inescapable presence of man, his dogs and machinery.

By comparison with other aspects of physiology, animal behavior is still in the ethologic, or observational stage; there are comparatively few aspects of behavior that can yet be adequately explained in physiologic terms. Much of the documented knowledge is derived from observations made under restricted environmental conditions, with animals of a single breed or strain, or a narrow age range and a unique background of early experience; conclusions about species differences should therefore be made with great caution.

I. ETHOLOGY OF SEXUAL BEHAVIOR

Various patterns of courtship, display, motor activities and postures are directed to bring the male and female gametes together to ensure fertilization, pregnancy and propagation of the species. The coordination of motor patterns leading to insemination of the female has been achieved by the evolution of an orderly series of responses to specific stimuli. Each response becomes a stimulus in turn, and thus leads to other responses and stimuli, a phenomenon known as a behavioral chain or sequence.

A. Social Structure, Home Range and Reproduction

The encounter of sexual partners is the first step of reproductive behavior. In free-living animals, this occurs largely under the influence of preexisting social structure and the territorial or home range behavior of males and females, and leads to an organized pattern of reproduction which varies with the socio-spatial or territorial characteristics of the species. In the roe deer and muntjak antelope, males and females live in a limited area, the boundaries of which are defended against any intruder of the same sex. The territories of males and females are overlapping, with permanent association between potential sexual partners. In other species, as in the wild rabbit and beaver, the territory is occupied by a permanent couple or harem, and the male avoids any encounter outside his territory. This pattern persists under artificial environments. For example, male rabbits breeding in cages display sexual behavior toward receptive females only after the male occupies the cage for a sufficiently long time to consider it as his territory. Territorial behavior is intensified during the season of reproduction and in fact in a number of species such as the seal it exists only at that time.

In some species, such as the Uganda kob, courtship is confined to a special area defended as a territory (Fig. 10–1). The estrous females visit small areas (15–30 m in diameter) each occupied by a male; nonterritorial males are virtually excluded from reproduction.

Under feral conditions, farm animals do not defend defined territories against intruders, but herds and flocks tend to

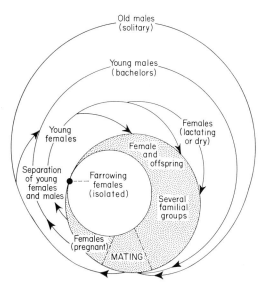

FIG. 10–2. Diagram of the evolution of the social organization in pigs.

occupy a "home range." The basic unit is matriarchal consisting of a female, her adult female offspring and their immature young. The males either congregate into herds, as in feral sheep and goats, or form isolated loose groups, as in the bison and pig (Fig. 10–2). During the breeding season the males leave the groups and range more widely; they may gather some estrous females, thus forming a temporary harem. This pattern gives rise to competition among males. In horses, each matriarchal herd is the permanent harem of a dominant stallion, whereas younger males form a permanent "bachelor" herd (Klingel, 1967).

B. Sequence of Sexual Behavior (Comparative Ethogram)

The motor patterns of courtship behavior are stereotyped and are not altered by experience which acts mainly on the latency and efficiency of mating. The components of copulatory patterns are sexual arousal, courtship (sexual display), erection, penile protrusion, mounting, intromission, ejaculation, dismounting and refractoriness. The duration of courtship and copulation varies with the

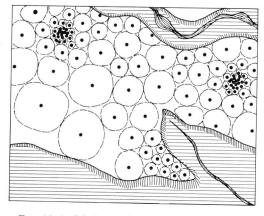

FIG. 10–1. Mating territories in Uganda kob. Note the irregular gradients in size and density of territories. Along the swamp (*upper part*) they are slightly concentrated; on a raised, level area (*lower right*) they form a cluster. The central part of the territories are marked with black dots; their exact boundaries are unknown (*broken outline*). (Leuthold, 1967. *Behaviour* 27, 215.)

Table 10-1. Species-specific Patterns of Courtship in the Male

Behavior	Cattle	Sheep	Goat	Swine	Horse
Sniffing		Sniffing to female's genitalia and urine			Sniffing to female's head	
Ritualized reaction to female's urine	"Flehmen"; male stands rigidly and holds his head in horizontal position which may move slowly from side to side with his neck extended and upper lip raised (Fig. 10–3); "Flehmen" lasts 10–30 seconds			Absent	Same as in cattle, sheep and goat
Urination	No	No	Frequent miction on foreleg during excitement	Rhythmic emission of urine during sexual excitement	Marks with urine the place where a mare has urinated
Vocalization	No	Courting grunts during sexual approach		Courting grunts	Neighing during sexual excitement
Tactile stimulation of female		Licks female's genitalia		Noses female's flanks	Licks female's body; bites back and neck
Postures during courtship	Head on female's back	"Nudging": ritualized approach of female, head turned on side with a motion of foreleg (Fig. 10–3)			
Postures during copulation	Presses head on female's back, "leap" at ejaculation	Rapid movement backwards of head at ejaculation		Motionless during ejaculation, scrotal contraction	Bites female's neck
Postcoital reactions	No	Stretches head and neck	Licks penis	No	No

species; both events are shorter in cattle and sheep than in swine and horses.

1. COURTSHIP

The sequence of courtship tends to be simpler and shorter in mammals than in birds, fishes or arthropods. The management of farm animals facilitates the contact of males and receptive females. Nevertheless, even in the wild, the patterns of courtship in mammals are simple, involving only a limited number of ritualized displays.

In the *male*, sniffing and licking the female are the most frequent patterns, suggesting an important function of chemical communication through olfac-

tion (Table 10–1). Except in swine, the male of domestic ungulates smells the female's urine and then raises his head, with lips curled, in the ritualized "Flehmen" reaction. In sheep, goats and cattle tactile stimulation of the female is made by nuzzling and licking the perineal region, whereas with the horse, the stallion often bites the female's neck and with swine the boar noses her flanks.

The characteristic odor of the male does not appear to be emitted through specific postures or displays. However, urination is used by the stallion for marking the place where an estrous female has urinated, and boars urinate in a rhythmic pattern during sexual activity.

Table 10-2. Species-specific Patterns of Courtship Behavior in the Female

Behavior	Cattle	Sheep	Goat	Swine	Horse
Sniffing	Sniffs male's body and genitals				
Urination	More frequent when the female is teased by the male; increased urination is not characteristic of estrus in sheep and pigs but is specific of sexual receptivity in horses				
Vocalization	Increased frequency of nonspecific bellows	Increased frequency of nonspecific bleats		Estrus grunts	?
Motor activity	Increased general motor activity. Attempts to sniff at the male's genitals leads to a reverse parallel position and circling movements				
	Frontal head-to-head contact; mounts other females	Tail wagging No mounts	Mounts other females	Head-to-head, sniffs; "mock fighting"	No mounts
Postures	Immobility when approached and teased by the male				
	Turning head back Tail deviation			Typical position of ears	Clitoris exposed by contraction of labia; extends hind legs, lifts tail sidewise; lowers croup
Postcoital reactions	Arching of back; elevation of tail	No	No	No	No

In most species, the estrous *female* shows increased motor activity, becoming restless and moving at the slightest disturbance. Receptive cows and goats exhibit increased frequency of nonspecific bellows or bleats, whereas the sow utters a typical estrous grunt. Cows, goats and sows tend to mount and to be mounted by other females but this is limited to some breeds of ewes and mares. In the presence of a male, the female sniffs at his perineum or scrotal region. Mutual sniffing leads both animals to circling motions in a reverse parallel position. Receptive sows also display interest in the boar's head. Frontal contact between the estrous cow or sow and the boar may be associated with "mock fighting." Estrous mares tend to urinate frequently in the presence of a stallion.

When approached and stimulated by the male, the female domestic ungulate assumes a mating posture. This entails immobilization, often accompanied by tail deviation, and some minor species-specific features such as turning the head back in the goat and ewe, cocking the ears in the sow, and exposing the clitoris in the mare (Table 10–2).

2. MATING

The posture of the sexually receptive female terminates courtship behavior by allowing mating to take place. The female stands immobile, and the male mounts and ejaculates (Table 10–3).

MOUNTING. In the presence of a proestrous female, the male attempts several mounts; the penis becomes partially erect and protrudes from the prepuce. These mounts are usually unsuccessful. During this activity, the male, especially the bull, excretes "dribblings" of accessory fluid, derived from the Cowper's gland and differing from the seminal plasma emitted from the vesicular glands during ejacula-

Table 10-3. **Patterns of Mating in Farm Mammals**

	Cattle	Sheep	Goat	Swine	Horse
Duration of estrus	15 hrs. (5–30)	24 hrs. (12–50)	32 hrs. (24–96)	50 hrs. (24–72)	7 days (2–10 days)
Time of ovulation	4–15 hrs. after onset of estrus	30 hrs. after onset of estrus	30–36 hrs. after onset of estrus	40 hrs. after onset of estrus	24–48 hrs. before end of estrus
Male anatomy: Penis	Fibroelastic	Fibroelastic with filiform process		Fibroelastic spiral tip	Vascular-muscular
Scrotum		Pendulous		Close to body	
Accessory glands	Vesicular gland large, prostate and Cowper's proportionally small			Vesicular and Cowper's gland large, important urethral gland, prostate relatively very small	Vesicular and prostate gland large, Cowper's gland small
Mating: Duration		Very brief (a second or less)		5 minutes	40 seconds
Site of semen ejaculation		Near os cervix		Cervix and uterus	Uterus
Volume of ejaculate	5–8 ml	1 ml	1 ml	150–200 ml	60–100 ml
Number of ejaculations to exhaustion, average	20	10	7	3	3
Maximum	60–80	30–40	14	8	20

tion. However, if the female is receptive, copulation may occur very rapidly. The male rests his chin on the female and she in turn responds by "standing." The male then mounts, "fixes" his forelegs around the female, grasps her firmly and performs rhythmic pelvic thrusts (Fig. 10–3). Some boars and stallions mount and dismount the female repeatedly before copulation whereas others mount once and copulate.

Intromission. At mounting, the abdominal muscles of the male, particularly the rectus abdominis muscles, contract suddenly. As a result, the pelvic region of the male is quickly brought into direct apposition to the external genitalia of the female. The boar, with the penis partially out of the prepuce, thrusts his pelvis until the tip of the penis penetrates the vulva; only then is the penis fully unsheathed and intromission accomplished. One intromission takes place per copulation in farm animals. The stallion oscillates the pelvis several times resulting in engorgement of the penis with blood and making it rigid for maximal intromission. In contrast ejaculation in several species of rodents is preceded by a series of mounts and intromissions.

The duration of intromission varies widely between species. Bulls and rams represent one extreme in that they typically ejaculate instantaneously on the first intromission. Boars, however, may maintain intromission for as long as 20 minutes with a single ejaculation.

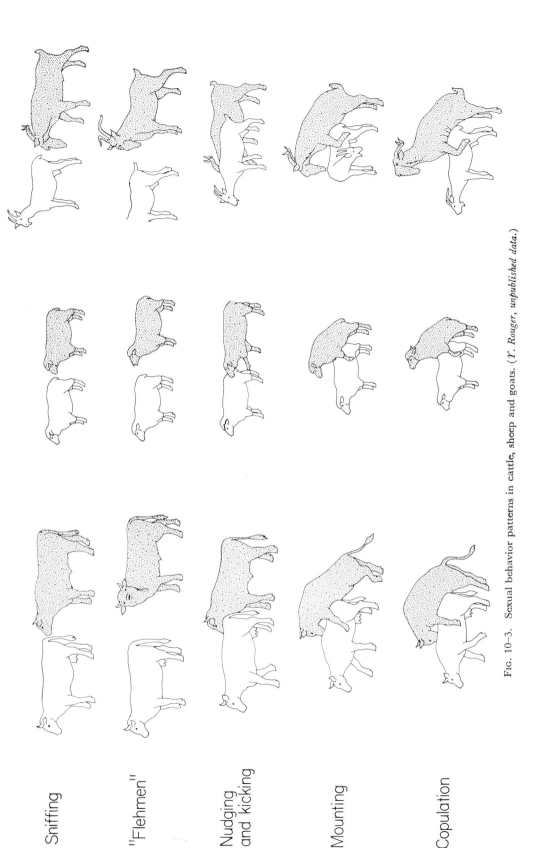

Sniffing

"Flehmen"

Nudging
and kicking

Mounting

Copulation

Fig. 10–3. Sexual behavior patterns in cattle, sheep and goats. (*Y. Rouger, unpublished data.*)

227

Ejaculation. Semen is ejaculated near the os cervix in the case of cattle and sheep, into the uterus in swine and horses. Abortive ejaculations may take place if the female refuses intromission or the penis fails to penetrate the vulva. In the ram, the goat and the bull, an intense generalized muscular contraction takes place at ejaculation. Often the force is so strong in the bull that the hind legs of the male leave the ground, giving the appearance of an active leap. In the ram and the goat the male's head is suddenly moved backward, whereas in the bull it is pressed down on the female's neck (Fig. 10–3). During ejaculation itself, the boar is quiet, presenting only slight rhythmic contractions of the scrotum; such periods of immobility are followed by some thrusts at irregular intervals.

After ejaculation, the male dismounts, and the penis is soon retracted into the prepuce. Postcoital displays are scarce in the domestic mammals: the cow arches her back and elevates the tail after copulation and keeps this posture for several minutes. The male goat usually licks the penis after ejaculation.

Refractoriness. Most males show no sexual activity immediately following copulation. The duration of the refractory period is extremely variable and increases gradually when several copulations are allowed successively with the same female. This period is modified by environmental stimuli.

Frequency of Copulation. The frequency of copulation varies with the species, the breed, the ratio of males and females present, available space, period of sexual rest, climate and nature of sexual stimuli. The maximal number of ejaculations is higher in bulls and rams than in stallions and boars: some bulls have been observed to copulate over 80 times within 24 hours or 60 times within six hours; an average of 21 copulations before exhaustion was observed (Wierzbowski, 1966).

After a long sexual rest, a ram may copulate up to 50 times on the first day after joining with the ewes, but this frequency is greatly reduced on subsequent days. The goat, the stallion and the boar reach exhaustion after a smaller number of ejaculations than in the ram and bull (Table 10–3). The number of copulations with the same estrous female is lower in natural mating than when the semen is collected in an artificial vagina. The bull will copulate only five to ten times with a free estrous female, the ram three to six times, the stallion and the boar two to four times daily.

Duration of Estrus. The duration of estrus is influenced by species, breed, climate and management. Estrus is limited to about a day in sheep and cattle but to longer periods in the sow and the mare (Table 10–3). In species in which the period of sexual receptivity is short, ovulation takes place after its end, but in species which remain receptive for long periods, ovulation occurs during estrus.

C. Measurement of Sexual Behavior

Measurement of mating behavioral responses involves both copulation and the whole sequence of courtship. However, the intensity of a response expressed at any specific time may not reflect the potential, since intensity is greatly dependent upon environmental factors. Hence conditions of testing should be highly standardized, with due regard to management of animals, to the individuality of sexual partners, to time of day and to other environmental factors.

The nature and time of any interaction between animals must be recorded during the test, and sufficient time must be allowed for complete expression of the whole sequence of activities. Accessory tests can be developed for the study of details of the sequence such as interattraction, postural responses to isolated stimuli and the control of specific and environmental stimuli.

Measurement of Male Sexual Behavior. Sexual behavior of male rats is classically measured under well-defined conditions, which included the arena, the duration of the test, the diurnal incidence (using red illumination at night), the ovariectomized and hormone-treated females, the patterns to be recorded, and the instrumentation for recording and computing the results. Similar techniques have been standardized for dogs and the rhesus monkey. The interest in farm animals has been generally limited to copulation; assessment of the intensity of their sexual behavior is based on the latency to ejaculate, the interval between successive ejaculations, the number of copulations required for satiation, environmental stimuli remaining unchanged and, finally, on the period required to recover after sexual satiation, when the same or different female partners are made available.

Measurement of Female Sexual Behavior. The intensity of estrus is subjectively classified according to the degree of sexual receptivity to the male. However, no objective measure of female sexual behavior is yet available for research, though there are some aspects that could be developed. Attraction of female to male or vice versa has been measured in a T maze (Lindstrom and Meyerson, 1969), and by using the frequency of crossing an electrified grid to reach a partner (Meyerson and Nordstrom, 1969). The frequency with which the female displays particular behavior patterns during courtship could also be measured; such patterns include sniffing and head turning but they are few and not highly specific. The willingness to adopt the mating posture as a response to the sexual approach of the male could be measured in terms of latency or by the percentage of accepted mountings.

II. MECHANISMS OF SEXUAL BEHAVIOR

The physiologic signal which originates sexual motivation is the gonadal steroid balance; it is clearly identified and well-known. Transmitted by the blood flow, the hormones activate the central nervous system. The humoral signal is transformed into sexual motivation or sex drive. The motor patterns of copulatory activity are programmed according to preexisting species-specific neuronal circuits. The sensory information allows the initial searching for sexual partners and the identification of their physiologic stages, and releases the appropriate motor reactions.

The following discussion deals with the sensory information organizing the sequence of courtship and copulation, and with the endocrine, neural and genetic control of the reproductive behavior.

A. Sensory Input

The behavioral interactions leading to copulation can be divided into four major phases: first, mutual searching for the sexual partner; second, identification of the physiologic state of the partner; third, the sequence of behavioral interactions resulting in the adoption of the mating posture by the female; and finally, the mounting reaction of the male leading to copulation.

1. Finding a Sexual Partner

Sex drive in the ram and goat shows seasonal variations but this is not so in the boar and bull. However, during the sexual season the male is always sexually active whereas the receptivity of the female is restricted to a few hours or a few days. Thus, the behavioral sequence appears to be initiated by the female. Her role during estrus can be passive in emitting specific signals to stimulate the male. The female may also play an active role by searching for the male or changing her motor or postural reactions to attract him.

The Role of the Female. During estrus, general activity of the female increases and is oriented toward exploration. In an attempt to study the role of attraction Lindsay and Robinson (1961) compared the mating efficiency of rams either free with the ewes in a pasture or tethered with a five-meter chain. One third of estrous ewes were not mated by the tethered rams; the other two thirds actively sought out the male.

The sow placed in a T maze is strongly attracted by the boar during estrus. The stimuli eliciting the approach of the estrous female appear to be essentially of olfactory nature. When the sow is unable to see the boar, the attraction is not modified, but the ability of the sow to discriminate between male and female animals is drastically impaired by the removal of her olfactory bulbs. The production of the attracting male pheromone is under the control of androgens; after daily injection of testosterone, both castrated boars and ovariectomized sows attract estrous females as efficiently as intact males. In the sheep, the impairment of any sensory information in estrous females reduces the percentage of animals mated, indicating that the senses of sight and hearing, as well as olfaction, help the estrous female to be oriented to the male.

The Role of the Male. A male placed with a group of females is initially very active, testing them intensively. However, the male may actively pursue an anestrous female, whereas an estrous one is temporarily neglected. Thus the male is unable to identify the estrous female with certainty in the absence of close physical interactions. In addition, the discrimination between potential sexual partners is more efficient in the sow than in the boar placed in a similar testing situation in a T maze. However odor does play a role in attracting males to females. In sheep, for instance, attracting pheromones appear to be emitted by estrous ewes; receptive and nonreceptive females are approached at random by anosmic rams but not by intact rams (Lindsay, 1965). Hearing and sight are not effective in the identification of estrous ewes.

Conclusion. Both males and females search for social contact with homospecific animals and the search is selectively oriented toward the potential sexual partner. Both participate to this orientation, but the female's role seems to be the most important. During estrus, the orientation to the male is very strong, highly selective, and results from the cumulative action of several long-range acting clues among which the pheromones appear to be of utmost importance. Conversely, the male is less sensitive to the stimuli from the estrous female. His approaches, although not random, tend not to be selectively oriented.

2. Identification of Female Receptivity by the Male

The identification of the receptive female is not made from a distance but takes place during the course of the courtship sequence. The patterns of interaction are the same when a male is testing an estrous or an anestrous female. It is possible that the male engages in courtship behavior upon contact with any female, and the essential clue appears to be the female's postural and motor reaction to the testing male. For example, the various interactions between the boar and the sow take place rapidly with a succession of nasonasal and nasogenital contacts, and tactile stimulation through nosing the flanks and mounting attempts (Signoret, 1970). The sequence is resumed after any of the patterns, provided that the sow has not "stood" when tested. Immobility or standing is the final clue that the female is receptive.

3. Stimulation of Reactions Allowing Copulation

The Female's Mating Posture. The mating posture is often adopted spontaneously when a female is approached by a male or by another female. It may appear under nonspecific tactile stimulation of the herdsman in sows or even without external stimulation; however, an intense amount of stimulation is necessary to elicit the response especially during the beginning and the end of estrus. Stimuli from the male are much more effective in facilitating and releasing the postural response. The "mating stance" of the sow—clear, long lasting and easy to release by an experimenter— is especially suitable for a study of releasing mechanism. During the "standing reaction" the receptive sow is absolutely immobile, arches her back and cocks the ears, and this reaction may be exhibited when an estrous female is touched on the back. However, only 48% of estrous gilts will "stand" in the absence of the male. The frequency of the response increases from less than 40% at the beginning of estrus to 60% between the 24th and 36th hours. Stimuli from the boar, especially olfactory and auditory, are effective in increasing the rate of response to 100% in estrous females; the standing reaction is observed in 90% of estrous gilts when the boar cannot be seen or touched. The addition of visual and tactile stimuli increase the number of gilts responding by 7% and 3% respectively.

Over 60% of estrous gilts, previously negative to the experimenter, exhibit the standing reaction when placed in the boar's pen, or presented to the odor of the boar's preputial secretion (Signoret, 1970) or to that of the compound, 5α-androst-16-ene 3-one, responsible for the boar odor (Melrose et al., 1971). Broadcasting of tape recorded "courting grunts" is similarly effective in 50% of previously negative females. Thus the stimuli emitted during precopulatory interactions facilitate the release of the female's postural response. This appears to be true for ungulates in general.

Mounting by the Male. The female in the mating stance is generally mounted immediately, and this reaction seems to be released mainly by visual and tactile clues. A restrained female, although not in estrus, is immediately mounted even by a sexually experienced bull or ram. Similarly, a ram does not copulate selectively with an estrous ewe when presented with two restrained anestrous females. Sexual reactions of

the male toward stimuli other than the female are common. For example, the bull or the boar react very rapidly to a restrained male or to a dummy. Some 90% of bulls mount a "dummy cow," and 75% of sexually naive young boars react to a very simplified dummy when presented to it for the first time (Signoret, 1970).

It is possible that the sexual releaser for mounting is the overall shape of the female and her immobility. The other visual, olfactory or acoustic information from the estrous female may be of minor but complementary importance. This scheme for the bull, termed a "Torbogen" or archway, seems to apply to naive as well as to experienced males. Other sensory information from the female enhances sexual responsiveness of the male. The need for different types and amount of stimulation according to the species, the breed and the individual, may explain the various reactions to dummies or anestrous females.

B. Endocrine Mechanisms of Sexual Behavior

The steroid hormones in the male and the female have close biochemical similarities, but the rhythm of their release into the blood stream is totally different. In the male, at least in bulls, a number of peaks are observed within 24 hours, but the average day-to-day secretion is practically constant. Any seasonal fluctuations are progressive and slow. In the female, however, estrogens are only present during a few days of the estrous cycle. The effects of the deprivation of the gonadal hormones and therapy in males are completely different from those in females.

1. HORMONAL MECHANISMS OF MALE SEXUAL BEHAVIOR

Castration of males is routinely used in animal husbandry but little information is available on its effects on sexual behavior. The depressing action of castration varies with the species, the individual and the physiologic and behavioral status of the animal at the time of operation.

Generally, some mounting activity is retained after prepubertal castration in bulls and rams, but the underdevelopment of the genital tract, due to the lack of androgen during ontogeny, drastically inhibits mating. There seems to be wide species variation in the effect of castration. Banks (1964) reported variable activity in rams castrated prior to puberty. In contrast, male rodents and cats rarely mount if castrated prepubertally.

Postpubertal castration causes decreased sexual activity. In the stallion, the bull or the ram, casual observations show that erection and intromission may persist for several years postoperatively, but with a marked decrease in frequency.

The loss of ejaculation potency may be due to modifications of androgen-dependent structures of the copulatory organs. In the rat, the cornified epithelial spines of the penis disappear within one month after castration and their reappearance under androgen therapy parallels the recovery of ejaculation. The persistence of the other behavioral patterns cannot be attributed to the presence of androgen from adrenal origin; for example, adrenalectomy does not change the sexual behavior of castrated dogs.

Precastration sexual experience considerably influences the persistence of libido in cats (Fig. 10–4). The patterns of sexual behavior gradually reappear under daily androgen treatment in an order inverse of their disappearance. With increasing doses of testosterone, the frequency of male responses are first dose-dependent, but as the precastration

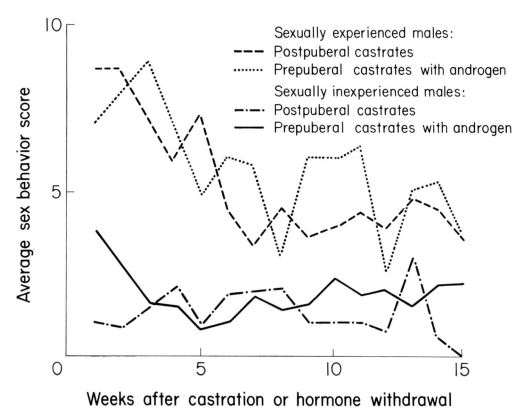

Fig. 10–4. Mean sex-behavior scores after castration or cessation of androgen treatment for males with and without sexual experience. (*Rosenblatt and Aronson, 1958. Behaviour 12, 285.*)

level of reaction is reached, further increases in the dose of androgen are without effect. The hormonal therapy allows the male to recover the preoperative level of copulatory activity, but the preexisting differences cannot be overcome by an extra dosage of hormone. The endocrine balance appears only to reveal the potential intensity of reaction without being able to modify it.

2. Hormonal Mechanisms of Female Sexual Behavior

Female sexual behavior depends on an appropriate endocrine balance resulting in the development of the ovarian follicles. Ovariectomy inhibits sexual behavior, but in the cow and sow sexual behavior is restored in ovariectomized females after the injection of a minimal dose of estrogen following 8–12 days of progesterone pretreatment. In the sow and ewe there is a linear dose-response relationship between the duration of heat and the logarithm of the dose of estrogen (Fig. 10–5). There is also a relationship between duration of natural estrus and the number of the ovulations in the sow and the ewe.

Progesterone inhibits the female's sexual reaction when injected after the appropriate hormonal sequence during the long lasting period of latency.

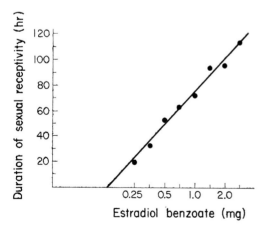

Fig. 10–5. Effect of the dose of estrogen on the duration of sexual receptivity in the sow. (*Signoret, 1967. Ann. Biol. Anim. Bioch. Biophys. 7, 1.*)

Sexual responses can be restored in ovariectomized females by injection of androgen: After one injection of testosterone propionate, the female responses are entirely normal without any male component, probably as a result of biochemical transformation of androgens into estrogens through aromatization.

The action of exogenous hormones in the intact female depends on her physiologic state at injection. When estrogens are injected during proestrus, sexual receptivity is hastened. During the luteal phase of the cycle, the inhibitory action of progesterone prevents an estrous response. The effects of hormonal treatment during anestrus are similar to those observed in spayed females.

C. Neural Mechanisms of Sexual Behavior

The physiologic signal which initiates sexual motivation is the secretion of steroid hormones. Once released in the blood stream, hormones are rapidly bound to receptor sites in the CNS. Maximal estrogen levels in the blood of the ewe and sow occur some 24 hours after onset of estrus. A similar latency is observed after intravenous or intramuscular injection of estrogens. When the animal is sexually motivated, behavioral events are initiated. Specific or unspecific sensory stimuli acting on the sense organs, through innate or acquired mechanisms, are integrated in the brain to elicit appropriate motor reactions.

Hypothalamus. Certain hypothalamic loci (centers) participate in sexual behavior independently of those involved in the regulation of the pituitary-gonadal axis. Several attempts were made to identify a possible "sexual center" in the hypothalamus. Electrolytic lesions in the anterior preoptic area prevented copulatory activity but did not inhibit the secretion of gonadotropins in the male rat (Soulairac, 1963). Clegg and associates (1958) and Domanski (1970) differentiated distinct but closely associated centers in the hypothalamus of the ewe: one abolishes sexual behavior and the other inhibits ovarian acitivity, presumably by influencing gonadotropin secretion (Fig. 10–6). The localization of estrogen-neurons obtained by autoradiography is in agreement with these results (Fig. 10–7).

Sexual behavior can be elicited in gonadectomized males and females by the implantation of steroid crystals in the hypothalamus. The activation of copulatory patterns seems to be under varying degrees of inhibitory control by the CNS. Lesions in the junction of diencephalon and mesencephalon cause an increase in the copulatory performance of male rats. It would appear that the hypothalamus may control sexual behavior in several ways: (*a*) fixation of sexual steroids and slow acting elaboration of sexual motivation; (*b*) direct control of sexual activity; and (*c*) sexual satisfaction.

Rhinencephalon. The phylogenically older parts of the telecephalon (olfactory bulbs and olfactory tract, septum, amygdalum, hippocampus, cingulum etc.) are derived from primary olfactory structures. They are involved in the specialized behaviors (e.g. feeding and sexual and emotional reactions such as arousal, aggression) (Karli, 1968). Specific EEG patterns are recorded in the rhinen-

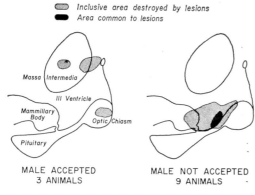

● Inclusive area destroyed by lesions
● Area common to lesions

Massa | Intermedia
III Ventricle
Mammillary Body
Optic/Chiasm
Pituitary

MALE ACCEPTED
3 ANIMALS

MALE NOT ACCEPTED
9 ANIMALS

Fig. 10–6. Effect of hypothalamic lesions on estrous behavior in the ewe. The male was accepted when the lesions were in the massa intermedia or in the optic chiasm of the hypothalamus. (*Clegg et al., 1958. Endocrinology 62, 790.*)

cephalon after coitus in the female rabbit. Estrogen-neurons exist in amygdala and hippocampus. Lesions in the amygdaloid and piriform cortex cause chronic hypersexuality in male cats; this response is androgen-dependent. MacLean and Ploog (1962) obtained erection in squirrel monkey, following electrical stimulation of various structures of the limbic system (hippocampus, septum, cingulate gyrus).

Cortex and Sensory Capacities. Deprivation of sensory capacity can limit sexual behavior, reduce the ability to detect the partner and/or impair orientation. Inexperienced males are impaired to a greater degree than experienced ones. If one sense is inhibited, another sense, that is ordinarily used to lesser degree, may be augmented. Thus elimination of the stimuli to visual receptors in males results in the use of tactile and olfactory receptors. Copulation in domestic mammals is not suppressed with the elimination of vision, smell or hearing provided contact with the partner has been established. Tactile stimuli are involved in the organization of postural responses of copulation: e.g. immobilization of the estrous sow and lordosis in the estrous rat in response to flank palpation.

Medulla. The postural reflexes of mating behavior are organized in the spinal cord of adult female mammals. In the female cat, elevation of the pelvis, treading of back legs and lateral deviation of the tail are observed after total spinal cord transection. Erection and ejaculation can be elicited in paraplegic male dogs (transection of spinal cord) as a response to genital stimulation. Sexual refractory period in the male following ejaculation seems to be partially due to refractoriness of spinal cord. The intensity of certain reactions of spinal males and females appear to be under the influence of sexual steroids, suggesting the possible effect of gonadal hormones on spinal neurons (Hart, 1971).

Neural Mechanisms of Erection and Ejaculation. Erection is predominantly under the influence of the parasympathetic sys-

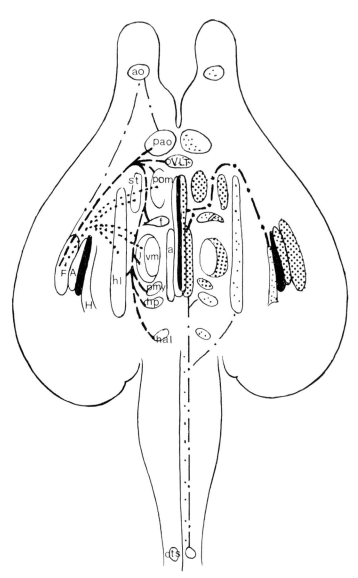

Fig. 10-7.   Estrogen-neuron systems in the rat brain.   Schematic drawing of a hypothetical horizontal plane of the rat brain showing the distribution of estrogen-neurons and probably related nerve fiber tracts.   Estrogen-neurons, which exist in selective areas of the periventricular brain, are indicated by *stipples* (*right* portion of the picture).   The intensity of the stipples reflects the frequency of the occurrence of estradiol-concentrating neurons in the different areas.   Few estrogen-neurons exist frontally in the n. olfactorius anterior and caudally in the ventrolateral portions of the substantia grisca centralis throughout the brain stem.   Few estrogen-neurons exist also in the n. commissuralis tractus solitarii.

Pathways: --- stria terminalis; ..... ventral amygdalofugal pathway; - · - ·- portion of the fornix, probably medial corticohypothalamic tract; —·—·— *left*: tractus olfactorius, *right*: tractus longitudinalis periependymalis.

Abbreviations: *A*, amygdala; *a*, n. arcuatus; *ao*, n. olfactorius anterior; *cts*, n. commissuralis tractus solitarii; *f*, n. paraventricularis; *H*, hippocampus; *hal*, n. habenulae lateralis; *hl*, lateral hypothalamus; *hp*, n. posterior hypothalami; *l vm*, n. ventromedialis hypothalami, pars lateralis; *OVLT*, organum vasculosum laminae terminalis (representing also the organon subfornicale and n. triangularis septi); *pao*, parolfactory region; *pom*, n. preopticus medialis; *pmv*, n. premamillaris ventralis; *st*, n. interstitialis striae terminalis. (The *black* areas indicate the ventricular system.) (*Stumpf, 1970. Am. J. Anat. 129, 207*.)

tem. The parasympathetic nerves in the bull, which supply the external genitalia, arise from the sacral segments of the spinal cord and are connected to the ventral roots of the spinal segments. These fibers are distributed via the pelvic nerves and pelvic plexus to all the reproductive organs except perhaps the testes. Thus drugs which affect the autonomic nervous system can be used to alter the ejaculatory process. Atropine reduces the volume of the ejaculate in blocking the secretion of the bulbo-urethral glands of the boar and the bull. Electrical stimulation of the sacral nerves causes erection and/or ejaculation, a phenomenon that has been put to practical use in collecting semen.

The copulatory patterns of the male are primarily governed by the neuromuscular anatomy and blood supply of the penis. The bull, ram and boar have a fibroelastic penis which is relatively small in diameter and rigid when nonerect. Although the penis becomes more rigid upon rapid erection, it enlarges very little and the amount of contractile tissue is limited. Protrusion is effected mainly by a straightening of the S-shaped flexure and the relaxation of the retractor muscle.

On the other hand, the stallion has a typical vascular penis with no sigmoid flexure. The function of the penis as an organ of intromission depends on the power of erection as a result of sexual excitement. The size, shape and length of the penis vary greatly between the flaccid and the erect state (Fig. 10–8).

Intromission and ejaculation are elicited by tactile stimuli (warmth of vagina and slipperiness of mucus) acting on the penile receptors. The penis of the bull and the ram is sensitive to temperature whereas that of the stallion is more sensitive to pressure exerted by the vaginal walls. In the boar, the corkscrew-shaped tip of the penis is engaged in the cervix during mating. The pressure exerted by this way is sufficient to elicit ejaculation even without any thermal stimulation.

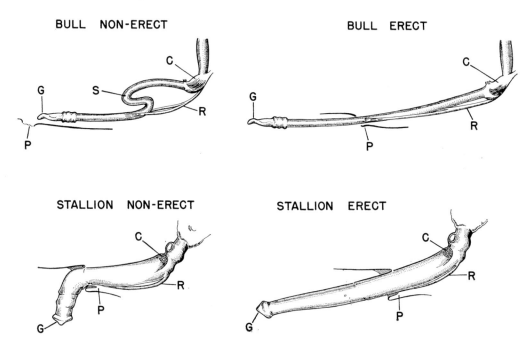

F IG. 10–8. Diagram of the anatomy of fibroelastic type penis (bull) and a vascular-muscular type penis (stallion) in the nonerect and erect positions. The anatomy of the penis determines, to a great extent, the ejaculatory responses of the species. C, cavernous muscle; G, glans; P, prepuce; R, retractor penile muscle; S, sigmoid flexure. (*Redrawn from Hafez, 1960. Cornell Vet. 50, 384.*)

Biochemical Mechanisms: Role of Biogenic Amines in Sexual Behavior. Biogenic amines are involved in behavioral activities as neurotransmitters. Recent advances of neuropharmacology allow the experimenter to interfere through specific inhibitors or precursors with different metabolic reactions of any neurotransmitter. Kobayashi et al. (1964) observed a marked decrease in brain monamine levels at estrus in the rat. Monoamine-oxidase inhibitors prevent such reduction and impair sexual receptivity.

III. FACTORS AFFECTING SEXUAL BEHAVIOR

The patterns and intensity of sexual behavior are affected by genetic, physiologic and environmental factors as well as previous experience.

A. Genetic Factors

Breed and strain differences in libido are not uncommon. Males of dairy breeds are more active than beef males whereas Brahman bulls are very sluggish. Yorkshire boars are easier to train for semen collections than Durocs. There are more differences in the pattern of sexual behavior between pairs of identical twin bulls than between members of the pair (Fig. 10-9). Breed differences in the duration of estrus in sheep and pigs may be partly due to differences in ovulation rates but similar differences occur in estrogen-induced estrus in ovariectomized females (Signoret, 1967b).

Individual differences in the amount of sexual stimulation required to elicit "immobilization reaction" in the sow are independent of sexual experience, of the female or the amount of estrogen injected (Signoret, 1967a). A genetic influence may account for such differences.

B. Sexual Differentiation of the Nervous System

Hormones secreted during fetal and neonatal life cause sexual differentiation of the nervous system. A female neonatally treated with androgen does not show cyclic patterns of gonadotropin secretion, characteristic of the female. The secretory activity of the hypophysis is more or less constant as in the male but males castrated in their neonatal life exhibit a cyclic pattern of pituitary secretion as in the female. Neonatal androgen appears to differentiate the hypothalamic structures controlling gonadotropin secretion from a neutral cyclic pattern of the female into a tonic type of the male.

Sexual differentiation is not limited to the diencephalic control of pituitary function but involves the structures organizing the behavioral sequence of copulation. If an adult female rat is ovariectomized she may react as a female after appropriate estrogen-progesterone

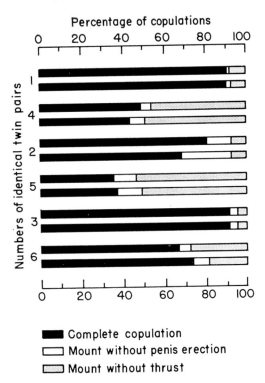

FIG. 10-9. Ejaculatory behavior in identical twin dairy bulls showing percentage distribution of complete copulations, mounts without penis erection and mounts without thrust. Note the great similarities in the ejaculatory pattern of the twin brothers and the great variability between the twin pairs. (*Adapted from Bane, 1954. Acta Agric. Scand. 4, 95.*)

treatment, or as a male after daily administration of testosterone. On the other hand if an adult male is castrated he never presents the lordosis reaction characteristic of the female. The secretion of androgen during neonatal life "defeminizes" the bisexual-potential nervous organization of sexual behavior. Similar mechanism may exist in ungulates. An adult ewe ovariectomized reacts as a male to continuous androgen treatment, and as a female to progesterone and estrogen injections.

C. Environmental Factors

The effect of external stimulation on sexual behavior is more pronounced in the male than in the female.

Effect of Novelty of Stimulus Females (Coolidge Effect). Sexual activity of the male increases when new females in the herd become receptive. If four receptive ewes are available, the ram mates three times as much as he does when only one ewe is in estrus. The enhancing effect of a new stimulus animal should be kept in mind under modern husbandry conditions. For instance changing the teaser cow is a very effective way to increase sexual behavior of a sluggish male.

Nonspecific Stimuli. Nonspecific external stimuli may lead to sexual activity during the refractory period which follows ejaculation in males of low libido. In the rat both painful stimuli such as an electric shock and gentle handling increase the frequency of ejaculations and reduce the postejaculatory interval. Changing the place of semen collection by moving the teaser animal, or "encouraging" the bull, are all effective in sluggish bulls.

Presence of Other Animals. The presence of other males while teasing a female or copulating improves sexual libido of the male. However, social hierarchy may interfere with sexual activity when several males compete for one receptive female. The dominant male performs most of the copulations and restricts the sexual performance of his subordinates. When females are in excess, however, dominant males cannot effectively control the activity of their inferiors. Adult rams usually dominate or "boss" yearling rams, and the degree of their dominance is greater than the dominance of yearling rams over other yearlings (Hulet et al., 1962). Unfortunately social dominance is not correlated to fertility of males. Thus a "bossy" male who is infertile or diseased may depress the conception rate of the entire herd. The size of the pasture also affects the competition among males and the number of copulations per female. The proportion of ewes bred by three rams compared to those bred to one ram was higher in a 1/5-acre pasture than 17-acre one (Lindsay and Robinson, 1961).

Season and Climate. Seasonal variations in sexual behavior of sheep, goats and horses are mostly due to seasonality of pituitary function controlling secretion of gonadal hormones. Seasonal changes are also reported in the responsiveness of ovariectomized ewes and sows to exogenous hormones showing a direct effect of the season on the sensibility of the nervous system. The intensity of sexual behavior is reduced in hot climates. The plane of nutrition per se does not seem to affect sexual behavior. Any physical trouble, however, may seriously affect sexual expression, e.g. inflammation of the hooves or joints, change of teeth, eczema, pains from accidents or certain diseases.

D. Effect of Experience

The efficiency of copulation of males and females is improved by experience. Individual contacts before puberty can have an organizing effect on subsequent sexual performance.

Social Contacts Before Puberty. Social deprivation during infancy drastically impairs adult sexual behavior in primates (Harlow et al., 1966). Sexually motivated male and female rhesus monkeys reared in isolation cannot perform postural adjustments for copulation. In other mammals, the female's sexual behavior does not appear modified by deprivation of social contacts. However, the male's sexual reactions can be seriously impaired. Only 16% of rams reared in isolation copulated when presented for the first time to an estrous ewe (Pretorius, 1967), suggesting an organizing action of heterosexual contacts during ontogeny. Conversely, boars reared in unisexual groups for performance testing have normal sexual responses. The effects of isolation or rearing in unisexual groups are still controversial in domestic mammals.

Sexual Experience. Young inexperienced males are usually awkward during their first contact with a receptive female: they approach hesitantly, spend a long time exploring the genitalia, mount without erection, descend and try to mount again. Erection and ejaculation are weak and the volume of semen is small. After the first ejaculation, the motor patterns are rapidly organized and a normal mating efficiency is reached. The search and identification of an appropriate sexual partner improves gradually by experience. On the other hand sexual reactions in the female are quite immediately adapted with little improvement due to sexual experience.

IV. EFFECTS OF SEXUAL INTERACTIONS ON REPRODUCTIVE PHYSIOLOGY

A. Female

The presence of the male may affect the physiologic events at the time of estrus and interfere with the mechanisms regulating the sexual season and puberty.

Effects of Male on the Duration of Estrus and Time of Ovulation. The presence of the male reduces the duration of the sexual receptivity of the ewe by 50%. This effect is not related to the possible changes in ovulation or steroid secretion since similar effects occur in ovariectomized ewes and sows treated with progesterone and estrogens. With the exception of the Alpaca and possibly some members of the Camelidae family (Fernandez-Baca et al., 1970), farm animals are known to be spontaneous ovulators. However, permanent association with the male hastens ovulation in the ewe and the sow.

Uterine Motility and Sperm Transport. Sterile matings with a vasectomized male stimulate sperm transport in the rabbit (Dandekar et al., 1972) but not in sheep (Hawk, 1972). However, stimulation of the genitalia, or precoital stimuli cause contraction of the cervix and uterus of the ewe (Lightfoot, 1970) and the cow, as a result of the release of oxytocin. Vaginal distension and precoital stimulation causes maximal oxytocin release in sheep and goats (McNeilly and Folley, 1970). MacNeilly and Alison-Ducker (1972) observed that oxytocin release often occurs before actual coitus has taken place. The exteroceptive factors which stimulate sexual behavior are in order of decreasing effectiveness: presence of the male, smell of the male, sounds emitted by the male, sight of the male.

Effect of Male on Anestrous Females. The introduction of the male into a flock of sheep or goat at the onset of the sexual season results in a synchronization of estrus 15–17 days later. An androgen-dependent pheromone from the male is responsible for similar synchronization of estrus in mice (Bronson and Whitten, 1968) and possibly in sheep. Neither sight nor contact is necessary for the synchronization of the first estrus of the breeding season in sheep.

Nursing the young delays estrus compared with hand-milking in cows and ewes. The presence of the ram results in an earlier postpartum estrus in nursing ewes, making the postpartum estrus similar in dry and milked females.

The introduction of the boar shortly before spontaneous puberty in a group of previously isolated gilts results in earlier onset of estrus.

B. Male

Precoital stimulation affects both composition of the ejaculate and androgen secretion. A period of restraint for 2–20 minutes causes an increase in semen volume, and concentration and number of sperm in bulls, with the sperm motility being unaffected. False mounts cause further increase in semen characteristics. The presence of another bull, changing the teaser, or using the bull as a teaser prior to collection has no such augmenting effect on semen characteristics despite a great increase in sexual excitement. Thus the stimuli which influence semen compositions differ from those which cause sexual excitement.

In the rabbit, copulation or the presence of a doe causes, within 30 minutes, a remarkable increase in blood testosterone (Haltmeyer and Eik-Nes, 1969). Sexual stimulation of bulls causes an immediate release of LH followed by a peak of testosterone, if the blood level of this hormone is low at the time of stimulation (Katongole et al., 1971) (Fig. 10–10).

V. ABNORMAL SEXUAL BEHAVIOR

Homosexuality, hypersexuality, hyposexuality and auto-erotic behavior are not uncommon. These syndromes may be due to genetic factors, disturbance in the endocrine or nervous systems or to faulty management. Unadapted sexual reactions are more frequent among domestic animals and under conditions of captivity in the zoo than in the wild. Homosexuality refers to sexual behavior among males particularly at puberty and

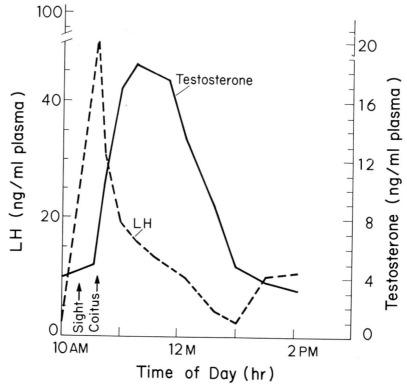

Fig. 10–10. The effect of coitus on concentrations of luteinizing hormone (LH) and testosterone in the peripheral blood of the bull Jambo, 7 January 1970. (*Katongole et al., 1971. J. Endocr., 50, 457.*)

when young males are housed together. The stimuli eliciting the male's sexual response are essentially visual. A releaser of an appropriate shape and size presented to a highly motivated male may elicit mounting. An immobile anestrous female, another male or an inanimate object may release sexual reactions. Most homosexual males in sex-segregated groups become heterosexual when placed with females and again homosexual when segregated.

Hypersexuality in males consists of increased sexual excitement, increased frequency of copulations and attempted copulations with young males and females of the same or different species. Hyposexuality is characterized by abnormalities in the ejaculatory pattern. Certain males may fail to ejaculate in spite of protrusion of erection whereas others cannot mount or exhibit no sexual desire for varying periods of time.

Auto-erotic behavior refers to the self-arousal of sexual responses which is called masturbation in males. The motor patterns vary with the species. The stallion rubs his rigid erected penis against the hypogastrium (anterior median of the abdomen) and lowers the loin region rapidly. This is followed by several forward movements of the pelvis, resulting in abortive ejaculation. The bull arches his back, performs pelvic movements and passes the penis in and out of the preputial orifice; such tactile stimulation causes ejaculation. Masturbation is less common in rams and is most commonly observed among bulls on high protein ration, e.g. bulls prepared for shows. As a result of such diets, the peripheral mucosa of the penis becomes more sensitive to tactile stimulation.

The most common abnormal female behavior is nymphomania in cattle.

Conclusions. It may be concluded that reproductive phenomena are influenced by social environment, sensory capacities and sexual stimuli. With a few exceptions most of our knowledge is based on research conducted on laboratory rodents. Extensive investigations are needed to understand the effect of hormonal, neural and sensory experiential factors on reproductive phenomena in farm animals.

VI. MATERNAL AND NEONATAL BEHAVIOR

Behavior concerned with mammalian birth and the establishment of the offspring as individuals commences with the prepartum period of any specialized activity such as selection of birth site; it covers the period of birth itself, the short critical period immediately after birth when the maternal-filial bond forms and when the young must learn to suck and adjust to the new environment; and it includes the much longer period of suckling, and the short period of weaning when the bond is disrupted (Table 10–4).

A. Maternal Behavior in Sheep

1. Prepartum Behavior

Within a few hours prior to lambing some ewes tend to cease grazing and wander about bleating as if searching for a lamb. As birth approaches, the frequency of urination and defecation increases markedly, probably due to the fetus pressing on the bladder and rectum. Some ewes remain with the flock for lambing, others become isolated and there is some evidence that sheep familiar with their environment seek a sheltered spot for lambing and so facilitate survival of the lamb.

Role of Fetal Fluids. Usually the birth site appears to be determined by where the placental membranes happen to rupture. Placental fluids are very attractive to ewes during the few hours prior to and after lambing, and once the membranes are ruptured, the ewe usually remains at the site, licking the ground and pawing so that hollows are sometimes scraped out. The placental fluids play a major role in attracting the ewe to her lamb, and ewes that have not yet lambed

Table 10-4.   Maternal and Neonatal Behavior in Farm Ungulates—
Semiquantitative Comparison of Species

	Sheep	Goats	Cattle	Horses	Swine
Prepartum behavior					
Restless	+		+	+	+
Seeks isolation	±	Probably the	±	?	+
Milk ejection	—	same as sheep	±	±	—
Builds nest	—		—	—	+
Time of most births	Variable	(?)	(?)	Night	Early hours of darkness
Duration of birth (approximate hours)	1	1 (?)	1 (?)	1 (?)	3 (from first to last piglet)
Delivery of placenta					
Delivery of young to delivery of placenta (approximate hours)	Several	Several	Several	1 (?)	Several
Eating of placenta by dam	±	± (?)	±	—	+
Time for dam to stand up after birth (if not already standing) (approximate minutes)	<1	<1 (?)	<1 (?)	10	3
Grooming					
Occurrence	+	+	+	+	—
Duration (approximate hours)	1	1 (?)	1 (?)	Several	Nil
Maternal solicitude for alien young	—	(?)	±	(?)	+
Abnormal maternal behavior					
Desertion of young	+		+	(?)	+
Moves from suckling	+	Probably the	+	+	NA
Butting young	+	same as sheep	+	+	NA
Cannibalism	—		—	—	+
Susceptibility of young to chilling	+	+	— (?)	— (?)	+
Progress of young					
Time to stand (approximate minutes)	15	Probably the	40	40	5
Time to first suckle (approximate hours)	1	same as sheep	2	2	0.3
Frequency of suckling (times per day approximately)					
First four days	30	(?)	4 (?)	(?)	10
Mid-lactation	15	(?)	3 (?)	(?)	25

Behavior usually present, +; behavior usually absent, —; behavior sometimes present, ±; behavior not documented, ?; not directly applicable, N.A.

Behavior, even with a group of homogeneous individuals of the one species, is extremely variable, so comparison between species must be made with great caution.

are attracted to the fluids and newborn of other ewes, leading to adoptions, sometimes referred to as "lamb stealing;" sometimes adoption ends with lambs being left without maternal care.

2. LAMBING

The time of lambing is not randomly distributed throughout the 24 hours of each day. The hours of peak births are affected by breed (George, 1969), and probably by weather, artificial illumination, disturbance or changes in nearby activity. Behavior of sheep giving birth depends largely on the speed and ease of the process, but generally the initial restlessness is broken by periods of lying with abdominal straining. Lambing occurs while the ewe is standing or lying but delivery is often completed by the act of standing. Most ewes are on their feet within a minute of birth, and the umbilical cord is broken by simple mechanical stretching. Parturition is completed by delivery of the afterbirth (fetal placenta) some two to five hours later; sometimes part is eaten by the ewe. Some lambs are born hind feet first, and these tend to have a high mortality during birth largely because of asphyxia.

The duration of the birth process varies widely (Fig. 10–11) even in a homogeneous flock. In some individuals delivery may be complete within a few minutes. Sometimes an extended birth process is associated with the lamb being too large for the birth canal—for example, when prenatal nutrition has been particularly good, when a sire of a large breed has been used, or when the lamb is a male with large horn buds as in the Dorset Horn breed. Protracted labor also results in exhaustion of the ewe with adverse effects on subsequent maternal behavior. Ewes lambing for the first time tend to have prolonged births and the process is usually protracted in ewes debilitated by disease or poor nutrition.

When a ewe gives birth to more than one lamb, the lambs are usually smaller than single lambs and the birth process is usually shorter. The interval between delivery of twins also varies widely from a few minutes to an hour or more, and the birth of the second lamb is accomplished with much less apparent effort than the birth of the first.

3. POSTPARTUM BEHAVIOR

Grooming and Maternal Imprinting. Ewes normally stand within seconds of giving birth, turn to face the newborn lamb, and commence vigorous licking (grooming) and eating of the amniotic and allantoic

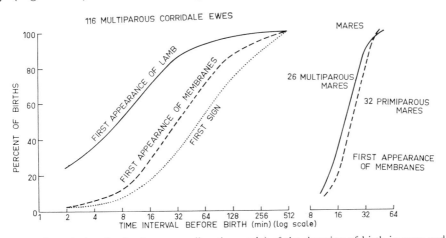

FIG. 10–11. Cumulative distribution curves (log time scale) of the duration of birth in ewes and mares measured from the first sign of impending birth, from the time of first appearance of the membranes or from the time the fetus first appeared at the vulva. Although the modal time for the interval from first appearance of the membranes to birth is similar in ewes and mares, the variability is much wider in ewes and includes some very long intervals. (*Drawn from data of Alexander [unpublished] and Rossdale [1967]. Brit. Vet. J. 123, 740.*)

membranes in which the lamb is sometimes enveloped. This brief period of intense olfactory and gustatory contact is sufficient for a ewe to learn to distinguish her lamb from alien lambs within very few hours of birth; aliens will be vigorously rejected by butting. It appears that during this time the intense initial attraction to the birth fluids is replaced by a more specific attraction to the lamb. This attraction appears to be largely olfactory, and this is confirmed by experiments with anosmic goats (Klopfer, 1971). However, maternal behavior is not abolished by abolishing the sense of smell; so other factors beside taste and smell must also be important in maintaining attraction of the mother to the newborn, and these may include warmth and movement since ewes rapidly lose interest in a stillborn or immobile chilled lamb.

In experiments where the young have been removed at various times after birth, five minutes of contact is sufficient to imprint the mother goat with the scent of her young (Klopfer, 1971), but if the kids are removed at birth, receptivity may be reduced an hour later and may be completely gone after six hours. Similar experiments have not been done with sheep, but the period in which receptivity in ewes is retained (termed the critical period) may exceed eight hours.

The coats of newborn lambs can contain about 0.5 kg of fluid at birth, and some of this is removed by the mother, so reducing evaporative cooling of the lamb, but licking also appears to play an important role in increasing neuroexcitability and facilitating rapid motor development of the newborn.

Finding the Teats and Imprinting of the Newborn. Most lambs stand within 15 minutes of birth and with increasing coordination make exploratory approaches to the ewe, moving from her head along her side, and nosing into body angles around the legs. Finding the teats is greatly facilitated by the initial experience of suckling. During the initial approaches, the ewe tends to move so as to maintain orientation with the lamb in front of the head, a position that facilitates grooming, but within an hour or so most ewes will allow the lamb to move toward the udder, and will continue grooming by turning the head (Plate 2; Fig. 10–12).

Both maternal orientation and grooming appear to facilitate the lamb finding the teats for the first time (Fig. 10–13) but they are not essential, because some lambs will find the teats in their absence, albeit more slowly. Maternal geometry appears to be important in cattle, but trial and error obviously play an important role. Lambs will follow a vertical surface and concentrate on corners. In the absence of the mother during the initial critical hours the lamb can be imprinted with a substitute maternal figure and will respond to it rather than to a sheep, even if the figure is not associated with feeding. During this initial phase the lamb also learns to follow the ewe as it moves from the birth site. For several

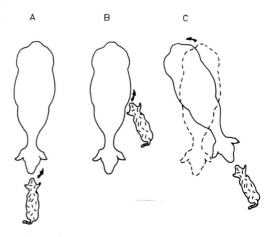

Fig. 10–12. Normal and abnormal maternal behavior in sheep. The well behaved mother orients with head toward the newborn (*A*) but stands quietly during the initial teat-seeking attempts (*B*). Young and inexperienced ewes tend to rotate away from the initial teat-seeking attempts and so maintain the initial relative orientation (*C*). This behavior may persist for some hours during which the teat-seeking activity of the lamb declines markedly. (*From Lynch and Alexander, 1973. In The Pastoral Industries of Australia. Alexander and Williams [eds.], Sydney, Sydney University Press.*)

days after birth the following responses can be elicited by large moving objects but within a few days this response to a foreign object is replaced by a fear response.

Aberrant Behavior. Ewes that are exhausted by a long and difficult parturition, or ewes that are debilitated by undernourishment or disease may remain prone for some hours after birth. Some ewes lambing for the first time display little or no interest in their newborn lambs; they may wander away without even smelling the lamb. Few return and become good mothers. Other forms of misbehavior common in ewes lambing for the first time include butting of the newborn lamb, as if in fear, when it attempts to stand or approach the ewe; also the tendency of ewes to move during the initial approaches of the lamb to the udder is sometimes prolonged. When these patterns persist for more than an hour or two, the lamb's chances of finding the teats and sucking are likely to be reduced since the teat seeking activity declines rapidly from about two hours after birth in the absence of the reward of milk (Fig. 10–14).

Lambs also show aberrant behavior patterns. Some lambs are excessively slow to stand and suck the teats or may even fail to stand at all. For example, activity is depressed by poor prenatal nutrition or by a long birth process or by birth injuries or by failure to maintain a normal body temperature.

Aberrant patterns are more common in twin births than in single births, because for several days after birth, most ewes appear unable to recognize that they have more than one lamb. Ewes distracted

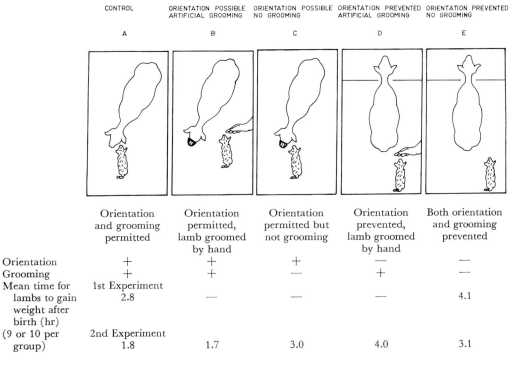

	CONTROL	ORIENTATION POSSIBLE ARTIFICIAL GROOMING	ORIENTATION POSSIBLE NO GROOMING	ORIENTATION PREVENTED ARTIFICIAL GROOMING	ORIENTATION PREVENTED NO GROOMING
	A	B	C	D	E
	Orientation and grooming permitted	Orientation permitted, lamb groomed by hand	Orientation permitted but not grooming	Orientation prevented, lamb groomed by hand	Both orientation and grooming prevented
Orientation	+	+	+	—	—
Grooming	+	+	—	+	—
Mean time for lambs to gain weight after birth (hr) (9 or 10 per group)	1st Experiment 2.8	—	—	—	4.1
	2nd Experiment 1.8	1.7	3.0	4.0	3.1

FIG. 10–13.   Experiment to show importance of orientation and grooming components of maternal behavior in sheep. Grooming was prevented by fitting a mask to the ewe (*B* and *C*); orientation was prevented by restraining the ewe in a bail (*D* and *E*); artificial grooming was supplied by rubbing the lamb with the hand (*B* and *D*). It is clear that prevention of maternal orientation or grooming, or both, hamper the progress of the lamb. (*From Lynch and Alexander, 1973. In The Pastoral Industries of Australia. Alexander and Williams [eds.], Sydney, Sydney University Press.*)

from the first-born by the second birth may drop the second lamb some feet from the first and thereafter give full attention to the second-born. The first may become lost or may be repelled as an alien if it remakes contact with its mother after the critical receptive period of the ewe has passed; imprinting with one lamb does not guarantee acceptance of others in the same litter. One of twins may also be lost when the ewe moves from the birth site; provided one lamb follows, most ewes appear satisfied.

Adoption and Fostering. Adoption, usually mothering of a wet newborn lamb by a receptive ewe that is not its mother, is largely restricted to experienced ewes. Adoptions result from onset of maternal behavior prior to parturition, from stillbirth or death of a lamb soon after birth, or from several ewes lambing at the same time in the same place; twins rather than single lambs tend to be involved. The outcome is variable; sometimes the adoption is permanent and successful, but in many instances the maternal bond

with the adopted lamb is weak or temporary with the ewe displaying interest in many lambs, and wandering further from the adopted lamb than normal, until finally it is permanently lost. Adoptions in sheep flocks could be of some consequence in genetic experiments where lambs are identified with ewes many hours or even days after birth.

Fostering lambs onto alien ewes presents few problems if attempted soon after birth when the ewe is still within the critical period, the lamb is still wet, and the teat-seeking activity of the lamb is still strong. Fostering is facilitated if lambs have been kept isolated and unstimulated from birth so that teat-seeking has not been suppressed. Even beyond the critical period, close confinement may result in successful fostering (Hersher et al., 1963) but the bond so formed may be readily disrupted.

Behavior and Thermoregulation. The temperature regulatory mechanisms of lambs are remarkably well-developed at birth (Alexander, 1970) but some lambs

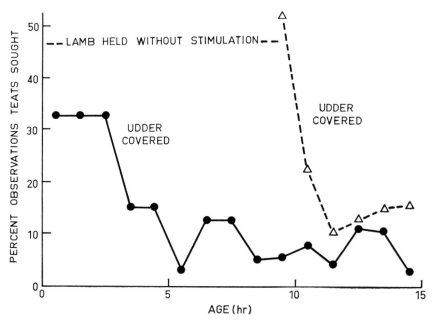

FIG. 10–14. Mean teat-seeking activity in lambs prevented from suckling by an udder cover. Ten lambs were left with their ewes from birth, and ten were first held in a dark box under conditions of minimal stimulation for nine hours after birth. The teat-seeking activity declines rapidly after birth but is preserved by depriving the lamb of stimulation. (*Adapted from Alexander and Williams* (1966). *Animal Behav.* 14, 166.)

will tend to chill at birth either because of an innate physiologic deficiency or through excessive heat loss. For example, many wet newborn lambs in a wind, would chill even at 10° C air temperature. Even when body temperature is normal, cold conditions suppress teat-seeking activity. These effects result in different rates of progress by lambs in different seasons of the year (Fig. 10–15).

4. The Suckling Phase

Ewes that have become separated from the flock return to it within a day or so of birth. The sheep is a classic example of a "follower" species, and for some days after birth the lamb and the ewe remain within earshot. Accidental separation results in considerable agitation of both ewe and lamb.

After the lamb has achieved satiety soon after birth the frequency of suckling stabilizes at about one or two bouts per hour, at least during daylight (Fig. 10–16). During the first week suckling bouts are usually initiated and terminated by the lamb. Twin lambs may be suckled individually, and their suckling bouts tend to be more frequent than those of single lambs, probably because less milk is available to each twin than to a single.

As the lamb becomes older, ewes and lambs stray further apart without agitation, and the frequency of suckling bouts declines, associated with an increasing tendency for the ewe to refuse to allow lambs to suckle ad libitum. Many of the bouts are initiated by the ewe approaching and calling the lamb. At this stage twins are not usually permitted to suckle unless both are present; so the ewe now recognizes that she has two lambs. The duration of suckling bouts also declines with advancing age.

By the time the lambs are three months old the frequency of suckling bouts has fallen to about one every three or four hours and occupy a few seconds. Under natural conditions these changes culminate in weaning, but weaning of domestic animals usually results from enforced separation of mother from young. Behavior during natural weaning has been little studied in domestic animals, though it is a distinct behavioral phenomenon in some wild species, maternal solicitude being suddenly replaced by antagonistic behavior (Lent, 1973).

During this suckling phase ewes become greatly distressed if the lamb dies, but there is no fierce maternal protective behavior in sheep; the simple presence of the ewes appears to act as a deterrent to predators. Maternal care consists pri-

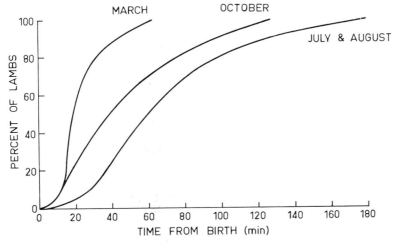

Fig. 10–15. Effect of seasonal climatic conditions in southwest Australia on the time between birth and first sucking in lambs. The weather is dry and mild in March, mild but wet in October and cold and windy in July and August. (*From McBride, et al., 1967. Proc. Ecol. Soc. Aust. 2, 133.*)

marily of the supply of nourishment and reservation of the milk supply for the ewe's own lamb.

Few behavioral abnormalities of this suckling phase have been recognized. A small minority of ewes will allow alien lambs to suck indiscriminately; perhaps because of disruption of the maternal bond soon after birth. Also, some lambs become adept at stealing milk from an alien ewe. During artificial rearing of lambs, certain lambs exhibit displaced sucking behavior and spend considerable time sucking the scrotum, the navel, or ears of other lambs. This leads to disruption of feeding by the sucked lamb, and sometimes to ingestion of feces and to scouring.

B. Maternal Behavior in Goats

Maternal and neonatal behavior in goats is similar to that of sheep (Blauvelt, 1954; Klopfer, 1971). However, goats appear less likely than sheep to reject one

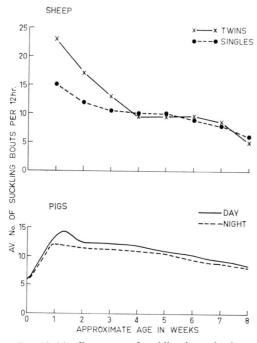

Fig. 10–16. Frequency of suckling bouts in sheep and swine. Peak frequency is reached later in swine than in sheep. (*From Ewbank*, [*1967*], *Animal Behav.* 15, 251 and data of Niwa et al., [1951], Bull. Nat. Inst. Agric. Sci. Tokyo [Ser. G] 1, 135.)

of twins. There is some debate about whether the goat is a "follower" or a "hider" species. Certainly goats that have escaped from domestication show a hiding phase of up to four days, at least in mountainous country (Rudge, 1970). This is associated with a clear preference for maternal isolation at the time of birth.

C. Maternal Behavior in Cattle

Studies with cattle have been largely restricted to animals in confined situations such as small paddocks on dairy farms or enclosed stalls (Selman et al., 1970a, b) so the expression of some aspects of behavior may be limited and some patterns may be artificial, but in general, cows show similar patterns of behavior to those described for ewes.

As early as the eighth month of the nine-month gestation period pregnant cows tend to avoid physical conflict with others of the herd during feeding and watering. A distinct restlessness and preference for isolation appear a week or two prior to calving, and the udder becomes sore and swollen. Within a day or two of calving, the restlessness becomes more pronounced, reactivity to noise is heightened, the tendency for isolation increases and the cow concentrates on a small secluded area which she defends against other cows.

1. Calving

Like the ewe, the cow is attracted to the spilled fetal fluids and often consumes soil and leaves contaminated with the fluids. During early labor the cow tends to stand up and lie down repeatedly and shows signs of slight straining. Longer periods of marked straining occur as the calf is forced down the birth canal. Most cows are lying for final delivery, and in some delivery is effected by the cow getting to her feet. Some cows calve without lying down.

2. Postpartum Behavior

Maternal Behavior. The cow usually stands immediately after birth but appears to take particular care to avoid stepping on the calf. Normally, vigorous grooming, accompanied by grunting calls, commences at once. The commencement of grooming can be much delayed by the cow remaining prone particularly when labor has been long. Cows that are slow to stand are usually stimulated to get up by movement and cries from the calf. The initial vigorous grooming persists for upward of an hour in normal experienced cows, but averages about 10 minutes in cows calving for the first time. The bovine placenta which is shed several hours after birth of the calf is often completely eaten.

Some cows back or sidle away from the teat seeking advances of the newborn calf and some inexperienced mothers show a fear response by butting or kicking when the newborn moves, but these phenomena are usually transient. More exaggerated forms of aberrant maternal behavior are also seen occasionally, including desertion sometimes associated with extremely violent attacks on the calf. On the other hand, behavior suggestive of maternal concern is often seen when a calf remains inert after birth; the calf is nudged with head or hoof as if to stir it into activity.

During this immediate postpartum period most cows will vigorously defend the area around the calf and will endeavor to remain between an intruder and the calf; cows are considerably more aggressive than sheep at this time. "Calf stealing" sometimes occurs in cattle.

Behavior of the Calf. The progress of newborn calves appears to be slower than that of lambs. Most calves have not successfully stood until about 45 minutes after birth, and many take upward of four hours to suck for the first time.

The search for the udder follows a similar pattern to that in sheep. The calf may be guided to the udder by its tendency to follow the line of the under belly of the cow towards its highest point (Selman et al., 1970*b*). In most species this is usually in the region of the udder; however, when the abdomen and udder are large and pendulous as in many dairy cows, the highest point is between the forelegs, and calves tend to be slow to suckle for the first time when the cow has this profile.

Maternal postural changes that appear to facilitate teat-seeking and suckling of calves include rotating the body, abducting the hind leg or moving forward bringing the udder closer to the calf. Cows tend to lick the perineal region of the calf during early suckling attempts and to lick and eat meconium.

The Suckling Phase. Cows suckle single calves about four times daily, and twins more frequently; the frequency tends to decline slightly with advancing lactation. Each suckling bout lasts about 10 minutes; all teats are sucked and suckling continues with frequent teat changes after milk let-down ceases. Suckling occurs most frequently at dawn, dusk and at the start of periods of grazing. Artificially reared calves, with milk available ad libitum, suckle much more frequently than naturally reared calves. Like lambs they may show nonnutritive sucking on other calves.

In contrast to sheep, unconfined cattle exhibit a short "lying out" phase during the first few days after birth at least in some environments. This "lying out" behavior can lead to exposure of calves to intense solar radiation and to the risk of heat stroke. However, most calves follow their mother within four or five days of birth and move to shade with the cow. On the other hand "lying out" behavior avoids the necessity for calves to travel long distances to water with their dams and perhaps run a greater risk of heat-stress through exercise. It is common to see a group of calves apparently being cared for by one or two cows, the other mothers being absent grazing or watering.

D. Maternal Behavior in Horses

Studies on horses have been confined to thoroughbreds in small paddocks or in loose boxes, and often in the close presence of man (Rossdale, 1968). Mammary development becomes noticeable within the last month of the $11\frac{1}{2}$ month mean gestation period, and milk ejection is sometimes seen days or even weeks before birth. The typical restlessness and unease, characteristic of the early stages of parturition, are also sometimes seen many days before birth.

1. FOALING

Parturition in mares has a clear diurnal incidence, with most births recorded between 7 PM and 7 AM, but it is not clear how much of the control is photoperiodic and how much is a result of routine husbandry procedures.

Early parturition is characterized by frequent changes of posture. Rupture of the membranes is followed by heavy straining with the mare usually recumbent and birth usually follows about 20 minutes later (Fig. 10–11). In contrast to the situation in sheep and cattle the duration of birth scarcely appears to be affected by whether or not the mare has foaled before. Delays in parturition have been attributed to the presence of spectator groups of mares as well as to human presence and interference.

Mares tend to remain recumbent for longer than sheep often not standing until 10 or more minutes after birth; the transfer of blood to the foal through the intact umbilical cord during this period may have significance for the well-being of the young animal. The placenta is usually delivered about an hour after birth and it is not usually eaten by the mare. Abnormally long parturition in horses is usually associated with malpresentation of the fetus, although there is the possibility of a behavioral problem, especially with old mares whose efforts to foal may not be sustained.

Postpartum Behavior. Upon standing, the mare immediately commences to lick the foal, still wet with amniotic fluid, and grooming usually continues for several hours. As in ruminants early maternal behavior in mares displays a range of variations, mostly transient; inexperienced mares, particularly, show aggression toward the foal and a tendency to rebuff early suckling attempts.

Most foals take upward of an hour to reach and maintain a stable standing position and most have suckled within two or three hours of birth. Premature parturition is much more common in horses than in ruminants, and the premature foal tends to be weak and slow to progress, though the sucking reflex is well-developed.

E. Maternal Behavior in Pigs

Most of the studies on pigs have been made with animals confined in small areas, often indoors (Jones, 1966). Behavior in the pig with its large litters of somewhat immature young contrasts markedly with the patterns in domestic ruminants and horses. Immaturity of the young pig takes the form of marked susceptibility to starvation (Mount, 1968), to chilling (Curtis, 1970) and, because of its small size, to injury by the sow.

1. PREPARTUM BEHAVIOR

The approach of parturition is indicated by characteristic nest-building activity, but in practice, sows are so restricted by modern farrowing pens, designed to reduce crushing of the piglets by the mother, that nest-building is virtually impossible and sows deprived of nesting material may be considerably disturbed. As parturition approaches, characteristic vocalizations become evident and the nest is defended as if piglets are present.

2. FARROWING

Parturition in the sow is more frequent during the early hours of darkness than at any other time of the day particularly in spring and summer. Though most domestic sows tolerate the presence of an observer at farrowing, some become highly disturbed if man is present. Sows normally farrow lying on one side, restlessness often being confined to the early stages of delivery. Delivery appears to be accomplished with very much less physical effort than required in other farm animals, perhaps because piglets are small in relation to the dimensions of the maternal pelvis.

The rupture of the membranes and the voiding of the fetal fluids is not well-defined as in other ungulates; a little blood-stained viscous fluid is passed intermittently. At the end of delivery the sow usually stands to urinate; the process of lying down again, both at this stage and during earlier bouts of restlessness, poses a major hazard to the newborn piglets which are prone to be injured if lain upon.

Most piglets are born partly covered with fetal membranes and, in contrast with the young of larger ungulates, the piglet must escape from these without maternal aid. Weak piglets may fail to escape and are suffocated. The umbilical cord is broken through the piglet moving away from the vulva.

The average duration of farrowing is about three hours but the duration can range from about half an hour to eight hours or more. The interval between birth of individual piglets ranges from less than a minute to three or more hours. Piglets in the last half of the litter to be born are prone to be stillborn, perhaps because of prolonged hypoxia (Fig. 10–17). About 70% of piglets are born with head foremost, and the 30% incidence of posterior births is much higher than in other farm ungulates. Piglet mortality tends to be greater with posterior than with anterior births.

3. Postpartum Behavior

Newborn piglets are almost immediately mobile. Usually they rapidly find their way to the region of the udder and may be suckling within five minutes. Most have suckled and obtained milk within half an hour of birth. At this very early stage milk appears available on demand, perhaps because the udder is full or perhaps because of continuous milk let-down due to circulating oxytocin associated with the birth process. Behavior of litters at this stage is very variable, and piglets may be attracted to infrared heaters before moving to the udder. The direction of the sow's hair patterns and warmth of the udder seem to be important in attracting newborn piglets to the udder but no critical experiments have been reported.

In some litters piglets tend to remain at the pectoral region of the udder for some hours (Fig. 10–18), competing with each other for the pectoral teats, though other functional teats are unoccupied. However, most litters soon space themselves out and piglets that do not find a functional teat soon after birth rapidly deplete their energy reserve in cold weather; they become hypothermic and die.

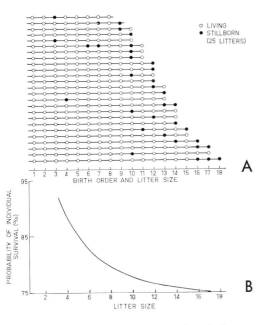

Fig. 10–17. Mortality in piglets in relation to birth order in 25 randomly selected litters (A) and in relation to litter size (B) (smoothed curve). Later born piglets are likely to be stillborn, and survival of live-born piglets during the first week after birth decreases with increasing litter size. (A drawn from unpublished data of P. English; B taken from Fraser, 1968, Reproductive Behavior in Ungulates, New York, Academic Press.)

4. The Suckling Phase

Suckling Behavior. Suckling bouts are initiated by the sow either spontaneously, often after feeding, or by the piglets squealing or attempting to suckle. There is a distinctive food call, a series of soft grunts, to which the litter rapidly becomes conditioned and which initiates suckling. The bouts can be stimulated by the piglets or sow being startled, and suckling by one sow may stimulate activity in neighboring litters so that all litters in the farrowing house may be suckling together. The position taken up by the sow for suckling tends to be characteristic of the individual. The standing posture tends to be more common in feral than in domestic pigs. Sows about to move use a "threatlike" warning call to which piglets respond by moving away and so reduce the risk of being crushed; sows that suckle lying down usually give the food call after they have lain down.

The food call initiates a period, lasting several minutes, of intense udder massage by the noses of the piglets; at the same time a recumbent sow rotates her body to expose the teats. The grunt frequency increases to a peak and the phase of active movement by the piglets is suddenly replaced by a quieter phase of milk let-down and suckling, which lasts no more than a minute or so. The frequency of suckling increases during the first week to about one per hour but thereafter begins to decline (Fig. 10–16).

Teat Preferences. The establishment of teat preferences usually occurs within a day or so of birth (McBride, 1963) but sometimes requires one or two weeks, and the established order is not completely inflexible. This variation may be due to differences in temperament of the sow or to husbandry conditions. The "teat-order" develops most rapidly in sows that remain lying on the one side for suckling during the first day or so. Changes in position unsettle the teat order and are accompanied by fighting between piglets a day or more old. Identification of teats appears to be based on sight, smell and recognition of neighboring piglets but this has not been examined experimentally. In small litters some piglets have the regular use of two adjacent teats, though one is usually preferred. There is a clear preference for anterior teats (Fig. 10–18). The apparent advantage of a greater milk supply of anterior teats appears to result from the preference for them rather than to be the cause of the preference.

Maternal Recognition. Sows appear to recognize their litters through smell, but

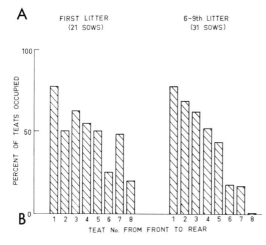

A   FIRST LITTER   6–9th LITTER
     (21 SOWS)   (31 SOWS)

PERCENT OF TEATS OCCUPIED

B   1 2 3 4 5 6 7 8   1 2 3 4 5 6 7 8
     TEAT No. FROM FRONT TO REAR

Fig. 10–18. Use of teats by piglets. *A,* Line drawing from photographs suggesting attraction of newborn piglets for anterior teats. (*Adapted from Fraser,* [1968]. *Reproductive Behavior in Ungulates. New York, Academic Press.*)

B, Occupation of teats during established lactation; there is a clear preference for anterior teats. (*Drawn from data of P. English, unpublished.*)

they are more tolerant of alien young than larger farm animals and will permit alien piglets to suckle, particularly during the first days of lactation. Piglets do not appear to discriminate between sows. Fostering is not difficult and litters can be reared together, with piglets suckling more than one sow. However, lactating sows are intolerant of intruders and may become markedly aggressive if piglets are disturbed or threatened.

Behavior and Thermoregulation. Newborn piglets with their small body size, sparse pelage and skin wet with fluids are prone to chill, despite a vigorous thermogenic response, in conditions as mild (for man) as 20 ° C air temperature and a 5 km per hour wind (Mount, 1968). Young piglets huddle closely together or against the sow to minimize heat loss; feral piglets appear to be more cold resistant than domestic piglets. The resistance of piglets to cold exposure markedly increases during the two or three days that they would remain in a nest. Suckling can be prevented or inhibited by chilling of piglets, and deaths from chilling are common in unheated farrowing houses.

5. Behavioral Abnormalities Contributing to Piglet Mortality

Sows occasionally kill and eat their piglets as they are born. Aggressive behavior in these sows is sometimes apparent several days before farrowing, but if piglets are removed as they are born and are returned to the sow after farrowing is complete they are usually cared for normally. Cannibalism also occurs occasionally in established lactation.

Starvation accounts for nearly half of the 10% mortality in live-born domestic pigs. Young piglets will suck anything that contacts their noses including the maternal vulva and other piglets, and excessive time may be spent in these activities while litter mates are occupying the teats. Sometimes there are more piglets in the litter than teats on the udder. The number of teats available is

effectively reduced in some sows, particularly in older animals, in which teats are hidden beneath the udder. Piglets that are displaced from their established position at the udder after the first day or two do not readily accept a vacant teat and may die. These deaths are concentrated among the smallest piglets in the litter (Fig. 10–17).

REFERENCES

Alexander, G. (1970). "Thermogenesis in Young Lambs." In *Digestion and Metabolism in the Ruminant.* A. T. Phillipson (ed.), Newcastle on Tyne, Oriel Press, pp. 199–210.

Banks, E. M. (1964). Some aspects of sexual behavior in domestic sheep, (vis aries.) 469–1 *Behav. Int. J. Comp. Ethol. 23*, Part 3–4, 249–279.

Blauvelt, H. (1954). "Dynamics of the Mother-Newborn Relationship in Goats." In *Group Processes; Transactions of the First Conference.* B. Schaffner (ed.), New York, Macy Foundation, pp. 221–258.

Bronson, F. H. and Whitten, W. K. (1968). Oestrus-accelerating pheromone of mice: Assay, androgen-dependency and presence in bladder urine. *J. Reprod. Fert. 15* (1), 131–134.

Curtis, S. E. (1970). Environmental-thermoregulatory interactions and neonatal piglet survival. *J. Anim. Sci. 31*, 576–587.

Dandekar, P., Vaidya, R. and Morris, J. (1972). The effects of coitus on transport of sperm in the rabbit. *Fertil. Steril. 23*, 759–762.

Domanski, E. (1970). Hypothalamic areas involved in the control of ovulation and lactation in the sheep. Final Report E.21.AH.13.FG.PO.202.

Fernandez-Baca, S., Madden, D. H. L. and Novoa, D. (1970). Effect of different mating stimuli on induction of ovulation in the Alpaca. *J. Reprod. Fert. 22*, 261–267.

George, J. M. (1969). Variation in the time of parturition of Merino and Dorset Horn ewes. *J. Agric. Sci. 73*, 295–299.

Haltmeyer, G. C. and Eik-Nes, K. B. (1969). Plasma levels of testosterone in male rabbits following copulation. *J. Reprod. Fert. 19*, 273–277.

Harlow, H. F., Joslyn, W. D., Senko, M. G. and Dopp, A. (1966). Behavioral aspects of reproduction in primates. *J. Anim. Sci.* Suppl. *25*, 49–67.

Hart, B. L. (1971). Facilitation by oestrogen of sexual reflexes in female cats. *Physiol. Behav. 7*, 675–678.

Hawk, H. W. (1972). Failure of vasectomized rams to improve sperm transport in inseminated ewes. *J. Anim. Sci. 35*, 69–72.

Hersher, L., Richmond, J. B. and Moore, A. U. (1963). Modifiability of the critical period for the development of maternal behavior in sheep and goats. *Behavior 20*, 311–320.

Hulet, C. V., Ercanbrack, S. K., Blackwell, R. L., Price, D. A. and Wilson, L. O. (1962). Mating behavior of the ram in the multi-sire pen. *J. Anim. Sci. 21*, 865–869.

Jones, J. E. T. (1966). Observations on parturition in the sow. Part II. The parturient and post parturient phases. *Brit. Vet. J. 122*, 471–788.

Karli, P. (1968). Systeme limbique et processus de motivation. *J. Physiol. 60*(1), 3–148.

Katongole, C. B., Naftolin, F. and Short, R. V. (1971). Relationship between blood levels of luteinizing hormone and testosterone in bulls and the effects of sexual stimulation. *J. Endocr. 50*, 457–466.

Klingel, H. (1967). Soziale Organization und Verhalten freilebender Steppenzebras. *Z. Tierpsychol. 34*, 580–624.

Klopfer, P. H. (1971). Mother love: What turns it on? *Amer. Scient. 59*, 404–407.

Kobayashi, T., Kato, J. and Minaguchi, H. (1964). Fluctuations in monoamineoxydase activity in the hypothalamus of the rat during oestrus cycle and after castration. *Endocr. Japon. 2*, 283–290.

Lent, P. (1973). "Mother-Infant Relationship in Ungulates." In *Behavior of Ungulates and Its Relationship to Management.* V. Geist and F. Walther (eds.), Morges, Switzerland, International Union for Conservation of Nature.

Leuthold, K. (1967). Variations in territorial behaviour of Uganda Kob (Adenota Kob Thomasi N. 1899). *Behaviour 27*, 215–258.

Lightfoot, R. J. (1970). The contractile activity of the genital tract of the ewe in response to oxytocin and mating. *J. Reprod. Fert. 21*, 376.

Lindsay, D. R. (1965). The importance of olfactory stimuli in the mating behaviour of the ram. *Animal Behav. 13*(1), 75–78.

Lindsay, D. R. and Robinson, T. J. (1961). Studies on the efficiency of mating in sheep. *J. Agric. Sci. 57*, 137–140, 141–145.

Lindstrom, L. H. and Meyerson, B. (1969). Sexual motivation in the female rat; II. The effect of oestrogen measured by a T-Maze choice method. (abstr.) *Acta Physiol. Scand.* Suppl. *440*, 65.

McBride, G. (1963). The "teat order" and communication in young pigs. *Animal Behav. 11*, 53–56.

MacLean, P. D. and Ploog, D. W. (1962). Cerebral representation of penile erection. *J. Neurophysiol. 25*(1), 29–55.

MacNeilly, A. S. and Folley, S. J. (1970). Blood levels of milk ejection activity (oxytocin) in the female goat during mating. *J. Endocr. 48*(1), IX-X.

MacNeilly, A. S. and Alison-Ducker, H. (1972). Blood levels of oxytocin in the female goat during coitus and in response to stimuli associated with mating. *J. Endocr. 54*, 399–406.

Melrose, D. R., Reed, H. C. B. and Patterson, R. L. S. (1971). Androgen steroids associated with boar odour as an aid to the detection of oestrus in pig artificial insemination. *Brit. Vet. J. 137*, 497.

Meyerson, B. and Nordstrom, E. B. (1969). Sexual motivation in the female rat. In: Effect of oestrogen and testosterone measured by an increasing barrier method. *Acta Physiol. Scand.* Suppl. *330*, 64.

Mount, L. E. (1968). *The Climatic Physiology of the Pig.* London, Arnold.

Pretorius, P. S. (1967). Libido and mating dexterity in rams reared and kept in isolation from ewes. *Proc. S. Afr. Soc. Anim. Prod.*, pp. 208–212.

Rosenblatt, J. S. and Aronson, L. R. (1958). The decline of sexual behaviour in male cats after castration with special reference to the role of prior sexual experience. *Behaviour 12*, 285–338.

Rossdale, P. D. (1968). Abnormal perinatal behaviour in the thoroughbred horse. *Brit. Vet. J. 24*, 540–551.

Rudge, M. R. (1970). Mother and kid behaviour in feral goats. (*Capra hircus L.*) *Z. Tierpsychol. 27*, 687–692.

Selman, I. E., McEwan, A. D. and Fisher, E. W. (1970*a*). Studies on natural suckling in cattle during the first eight hours post-partum. I. Behaviour studies (dams). *Animal Behav. 18*, 276–283.

Selman, I. E., McEwan, A. D. and Fisher, E. W. (1970*b*). Studies on natural suckling in cattle during the first eight hours post-partum. II. Behavioural studies (calves). *Animal Behav. 18*, 284–289.

Signoret, J. P. (1967*a*). Attraction de la femelle en oestrus par le male chez les porcins. *Rev. Comp. Anim. 4*, 10–22.

Signoret, J. P. (1967*b*). Durée du cycle oestrien et de l'oestrus chez la Truie; action du benzoate d'oestradiol chez la femelle ovariectomisée. *Ann. Biol. Anim. Bioch. Biophys. 7*, 1.

Signoret, J. P. (1970). Reproductive behaviour of pigs. *J. Reprod. Fert.*, Suppl. *11*, 105–117.

Soulairac, M. L. (1963). Étude éxperimentale des régulations hormono nerveuses du comportement sexuel du rat male. *Ann. Endocr. 24*(3) Suppl., 5–98.

Wierzbowski, S. (1966). The scheme of sexual behaviour in bulls, rams and stallions. *Wld. Rev. Anim. Prod. 2*, 66–74.

III. Reproductive Cycles

Chapter 11

Cattle

W. D. Foote

I. THE BULL REPRODUCTIVE SYSTEM—BIRTH TO PUBERTY

At birth the testicles in the calf have descended into the scrotum in most cases, but they may return through the partially open inguinal canal into the body cavity if manipulated or if the calf is placed on his back. Closing of the inguinal canal within the first few weeks prevents this. Spermatogonia and a small number of primary spermatocytes are present in the seminiferous tubules at 63 days of age. Spermatozoa first appear at 224 days (Phillips and Andrews, 1936). The cells of Leydig are formed by $3\frac{1}{2}$ months (Hooker, 1944) and androgen is being secreted before sperm are produced (Lindner and Mann, 1960). Testosterone production and release from the testicles generally increase from birth to near puberty. Several additional morphologic developments of the reproductive system take place between birth and puberty. These include development of complete patency of the genital tract and freedom of the penis from the sheath. Lack of completion of the latter by puberty sometimes results in a frenulumlike tissue attachment of the penis to the sheath which prevents mating. Age at puberty is influenced by both breed and environment.

Bulls usually reach puberty by 10 to 12 months of age. Most bulls have considerable sexual desire before reaching puberty. Nursing bull calves six months old or older attempt to mount cows, heifer calves or sometimes other bull calves. They have keen ability to sense estrus in females. Often a bull calf will follow a cow two or three days, starting at least one day before standing estrus and continuing that long after the end of heat.

MacMillan and Hafs (1968) reported an increase in plasma LH and hypothalamic LH releasing factor until age of puberty while pituitary FSH declined after six months of age. Most of the endocrine changes leading to puberty occur between 2 and 10 months of age.

II. BULL REPRODUCTION AFTER PUBERTY

A. Sperm Production and Release

Although testicular function is considered to be rather constant after puberty is reached, research suggests that intensity of sexual activity may follow a cyclic pattern. Although semen production is influenced by various factors, bulls maintain spermatogenesis under considerable nutritional deficiencies and other stresses. A mature bull can be expected to ejaculate 2–8 ml of semen per ejaculation with a concentration of 1.0×10^9 to 1.5×10^9 sperm per ml. The length of interval between collections affects semen output and characteristics.

Continued frequent ejaculations often result in reduced volume and concentration per ejaculate. Bulls maintained with groups of cows or heifers often breed one or more females daily for several weeks with high conception rates and no adverse effects on the bull. However, bulls studs in artificial insemination are usually collected from no more often than twice weekly with one or two ejaculations per collection. Sexual preparation by restraining the bull near an estrous cow or allowing him to make one or more false mounts before ejaculation increases the seminal volume and number of sperm per ejaculate for dairy bulls (Hale and Almquist, 1960; Signoret, 1961; Convey et al., 1971). Foster et al. (1970) observed that this type of sexual preparation did not cause this effect in Angus bulls. Although not all breeds of dairy or beef bulls have been tested for this effect, the general temperament differences between dairy and beef bulls may be correlated with epididymal and sex accessory gland response to this neural stimulation.

Genetic factors influence bull performance. Heritability estimates for traits such as ejaculate volume, sperm concentration, sperm motility and sperm viability range from .35 to .60.

B. Endocrinology

The reproductive hormones do not show abrupt changes in gland or serum levels after puberty. Serum gonadotropins do not appear to be affected by age (Convey et al., 1971). Testosterone production increases as the bull becomes older (Lindner and Mann, 1960; Rawlings et al., 1972). Bulls usually reach maximum reproductive capacity when three or four years old. Decreased libido and testicular atrophy are usually evident by 10 to 12 years of age. Some bulls are active and fertile when several years older than this. Most bulls used for semen collection remain in service until 7 to 10 years old. Testosterone has been detected in the testes of a Shorthorn bull $17\frac{1}{2}$ years old (Lindner and Mann, 1960).

III. THE HEIFER REPRODUCTIVE SYSTEM—BIRTH TO PUBERTY

The duct system is patent at birth. The pituitary gland is producing gonadotropins and in many cases the ovaries contain small antral follicles (Casida et al., 1935). Oocyte mitosis is complete so that the total number of gametes for the individual are present at birth. This has been estimated to be 600,000 or more.

Follicular development and degeneration occurs throughout prepuberal life. The degree of follicular development obtained before degeneration occurs becomes greater as the heifer approaches puberty. Follicular development to maturation and ovulation usually occurs by 10–12 months of age, although many heifers reach puberty younger and some older than this. Some heifers reach puberty before being weaned at eight months of age. Occasionally such heifers are bred and conceive to early maturing bull calves or herd bulls.

Inadequate rations for growing heifers result in both reduced body weight and increased age at puberty. Conversely, providing high levels of nutrition results in heifers reaching puberty at a younger age and heavier weight. Breed differences exist for this trait. The averages range from 319 days for the Jersey to 390 days for Hereford heifers (Laster et al., 1972). Crossbreeding often decreases age at puberty in addition to the effect of heterosis expressed through daily gain in body weight (Wiltbank et al., 1969).

Heifers show no sexual behavior until puberty unless stimulated by hormone treatments. The ovaries are responsive to gonadotropins and can be stimulated to develop numerous follicles, ovulate fertile ova and form corpora lutea at various prepuberal ages. Beef calves four months old have demonstrated estrus, superovulation and some ova cleavage following treatment with PMSG and LH and insemination. It is not expected that estrous cycles or successful pregnancy, the criteria for puberty, can be induced in calves at this age. However the age at puberty can be decreased several weeks by treating heifers with progesterone and PMSG if their nutrition is adequate.

IV. THE ESTROUS CYCLE

A. Endocrinology

Mature follicles ovulate when 16–18 mm in diameter and the corpora lutea reach a maximum diameter of 18–24 mm. During the estrous cycle the luteal phase lasts until about the 16th day when the corpus luteum begins regression. Often follicles develop to near ovulatory size during the luteal phase and regress making it impossible to determine by follicle size which one will ovulate at the next estrus. However, during the last three days of the follicular phase, the largest follicles complete maturation and ovulate (Dufour et al., 1972).

Blood levels of estrogen are relatively low during the luteal phase of the cycle but increase markedly one to three days before estrus and decrease rapidly following ovulation. Progesterone levels in the blood are low at estrus, increase starting about the second day after estrus with the development of the corpus luteum and decrease at the end of the luteal phase (Hendricks et al., 1971; Wettemann et al., 1972; Swanson et al., 1972; Hendricks et al., 1970). LH is released from the pituitary gland near the beginning of estrus and reaches the highest level in the blood within a few hours. This ovulatory peak precedes ovulation approximately 24 hours (Hendricks et al., 1970; Carr, 1972). FSH levels are less well-defined but appear not to follow as distinct a cyclic pattern as LH does.

The endocrine picture during the estrous cycle is similar between cows and heifers although gonadotropic activity may be somewhat higher in the cow (Carr, 1972). The estrous cycle can be lengthened by exogenous progesterone treatments or shortened by treatments with various materials to cause early corpus luteum regression. These include estrogen, oxytocin and prostaglandins. Prostaglandin $F_2\alpha$ ($PGF_2\alpha$) is a very effective luteolytic agent when given after the fifth day of the estrous cycle. Marked corpus luteum regression occurs within two days when $PGF_2\alpha$ is injected systemically. Uterine infusion of adequate levels into the uterine horn ipsilateral to the corpus luteum results in substantial luteal regression and a decrease in progesterone production within 24 hours. These treatments, therefore, can cause pregnant heifers to abort and return to estrus, and nonpregnant females to return to estrus after a shortened luteal phase of the estrous cycle.

Extirpation of the corpus luteum causes return to estrus and ovulation, usually within two to four days in Holstein heifers. However Hereford heifers initiate new cycles at longer and more variable intervals after such treatment. Also, corpus luteum removal causes an increase in subsequent twin ovulations in dairy heifers (Foote et al., 1959) while this effect has not been observed in beef females. Corpus luteum removal in conjunction with gonadotropin treatment to induce superovulation seems to cause a greater incidence of follicular luteinization than when gonadotropins are given with a waning corpus luteum present.

B. Estrous Cycle Characteristics and Breeding

The estrous cycle length, usually 20 to 21 days, varies both between and within animals. Heifers often have shorter cycles than do cows. Adverse environment such as poor nutrition or inclement weather may cause estrual hiatus with or without ovulation interruption. Continued ovarian function during behavioral estrus suppression suggests different thresholds for these two phenomena.

Over half of all ovulations are from the right ovary in the bovine. The reason for this is not clear. Conjectures include asymmetric blood distribution between the two ovaries and crowding in the left side of the body cavity by the rumen. Cows and heifers usually have estrous cycles throughout the year when not pregnant or in postpartum anestrus. Females in estrus attract other females as well as bulls and will stand to be mounted by either. Other signs of estrus are mucous discharge from the vagina, nervousness and attempting to mount other animals. Although these signs, especially the first one, are quite objective, inadequate detection of estrus is one of the greatest problems in artificial insemination programs. This is because animals have varying intensities of behavioral estrus. Estrus manifestation is often more pronounced in dairy than in beef females. Beef cows which are not in estrus are less aggressive in mounting estrous cows than diestrous or pregnant dairy cows are. Sometimes cows in beef breeding herds appear to come into estrus two or more at a time only, because a single estrous cow is ignored by other cows and therefore undetected, while two or more synchronized females will mount each other. Estrus usually lasts 12–14 hours but has considerable variation and tends to be shorter for heifers than for cows. Ovulation usually occurs 25–30 hours after onset of estrus. This corresponds to the time required for oocyte maturation preparatory for fertilization. The oocyte in the follicle which will ovulate next begins maturation division near the onset of estrus. The oocyte reaches metaphase II shortly before completion of follicle maturation and ovulation. Superovulation, by development of small follicles, can be induced by treatment of females with PMSG and FSH followed by HCG or LH injection. Oocyte meiosis within these stimulated follicles follows the same time schedule as in untreated animals and oocytes in small follicles that do not mature do not mature whether hormone treatment is given or not (Hafez and Ishibashi, 1964).

Some females show a slight bloody discharge from the vagina about the third day after estrus. The exact source of this discharge is not known, although it is believed to be related to ovulation. The optimum time for artificial insemination is near the end of estrus.

Fertility is affected by various factors. It is not possible to distinguish between failure to conceive and very early embryo loss except under experimental conditions. Cows returning to estrus after longer than normal estrous cycles indicate that conception may have occurred but that the embryo did not survive beyond the early stages. Females on a low level of nutrition or improper quality ration during or just preceding the breeding season require more services per conception for successful pregnancy. This, together with irregular estrous cycles, results in later successful conception dates. Long periods of extremely cold or stormy weather may have the same effect. Hot climates also decrease fertility by decreasing the chance of conception at each service and by delaying onset of estrus. Although most causes of low fertility are believed to be mediated through suboptimal endocrine functions, heat stress may also have the direct effect of high temperature on the gametes or young embryo.

Although heritability estimates for reproductive traits are low, cows which require several services to conceive often produce heifers with the same problem. Conception rates of 60–75% at first service can be expected by either natural mating or artificial insemination under most conditions. Intensive selection by eliminating all "slow breeders" can increase this percentage substantially. One purebred Angus herd achieved over 90% pregnancy with a 45-day breeding period.

C. Anomalies

Several anatomic and physiologic characteristics develop through natural processes which limit or prevent reproduction by an individual. They are referred to as abnormal because of their interference with desired fertility. Occasionally, follicles fail to ovulate and continue to develop to 8 cm or more in diameter. Cows with this condition of cystic follicles often are in constant or intermittent estrus but are sterile until a fertile ovum is released. In some cases, normal ovarian function is resumed with no treatment, but usually with a delay in breeding dates and an increase in calving interval. Cystic follicles can be ruptured by pressure with the fingers per rectum or sometimes by injections of LH or HCG. These treatments do not correct the endogenous cause of the problem and new cysts soon develop. Ovarian cysts are a main cause of nymphomania. However, the development of malelike morphologic and behavioral characteristics associated with this disease indicate endocrine involvement other than increased release of follicular estrogen. The incidence of cystic follicles is greatest in high-producing dairy cows.

Some cows and heifers with no detectable infection or morphologic deterrent to fertility repeatedly return to estrus after natural or artificial inseminations with fertile semen and are referred to as repeat breeders. Some conceive after several services and some never do. Studies to determine the cause of this problem have been largely unsuccesssful and so no satisfactory treatment is known. Probably many different causes exist which involve both failure to conceive and early pregnancy loss. Some repeat breeders maintain normal estrous cycles and others have irregular or lengthened intervals. Many repeat breeder animals are kept in a breeding herd because of the chance that conception may occur at a future service. Barrett et al. (1948) found that 54% of apparently normal dairy cows conceived at first service. Based on their work, the predicted chances of a cow conceiving decreases about 4.5% for each previously unsuccessful service.

Low reproductive efficiency due to delayed conception or early pregnancy loss probably causes greater economic loss in breeding herds than the more obvious conditions of complete sterility which are readily identified and eliminated.

V. GESTATION AND PARTURITION

Most breeds have gestation lengths within 278 to 293 days (average: 283). In addition to genetic influences evidenced by breed differences, specific genes for prolonged gestation exist in some breeds. The genotype of the fetus is as important as that of the dam in determining the duration of pregnancy. Fetal influence upon gestation length is probably manifest through body size and placental endocrine function. Bull calves tend to be carried about one day longer than heifers.

Most prenatal deaths occur within the first 34 days of gestation (Hawk et al., 1955). During this time, an endocrine transition from the estrous cycle to pregnancy occurs and placentation develops to the point of initial attachment to the uterine caruncles. Prenatal death may result from many causes. In addition to animals afflicted with disease, injury or improper nutrition, animals in good health lose pregnancies because of genetic or other developmental deficiencies in embryonic or maternal function. Undoubtedly, many embryos which are lost are incapable of developing into normal offspring. This is evident in increased prenatal loss in inbreeding.

Some evidence suggests that Holstein calves carried in the right uterine horn are slightly heavier at birth than those from the left side. This difference was not found in beef cows (Foote et al., 1960).

Occasionally, cows come into estrus while pregnant, especially during late pregnancy. This may result from a higher than normal estrogen level, establishment of an estrogen-progesterone ratio similar to that inducing nonpregnant estrus or a lower neural threshold for behavioral estrus. Follicular development can usually be detected throughout pregnancy and the placenta produces a relatively high level of estrogen during the later stages.

Pregnancy can be maintained without the corpus luteum or exogenous progesterone during the last trimester, although the incidence of retained placenta tends to increase (McDonald et al., 1952). The placenta contains both estrogen and progesterone and is believed to be the extra ovarian source of these steroids during pregnancy. LH-like activity has also been detected in placental cotyledons. Blood levels of LH and prolactin appear to be quite constant during pregnancy (Oxender et al., 1972).

Mature cows seldom require help at calving. Dystocia occurs more frequently in heifers, especially when bred young or to bulls which tend to sire large calves. Breeding only heifers 600 pounds or heavier minimizes this problem. In addition to size and age of the dam, the frequency and extent of dystocia is also influenced by breed and the sex of the calf. Bull calves are usually larger and require help at birth more often than heifer calves do.

Methods are being developed to induce parturition early at a controlled convenient time using various agents, including methasone and prostaglandin.

VI. POSTPARTUM INTERVAL

Cattle have an indefinite postpartum anestrus. The reproductive system undergoes great changes from the status of pregnancy termination to a condition capable of initiating and maintaining a new pregnancy. While the uterus is involuting, the ovaries are changing from a relatively quiescent state to a cyclic endocrine function. Although the corpora lutea of pregnancy are detectable at parturition, they contain little or no progesterone. Follicular development is usually limited at time of parturition and increases until ovulation occurs and estrous cycles are resumed. The first postpartum ovulation is not always accompanied by estrus, although the incidence of "quiet ovulations" decreases as estrus detection is improved.

Postpartum intervals to first ovulation, first estrus and uterine involution are approximately 50, 60 and 45 days, respec-

tively, for beef cows. The first two intervals are usually shorter for dairy cows. Cattle are usually bred to calve at 12-month intervals. Conception, therefore, ideally takes place between two and three months after calving. The interval to conception is about 90 days when cows nursing calves are bred at each estrus after calving (Saiduddin et al., 1968). The intervals to estrus and ovulation are increased in cows nursing calves. This may be due to both the increased energy requirement of lactation and a neural stimulus of the calf sucking its dam. Cows with calves removed and milked twice daily have intervals to ovulation and estrus up to three weeks shorter than similar cows nursing calves. These intervals are further reduced by mastectomy. The suckling stimulus appears to delay ovarian function by limiting pituitary gonadotropin release.

The level of nutrition also affects postpartum reproductive activities. Nutrient requirements are elevated substantially during the last trimester of pregnancy and during lactation. Reproductive functions have secondary priority in the individual and ovarian follicular development and ovulation are delayed in cows on low energy rations either near the end of pregnancy or during the postpartum interval (Dunn et al., 1969; Oxenreider and Wagner, 1971). Inadequate nutrition presents a serious fertility problem in the breeding herd because of a necessity of a minimal interval from parturition to conception to maintain the desired calving interval.

Another cause of delayed fertility after calving in some cows is dystocia. Calving difficulty increases the interval to estrus and decreases fertility at ensuing breedings. This effect of stress may be due to trauma, physical damage or resulting infection.

The postpartum anestrus may be due to inadequate pituitary gonadotropin release and low ovarian sensitivity. The pituitary LH activity is low at calving and increases throughout anestrus until ovulation occurs and a cyclic pattern is

Table 11-1.   Average Postpartum Intervals, in
Days, of Beef and Dairy Cows
(20 Cows Per Group)

Postpartum Interval	Hereford	Holstein
Uterine involution	46	44
First ovulation	49	26
First estrus	60	33
First to second ovulation	23	20

Data from University of Nevada Experiment Station.

reestablished. FSH activity in the pituitary gland appears to be inverse to this and to ovarian follicular growth, suggesting that this hormone is being released, causing increased follicular development until LH is released and estrus and ovulation occur, ending anestrus. An interrelationship exists between the functions of the various organs of the reproductive system in overcoming anestrus as well as controlling the estrous cycle. Removal of the ovaries from beef cows the day of calving decreases the LH activity of the pituitary gland three weeks later, as compared with the pituitary glands in cows with ovaries intact (Foote, 1971). The anestrous interval can be shortened to some extent by treating cows with ovarian hormones or gonadotropins. Dairy cows appear to be more responsive than beef cows to hormone treatment during postpartum anestrus. Treatments with estradiol or HCG 12 days after calving caused ovulation in a high percentage of Holstein cows but had less effect in Hereford cows (Foote, 1971). Normal estrous cycles usually follow these induced ovulations.

Differences between beef and dairy cows in endocrine function may be due to genetic factors, milk production level, method and frequency of milking and nutrition.

REFERENCES

Barrett, J. W., Lloyd, C. A. and Carpenter, R. A. (1948). Order number of insemination and conception rate. *J. Dairy Sci., 31*, 683.

Carr, W. R. (1972). Radioimmunoassay of luteinizing hormone in the blood of Zebu cattle. *J. Reprod. Fert. 29*, 11–18.

Casida, L. E., Chapman, A. B. and Rupel, I. W. (1935). Ovarian development in calves. *J. Agric. Res. 50*, 953–960.

Convey, E. M., Bretschneider, E., Hafs, H. D. and Oxender, W. D. (1971). Serum levels of LH, prolactin and growth hormone after ejaculation in bulls. *Biol. Reprod. 5*, 20–24.

Dufour, J., Whitmore, H. C., Ginther, O. J. and Casida, L. E. (1972). Identification of the ovulating follicle by its size on different days of the estrous cycle in heifers. *J. Anim. Sci. 34*, 85–87.

Dunn, T. G., Ingalls, J. E., Zimmerman, D. R. and Wiltbank, J. N. (1969). Reproductive performance of 2-year-old Hereford and Angus heifers as influenced by pre- and post-calving energy intake. *J. Anim. Sci. 29*, 719–726.

Foote, W. D. (1971). Endocrine changes in the bovine during the postpartum period. *J. Anim. Sci. 32*, (Suppl. I), 73–77.

Foote, W. D., Hauser, E. R. and Casida, L. E. (1960). Effect of uterine horn pregnant, parity of dam and sex of calf on birth weight and gestation length in Angus and Shorthorn cows. *J. Anim. Sci. 19*, 470–473.

Foote, W. D., Zimbelman, R. G., Loy, R. G. and Casida, L. E. (1959). Endocrine activity of corpora lutea from first-service and repeat-breeder dairy heifers. *J. Dairy Sci. 42*, 1944–1948.

Foster, J., Almquist, J. O. and Martig, R. C. (1970). Reproductive capacity of beef bulls. IV. Changes in sexual behavior and semen characteristics among successive ejaculations. *J. Anim. Sci. 30*, 245–252.

Hafez, E. S. E. and Ishibashi, I. (1964). Maturation division in bovine oocytes following gonadotrophin injections. *Cytogenetics 3*, 167–183.

Hale, E. B. and Almquist, J. O. (1960). Relation of sexual behavior to germ cell output in farm animals. *J. Dairy Sci. 43* (Suppl.), 145–169.

Hawk, H. W., Wiltbank, J. N., Kidder, H. E. and Casida, L. E. (1955). Embryonic mortality between 16 and 34 days post-breeding in cows of low fertility. *J. Dairy Sci. 38*, 673–676.

Hendricks, D. M., Dickey, J. F. and Niswender, G. D. (1970). Serum luteinizing hormone and plasma progesterone levels during the estrous cycle and early pregnancy in cows. *Biol. Reprod. 2*, 346–351.

Hendricks, D. M., Dickey, J. F. and Hill, J. R. (1971). Plasma estrogen and progesterone levels in cows prior to and during estrus. *Endocrinology 89*, 1350–1355.

Hooker, C. W. (1944). The postnatal history and function of the interstitial cells of the testis of the bull. *Am. J. Anat. 74*, 1–37.

Laster, D. B., Glimp, H. A. and Gregory, K. E. (1972). Age and weight at puberty and conception in different breeds and breed crosses of beef heifers. *J. Anim. Sci. 34*, 1031–1036.

Lindner, H. R. and Mann, T. (1960). Relationship between the content of androgenic steroids in the testes and the secretory activity of the seminal vesicles in the bull. *J. Endocr. 21*, 341–360.

McDonald, L. E., Nichols, R. E. and McNutt, S. H. (1952). Studies on corpus luteum ablation and progesterone replacement therapy during pregnancy in the cow. *Am. J. Vet. Res. 13*, 446–451.

MacMillan, K. L. and Hafs, H. D. (1968). Pituitary and hypothalamic endocrine changes associated with reproductive development of Holstein bulls. *J. Anim. Sci. 27*, 1614–1620.

Oxender, W. D., Hafs, H. D. and Edgerton, L. A. (1972). Serum growth hormone, LH and prolactin in the pregnant cow. *J. Anim. Sci. 35*, 51–55.

Oxenreider, S. L. and Wagner, W. C. (1971). Effect of lactation and energy intake on postpartum ovarian activity in the cow. *J. Anim. Sci. 33*, 1026–1031.

Phillips, R. W. and Andrews, F. N. (1936). The development of the testes and scrotum of the ram, bull and boar. *Miss. Agr. Exp. Sta. Bull. 331*.

Rawlings, N. C., Hafs, H. D. and Swanson, L. V. (1972). Testicular and blood plasma androgens in Holstein bulls from birth through puberty. *J. Anim. Sci. 34*, 435–440.

Saiduddin, S., Quevedo, M. M. and Foote, W. D. (1968). Response of beef cows to exogenous progesterone and estradiol at various stages postpartum. *J. Anim. Sci. 27*, 1015–1020.

Signoret, J. P. (1961). A study of the influence of various elements of sexual behavior in the bull on the characteristics of semen. *Proc. 4th Internat. Congr. Anim. Reprod.* The Hague 2, 166–170.

Swanson, L. V., Hafs, H. D. and Morrow, D. A. (1972). Ovarian characteristics and serum LH, prolactin progesterons and glucocorticoids from first estrus to breeding size in Holstein heifers. *J. Anim. Sci. 34*, 284–293.

Wettemann, R. P., Hafs, H. D., Edgerton, L. A. and Swanson, L. V. (1972). Estradiol and progesterone in blood serum during the bovine estrous cycle. *J. Anim. Sci. 34*, 1020–1024.

Wiltbank, J. N., Casson, C. W. and Ingalls, J. E. (1969). Puberty in crossbred and straightbred beef heifers on two levels of feed. *J. Anim. Sci. 29*, 602–605.

Chapter 12

Sheep

C. E. Terrill

Domestic sheep (*Ovis aries*) belong to the family Bovidae, the hollow-horned ruminants. They are even-toed, hoofed mammals of the order Artiodactyla.

I. SEXUAL MATURITY

A. Male

Differentiation of sexual organs in sheep commences about the 35th day following conception and the scrotum is apparent in the 50- to 60-day fetus. The testes of the ram lamb are generally descended at birth. However, full development of the reproductive organs is not reached until puberty at an age of about 100–150 days or longer.

At about 8 to 10 weeks of age at body weights of 16–20 kg, there is a marked increase in testis weight and volume (two to three times to 10–12 gm). This coincides with the appearance of primary spermatocytes and an enlargement (two times) in the average diameter of seminiferous tubules. By 15–24 weeks spermatids appear at testicular weights of 30 gm or more. Live sperm may be ejaculated with resulting fertility as early as 112 to 185 days with testicular weights of 65 gm or more as compared with an adult testicle weight of 200 gm. The weight of the epididymis is more closely related to testis weight than to age or body weight.

The penis remains infantile with preputial adhesions and can be protruded only slightly until just before puberty. The breakdown of the preputial adhesions to the mature form appears to be dependent on the testis hormone and may be used as an indication of onset of sexual maturity.

Sexual maturity of ram lambs seems to be more closely related to body weight than to age, occurring when the body weight is 40–60% of mature weight. Sexual maturity varies with breed; it is earlier for fast-growing breeds (e.g., Hampshire and Suffolk) and later in slower growing breeds (e.g., Merino). The highly prolific Finnsheep reaches sexual maturity early—at or before six months in both sexes. Crossbred lambs generally mature earlier than the average of their purebred parents. Sexual maturity may be delayed even beyond one year of age on a low plane of nutrition or under unfavorable climatic conditions.

From 10 to 50% of rams at six to seven months of age can be expected to be highly fertile and the proportion might be higher under favorable feed and climatic conditions. Sperm production of lambs is generally less than for mature rams and abnormal sperm, particularly immature types, may be more frequent. While ram lambs may be used successfully in breeding, they should be used for a limited mating season. Rams are more commonly mated first at 18 to 20 months of age and fertility at this age is generally equal to that of mature rams, with 70–95% of ewes lambing from a 35-day breeding period.

B. Female

Age and weight at puberty in ewe lambs are similar to those for ram lambs, although the first estrus may be expected slightly later in life than first sperm production. In general, the first estrus would occur at 4 to 10 months of age with live weight of 40–60% of mature weight. However, many ewe lambs do not have even one estrus until in their second year (Hafez, 1952). Season is an important factor in age of sexual maturity. Ewe lambs which do not show estrus during the first breeding season (fall of the year) will probably not do so until the same time of year after yearling age. With slower maturing breeds this may not occur until after two years of age for at least part of the ewes. A low plane of nutrition may delay sexual development of ewe lambs and the age of onset of first estrus and development of the reproductive organs is earlier in faster growing, heavier lambs. Puberty was reached in 158 ewe lambs of five breeds at an average age of 212 days (163–241) and an average weight of 45.5 kg (31–62) (Southam et al., 1971). Ovulation without heat generally precedes the first estrus by the length of one estrous cycle. Ovulation may be induced artificially as early as eight weeks of age.

Under favorable conditions about 50–95% of ewe lambs exposed will become pregnant. Ewes which reach puberty as lambs will possess a higher production potential than other ewe lambs and if bred first as lambs will generally produce more lambs during their lifetime than ewes bred first as yearlings. The practice of breeding ewes first as lambs is increasing but many are still bred first at the breeding season following one year of age.

II. BREEDING SEASON

A. Male

The male of domestic sheep does not show a restricted breeding season so common in the female, but seasonal variations in semen production and characteristics are evident. Sexual activity of the ram tends to be highest in the fall and often declines in later winter, spring and summer. Various aspects of semen quality, such as total sperm, motility, proportion of live and normal sperm and metabolic activity of the sperm, are generally highest in the fall. These often decline in the spring and summer. These vary not only with temperature but also with day length as semen quality tends to decline as the days become longer and to improve as they become shorter. High ambient temperatures above 27° C tend to depress semen quality and may result in summer sterility particularly in the black-faced British breeds. Continuous heat may be more detrimental than intermittent high temperatures. Artificial cooling of the rams will correct or prevent summer sterility and shearing of rams before hot weather is often helpful (Lodge and Salisbury, 1970).

The testes of rams are normally kept at temperatures below those of the body and thus testicular temperatures tend to be intermediate between scrotal and rectal temperatures. Under cold conditions the testicles are held close to the body. As air and body temperatures rise, the tunica dartos muscle relaxes and the scrotum elongates. Sweating of the scrotal skin also aids in lowering scrotal and testicular temperatures. Raising of testicular temperatures to about 36° C or higher results in degeneration of the germinal epithelium. Sperm in the seminiferous tubules and in the epididymis are also damaged. Abnormal sperm are evident within two to three weeks and the damage is related to the duration of the high temperatures. Raising of testicular temperatures may occur from high ambient or body temperatures, from insulation of the scrotum, or from natural or artificial cryptorchidism.

B. Female

The ewe of wild or primitive sheep normally breeds during a restricted season of the year so that the young tend to be born at the most favorable time, the late spring. Breeding seasons of domestic sheep vary from only a few estrous periods per year to year-round breeding in some breeds. There is also considerable variation among individuals and among years in the date of onset and the duration of the breeding seasons. Seasonal fluctuation in day length is an important factor affecting the length of the breeding season in different parts of the world (Fig. 12–1). The breeding season usually commences as the days become shorter, although sometimes earlier. Furthermore, the breeding season may be shorter or begin later at latitudes close to the poles. In the tropics and subtropics the ewes may tend to show estrus throughout the year, i.e., there is no restricted breeding season. Romney and Hampshire ewes at the equator lambed each month of the year although peaks occurred in April and October (Anderson, 1965). The breeding season is also affected by ambient temperature. Year-to-year variations are not easily explained by climatic differences. The presence of the ram may hasten the onset of first estrus near the beginning of the breeding season or may prolong the season. Mountain breeds tend to have a shorter breeding season than the Dorset Horn, Merino and Rambouillet.

III. ESTRUS AND OVULATION

A. Detection of Estrus

Sexual behavior of ewes is relatively inconspicuous and is not evident in the absence of the ram. The ewe in estrus

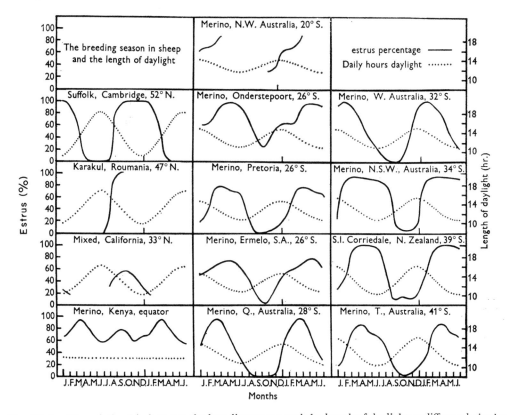

FIG. 12–1. The relationship between the breeding season and the length of daylight at different latitudes. (Breed, locality and latitude are shown in that order in each graph.) (*From Hafez, 1952. J. Agric. Sci. 42, 189.*)

may seek out the ram, but permitting the ram to mount and serve is the real evidence of estrus. Vasectomized or aproned rams must be used to detect ewes in heat. The use of marking harness or marking material on the ram's brisket is subject to some error unless it is supplemented with direct observation.

Coincident with estrus is the occasional enlargement of the vulva, high epithelium of the vagina and the liquefaction and flow of mucus from the cervix. These are often unnoticed externally. Cornification and sloughing of the vaginal epithelium occurs following estrus. The vaginal smear is much less diagnostic of the stages of the estrual cycle than in many other species, although it is of considerable help to the trained observer. Arborization of the cervical mucus is of help in detecting estrogenic effects.

B. Duration of Estrus

The duration of estrus varies from a few hours up to three to four days or more with an average of 24–48 hours. Estrus is generally shortest for ewe lambs with yearlings being intermediate. It is often shorter near the beginning and end of the breeding season. Estrus may be shorter when rams are with the ewes continually rather than at intervals. Breed differences in duration of estrus are not clearcut, although wool breeds may tend to have longer estrous periods than meat breeds.

C. Ovulation

Ovulation normally occurs near the end of estrus, although it may take place from several hours before to several hours after the end of estrus. Ovulation occurs before the end of estrus in a majority of cases and is probably more closely related to the end than the onset of estrus. Ovulation has been observed over a range of 12–41 hours after the onset of estrus. Mean ovulation times for various breeds vary from 21 to 33 hours. Twin ovulations may be separated for more than seven hours with a mean difference of almost two hours. The time of ovulation is independent of copulation, although it does tend to vary with other factors affecting the duration of estrus. Ovulations follow no regular pattern between both ovaries, although they occur more often in the right ovary.

D. Length of Estrous Cycle

The normal estrous cycle in the ewe ranges from 14 to 19 days with a mean of 17 days. Multiple cycles resulting from ovulation without estrus occur occasionally and are more common in some mountain breeds. Other causes of abnormal cycle lengths include failure of ovulation and luteinization, early regression of the corpus luteum and early prenatal death of the embryo. Abnormal cycle lengths are more common early and late in the breeding season. The length of the estrous cycle tends to be shortest at the peak of the breeding season and increases toward the end. It is also longer on a low as compared to a high plane of nutrition but breed differences are not clear-cut. Meat breeds may have slightly shorter cycles than wool breeds.

The occurrence of estrus and ovulation during lactation is quite variable and often infrequent. Ewes may show postpartum estrus from 15 to 27 hours after parturition especially when lambing during the normal breeding season. This estrus may last 12–50 hours but is unaccompanied by follicular development or ovulation.

Lactation anestrus may vary from a few days up to 10 months although it generally varies from about 4 to 10 weeks. A high proportion of ewes will rebreed successfully within two to three months after lambing (Hunter, 1968). Ewes which lamb early and those from breeds with long breeding seasons may have short postpartum anestrous periods. Nonsuckling ewes return to estrus earlier and are more likely to conceive at the first estrus.

E. Synchronization of Estrus and Induced Ovulation

Progesterone or a derivative can be given to ewes for about two weeks by injection, vaginal pessary, subcutaneous implant or by feeding to inhibit estrus. At the cessation of treatment the return to estrus will be concentrated at about 48 hours. To insure follicle development and ovulation, stimulating hormones are given when the progesterone treatment such as pregnant mare serum or FSH then HCG is stopped (Dziuk, 1972). Control of the regression of the corpus luteum during the breeding season by a single injection of prostaglandin, possibly followed by a single injection of estrogen to control the time of release of LH, has been suggested by Inskeep (1972). Fertility after treatment is still variable and timing as well as kinds of treatment may require further refinement and adjustment to the situations of recipient ewes in order to obtain consistently high success. Success of synchronization varies with season being high (82.5% pregnancies) in the fall and low (13.6% pregnancies) in the spring (Robinson, 1971). In general some reduction in fertility often results from synchronization of estrus.

IV. MATING AND CONCEPTION

A. Sperm Production

The production of sperm in the ram requires a total of about 50–53 days. Ortavant et al. (1969) have divided the process into the seminiferous epithelial cycle of 10 days, meiotic prophase—15 days, spermiogenesis—14 days followed by about 11 to 14 days in the ductus epididymis. Ortavant calculated the average daily production of sperm of Ile-de-France rams to be 5.5 billion or 12.2 million per gm of testicular tissue. Sperm production may improve with age up to three or four years but little difference has been found in fertility of rams of different ages.

Frequency of ejaculation, motility of sperm, proportion of live sperm and absence of abnormalities, particularly those of the sperm head, broken necks, or tailless heads have generally shown highest predictive value for subsequent ram fertility. High quality ram semen should have at least 1 to 2 billion sperm per cc, with rapid swirling motility, a pH of 7.0 or more acid and with less than 25% abnormal sperm or less than 5% abnormal heads. Semen from electro-ejaculation will usually have lower sperm concentration and more alkaline reaction than normally ejaculated semen.

B. Mating Behavior and Copulation

Rams vary widely in frequency of copulation. Rams observed continuously with one or more ewes in heat mated an average of once every three hours, but varied from one in ten hours to over once an hour (Hulet, 1966). Rams may copulate two or three times in a few minutes when first turned with a ewe. They generally mate more frequently when with more than one ewe in heat. Rams may have preference for individual ewes or for ewes of the same breed but this is usually not noticeable in practice.

Partner-seeking is affected most by vision, although olfaction is important and hearing is least important (Fletcher and Lindsay, 1968). Dominant rams are more successful in mating when only one or a few ewes are in estrus but with large numbers of ewes in heat competition stimulates mating.

C. Sperm in the Female Tract

Sperm travel through the reproductive tract of the ewe at an average of 4 cm per minute although considerable variation occurs and one to two hours or more may be required to reach the upper uterus and fallopian tubes. Sperm tend to survive longer in the cervix (up to three days) and flow out from there. They may survive up to 12 hours in the vagina and up to 30 hours in the uterus and oviducts. Sperm transport is often aided by motility of the female tract although sperm motility is important in the cervix (Lightfoot and Restall, 1971). Synchronization of estrus may interfere with sperm transport as also may grazing on estrogenic pastures or foreign devices inserted into the uterus. The latter effect appears to be caused by reverse uterine contractions (Warren and Hawk, 1971).

D. Fertilization

The eggs may remain viable for 10 to 25 hours but abnormal fertilization may increase with age of the sperm or egg. The cumulus oophorus breaks down early in the ewe so that tubal eggs may have few or no follicle cells. The slit by which the spermatozoon enters the egg has been identified in the sheep. Sperm penetration begins in about three hours, pronuclei are formed in about three to nine hours and the fertilized egg undergoes its first cleavage about 19 to 20 hours after ovulation. Sheep eggs have been fertilized in vitro but with low incidence. Fertilized ova may survive culture, and some when transferred to recipient ewes have developed into lambs (Moore, 1970). Mean cleavage stages from onset of estrus were 2-cell at about 53 hours, 4-cell at 62, 6-cell at 65, 8-cell at 75 and 12-cell at over 80 hours.

V. PREGNANCY

Sexual activity in the ewe occurs occasionally during pregnancy, particularly at the time of the normal breeding season. Ovulation generally does not occur but fertilization during pregnancy and superfetation have been reported. Scanlon (1972) reported two sets of twins born 15 days apart, both from apparently normal gestation lengths but from different breeds of sire.

Prenatal mortality is usually 20–30% (Edy, 1969). There is higher loss of twin than of single ova or embryos. High temperatures contribute to prenatal mortality as well as to low birth weights. Some poisonous plants interfere with fetal development.

A. Length of Pregnancy

The mean length of pregnancy for various breeds varies from 144 to 151 days, although individual normal pregnancies may vary from 138 to 159 days. Extremely short and long periods may be of questionable normality. The early maturing meat breeds (e.g., Southdown or Hampshire) and the highly prolific breeds (e.g., Finnsheep or Romanov) have short gestation periods averaging about 144–145 days. Slow maturing fine wool breeds (e.g., Merino or Rambouillet) have long periods of 150–151 days. Crossbred types (e.g., Columbia or Targhee) have intermediate periods of 148–149 days. Individual gestation periods within a breed will vary up to 13 days with a standard deviation of about 2.2 days.

Hereditary factors play an important role in the duration of pregnancy as would be expected from breed differences (Prud'Hon et al., 1970). Heritabilities of pregnancy duration from intraclass correlations of paternal half-sibs were in the confidence interval of 0.34 to 0.65. The maternal effect constituted 2.6 to 10.9% of the variance of pregnancy duration. Repeatability of gestation length for two or three pregnancies was 0.17 and 0.23 respectively. Phenotypic and genetic correlations between gestation length and birth weight were 0.65 and 0.86 respectively.

Pregnancy is shorter for twin lambs than singles, is sometimes longer for ram than ewe lambs and increases with age of dam. A low plane of nutrition tends to reduce gestation length, particularly in late pregnancy and for twins.

B. Pregnancy Diagnosis

A great variety of pregnancy tests have been studied with sheep (Richardson, 1972). Vaginal biopsy and radiograph (Hulet and Foote, 1968) and the peritoneoscope give early reliable results but require surgery. Ultrasonic techniques (Lindahl, 1969, 1971, 1972) are both reliable and easy to perform but instruments are required. The rectal-abdominal palpation technique (Hulet, 1972) is easy, rapid and quite reliable after about two months of pregnancy and may prove most useful in practice. Detection of number of fetuses can be accomplished by ultrasonics (Lindahl and Brobeck, 1972), radiography, fetal electrocardiography, rectal abdominal palpation, laparoscopy, laparotomy and others but a really satisfactory method is not yet available.

VI. PARTURITION

Near the end of pregnancy there is a relaxation of the pelvic ligaments, vagina and cervix. This is accompanied by increased irritability of the uterus and activity of the fetus. Use of radio transmitters shows that intrauterine pressure begins to increase about 12 hours before delivery (Hindson et al., 1965). The pressure waves increase in force with the approach of delivery but do not show any pattern of frequency or duration. The cervix dilates slowly at first with a rapid change in the last hour before delivery. Abdominal contractions occur at the peak of the intrauterine pressure and increase in intensity when the cervix is dilated through delivery. High intrauterine pressure persists after delivery.

The placenta is generally expelled from two to four hours following birth. Involution of the uterus, involving necrosis beginning at the base of the cotyledonary crypts followed by liquefaction and exfoliation of the necrotic crypt mass, follows rapidly and is usually complete by one month. The uterus returns to nonpregnant size in about two weeks. The ewe usually licks the newborn lamb. At this stage she will generally accept any lamb presented to her.

VII. REPRODUCTIVE EFFICIENCY

Reproductive efficiency depends on the proportion of ewes mated which become pregnant and the lambing rate. Lambing rate (number of lambs born of ewes lambing) depends on the number of ovulations minus those ova which failed to become fertilized or embryos which died and were resorbed before birth. The proportion of lambs born of ewes lambing is markedly affected by age and by breed.

Ovulation rate, or the number of eggs shed per heat period, through its important relationship to the number of lambs weaned per ewe bred, may be the most important aspect of reproduction in sheep. Significant differences are noted in ovulation rate due to breed, age, year, season and nutrition. Ovulation rate, often deduced from the number of lambs born at one time, is most commonly one, often two, occasionally three and rarely 4, but has been reported as high as seven to nine. Six normal lambs have been born together without any special treatment of the mother. It increases to a maximum from three to six years of age and then may decline slightly with advancing age (Fig. 12–2). The ovulation rate tends to be high from early to near the middle of the breeding season and decreases toward its end. The changes are illustrated by lambing frequency over the seasons (Fig. 12–3). Ovulation rate is quite repeatable both within a season and between years so that individual ewes with higher ovulation rates tend to have higher lifetime lamb production than those with lower ovulation rates (Hulet and Foote, 1967). Year differences are often confounded with nutrition and are not well-understood. The tendencies for the ovulation rate to be higher as sheep are kept at latitudes closer to the poles and to be higher for sheep with faces free of wool covering, as has been shown in a number of breeds, may be related to seasonal or nutritional effects but are not well-understood.

The restricted breeding season of many breeds of sheep has led to attempts to alter the breeding season so that lambs can be produced at any season of the year. This often has marked economic advantages. The fact that some breeds of tropical or subtropical origin will

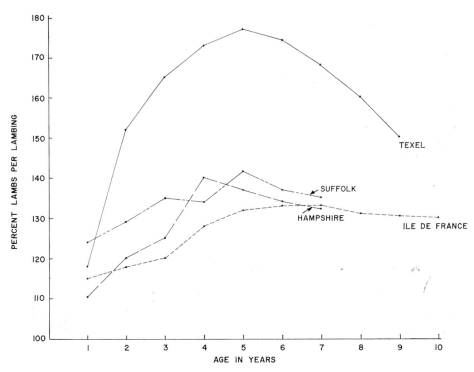

FIG. 12–2.   Effect of age on lambs per lambing. (*Institut Technique d' l'Elevage Ovin et Caprin, Paris, France, 1972.*)

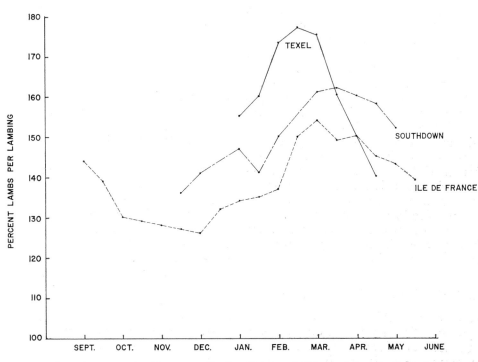

FIG. 12–3.   Effect of season on lambing rate. (*Institut Technique d' l'Elevage Ovin et Caprin, Paris, France, 1972.*)

breed at any time of the year indicates that the breeding season could be extended by genetic means. The heritability of the length of breeding season is appreciable and selection experiments are now underway to eliminate anestrus of some breeds or strains through breeding. Rate of twinning and milk production are also heritable although to a low or moderate degree. Selection for lambing of offspring of ewe with lambs every eight months or more often is also being practiced.

The use of artificial treatment to extend the breeding season offers hope of more rapid gains if practical methods can be developed. Shortening of the length of day and cooling have both been shown to be effective in permitting ewes to be bred out of season but the interactions between these treatments have not been established.

Estrus, ovulation and fertility may be induced in the anestrous ewe by use of hormones. Treatment with progesterone for 10–12 days followed by gonad stimulating hormone (750 to 1000 i.u. of PMS or HCG) in two or three days may be successful. A second treatment sequence may be advantageous. Prostaglandin may increase precision of timing of the induced ovulation which will be particularly helpful for artificial insemination. However, sheep often vary in their response to these treatments and high fertility is not always obtained.

The use of gonadotropins on about the 11th to the 13th day of the normal cycle to increase the ovulation rate (mild superovulation) is often effective if the proper dosage (500 to 750 i.u. PMS) is given but considerable labor is involved in checking for estrus and timing the hormone injections properly. Therefore, it may be questioned if the procedure is practical but under certain conditions it may be very useful. Lambing percentages may be increased by about 30% by such treatment.

The stimulation of lactation in the non-lactating ewe, though possible, is generally of little practical use. The development of practical milk replacers and methods of artificial rearing of lambs (Glimp, 1972) lessens the need to increase milk production as litter size is increased.

REFERENCES

Anderson, J. (1965). Reproduction in imported British breeds of sheep on a tropical plateau. *Proc. 5th Internat. Congr. Anim. Reprod. A. I. Trento 3*, 465–469.

Dziuk, P. (1972). The importance of reproductive performance in intensified sheep production. *Proc. Symp. on Intensive Sheep Management*. SIDP, Denver, Colorado.

Edy, T. N. (1969). Prenatal mortality in sheep: A review. *Anim. Breed. Abstr. 37*, 173–190.

Fletcher, I. C. and Lindsay, D. R. (1968). Sensory involvement in the mating behavior of domestic sheep. *Animal Behav. 16*, 410–414.

Glimp, H. A. (1972). Effect of diet composition on performance of lambs reared from birth on milk replacer. *J. Anim. Sci. 34*, 1085–1088.

Hafez, E. S. E. (1952). Studies on the breeding season and reproduction of the ewe. *J. Agric. Sci. 42*, 189–265.

Hindson, J. C., Schofield, B. M., Turner, C. B. and Wolff, H. S. (1965). Parturition in sheep. *J. Physio. 181*, 560–567.

Hulet, C. V. (1966). Behavioral, social and psychological factors affecting mating time and breeding efficiency in sheep. *J. Anim. Sci.* (Suppl.), *25*, 5–16.

Hulet, C. V. (1972). A rectal-abdominal palpation technique for diagnosing pregnancy in the ewe. *J. Anim. Sci. 35*, 814–819.

Hulet, C. V. and Foote, W. C. (1967). The relationship between ovulation rate and reproductive performance in sheep. *J. Anim. Sci. 26*, 563–566.

Hulet, C. V. and Foote, W. C. (1968). A rapid technique for observing the reproductive tract of living ewes. *J. Anim. Sci. 27*, 142–145.

Hunter, G. L. (1968). Increasing the frequency of pregnancy in sheep. I. Some factors affecting rebreeding during the post-partum period. *Anim. Breed. Abstr. 36*, 347–378.

Inskeep, E. K. (1972). Artificial insemination, semen handling and pregnancy diagnosis. *Proc. Symp. on Intensive Sheep Management*. SIDP, Denver, Colorado.

Lightfoot, R. J. and Restall, B. J. (1971). Effect of site of insemination, sperm motility and genital tract contractions on transport of spermatozoa in the ewe. *J. Reprod. Fert. 26*, 1.

Lindahl, I. L. (1969). Comparison of ultrasonic techniques for the detection of pregnancy in ewes. *J. Reprod. Fert. 18*, 117–120.

Lindahl, I. L. (1971). Pregnancy diagnosis in the ewe by intrarectal doppler. *J. Anim. Sci. 32*, 922–926.

Lindahl, I. L. (1972). Early pregnancy detection in ewes by intrarectal reflection ultrasound. *J. Anim. Sci. 34*, 772–775.

Lindahl, I. L. and Brobeck, S. L. (1972): Ewe pregnancy diagnosis by ultrasonic scanning. *J. Anim. Sci. 34*, 354.

Lodge, J. R. and Salisbury, G. W. (1970). "Seasonal Variation and Male Reprodutive Efficiency." In *The Testis*. Vol. 3. A. D. Johnson, W. R. Gomes and N. L. VanDemark (eds.), New York, Academic Press, pp. 139–147.

Moore, N. W. (1970). Preliminary studies on in-vitro culture of fertilized sheep ova. *Austr. J. Biol. Sci. 23*, 721–724.

Ortavant, R., Courot, M. and Hochereau, M. T. (1969). "Spermatogenesis and Morphology of the Spermatozoon." In *Reproduction in Domestic Animals*. H. H. Cole and D. T. Cupps (eds.), New York, Academic Press.

Prud'Hon, M., Desvignes, A. and Devoy, I. (1970). (Results of six year breeding of "Merinos D'arles" ewes, Domaine du Merle. IV. Duration of pregnancy and birth weight of lambs). *Ann. Zootech. 19*, 439–454.

Richardson, C. (1972). Pregnancy diagnosis in the ewe. *Vet. Rec. 90*, 264–275.

Robinson, T. J. (1971). The seasonal nature of reproductive phenomena in the sheep. II. Variation in fertility following synchronization of oestrus. *J. Reprod. Fert. 24*, 19–27.

Scanlon, P. F. (1972). An apparent case of super-foetation in a ewe. *Aust. Vet. J. 48*, 74.

Southam, E. R., Hulet, C. V. and Botkin, M. P. (1971). Factors influencing reproduction in ewe lambs. *J. Anim. Sci. 33*, 1282–1287.

Warren, Jr., J. E. and Hawk, H. W. (1971). Effect of intrauterine device on sperm transport and uterine motility in sheep and rabbits. *J. Reprod. Fert. 26*, 419.

Chapter 13

Pigs

L. L. ANDERSON

The pig (*Sus scrofa, domestica*) is a member of the family Suidae, in the order Artiodactyla. It is in the class of monogastric ungulates which have even-toed hoofs. The productivity of the pig is high when compared with other farm mammals. Reproductive potential depends upon early sexual maturity, a comparatively high ovulation rate, relatively short periods of gestation and lactation, as well as the capability of repeating the pregnancy cycle soon after weaning a litter.

I. SEXUAL MATURITY

Although the sex ratio at birth (secondary) is near 100, there are more male than female embryos and fetuses which perish during prenatal development.

A. Male

Boars reach puberty at about 110–125 days of age. Spermatozoa are found in the testes but there may be further delay before they are capable of fertilizing ova. Mounting activity occurs early (e.g., as early as 10 days of age); erection by four months; but sequential patterns of sexual behavior culminate after five months. First ejaculations occur at five to eight months of age. The number of spermatozoa and semen volume continue to increase during the first 18 months of life.

B. Female

Sexual maturity (puberty) occurs between six and eight months of age with the first litter being produced at about one year of age. During the prepubertal period the ovaries contain numerous small follicles (2–4 mm diameter) and several (e.g., 8–15) medium-sized follicles (6–8 mm). The uterine development responds to increasing ovarian steroidogenic activity during late stages of the prepubertal period; the uterus weighs 30–60 gm during infantile and impuberal stages as compared with 150–250 gm in prepubertal gilts. Puberty is characterized by first estrus, ovulation of follicles and release of ova capable of fertilization.

Age at puberty may be influenced by level of nutrition, social environment, body weight, season of year, breed, disease or parasite infestation and management practices. Pubertal age in six linebred breeds indicates a considerable range (183–247 days) between these breeds, the average being 209 days (Table 13–1). Inbreeding increases (e.g., 243 days) whereas crossing inbred lines usually lowers pubertal age (e.g., 228 days). Utilizing boars to check for signs of estrous behavior in prepubertal gilts reduces age at onset of the first estrus. The domestic pig receives an intake of dietary energy sufficient not only for body maintenance but also usually for

Table 13-1.   Age at Puberty in Gilts

Classification	Breed	Age of Puberty (Days)	Reference
Linebred	Landrace	183	1
	Lacombe	197	2
	Yorkshire	199	2
	Poland China	201	3
	Chester White	204	3
	Large White	218	4
	Poland China	226	5
	Yorkshire	247	5
Inbred (Year 1)	Yorkshire; Chester White	236	6
(Year 2)	Yorkshire; Chester White	251	6
Inbred lines crossed	Yorkshire × Chester White	228	7
Crossbred	Landrace × Poland China	202	1
(Boar absent)	—	190	8
(Boar present)	—	163	8
	Poland China × Yorkshire	213	5
	Poland China × Yorkshire	231	5

1. Rathnasabapathy et al., 1956. *Mo. Agr. Exp. Sta. Res. Bull. No. 615.*
2. Dyck, 1971. *Can. J. Anim. Sci. 51,* 135.
3. Robertson et al., 1951. *J. Anim. Sci. 10,* 647.
4. Burger, 1952. *Onderstep. J. Vet. Res.* Suppl. 2, 1.
5. Clark et al., 1970. *J. Anim. Sci. 31,* 1032.
6. Warnick et al., 1951. *J. Anim. Sci. 10,* 479.
7. Foote et al., 1956. *J. Anim. Sci. 15,* 959.
8. Zimmerman et al., 1969. *J. Anim. Sci. 29,* 303.

Table 13-2.   Physiologic Development of Reproductive Organs in Gilts Weighing 90 kg

Stage	Classification of Largest Ovarian Follicles (mm)	Number of Corpora Lutea	Mean Ovarian Weight (gm)	Uterine Weight (gm)
Prepubertal	<4	0	2.8	26
	4–6	0	2.8	34
	6–8	0	2.4	61
	8–10	0	3.1	106
Estrous cycle	<4	11	4.7	190
	4–6	12	5.1	186
	4–10	11	4.2	203

(From Shaw et al., 1970. *Can. J. Anim. Sci. 50,* 185.)

an optimum growth rate. Limiting energy intake to half that of full-fed controls delays puberty more than 40 days. Restricting energy intake to 60–70% that of ad libitum feeding for a few days to several weeks has resulted in hastening onset of puberty (e.g. −11 days) in some trials and delaying puberty (e.g. +16 days) in other trials.

Body weight seems a poor indicator for age of puberty; full-fed gilts may attain puberty at earlier age and thus weigh less than gilts with limited food intake. Even though prepubertal gilts may be of similar body weight, there are marked changes in the morphology of the reproductive organs which occur with onset of estrous cycles (Table 13–2). As ovarian follicles develop in prepubertal gilts there are corresponding increases in ovarian and uterine weights. In cycling gilts, ovarian weight increases further, reflecting primarily the presence of corpora lutea; uterine weight about doubles. Estrogens, secreted by the developing follicles, stimulate endometrial and myometrial growth, whereas progesterone secreted by the corpora lutea induces proliferation of endometrial glands and stroma as well as uterine vascularity.

C. Growth

Both male and female pigs grow to about 160 kg body weight during the first year (Fig. 13–1). After the first year growth rate in male pigs continues but at a rate higher than found in females. Although the absolute body weight continues to increase in both sexes, it does so at a decreasing rate. The increase in body weight of the dam through several reproductive cycles is caused by anabolic effects of pregnancy (Fig. 13–2). Weight

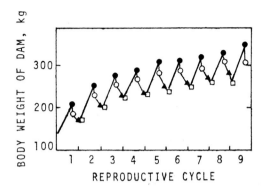

WEIGHT AT BREEDING
WEIGHT BEFORE PARTURITION
WEIGHT AFTER PARTURITION
WEIGHT TO WEANING

FIG. 13–2. Body weight changes in sows in relation to reproductive cycles. (*Adapted from Salmon-Legagneur et al., 1966. Ann. Zootech. 15, 215.*)

increases during pregnancy and decreases during lactation, but the overall weight change is positive during each reproductive cycle.

II. ESTRUS AND ESTROUS CYCLE

A. Estrus

Onset of estrus is characterized by gradual changes in behavioral patterns (e.g., restlessness, mounting other animals, lordosis response), vulva responses (e.g., swelling, pink-red coloring) and occasionally a mucous discharge. Sexual receptivity lasts an average of 40–60 hours. The pubertal estrous period usually is shorter (47 hours) than later ones (56 hours), and gilts usually have a shorter period of estrus than sows. Breed, seasonal variation (e.g., longest in summer, shortest in winter) and endocrine abnormalities affect the duration of heat.

B. Ovulation

Ova are released 38–42 hours after onset of estrus and the duration of this

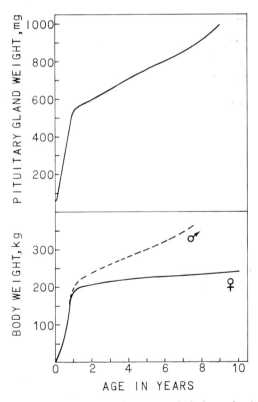

FIG. 13–1. Body development and pituitary gland weight in relation to age of the pig. (*Adapted from Das et al., 1971. Exp. Geront. 6, 63.*)

ovulatory process requires 3.8 hours (du Mesnil du Buisson et al., 1970). Ovulations occur about four hours earlier in mated than in unmated animals (Signoret et al., 1973).

C. Estrous Cycle

The length of the cycle is about 21 days (range: 19–23 days). The pig is polyestrous throughout the year; only pregnancy or endocrine dysfunction interrupts this cyclicity.

D. Cyclic Changes

Throughout the estrous cycle the interdependence of the ovary, hypothalamus, pituitary gland and uterus is reflected in their secretory functions as indicated by morphologic and hormonal changes. The estrous cycle may be categorized into a follicular phase (proestrus and estrus) and a luteal phase (metestrus and diestrus).

E. Ovarian Morphology

There are about 50 small follicles (about 2–5 mm in diameter) per animal during the luteal and early follicular phases of the cycle. During the proestrous and estrous phases about 10–20 follicles approach preovulatory size (8–11 mm) while there is a decline in the number of smaller follicles (those < 5 mm). An index of ovarian follicles (sum of number of follicles × diameter) illustrates this pattern of follicular development during the cycle (Fig. 13–3). The low index during the first four days after estrus accounts for a relatively small number of small follicles. Between days 5 and 16 the index increases as a result of increasing numbers of follicles 2–5 mm in diameter (with a few up to 7 mm), whereas the high index after day 18 reflects primarily growth of the preovulatory follicles (those ≥8 mm diameter).

Soon after ovulation there is rapid proliferation of primarily granulosa and a few theca cells lining the follicle wall.

These cells become luteinized to form luteal tissue, thus the corpus luteum. Initially the corpus is considered a corpus hemorrhagicum because of the blood-filled central cavity, but within six to eight days the corpus luteum is a solid mass of luteal cells with an overall diameter of 8–11 mm. The growth and decline of the corpus luteum is shown in Figure 13–3. The relatively long luteal phase (about 16 days) is characterized by rapid development to maximal weight

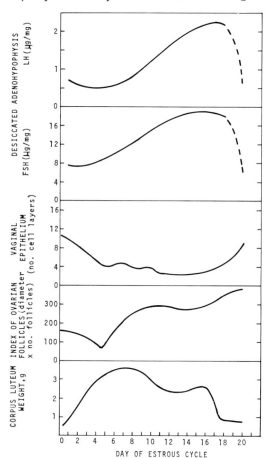

Fig. 13–3. Morphologic changes in the ovaries and vagina are indicated for different stages of the estrous cycle in the pig. Concentrations of follicle stimulating hormone (FSH) and luteinizing hormone (LH) found in the adenohypophysis are shown for different stages of the estrous cycle. (*Adapted from Robinson and Nalbandov, 1951, J. Anim. Sci. 10, 469; Parlow et al., 1964, Endocrinology 75, 365; Masuda et al., 1967, Endocrinology 80, 240; Wrathall, 1970, J. Reprod. Fert. 21, 127.*)

by days 6–8, maintenance of cellular integrity and secretory function to day 16, and then rapid regression to a non-secreting corpus albicans.

Some characteristic cytologic features of a steroid-secreting cell include a large Golgi complex, few cisternal profiles of granular endoplasmic reticulum and extensive agranular (smooth surface) endoplasmic reticulum, whereas a protein-secreting cell contains prominent granular endoplasmic reticulum with well-developed cisternae (Fawcett et al., 1969). Fine structural changes in the lutein cell indicate a close correlation between its morphology and steroid secretion during the cycle (Cavazos et al., 1969). During luteinization (day 1) granulosa cells at the periphery of the ruptured follicle are cuboidal to columnar and separated by irregular extracellular spaces which contain precipitated liquor folliculi. The cytoplasm in these peripheral cells contains granular endoplasmic reticulum and free polysomes. Deeper cells within the corpus are hypertrophied with an eccentrically located nucleus; the cytoplasm contains abundant granular endoplasmic reticulum. By day 4, luteinization is essentially complete; the cells are hypertrophied with masses of agranular endoplasmic reticulum. These cells typify the secretory phase (days 4–12) by their protein and steroid production. Small, coated vesicles are found near the Golgi complex and larger ones are found in peripheral locations near the membrane wall; these vesicles may be related to cellular transport. During cell regression (days 14–18) there is an increase in cytoplasmic lipid droplets, cytoplasmic disorganization and vacuolation of the agranular endoplasmic reticulum. At the terminal phase of the cycle there is an increase in the number of lysosomes, a marked vacuolation of agranular endoplasmic reticulum and invasion of connective tissue; these events result in formation of the corpus albicans.

The vaginal epithelium responds to steroids from the ovaries. The number of cell layers of vaginal epithelium is highest at estrus, declines to lowest levels during the luteal phase, and then increases during proestrus (Fig. 13–3). Vaginal smears also indicate cyclic variations in the distribution of leukocytes, epithelial cells and cornified cells (McKenzie, 1926).

Steroid-secreting activity of the corpora lutea is indicated by the levels of progesterone in ovarian venous plasma throughout the cycle (Fig. 13–4). Progesterone levels are low after estrus, steadily in-

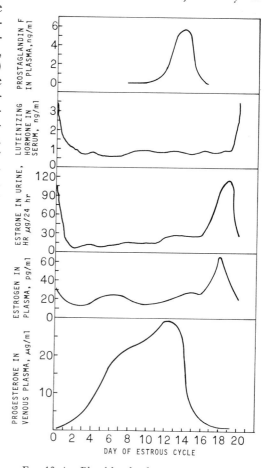

Fig. 13–4. Blood levels of progesterone, estrogen, luteinizing hormone and prostaglandin during the estrous cycle in the pig. (*Adapted from Masuda et al., 1967, Endocrinology 80, 240; Guthrie et al., 1972, Endocrinology 91, 675; Niswender et al., 1970, Endocrinology 87, 576; Gleeson and Thoburn, 1973, J. Reprod. Fert. 32, 343.*) Estrone levels in 24-hour urine collections throughout the cycle in gilts. (*Adapted from Bowerman et al., 1964. Iowa State J. Sci. 38, 437.*)

crease to peak values by days 12–14 and then decline precipitously by days 16–18. These steroid levels follow a pattern similar to the morphologic development and decline of the corpus luteum as well as ultrastructural changes in luteal cells.

Estrogen levels in blood plasma and estrone levels in 24-hour urine collections show a marked decline to low levels soon after estrus, remain low during the luteal phase and increase to peak values with growth and maturation of preovulatory follicles, and finally onset of estrus (Fig. 13–4).

Ovarian follicles depend upon secretion of adenohypophysial gonadotropins for their growth and maturation; hypophysectomy (du Mesnil du Buisson and Léglise, 1963) or hypophysial stalk transsection (Anderson et al., 1967) results in abrupt regression of these follicles. The adenohypophysis synthesizes but secretes little luteinizing hormone during the luteal phase of the cycle (Fig. 13–3). FSH levels increase in the pituitary gland during the luteal phase, a time when follicles are growing. Adenohypophysial LH and FSH levels peak at proestrus and estrus and are low again during the early luteal phase. Serum levels of LH show one sharp peak at estrus and drop to low levels during the remainder of the cycle (Fig. 13–4). Concentrations of LH in blood and adenohypophysial tissue suggest a pattern of synthesis, storage and release.

Relaxin activity remains extremely low in luteal tissue throughout the estrous cycle and shows no relationship to steroid secretion during this brief period.

Endogenous levels of prostaglandin F found in blood plasma indicate a marked increase to peak values by day 14, a time just preceding onset of luteal regression (Fig. 13–4).

III. OVULATION RATE

Ovulation rate is associated with breed (lines or crosses), amount of inbreeding, age at breeding and weight at breeding. In inbred lines there is an average in-

crease of 0.8 ova from the first to second estrous period; ovulation continues to increase (1.1 more ova) at the third estrus but little, if any, additional increase beyond the fourth postpubertal estrus. Reproductive experience correlates with ovulation rate; ovulations increase with parity to seven and more litters (Fig. 13–5). Inbreeding reduces ovulation rate whereas crossing inbred lines increases markedly the number of ovulations. For example, ovulation rates in crossbred pigs increase 0.55 ova for each 10% of inbreeding of their parent strains. The age at breeding in young gilts is positively correlated with ovulation rate. For each additional 10 days

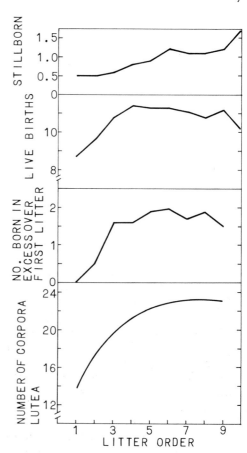

Fig. 13–5. Reproductive performance in relation to parity in the pig. (*Adapted from Perry, 1954, J. Embryol. Exp. Morph. 2, 308; Lush and Molln, 1942, Tech. Bull. U. S. Dept. Agric., 836; Rasbech, 1969, Brit. Vet. J. 125, 599.*)

of age at breeding there are about 0.5 additional corpora lutea. Weight at breeding is also positively associated with ovulation rate when compared with weaning weight or weight at 154 days. Selection experiments based on a controlled gene pool over five generations indicate that heritability of ovulation rate is 0.5 (Cunningham and Zimmerman, 1973).

A. Methods to Increase Ovulation Rates

Ovulations can be induced just before puberty by an injection of PMSG or PMSG followed by an injection of HCG. The ovulatory response is dependent primarily upon the dosage of PMSG. For example, 750–1000 IU PMSG yield from 12 to 25 ovulations.

HCG induces ovulation in cycling gilts but causes little, if any, increase in ovulation rate. After an intramuscular injection of HCG (e.g. 500 IU) during the proestrous period, ovulations occur in most of the animals 44–46 hours later; genetic and environmental factors as well as the hormone preparation may modify this response (Hunter, 1972).

The injection of PMSG induces superovulatory responses when given on days 15 or 16 of the cycle. Hunter (1964) calculated a regression coefficient of 1.9 ± 0.5 more ovulations for each additional 100 IU PMSG.

These gonadotropins usually reduce the length of the cycle, increase the duration of estrus and may increase the incidence of cystic follicles. Ova shed from induced follicles are capable of acceptable fertilization rates, and embryo survival rates are similar to those found in controls.

The onset of estrus in prepubertal gilts or anestrous sows and gilts is induced by a low level of PMSG (e.g. 400 IU) and HCG (e.g. 200 IU), without causing a superovulatory response (Schilling and Cerne, 1972).

B. Nutrition and Ovulation Rate

High-energy diets induce a higher ovulation rate in the pig when the diets are fed for a restricted duration. The number of ovulations is predominantly affected by genetic background, nevertheless ovulation rate is usually affected in a positive way with increasing levels of energy intake. The level of energy restriction before feeding the pigs a high-energy diet is an important factor influencing ovulation rate. A low level of energy intake (e.g., 3000–5000 kcal) is usually given before high-energy (e.g., 8000–10,000 kcal) diets. Durations for feeding the high-energy diets have ranged from 1 to >21 days. The optimum duration of a high-energy regimen seems to be 11–14 days before the expected estrus or mating. Results from 14 trials indicated an additional 2.2 ova shed as compared with ovulation rates in pigs on restricted diets. Protein levels in high- and low-energy diets are usually similar; there is little evidence that increased protein intake during brief periods increases ovulation rate.

IV. CONCEPTION RATE

Fertilization rate in pigs is usually high (>90%). Low or high ovulation rates have little or no effect on fertilization rate; superovulation (e.g., >25 ova) usually results in continued high fertilization rates.

Loss of the whole litter may result from failure of fertilization or death of all the embryos. Estimates indicate that about 5% of the litters are lost during the remainder of gestation. Early embryonic death results in resorption of the conceptus, whereas losses occurring after day 50 may result in abortion, fetal mummification or delivery of stillborn fetuses at term.

V. EMBRYO SURVIVAL

Embryo loss is an important factor of considerable size in the pig. At least 40% of the embryos are lost before parturition and a major part (85%) of this loss occurs during the first half of gestation. Fertilization failure accounts for only

5–10% of this loss. By day 25 about 33% of the embryos die and this increases to 40% by day 50. Sows have greater fecundity than gilts but they lose a greater proportion of their embryos during the first 40 days. With each 10% inbreeding in the dam there results 0.55 to 0.76 fewer ova, loss of 0.53 more fertilized ova, and 0.8 fewer embryos by day 25. Crossing these inbred lines results in 0.55 more ova, 0.33 increase in number of fertilized ova and 0.80 more embryos at day 25 of gestation.

A. Uterine Capacity and Embryo Survival

Embryo mortality rates represent a major loss of production potential in the pig and most of these embryos die in the early part of pregnancy. The relationship between uterine capacity and embryo development has been tested experimentally by several methods; these include superovulation, embryo transfer, superinduction and compensatory ovulation in unilaterally ovariectomized-hysterectomized animals.

By transferring either 12 or 24 embryos to recipient gilts and determining embryo survival at days 26 to 29, Pope et al. (1972) found that the number (and percentage survival) was 6.8 (57%) and 16.3 (68%). Thus, overcrowding during early pregnancy is not a major factor limiting litter size in the pig. By utilizing the technique of unilateral ovarectomy-hysterectomy to accomplish a compensatory ovulation rate in the remaining ovary and a limited amount of uterus, Fenton et al. (1970) showed that fetal survival by day 105 is dependent upon the amount of available uterine space. Only half as many fetuses survived in the experimental group as compared with intact controls, even though ovulation rates were similar in both groups.

A summary of experimental results from several investigators indicates that as the ovulation rate (or potential embryos by embryo transfer or superinduction) increases the survival rates of

embryos remain higher during early as compared with late pregnancy (Fig. 13–6).

VI. LITTER SIZE

Reproductive performance is measured primarily by the number of living pigs at birth or by the total farrowing or weaning weight of pigs produced by the dam within one year. Ovulation rates continue to increase with subsequent gestations but litter size reaches maximal levels

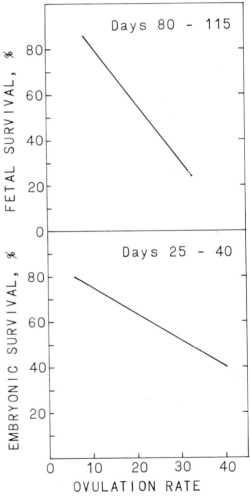

Fig. 13–6. Relationship between ovulation rate and percentage of embryos and fetuses surviving during early and late pregnancy. (*Compiled by Wrathall, 1971. Prenatal Survival in Pigs. Commonwealth Agriculture Bureaux Series 9, 1.*)

by the fourth or fifth parity (Fig. 13–5). The number of pigs farrowed increases markedly between the first and fourth litter, but by the eighth litter the number of live births declines while the number of stillborn increases. When litter size is related to the age of the dam, reproductive performance begins to decline after 4.5 years. The genetic contribution (heritability) to litter size is estimated as 0.17. Thus most of the variation is subject to environmental factors. Breed of sire has been found to affect in a positive as well as negative way litter size, birth weight, stillbirth rate and rate of preweaning mortality.

A. Pig Survival

The most critical period for survival of the neonate is the first three weeks of life. Factors involved in the cause of death include inadequate nutrition, maternal behavior, disease, herd management, season, litter size, degree of inbreeding, age of dam and sex of pig. Although male pigs are heavier than females at birth, the females have a higher survival rate at weaning (Table 13–3). The female pig retains this advantage as litter size increases. Maximum survival rate, regardless of sex, results when the newborn weighs 1.86 kg; large fetuses are not detrimental to neonate survival in this polytocous species. The number of pigs born alive and the sex of the pig are factors of lesser importance to survival. For each 10% in litter inbreeding the neonate survival declines about 1.2%.

B. Nutrition and Litter Size

Although a considerable proportion of embryos die during the course of gestation, the nutritional status of the dam seems of doubtful significance to this loss. During the first 25 days after mating, pregnancies are maintained on protein-free diets, on low-energy diets or during brief periods of complete inanition. Restricting metabolizable energy intake (e.g., < 5000 versus > 9000 kcal/

Table 13-3. Litter Values of Traits Related to Survival

Variable*	Mean	Percentage
Pigs born alive	7.8	94
Pigs born dead	0.5	6
Pigs weaned	5.6	71
Males		51
Pigs living at 154 days	5.3	66
Inbreeding of dam		17
Inbreeding of litter		16
Age of dam, months	12.7	
Birth weight of ♂ pig, kg	1.2	
Birth weight of ♀ pig, kg	1.2	

* Values based on 7018 litters.
(From Bereskin et al., 1973. *J. Anim. Sci. 36*, 821.)

day) throughout gestation has no effect on embryo survival rates (10.4 versus 10.5 live embryos) and only a slight decrease results in piglet birth weight (Table 13–4). There is an increase in litter size of about two pigs between the first and third gestation, but this is unrelated to the nutritional status of the dam. The results indicate that dams given restricted energy levels deliver more pigs during the second and third gestation than do full fed controls. Thus high energy intake during pregnancy has no beneficial effect on fetal survival. Long-term effects of energy intake on reproductive performance through three consecutive gestation periods indicate that the total number of pigs farrowed declines but birth weight of living pigs increases with increasing levels of metabolizable energy for the dam (Table 13–5). Body weight change from breeding to 24 hours postpartum is markedly affected by level of dietary energy intake.

Daily intake of protein during gestation is not an important factor affecting litter size or birth weight. Nutritionally adequate gestation rations include levels of protein of 200–300 gm/day. Protein restriction (e.g., 90 gm/day) seems to have little effect on total litter size, number of live pigs born or birth weight. Dams given protein-free (e.g., 9 gm/day) diets throughout pregnancy maintain litter size, but birth weight is reduced

Table 13-4.   Metabolizable Energy Intake and Reproductive Performance in the Pig

Number of Trials	Performance on Restricted Diet Gestation			Difference in Pigs Born to Dams on High Energy Diet
	1	2	3	
	Live Pigs Born			
28	9.1			+0.2
16		10.9		−0.2
12			11.8	−0.3
	Piglet Birth Weight, kg			
26	1.19			+0.11
10		1.29		+0.09
6			1.17	+0.18

(From Anderson and Melampy, 1972. In *Pig Production*. D. J. A. Cole [ed.], London, Butterworths.)

Table 13-5.   Effects of Energy Intake on Reproductive Performance

	Metabolizable Energy Intake, kcal/day*			
	3000	4500	6000	7500
Pigs born	10.6	10.7	10.4	8.7
Live pigs born	9.8	10.1	9.4	8.2
Birth weight (kg)	1.2	1.3	1.3	1.5
Dam weight change (kg)	5	17	27	38

* Mean values from three consecutive gestation periods on same dams.
(From Frobish et al., 1973. *J. Anim. Sci. 36*, 293.)

(e.g., 0.8 versus 1.1 kg) (Pond et al., 1969). Postnatal performance of pigs from protein-deprived dams is reflected primarily in reduced protein synthesis in the cerebrum and *longissimus* (e.g., differences in tissue RNA and RNA/DNA ratios) but little or no effect on growth rate. Severe protein restriction during gestation and lactation impairs subsequent reproductive efficiency as indicated by an increased number of days from weaning to estrus, reduced ovulation rate and lower uterine weight (Svajgr et al., 1972). These detrimental effects from previous protein restriction are more severe on younger animals but estrus and ovulation can be induced in such noncyclic animals by injecting PMSG and HCG.

When pigs are deprived of a diet adequate for growth and development for prolonged periods, they are capable of reproduction if allowed to consume adequate diets later in life. Pigs limited to 4–14 kg body weight during their first two years and then allowed to grow on adequate diets eventually exhibit reproductive cycles and produce litters. Comparison of pigs severely undernourished from 10 days to the first year of age with full-fed littermate controls indicates that body weight in the undernourished animals is held to 3%, whereas the brain weighs 80% that found in the well-fed littermates. When these undernourished animals are then given full diets after the first year they begin to grow rapidly until two years of age but they are slightly smaller than their littermate controls (Widdowson, 1972).

VII. PREGNANCY AND LACTATION

Embryos are usually in the 4-cell stage when they enter the uterus. Cleavage advances to the morula stage by day 5 and then blastocyst formation by days 6–8. The blastocysts are unevenly distributed throughout the uterine horns. By day 11 half the blastocysts have undergone elongation (e.g., 10–100 cm) and by day 13 most of the embryos have completed this process. Intrauterine spacing of embryos is virtually complete even though implantation occurs later (days 14–18).

Pregnancy lasts about 115 days in the pig and its successful completion is dependent upon functional corpora lutea; ovariectomy as late as day 110 results in abortion within a few hours (Belt et al., 1971).

As pregnancy advances urinary levels of estrone increase to day 30, then decline to low levels until about days 60–70. Thereafter estrone levels increase markedly to peak values by the last days of pregnancy and then drop to low levels following parturition. Relaxin activity in luteal tissue accumulates to peak values during late pregnancy (days 105–110). Between 44 and 26 hours before parturition relaxin levels decline by half and the last day before delivery the level of relaxin in corpora lutea becomes minimal.

The conceptus seems to maintain luteal function during the critical phases of early pregnancy in the pig by overcoming luteolytic action of the uterus. A sufficient amount of nongravid uterus in these pregnant animals can cause a local luteolytic effect. For example, in Figure 13–7, let A represent the estrous cycle and D pregnancy. When one uterine horn is removed (C), the pregnancy continues and the corpora lutea are maintained in both ovaries throughout the gestation. When only a small part of one nongravid horn remains and the animal is mated (B), the corpora lutea near the gravid horn remain functional whereas those adjacent to the nongravid segment regress within 20 days (du Mesnil du Buisson, 1961). As the amount of nongravid horn increases the luteolytic action becomes dominant and the pregnancy fails (Anderson, 1966). Niswender et al. (1970) confirm previous findings by observing maintenance of corpora only in the ovary ipsilateral to the gravid horn in animals made unilaterally pregnant on day 2. The corpora lutea are maintained in both ovaries, however, when the animal is made unilaterally pregnant after day 12. Thus the first 12 days are the critical periods in the pig for uterine luteolytic action on a local or systemic basis.

If the uterus is removed (hysterectomy) the corpora lutea are maintained for a period exceeding that of gestation. The corpora lutea in these pigs secrete progesterone and relaxin, but urinary levels of estrone remain low and unlike those found in pregnant dams.

A. Postpartum

Estrus frequently occurs within one to three days after parturition. If mated, the sow fails to conceive at this estrus because the ovarian follicles are immature (e.g., 3–4 mm diameter) and ovulation usually does not occur.

B. Lactation

With the exception of the postpartum estrus, sows rarely exhibit estrus during lactation. Ovarian morphology during this anestrous period is indicative of an absence of gonadotropic stimulation. Average diameters of the ovarian follicles decrease (e.g., from 4.6 to 2.7 mm) during the first week after parturition and then gradually increase (e.g., >5 mm)

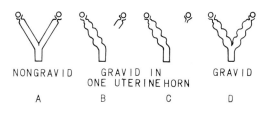

NONGRAVID   GRAVID IN   GRAVID
               ONE UTERINE HORN

A   B   C   D

Fig. 13–7. Local uterine luteolytic action in the pig. (*From Anderson, 1972. In Biology of Mammalian Fertilization and Implantation. Moghissi and Hafez [eds.], Springfield, Charles C Thomas.*)

by the fifth week of lactation (Palmer et al., 1965). Uterine weight and length decline rapidly for 21–28 days following parturition; thereafter, both remain constant. The endometrium is thinner and the uterine glands are less numerous, particularly in the basal region near the myometrium.

Pituitary FSH activity is high during early and late lactation while LH activity remains low throughout lactation. The lactational anestrus may be a period of depressed FSH release and reduced LH synthesis.

By the third week of lactation, estrus and ovulation can be induced by separating the sow from the litter for periods of 12 hours for three consecutive days and injecting PMSG (Crighton, 1970).

C. Weaning

After weaning, pituitary FSH and LH decline and there is an increase in the average diameter of the follicles as well as an increase in the length and weight of the uterus (Crighton and Lamming, 1969; Palmer et al., 1965). When pigs are removed on days 10, 21 and 56 of lactation the average interval from weaning to estrus is nine, six and four days (Self and Grummer, 1958). As the lactation period progresses, the interval from weaning to onset of estrus is shortened and less variable.

REFERENCES

Anderson, L. L. (1966). Pituitary-ovarian-uterine relationships in pigs. *J. Reprod. Fert.* Suppl. *1*, 21–32.

Anderson, L. L., Dyck, G. W., Mori, H., Henricks, D. M. and Melampy, R. M. (1967). Ovarian function in pigs following hypophysial stalk transection or hypophysectomy. *Amer. J. Physiol.* 212(5), 1188–1194.

Belt, W. D., Anderson, L. L., Cavazos, L. F. and Melampy, R. M. (1971). Cytoplasmic granules and relaxin levels in porcine corpora lutea. *Endocrinology 89*, 1–10.

Cavazos, L. F., Anderson, L. L., Belt, W. D., Henricks, D. M., Kraeling, R. R. and Melampy, R. M. (1969). Fine structure and progesterone

levels in the corpus luteum of the pig during the estrous cycle. *Biol. Reprod. 1*, 83–106.

Crighton, D. B. (1970). The induction of pregnancy during lactation in the sow: The effects of a treatment imposed at 21 days of lactation. *Anim. Prod. 12*, 611–617.

Crighton, D. B. and Lamming, G. E. (1969). The lactational anoestrus of the sow: The status of the anterior pituitary-ovarian system during lactation and after weaning. *J. Endocr. 43*, 507–519.

Cunningham, P. J. and Zimmerman, D. R. (1973). Selection for ovulation rate in swine. *J. Anim. Sci. 37*, 231–232.

Fawcett, D. W., Long, J. A. and Jones, A. L. (1969). The ultrastructure of endocrine glands. *Rec. Prog. Hormone Res. 25*, 315–380.

Fenton, F. R., Bazer, F. W., Robison, O. W. and Ulberg, L. C. (1970). Effect of quantity of uterus on uterine capacity in gilts. *J. Anim. Sci. 31*, 104–106.

Hunter, R. H. F. (1964). Superovulation and fertility in the pig. *Anim. Prod. 6*, 189–194.

Hunter, R. H. F. (1972). Ovulation in the pig: Timing of the response to injection human chorionic gonadotrophin. *Res. Vet. Sci. 13*, 356–361.

McKenzie, F. F. (1926). The normal oestrous cycle in the sow. *Mo. Agr. Exp. Sta. Res. Bull, 86*, 1.

du Mesnil du Buisson, F. (1961). Régression unilatérale des corps jaunes après hysterectomie partielle chez la truie. *Ann. Biol. Anim. Bioch. Biophys. 1*, 105–112.

du Mesnil du Buisson, F. and Léglise, P. C. (1963). Effet de l'hypophysectomie sur les corps jaunes de la truie. Résultats préliminaires. *Compt. Rend. 257*, 261–263.

du Mesnil du Buisson, F., Mauleon, P., Locatelli, A. and Mariana, J. C. (1970). Modification du moment et de l'étalement des ovulations après maîtrise du cycle sexuel de la truie. *Colloque Sté Nat. Etude Steril-Fertil* "L'inhibition de l'ovulation," Paris-Masson, éd., 225–234.

Niswender, G. D., Dziuk, P. J., Kaltenbach, C. C. and Norton, H. W. (1970). Local effects of embryos and the uterus on corpora lutea in gilts. *J. Anim. Sci. 30*, 225–228.

Palmer, W. M., Teague, H. S. and Venzke, W. G. (1965). Histological changes in the reproductive tract of the sow during lactation and early postweaning. *J. Anim. Sci. 24*, 1117–1125.

Pond, W. G., Strachan, D. N., Sinha, Y. N., Walker, Jr., E. F., Dunn, J. A. and Barnes, R. H. (1969). Effect of protein deprivation of swine during all or part of gestation on birth weight, postnatal growth rate and nucleic acid content of brain and muscle of progeny. *J. Nutr. 99*, 61–67.

Pope, C. E., Christenson, R. K., Zimmerman-Pope, V. A., and Day, B. N. (1972). Effect of number of embryos on embryonic survival in recipient gilts. *J. Anim. Sci. 35*, 805–808.

Schilling, E. and Cerne, F. (1972). Induction and synchronisation of oestrus in prepuberal gilts and anoestrous sows by a PMS/HCG-compound. *Vet. Rec. 91*, 471–474.

Self, H. L. and Grummer, R. H. (1958). The rate and economy of pig gains and the reproductive behavior in sows when litters are weaned at 10 days, 21 days, or 56 days of age. *J. Anim. Sci.* *17*, 862–868.

Signoret, J. P., du Mesnil du Buisson, F. and Mauleon, P. (1973). Effect of mating on the onset and duration of ovulation in the sow. *J. Reprod. Fert.* *31*, 327–330.

Svajgr, A. J., Hammell, D. L., Degeeter, M. J., Hays, V. W., Cromwell, G. L. and Dutt, R. H. (1972). Reproductive performance of sows on a protein-restricted diet. *J. Reprod. Fert.* *30*, 455–458.

Widdowson, E. M. (1972). "Effects of Early Malnutrition on General Development in Animals." In *Normal and Abnormal Development of Brain and Behaviour.* G. B. A. Stoeling and J. J. Van der Werff Ten Bosch (eds.), Baltimore, Williams & Wilkins.

Chapter 14

Horses

Y. Nishikawa and E. S. E. Hafez

The present varieties of the domestic horse *(Equus caballus)*—draft, light and pony—are members of the family Equidae, which belongs to the order Perissodactyla. Many aspects of reproductive endocrinology and pregnancy in horses are of particular interest to the student of reproduction.

The mare has small uterine horns, which join the relatively large uterine body almost perpendicularly, thus giving the flattened dorsoventrally located organ a T-shaped form. The cervix is shorter than that in cattle and has the form of a flat cone. The structure of the cervix is simpler than that of ungulates. The cervical canal is open throughout the estrous cycle, whereas it is closed during pregnancy.

I. SEXUAL MATURITY

A. Male

In the early fetus, the testes of the horse are yellow-white but at birth become dark brown or black. The testes descend in the scrotum at the age of two to three weeks; in a few cases the testes are already down in the scrotum at birth. Postnatal growth of the testes begins during the 11th month, with the left testis usually developing earlier and growing more rapidly than the right. At this time, there is also a gradual outward development of the seminiferous tubules around the rete testis (Fig. 14–1).

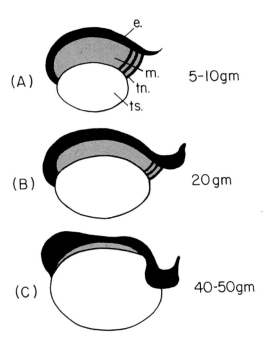

Fig. 14–1. Developmental changes in the attachment of epididymis with the testis in relation to sexual maturity. Figures on right indicate testis weight. *e.*, epididymis; *m.*, thin membrane; *tn.*, tendon; *ts.*, testis.

A, Loose attachment between testis and epididymis; note tendon attachment at head of epididymis; *B,* elongation of epididymis; *C,* epididymis fully developed and completely attached to testis. (*Adapted from Nishikawa, 1959. Studies on Reproduction in Horses. Tokyo, courtesy of Japan Racing Association.*)

At the age of one year, the sperm is first produced in the testes. Stallions may attain sexual maturity by the age of two years and show intense sexual desire when approached by estrous mares. Young stallions begin ejaculating into the artificial vagina 4–14 weeks before the appearance of spermatozoa in the ejaculate at 13 months of age (Skinner and Bowen, 1968). Although semen from pubertal stallions can be used for artificial insemination, the general practice is *not* to use males for natural breeding until they are three to four years old. In practice the age at which stallions are first used for breeding is determined primarily by managerial factors; for example, the use of thoroughbred stallions for breeding at the age of five years is probably much more dependent upon the fact that they are raced until that time, on the average, rather than upon stage of sexual maturity.

While androgen secretion is greatly influenced by day length, spermatogenesis is not. Following the second winter period, a second rise in androgen output occurs with increasing day length, after which the stallion can attain his inherent potential sexual function (Skinner and Bowen, 1968). The second rise in androgen output coincides with the rise in urinary estrogen excretion believed to accompany sexual maturation in the stallion.

B. Females

Fillies attain puberty at 12–18 months of age, but this is influenced by season of birth. When approached by the stallion at this time, the pubertal mare does not resent courtship as do adults. After two to three estrous cycles, the sexual behavior of the pubertal animal becomes similar to that of the adult.

II. BREEDING SEASON

Sexual activity in both sexes is influenced by the season of the year and is related to day length. Activity is usually greatest during the spring and summer period, when the days are long. The length of the breeding season is shorter near the poles than in tropical and subtropical regions, where the breeding season may extend through the whole year. Breed differences in the length of the breeding season have not been established but thoroughbreds tend, probably as the result of artificial selection, to have an earlier breeding season than other horses. Race horses are usually aged from January 1 in the year that they are foaled, and it has been the practice to breed them as early in the year as possible so that, in racing as a two-year-old, they have maximum physical advantage.

A. Male

The breeding season of the stallions is not well-marked and semen can be collected throughout the year. Remarkable seasonal variations are noted in reaction time, number of mounts per ejaculate, volume of gel-free semen, total number of spermatozoa per ejaculate, sperm agglutination and motility in raw and in diluted semen. The total sperm output for first ejaculates in Colorado varies from 6.6 $\times$ 10^9 in November to 12.7 $\times$ 10^9 in May (Pickett et al., 1970). Sperm agglutinations are common during the transition from the nonbreeding to the breeding season. The volume of semen is highest during the spring and summer. The concentration of sperm and ergothioneine and citric acid concentration of semen (Mann et al., 1965) tend to increase during the fall and winter.

The ejaculate normally contains a large fraction of gelatinous material; in some cases the ejaculate is devoid of such material. In Japan (latitude of 37° N.) there are marked seasonal variations in volume of semen ejaculate and of gelatinous fraction. The vesicular glands (commonly called seminal vesicles) are the source of gelatinous material. The absence of gelatinous material from the semen does not affect the motility and fertilizing capacity of the sperm.

An additional five hours of artificial light daily after sunset during November enhances sex drive and improves the semen.

B. Female

The literature pertaining to the nature of the breeding season in the mare is conflicting and inconsistent. This apparent inconsistency probably reflects real differences among the samples studied with regard to heredity and environment. Mares can be classified into three categories: (*a*) defined breeding season: the wild breeds of horses manifest several estrous cycles during a restricted breeding season coinciding with the longest days of the year; the foals are born during a restricted period of the year; (*b*) transitory breeding season: some domestic breeds and some individual mares manifest estrous cycles throughout the year, but ovulation accompanies estrus only during the breeding season, the foals are born during a limited foaling season; (*c*) year-round breeding: some domestic breeds and some individual mares exhibit estrous cycles accompanied by ovulation throughout the year. Thus it is evident that although some mares, at certain latitudes, may show estrous cycles throughout the year, they do not necessarily conceive at all periods. For example, the conception rate of mares in South Africa is markedly influenced by the season of breeding (Fig. 14–2).

In localities where there is a breeding season, the two transitory periods preceding and following the breeding season are characterized by extreme variability of ovarian activity and sexual behavior. At this time the ovarian follicles develop only to a limited degree and then undergo atresia. Also there is a high frequency of prolonged estrus or estrus of short duration as well as irregular estrous cycles.

Near the equator, there is little seasonal variation in the length of the estrous cycle. At latitudes of the British Isles the cycle length is very long during the spring and gradually shortens thereafter. At the same latitude, about 50% of the mares are anestrus during the winter.

Data by Burkhardt (1947) and by Nishikawa (1959) suggest that the exposure of mares to additional hours of light during winter will induce estrus and advance the onset of the breeding season.

The ovaries of the anestrous mare cannot be activated, even by the injection of fairly large doses of serum gonadotropin, PMS, or chorionic gonadotropin, HCG, in contrast to the anestrus ewe, which responds to either hormone. Estrogen therapy may, however, be useful; daily injections of 5–10 mg of stilbestrol for 10–20 days during the luteal phase of the cycle, given in the later part of the breeding season, will inhibit follicle development for two to four months after the last injection. After this period, ovarian functions resume and estrus occurs at regular intervals throughout the nonbreeding season (Nishikawa, 1959).

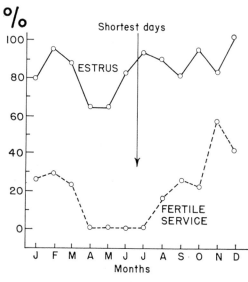

Fig. 14–2. The breeding season of the mare in Onderstepoort, South Africa (25°S). Note the lowest frequency of estrus and of fertile service coincide with the shortest days of the year. (*Adapted from Quinlan et al., 1950. Onderstep. J. Vet. Res. 25, 105.*)

III. SEMEN CHARACTERISTICS

Swiersta and associates estimated that the duration of the cycle of the seminiferous epithelium in the stallion is 12–13 days. The volume of gel, about one-third of the ejaculate, varies considerably and is not characteristic of the individual stallion. This is in contrast to gel obtained from boar ejaculates, which is the most constant feature among ejaculates. Ejaculates containing gel seem to require fewer mounts, a shorter reaction time and possess a slightly larger volume of gel-free semen than ejaculates without gel (Pickett et al., 1970).

Fructose is present in negligible quantities in stallion semen, while relatively large quantities of ergothioneine and citric acid are present. Although the majority of the lipids of stallion spermatozoa are phospholipids, the percentage of phospholipid is much lower than that reported for the bull and boar considering the functional role which phospholipids have in relation to cell membranes. The lower percentages of phospholipid may be related to the great sensitivity of stallion spermatozoa to stress as compared with other species (Komarek et al., 1965).

Byers and associates reported that complete removal of seminal plasma from stallion ejaculates prior to resuspension of sperm in extender caused severe loss of sperm motility. The optimal pre- and postfreeze dilution rate was four parts dried-skim milk extender to one part stallion semen.

IV. ESTRUS AND OVULATION

A. Estrus

During estrus the vulva becomes large and swollen; the labial folds are loose and readily open on examination. The mucous membrane of the vulva is congested, scarlet or orange, wet, glossy and covered with a film of transparent mucus. The vaginal mucosa is highly vascular, and thin watery mucus may accumulate in the vagina. Such mucus spreads uniformly on a glass slide and leaves very little residue on drying. The cellular components of the vaginal smear have no value for detecting estrus in the mare.

In estrus, the cervix dilates enough to admit two to four fingers; during diestrus one finger only can be inserted.

Mares do not show homosexual desire during estrus. They will, however, seek the company of other mares and geldings, crowd them and behave in a similar manner to that shown in the presence of a stallion or teaser. The estrous mare repeatedly assumes a stance characteristic of urination. The tail is raised, urine is expelled in small amounts and the clitoris is exposed by prolonged rhythmic contractions and congested mucous membrane of the moist vulvar lips. A vaginal discharge is observed in the form of soiled edges of the vulvar lips and mucous smears on the tail and between or on the thighs.

1. Intensity

The intensity of behavioral estrus varies both throughout the estrous period and among individual mares at comparable stages of the period. Responses have been graded in eight phases ranging from *very receptive* to *very actively resistant* (Andrews and McKenzie, 1941). The intensity increases slowly at the onset of estrus, reaches its maximum within two or three days, is maintained at this level for two or three days, and declines following ovulation. There is no correlation between intensity of estrus and degree of follicular development (Mahaffey, 1950).

2. Duration

The duration of estrus varies between individuals and also between estrous cycles of the same mare (Fig. 14-3). In most cases, it ranges from two to eleven days with an average of six days. The long duration of estrus in the mare may be due to the following factors: (*a*) the ovary is surrounded for the most part by a serous coat and some time is necessary for follicles to grow large enough to reach the ovulation fossa and rupture. The long duration of estrus reflects the length of the follicular phase in this species; (*b*) the ovary is less sensitive to exogenous

FSH than other species (e.g. cattle, sheep) and it may take a long time for the follicle to reach maximum size immediately before ovulation; and (c) the level of LH is low compared with FSH and this delays ovulation.

The duration of estrus is generally prolonged in old mares, in underfed mares and during the early parts of the breeding season. It is very rare for estrus to continue for more than 48 hours after ovulation occurs, except when there are twin ovulations with the separation of some days between them. This interval may vary from 1 to 10 days.

B. Ovulation

The ovary of the mature mare is kidney-shaped and larger than in other farm mammals. Right and left ovaries do not necessarily ovulate alternately at successive estrous periods; the left tends to be more active than the right. The frequency of twin ovulations may range from 3 to 30% depending on the breed and season of year, being rare in pony breeds and during spring. The incidence of twin pregnancy varies from one to five percent. Usually one member of the twins dies before birth and twin births occur only in very few cases.

During the normal breeding season, the onset of estrus is associated with the appearance of one or several follicles in either or both ovaries. Ordinarily one follicle undergoes a marked increase in size before ovulation. Rarely can ovulation occur from follicles less than 20 mm in diameter.

1. OVULATION AND CORPUS LUTEUM

Most ovulations occur on day 3, 4 or 5 of estrus, 24–48 hours prior to the end of true estrus, i.e. the time of ovulation is more closely related to the end than the onset of estrus; follicle size and the day of ovulation are consistent in individuals from one cycle to another and most ovulations occur between 11 PM and 7 AM (Witherspoon and Talbot, 1970a, b). Ovulation takes place after the first meiotic division. The fertility of mating gradually rises to a peak about two days before the end of estrus and then falls sharply on the final day.

Due to the lengthy estrus in mares copulation must be synchronized with ovulation in order to insure fertilization. The time of ovulation can be detected by repeated rectal palpations. Prior to ovulation, one of the ovaries enlarges and the developing follicles occupy much of the ovarian stroma. After ovulation

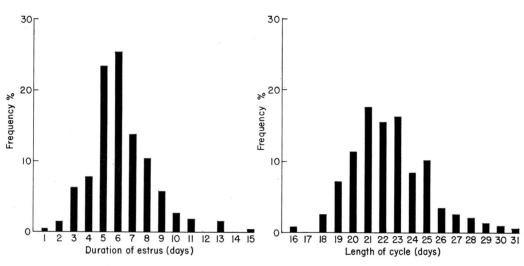

FIG. 14–3. Frequency distribution of the duration of estrus and the length of the estrous cycle in the mare. (*Adapted from Nishikawa, 1959. Studies on Reproduction in Horses. Tokyo, courtesy of Japan Racing Association.*)

the volume of the ovary decreases markedly and becomes very soft and flaccid.

The average volume of oviductal secretions in Shetland mares is 5 ml/day; maximal secretions occur one to four days before ovulation (Engle et al., 1970).

The ovary is less sensitive to FSH than that of the cow, ewe and nanny goat. Injection of massive doses of PMSG during the nonbreeding season was ineffective to induce ovulation of follicles. Injection of PMSG toward the end of the estrous cycle was also ineffective to promote follicular development (Nishikawa and Sugie, 1972). On the other hand, the injection of 2000 i.u. of HCG hastens the time of ovulation and shortens the duration of estrus (Loy and Hughes, 1966). Repeated injection of HCG does not appear to cause refractoriness to subsequent ovulating doses of the hormone.

The developing corpus luteum cannot be detected by rectal examination 48 hours after ovulation because it develops within the ovarian stroma (Plate 14).

The corpus luteum reaches only one-half to three-fourths the size of the follicle at the time of ovulation. The maximal size is attained at 14 days, when luteal cells are very large and have a peripheral vacuolation (Harrison, 1946).

The daily intramuscular injection of 100 mg or more of progesterone during the midcycle prevents estrus and ovulation; 50 mg per day inhibits only estrus (Loy and Swan, 1966). The interval between termination of treatment and estrus appears to depend upon dosage. These effects were, however, confounded by seasonal variations. Neither estrus nor ovulation would be prevented by 50 or 100 mg/day if treatment was started on day 1 of estrus. Postpartum estrus and ovulation could be delayed with 100 mg/day but 50 mg/day was ineffective.

2. Ova and Ova Transport

At ovulation the ovum is without corona radiata but is enclosed in a large irregular gelatinous mass of ovarian origin which is separated within two days. Fertilized ova are transported in the uterus, whereas unfertilized ova are trapped for several months in the isthmus of the oviduct (Fig. 14–4). During this time the ova undergo degeneration and fragmentation during the ensuing months (van Niekerk and Gerneke, 1966). If a mare has a succession of sterile estrous cycles, followed by a fertile mating, the developing embryo may outrun the unfertilized eggs trapped in the oviduct and enter the uterus. The physiologic mechanism of this phenomenon is not understood.

The time at which fertilized ova arrive in the uterus in the mare is much later (>144 hr) than in the cow, and the cleave stage of equine ova at arrival is more ad-

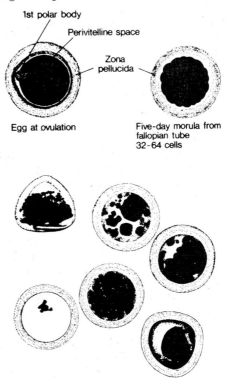

Fig. 14–4. A normal horse egg before and after fertilization, and degenerating oviductal eggs ranging in age from 1 day to 7.5 months. The mechanisms by which the equine unfertilized ova are trapped in the oviduct are unknown. (*From van Niekerk and Gerneke, 1966. Onderstep. J. Vet. Res., 33, 195; redrawn by Short, 1972. In Reproduction in Mammals. Austin and Short [eds.], Cambridge, Cambridge University Press.*)

vanced than in cattle. Transuterine migration of ova occurs in 50% of the cases. Equine ova were successfully recovered nonsurgically from the uterus using a three-way system apparatus, the recovery rate being 45% (Oguri and Tsutsumi, 1972).

C. Estrous Cycle

1. LENGTH OF CYCLE

In general the length of the normal estrous cycle varies from 16 to 24 days, with an average of 22 days. The length of the cycle is prolonged under poor nutritional conditions and during the early spring.

The length of diestrus, the interval from the end of estrus to the onset of the next, is 14–19 days in most cases. When ovulation does not occur the length of diestrus is usually 7–10 days. Uterine flushing during the follicular phase of the cycle shortens the length of the cycle while flushing during the luteal phase delays subsequent ovulation (Oguri and Tsutsumi, 1972).

2. CYCLIC CHANGES

The cyclic changes that occur in the reproductive organs are summarized in Plate 14 and Plate 15. A few days before the onset of estrus a developing follicle of about 2 cm in diameter can often be detected by palpation of the ovary per rectum. As estrus progresses, the follicle markedly increases in size. Other follicles, an average of 22 per ovary, also grow during estrus. The maximum diameter of the corpus luteum never reaches the maximum size of the mature follicle.

The characteristic changes in the vulva and vagina during diestrus reach a maximum about 10 days after the end of estrus and disappear one to two days before the next estrus. During the entire luteal phase of the cycle the mucus is tenacious and adheres to a glass slide, giving a spotted pattern. The cellular components of the vaginal mucus are: leucocytes (chiefly neutrophils, rarely eosinophils),

vaginal epithelial cells and ciliated epithelial cells. The occurrence of leucocytes and epithelial cells is not closely related to a particular stage of the cycle, but ciliated cells characterize the luteal phase.

IV. GESTATION AND PARTURITION

A. Secretion of PMSG

The primary corpus luteum of pregnant mares has a functional life of some 40 days, after which time it degenerates. The allantois develops into a primitive yolk sac placenta which in turn forms allantochorionic placenta with an equatorial band (allantochorionic girdle) around the embryo (Fig. 14–5). Endometrial cups (decidual cells) develop in this region and secrete pregnant mare serum gonadotropins (PMSG) which appear in the maternal circulation on day 40 of pregnancy, reach a maximum on day 60 then decline at 120 days of gestation (Allen and Moor, 1972). The maternal ovaries are thus stimulated, forming ovarian follicles. In Welsh ponies, the first ovulation occurs at 53 days and the second ovulation at 66 days of gestation

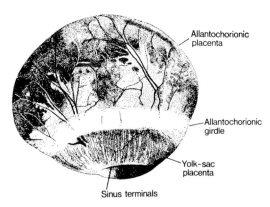

FIG. 14–5. A drawing of an equine embryo and its membranes in section and in surface view on the 35th day of gestation. The conceptus is about the size of an orange, and cells become detached from around the allantochorionic girdle at this stage to invade the endometrium and from the endometrial cups. (*Short, 1972. In Reproduction in Mammals. Austin and Short [eds.], Cambridge, Cambridge University Press.*)

PLATE 14

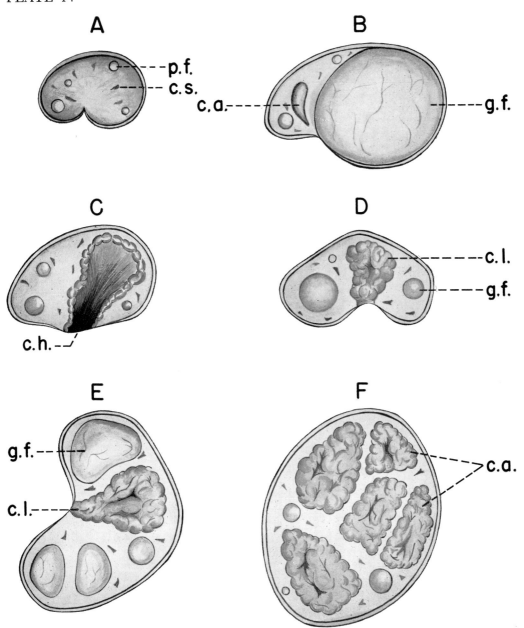

Microdrawing of cross section of ovaries (drawn to scale) of the mare at different stages of reproduction.

A, Nonbreeding season: the ovary is small and contains primary follicles (*p.f.*) and scars (*c.s.*) from degenerating corpora lutea.

B, Breeding season: during estrus, the ovary contains a mature graafian follicle (*g.f.*) and corpus albicans (*c.a.*) from previous ovulation.

C, Three days after ovulation: the corpus haemorrhagica (*c.h.*) develops from the walls of ruptured graafian follicle.

D, Ten days after ovulation: fully developed corpus luteum (*c.l.*). Graafian follicles (*g.f.*) start to develop for subsequent cycle.

E, Pregnancy—60 days: the corpus luteum of pregnancy (*c.l.*) is maintained and graafian follicles (*g.f.*) develop as a result of circulating PMS.

F, Pregnancy—80 days: accessory corpora lutea (*c.a.*) develop from the unruptured follicles.

PLATE 15

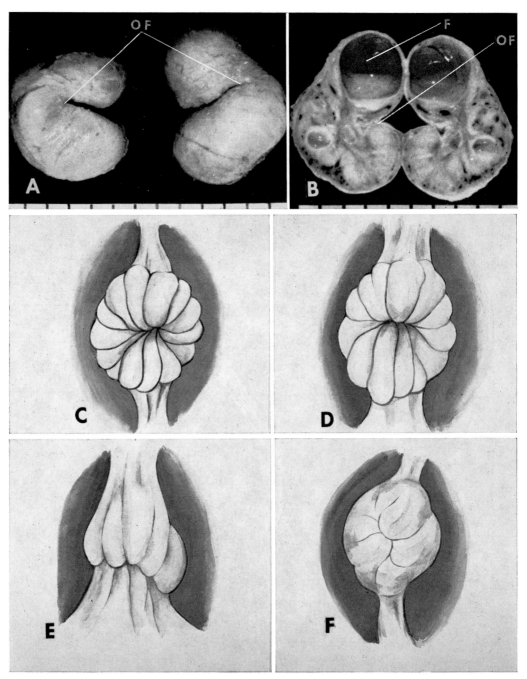

A, *B*, Ovaries of nonpregnant mare during the breeding season. (*OF*), Ovulation fossa; (*F*), graafian follicle. *A*, Note the protruding follicle on the right ovary. *B*, Ovary cut in half.

C,*D*,*E*,*F*, Microdrawings of the os cervix in the mare as seen by clinical examination during estrous cycle and pregnancy. *C*, Diestrus, 10 days after ovulation: the cervix is hard and the folds are well-defined. *D*, Onset of estrus (1st day): the cervix is somewhat swollen, the folds are shallow and less defined, and the orifice of the cervix is opened. *E*, End of estrus (6th day): the cervix is markedly swollen and relaxed and the upper folds hang down in a "membranous" appearance and cover the orifice. *F*, Pregnancy (four to six months): the cervix is hard and budlike in appearance and is covered with pasty mucus and the orifice is tightly closed.

(Allen, 1971). Many of these follicles ovulate and form accessory corpora lutea, while others luteinize without ovulating. The accessory corpora lutea persist until about 180 days of gestation and then degenerate. During the remainder of gestation, no lutein tissue is formed. If a mare aborts about the 35th day of gestation, when the fetal cells have already invaded the endometrium, the endometrial cups continue to develop normally and secrete PMSG, and the corpora lutea that are present may be maintained (Short, 1972). A similar phenomenon occurs when fetal death occurs after the appearance of PMSG.

PMSG also stimulates the fetal gonads of both sexes to increase in size. In fact, the fetal ovaries become larger than those of the mother, and the testes become larger than the testes of newborn foals. With the decline of circulating gonadotropin, this influence diminishes. At this stage, the levels of estrogen and progesterone in the maternal blood increase, causing the female fetal uterus and the male fetal vesicular glands to enlarge. After parturition, however, the reproductive tract and the accessory glands of the foals decline dramatically in size.

A high level of PMSG occurs in the serum of a mare carrying twin fetuses if a set of endometrial cups develops in both uterine horns (Rowlands, 1949). The sudden fall of this blood hormone in an aborted mare compared to the maintenance of a high level in another in which resorption had taken place suggests that the conceptus may be necessary for the continued function of this hormone. The fetal genotype also affects gonadotropin secretion during gestation. The concentration in mare-jack crosses is only 10% of that found in mare-stallion crosses (Clegg et al., 1962).

B. Pregnancy Diagnosis

Clinical diagnosis, done on 30–40 days of gestation, is based on the examination of the vulva, vagina and cervix with the speculum and inspection lamp, and palpation of the ovaries and the uterus per rectum. During pregnancy the vulva becomes small, contracted and difficult to open. The exposed mucous membrane of the vulva and vagina becomes pale and dry. Insertion of an unlubricated speculum into the vagina is difficult because of the adhesive mucus. The cervix becomes small and dry and the cervical orifice is closed. The neck of the cervix resembles a closed flower bud and it protrudes into the vagina (Plate 15). The cervical mucus becomes adhesive and pasty; it contains mucous globules as it does during diestrus.

Forty to fifty days after conception pregnancy is successfully diagnosed by palpating the ovaries and the uterus per rectum.

PMSG can be detected in the blood 40–120 days of gestation. A high amount of estrogen, which can be detected by biologic or chemical methods, appears in the urine at the fourth month of pregnancy and reaches a maximum at the seventh to ninth months.

C. Foaling

The time of foaling which usually takes place between nightfall and daybreak seems to be influenced by light intensity and quietness in the stable. The imminence of foaling is suggested by the degree of mammary hypertrophy, waxing of the teats and possibly the discharge of milk from the udder. The best indication that the first stage has begun is the onset of patchy sweating behind the elbows and about the flanks. This sweating commences about four hours before foaling and increases as the stage progresses. The tail is frequently raised or held to one side. The mare may also swish or slap it against the anus. This stage is followed by crouching, straddling of the hind limbs, kneeling on the knees or breast and rising again and glancing at the flank (Arthur, 1964).

D. Twinning

Twinning is rare. In most dizygotic twin pregnancies there is invagination of the adjacent allantochorions. Identical blood groups are found in twin foals, indicating chimerism and macroscopic or microscopic anastomosis between both chorions. Stillbirth is frequent in twin pregnancies and only one-half of the born foals survive. Twin pregnancies are not desirable in view of the high peri- and postnatal losses, and the poor viability and racing performance of twins. Subsequent conception rate is not affected by twin birth if the mare foals at full-term, but it decreases following abortion.

E. Uterine Involution

Involution of the uterus after normal foaling is extremely rapid. Regression in size is almost complete by the first day of "foal heat" (11 days postpartum). The relatively low conception rate from copulations during this period appears to indicate that involution of the endometrium is not complete in all mares.

F. Postpartum Estrus (Foal Heat)

Postpartum estrus usually occurs 5–15 days after foaling. Some mares, however, may show estrus as late as 45 days after parturition; such estrus may have been preceded by a quiet ovulation. The interval between postpartum estrus and the following estrus may be affected by the milk yield.

Breeding at the postpartum estrus may cause an increased percentage of abortions, dystocias, stillbirths and retained placentae (Jennings, 1950). This may be due to the introduction of bacteria into the uterus before it is completely involuted and while it still lacks contractility. Frequently, the uterine epithelium is not completely restored for some time after foaling.

V. LOWERED FERTILITY AND INCREASING FERTILITY

Reproductive efficiency in horses is low as compared to that in other farm mammals. The incidence of lowered re-productive efficiency varies with management and may often be due to improper detection of estrus in the mare and failure to breed at the right time. Conception rates range from 60 to 65% and the foaling percentage is no more than about 50%. In well-managed ranches, conception rate may reach 90% and foaling percentage varies from 80 to 85%.

Flushing of mares is recommended to improve conception rate. For a stallion serving 60–80 mares per season, an 80% conception rate is considered satisfactory.

A. Male

Overuse of popular stallions causes infertility or sterility. Defects in the penis, ejaculatory disturbances, small volume of ejaculate, oligospermia, azospermia and poor quality semen are common causes of lowered fertility in stallions. Day described a simple way to insure that ejaculation has taken place after intromission by holding the hand under the base of the penis. In the absence of ejaculation a few weak urethral waves may be felt, but when ejaculation takes place it feels like the contents of a 10-ml syringe going along the urethra. In a stallion of good fertility it is usual to feel about five of these waves; in the lower fertility stallion, it is sometimes only about $1\frac{1}{2}$.

B. Female—Abnormal Estrus

"Quiet ovulation," anovulatory estrus, "split estrus" and prolonged estrus are not uncommon in the mare. Anovulatory estrus is due to the low LH output in the species, for in other farm animals, especially cattle and sheep, it is rarely encountered.

Split estrus commonly observed in healthy mares and during the early spring occurs at all stages of maturation of the graafian follicle. In such cases, behavioral estrus is interrupted by an interval of sexual nonreceptivity. Fairly frequently the follicle which is present at the beginning of the first part of the split estrus will continue its growth normally throughout the period and will ovulate.

Prolonged estrus, lasting 10–40 days, may also occur in early spring and in mares used for heavy draft.

1. Abortion

The average rate of abortion in mares is high (10%). This may possibly be due to peculiarities in the hormonal balance of the mare during pregnancy. Abortion is lowest between three and six years of age. Conception rate is low following abortion expecially in older mares.

During the fifth and tenth months of pregnancy, the mares are endocrinologically susceptible to abortion owing to hormonal deficiencies. It is a wise precaution to avoid sudden changes in the diet or the amount of physical exercise at these times. Nishikawa (1959) has reported that abortion may be prevented in mares which have experienced a previous failure by continual administration of estrogen. Day has recommended 300 mg of progesterone implanted subcutaneously after 120 days of conception.

2. Prenatal Mortality

Prenatal mortality is frequent in lactating mares mated early in the season or mated after foal heat. Prenatal death is also common in certain horse families.

3. Neonatal Mortality

Neonatal mortality may be a result of weakness of the mother or the foal or bacterial infection through the umbilical cord of the young. Proper management, clean stables for foaling and sanitary precautions at foaling are the common preventive methods of neonatal mortality.

C. Recommendations for Breeding Techniques

1. Stallions

Stallions used for breeding should possess well-developed testes fully descended in the scrotum. The semen should be evaluated before it is used for breeding.

The detrimental effect of frequent ejaculations on the number of sperm per ejaculate is pronounced in the stallion. In natural breeding, where several mares may exhibit estrus simultaneously, a stallion may copulate several times on one day; this causes a decline in fertility. The use of artificial insemination during such periods of mating will improve the conception rate. The stallion ejaculates directly into the uterus and in most cases there is very little semen left in the vagina. The transfer of semen from the vagina to the uterus following insemination is seldom necessary, but it is recommended with wriggling mares or if a stallion is apt to come off the mare with the penis still erect, as this pulls semen back into the vagina.

2. Mares

Careful testing for estrus with the stallion, routine examination of the vagina and rectal palpation of the reproductive organs may help to improve conception rate. Whenever possible, the time of ovulation should be predicted since the duration of estrus and the time of ovulation from the onset of estrus may differ markedly between individuals. Conception rate depends primarily on the time and number of inseminations.

On occasion, the mare may strain (as in micturition and defecation) after mating and evacuate most of the semen from the uterus. This may be prevented by having the mare walk for a while after mating. The mare is susceptible to endometritis, especially after foaling, since the cervix of the mare is not a strong barrier to the introduction of bacteria. The mare is more prone to a deficiency in LH than other farm mammals; such deficiency may be alleviated by the use of exogenous LH. The intravenous injection of 1500 to 3000 i.u. of HCG will cause ovulation during anovulatory estrus, provided the ovarian follicle is at least 3 cm in diameter.

REFERENCES

Allen, W. E. (1971). The occurrence of ovulation during pregnancy in the mare. *Vet. Rec. 88*, 508–509.

Allen, W. R. and Moor, R. M. (1972). The origin of the equine endometrial cups. I. Production of PMSG by fetal trophoblast cells. *J. Reprod. Fert. 29*, 313.

Andrews, F. N. and McKenzie, F. F. (1941). Estrus, ovulation and related phenomena in the mare. *Mo. Agr. Exp. Sta. Res. Bull.* No. 329.

Arthur, G. H. (1964). *Wright's Veterinary Obstetrics*. London, Bailliere, Tindall & Cox.

Burkhardt, J. (1947). Transition from anoestrus in the mare and the effects of artificial lighting. *J. Agric. Sci. 37*, 64–68.

Clegg, M. T., Cole, H. H., Howard, C. B. and Pigon, H. (1962). The influence of fetal genotype on equine gonadotrophin secretion. *J. Endocr. 25*, 245–252.

Engle, C. C., Witherspoon, D. M. and Foley, C. W. (1970). Technique for continuous collection of equine oviduct secretions. *Am. J. Vet. Res. 31*, 1889–1896.

Harrison, R. J. (1946). The early development of the corpus luteum in the mare. *J. Anat.* (London) *80*, 160–166.

Jennings, W. E. (1950). Twelve years of horse breeding in the army. *J. Amer. Vet. Med. Assn. 116*, 11–19.

Komarek, R. J., Pickett, B. W., Gibson, E. W. and Lanz, R. N. (1965). Composition of lipids in stallion semen. *J. Reprod. Fert.*, *10*, 337–342.

Loy, R. G. and Hughes, J. P. (1966). The effects of human chorionic gonadotrophin on ovulation, length of estrus, and fertility in the mare. *Cornell Vet. 56*, 41–50.

Loy, R. G. and Swan, S. M. (1966). Effects of exogenous progestogens on reproductive phenomena in mares. *J. Anim. Sci. 25*, 821–826.

Mahaffey, L. W. (1950). Studies on fertility in the post-partum oestrus ("foal heat"). *Aust. Vet. J. 26*, 267–273.

Mann, T., Leone, E. and Polge, C. (1965). The composition of the stallion's semen. *J. Endocr. 13*, 279–290.

Nishikawa, Y. (1959). *Studies on Reproduction in Horses*. Tokyo, Japan Racing Association.

Nishikawa, Y. and Sugie, T. (1972). Low sensitivity of horse ovary to gonadotrophin, with special reference to comparison between species of farm animals. *Ann. Kyoto Univ.* No. 101, 45–55.

Oguri, N. and Tsutsumi, Y. (1972). Non-surgical recovery of equine eggs, and an attempt at non-surgical egg transfer in horses. *J. Reprod. Fert. 31*, 187–195.

Pickett, B. W., Faulkner, L. C. and Sutherland, T. M. (1970). Effect of month and stallion on seminal characteristics and sexual behavior. *J. Anim. Sci. 31*, 713–728.

Rowlands, I. W. (1949). Serum gonadotropin and ovarian activity in the pregnant mare. *J. Endocr. 6*, 184–191.

Short, R. V. (1972). "Species Differences." In *Reproduction in Mammals*, Vol. 4. C. R. Austin and R. V. Short (eds.), Cambridge, Cambridge University Press.

Skinner, J. D. and Bowen, J. (1968). Puberty in the Welsh stallion. *J. Reprod. Fert. 16*, 133–135.

Van Niekerk, C. H. and Gerneke, W. H. (1966). Persistence and parthenogenetic cleavage of tubal ova in the mare. *Onderstep. J. Vet. Res. 33*, 195–231.

Witherspoon, D. M. and Talbot, R. B. (1970a). Ovulation site in the mare. *J. Amer. Vet. Med. Assn. 157*, 1452–1459.

Witherspoon, D. M. and Talbot, R. B. (1970b). Nocturnal ovulation in the equine animal. *Vet. Rec. 87*, 302–304.

Chapter 15

Poultry

A. B. Gilbert

In poultry the fertile egg is the sum of the physiologic processes in both male and female. The bird differs from the mammal, in that it has no well-defined estrous cycle and the breeding season is not divided into an "estrous" and a "pregnancy" phase, where one physiologic process supersedes and suppresses what went before. In poultry follicular development and oviductal function continue at the same time.

Modern breeds also differ from the wild species; for example, "broodiness"—the physiologic mechanism which results in the hen incubating the eggs—has disappeared from many breeds of chicken. The modern hen is an "indeterminate layer" and will continue to lay eggs for up to two years, given the right conditions. Similar improvements are being made with other species of domestic poultry.

I. THE FEMALE

A. The Egg

Large by mammalian standards, the poultry egg is formed basically of three

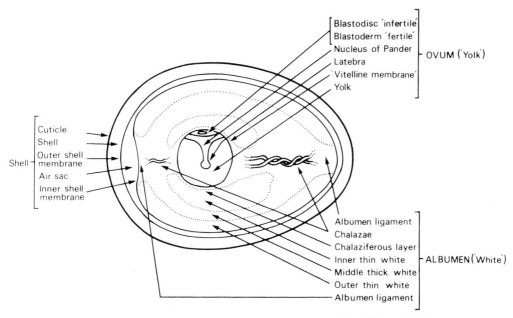

FIG. 15–1. Diagram of the components of the egg. (*From Gilbert, 1971a.*)

components: "yolk" (ovum), "white" (albumen) and "shell" (Fig. 15–1), although none is homogenous either structurally or chemically (Table 15–1); generally it is better to avoid the common names.

1. THE OVUM ("YOLK")

The ovum is bounded by the perivitelline membrane and weighs about 19 gm in the chicken although its size is age-dependent; lower weights are associated with pullets. The living cytoplasmic part of the cell, the *blastodisc*, or the *blastoderm* if the egg is fertile, occupies a small fraction of the ovum. It is a grayish-white spot, 3 mm in diameter, containing in the infertile egg the female haploid pronucleus which has the potential to produce either a male or female embryo; in the bird, unlike the mammal, it is the female which is the heterogametic sex. In a fertile egg the area is covered by a double layer of cells at the time of laying.

The germinal disc floats on a cone of light-colored yolk ("white yolk") which extends downward to end in a ball, the *latebra* (Fig. 15–1). The chemical composition of white yolk differs from that of the yellow: it contains a greater proportion of protein and its structure has certain peculiarities.

The main mass of the ovum (the yellow yolk) is composed of the familiar orange-yellow viscid fluid, which has been described as an oil-water emulsion with the continuous phase an aqueous protein. Scattered throughout this are various droplets and yolk spheres (Bellairs, 1964).

Chemically, the ovum is a heterogeneous mass containing proteins, lipids, pigments and a variety of minor organic and inorganic substances (Fig. 15–2). The low density fraction ("vitellenin" in older literature) contains nearly all the lipid fraction as lipoproteins and much phosphorus. The water-soluble fraction (livetin), which comprises 10% of the ovum, contains about 30% of the ovum protein mainly as plasma albumin, one of the plasma glycoproteins, and plasma γ-globulin. The granular fraction is formed of phosvitin and includes most of the remaining phosphorus of the ovum

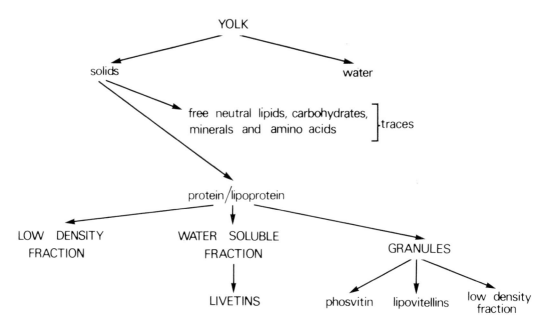

FIG. 15–2. Simplified breakdown of the major components of yolk.
(*Redrawn from Gilbert, 1971a.*)

Table 15-1. Names of the Main Components of the Egg

Common Name	Specific Name	Components

"Yolk"	Ovum	Blastodisc (infertile) or blastoderm (fertile)
		Nucleus of Pander
		Latebra ("white" yolk)
		Yolk ("yellow" yolk)
		"Vitelline" membranes:
		  (a) Primary membrane, the true vitelline
		     membrane
		  (b) Secondary membrane, the perivitelline
		     membrane
		  (c) Two tertiary membranes, formed
		     from the oviduct

"White"	Albumen	"Albumen ligaments" Chalazae Chalaziferous region Inner thin white Middle thick white Outer thin white
"Shell"	Shell membranes	Inner membrane (Air space) Outer membrane
	Mineralized part	Organic matrix Inorganic crystals
	Cuticle	

(From Gilbert, 1971. In *Physiology and Biochemistry of the Domestic Fowl.* D. J. Bell and B. M. Freeman [eds.], New York, Academic Press.)

as well as the majority of the ovum calcium and lipovitellins.

The function of the ovum is to provide the initial material for embryogenesis but it is difficult to distinguish between substances essential for embryonic development and those which may be present as accidents of the mechanisms responsible for ovum formation.

2. ALBUMEN ("WHITE")

Surrounding the ovum and constituting about two-thirds of the egg in weight, the layer of albumen has at least seven major regions, possibly produced by the rotation of the egg in the oviduct. It is amost pure aqueous protein and contains about 40 proteins (Table 15–2).

Since the albumen surrounds the developing embryo and ovum, it forms an aqueous mantle preventing desiccation. It is also known to contribute material to the embryo during later development.

Other functions have been ascribed to specific proteins (such as forming a barrier to bacterial contamination) but there is no conclusive evidence that their properties in vitro are related to their biologic functions in vivo; however many of the proteins in vitro do have bactericidal properties, while others have enzymatic activity and some are enzyme inhibitors.

3. THE "SHELL"

The "shell" is composed of three structures: the membranes, the mineralized part and the cuticle (Simkiss and Taylor, 1971). Two membranes may exist: they are toughened sheets of fibrous protein, together about 70 μm thick; at one end they separate to form the air sac. Functionally the membranes provide the surface on which mineralization can occur; fibers from them penetrate outward to produce the organic matrix of

11

Table 15-2.   Summary of Characteristics of Egg White Proteins from the Hen

Component	Percent	Molecular Weight	Biologic Properties
Ovalbumin	54.0	46,000	
Ovotransferrin	12–13	76,600–86,000	Binds iron, copper, manganese, zinc; may inhibit bacteria
Ovomucoid	11.0	28,000	Inhibits trypsin
Globulins	8.0	Between 36,000 and 45,000	
Lysozyme	3.4–3.5	14,300–17,000	Splits specific β-(1-4)-D-glucosaminides; lyses bacteria
Ovomucin	1.5–2.9	?	Antiviral haemagglutination
Flavoprotein	0.8	32,000–36,000	Binds riboflavin
Ovomacroglobulin	0.5	760,000–900,000	
Ovoglycoprotein	0.5–1.0	24,400	
Ovoinhibitor	0.1–1.5	44,000–49,000	Inhibits proteases, including trypsin and chymotrypsin
Avidin	0.05	68,300	Binds biotin
Papain inhibitor	0.1	12,700	Inhibits proteases including papain and ficin

the shell. They may reduce the speed of bacterial entry, allowing the bactericidal properties of egg white to act more effectively.

The mineral of the shell is almost pure calcite (calcium carbonate), with some magnesium, formed of radiating crystals. Running vertically through the shell are pores which allow gases to pass. The shell, apart from the pores, is impervious to most substances and forms a physical barrier to substances which may adversely affect the microenvironment of the embryo. It also provides mechanical strength and a rigid support to maintain the orientation of the heterogeneous internal components. Its strength is determined mainly by its curvature and thickness, although other factors are involved (Carter, 1970). It also provides calcium for the developing embryo.

The outer covering of the egg, the proteinaceous cuticle, reduces water loss and bacterial contamination. It contains the characteristic pigmented spots.

B. Egg Formation

The egg is a complex structure chemically and physically and its complexity is determined by the requirements to protect and to provide for the embryo living in a hostile environment. Its formation involves the transport of large quantities of material across numerous biologic membranes and the formation of many new substances, particularly specific proteins and lipids.

1. The Ovary

In birds as in mammals, two ovaries and oviducts are formed during embryogenesis but a characteristic feature of birds is the usual suppression of further development of those on the right side.

The suppression is probably brought about by the early production of steroid hormones by the left ovary; removal of the lefty ovary results in development of the right gonad into an ovo-testis with both male and female tissue.

PLATE 16

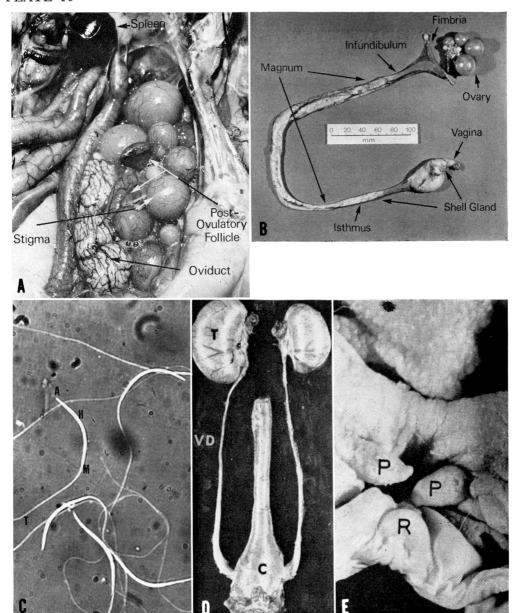

A, The ovary of the domestic hen in situ.

B, The ovary and oviduct of the domestic hen.

C, Characteristic appearance of chicken spermatozoa. Acrosome (A); head (H); midpiece (M); tail (T).

D, Reproductive organs of the male domestic chicken. The vas deferens (VD) leads from the testis (T) to the cloaca (C) ($\times$ 0.48).

E, Copulatory apparatus located in the posterior ventral cloaca of the male chicken showing papillae (P) and the rudimentary copulatory organ (R) ($\times$ 5.6).

PLATE 17

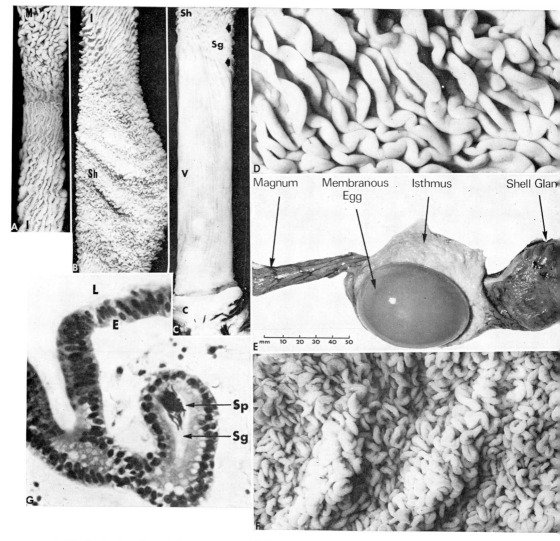

A, The luminal surface of the magnum and isthmus.
B, The luminal surface of the shell gland and its junction with the isthmus.
C, The luminal surface of the vagina.
D, Higher power views of the mucosal folds of the magnum.
E, An egg in situ in the isthmus. Note the membranous covering only and the typical egg-shape.
F, Higher power view of the mucosal folds of the shell gland.
G, "Sperm host" glands of the junction between the shell gland and the vagina.

C, Cloaca; E, epithelium; I, isthmus; L, lumen of oviduct; M, magnum; Sg, "sperm host glands"; Sh, shell gland; Sp, spermatozoa; V, vagina.

The functional left ovary (Plate 16) acts in the same way as those of mammals: it produces the female gametes, it releases them and it acts as an endocrine organ producing the steroid hormones (Gilbert, 1971e).

The ovary consists of a medulla, derived from the primary sex cords, which contains connective tissue, blood vessels and nerves, and a cortex derived from the secondary sex cords. The cortex contains the oogonia which give rise to the oocytes and the ova. The pear-shaped, immature ovary is about 15 mm long by 5 mm wide lying in the body cavity, ventral to the aorta and cranial to the kidney, and close to the two adrenal glands. Although variable, the blood supply usually arises from the gonadorenal artery. Two veins drain blood from the ovary. It is extensively innervated from the sympathetic chain via the adrenal/ovarian plexus.

Follicular Development and Gametogenesis. At the onset of sexual maturity there is an increase in ovarian weight from 0.5 to between 40 and 60 gm; most of this comes from the four to six developing follicles, the largest of which weighs about 20 gm with a diameter of about 40 mm; the main mass of the ovarian tissue only increases to about 6 gm.

Of the thousands of oocytes present only one starts its development at any given time. The interval between the onset of development of successive follicles appears to be about 24 hours or multiples of this, brought about possibly by the follicular stimulating hormone (FSH) and the luteinizing hormone (LH).

Maturation of the female pronucleus begins in the follicle and is completed in the oviduct. The first (reduction) division occurs two hours before ovulation, possibly under control of LH; the second (maturation) division occurs in the oviduct and may be initiated by the penetration of the sperm.

Development of the oocyte and yolk deposition can be considered in three phases (Marza and Marza, 1935). The first, lasting for many months or years, is characterized by a slow deposition of material consisting mainly of neutral fat. It is superseded by the second period of several weeks during which time the size of the follicle increases from about 1 mm to about 6 mm. During the last stage (the "rapid-growth phase") of seven to eight days the main mass of yolk material is laid down and follicular weight increases almost linearly until it reaches 18 or 19 gm.

None of the material of the yolk is formed directly within the ovary, though some rearrangement and combination may occur inside the oocyte. The liver is the major source of yolk proteins and phospholipids controlled by the ovarian estrogens (McIndoe, 1971). Since the ovarian follicle only accumulates yolk, the blood supply within the follicular wall is peculiarly complex, terminating in a capillary network adjacent to the basement membrane of the granulosa layer, to aid in the transport of material from the blood to the ovum.

Of importance is the accumulation of specific components of the yolk: although a uniform amount of material is deposited each day, the proportions of the fractions vary from time to time. More is becoming known of this (Gilbert, 1971c), but the functional significance of the differen-

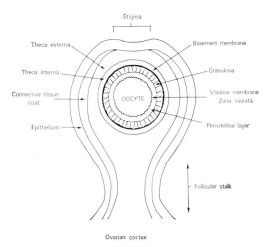

Fig. 15–3.   Cross section of a maturing follicle. *(From Gilbert, 1971c.)*

tial accumulation is not understood. Since follicles at different stages of development are present in the ovary at the same time, variations in liver output are not the cause: the follicle or the oocyte itself must control the accumulation of material in this way.

There has been some doubt whether or not yolk deposition continues indefinitely, to be terminated only by ovulation. Modern evidence suggests that yolk deposition often follows a fixed time course within certain limits, whether or not ovulation occurs. Only two courses of development are available to the follicle: more commonly it continues until such time as ovulation occurs, with the release of the oocyte, but it may become atretic. The changes which occur in atresia are not precisely known but the result is a general liquefaction of the yolk material and its resorption.

Ovulation. Ovulation occurs through rupture of the stigma (Fig. 15–3), a specialized region of the follicle. The hormone involved—ovulation-inducing hormone (OIH)—may be LH, as in mammals, for mammalian LH will cause ovulation in the chicken and fowl LH is a powerful ovulation-inducing hormone (Fraps et al., 1947). However LH per se may be incapable of causing ovulation and may have to be changed in the blood. In general the follicle must be mature enough to react to the ovulation-inducing stimulus (Fig. 15–4); however the follicle may be capable of reacting to the stimulus for several days since premature (17–30 hours) ovulations can be induced and some follicles ovulate 24 hours after they would normally be expected to.

The interval between pituitary release of LH and ovulation appears to be between four and eight hours (Cunningham and Furr, 1972) and there is evidence that its release may occur on at

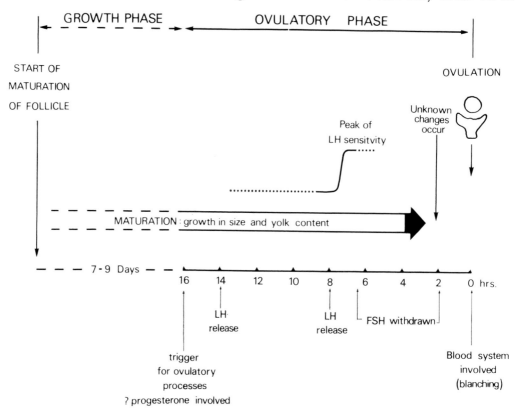

Fig. 15–4.   Stages in the maturation of the follicle and ovulation.   (*From Gilbert, 1971c.*)

least two occasions, the second occurring between 14 and 20 hours (Gilbert, 1971*d*). The ovarian steroids are involved, as in mammals, but whether estrogen or progesterone is the main one has yet to be resolved. However, unlike in the mammal, evidence at present favors the hypothesis that progesterone is related to pituitary regulation of LH. Whatever steroid is involved, it probably acts through the hypothalamus: progesterone is inactive when injected into the pituitary (Ralph and Fraps, 1960).

The physical events which lead to the rupture of the stigma are unknown; none of the many suggestions is tenable in the light of modern research but intrafollicular pressure is not involved, for premature ovulations occur. Nor does it seem likely that necrosis occurs.

2. THE OVIDUCT

The oviduct, about 700 mm in length, is suspended by the peritoneal dorsal ligament which continues around it to form the ventral ligament (Plate 16) (Aitken, 1971). It is highly vascular and the muscular layers are well-supplied with nerves from the autonomic system.

The functions of the oviduct are production of the remaining formed elements of the egg (Fig. 15–5) and transport of the developing egg. The oviduct consists of a glandular folded mucosa and of muscle arranged into an outer longitudinal layer and a thinner circular layer. The remainder of the formed elements of the egg are dissimilar and each is produced by a specialized region of the oviduct (Fig. 15–6): thus five major regions can be distinguished, each with a characteristic distribution of mucosa and a muscularis (Plate 17) (for more detailed descriptions see Aitken, 1971).

The Infundibulum. The infundibulum actively engulfs the shed ovum. In rare cases unovulated follicles attached to the ovary may pass down the oviduct and subsequent ovulation may occur within the oviduct. Its activity is due to the fimbriae, projections of the cephalic surface of the infundibulum, which become engorged with blood and also activity of the smooth muscle of both the infundibulum and the ventral ligament.

After passing through the initial funnel-shaped region of the infundibulum the ovum reaches the "chalaziferous region," a narrower, glandular portion (Plate 16). This region may contribute to the formation of the chalazae, for albumen is added, and to the perivitelline membrane. However, the formation of the

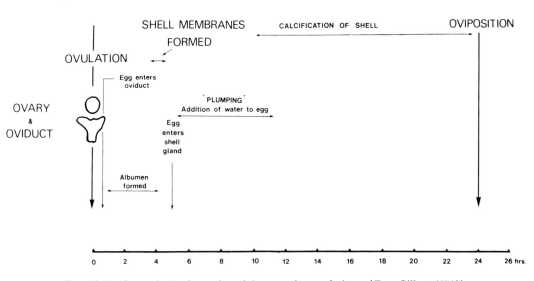

FIG. 15–5. Stages in the formation of the egg after ovulation. (*From Gilbert, 1971f.*)

perivitelline membrane and the chalazae is a continuous process starting in the infundibulum and ending in the distal oviduct.

The Magnum. The albumen-secreting region is longer than any other part, distinguishable from the infundibulum and isthmus by its creamy-white color, its greater external diameter and its thicker walls (Plate 16). The lining is thrown into marked folds filled with secretory cells (Plate 17) (Aitken, 1971).

Unlike yolk proteins, which are not formed in the ovary, the majority of the proteins of albumen are formed in the oviductal tissues; a possible exception is ovotransferrin (conalbumin) which appears to be similar to serum transferrin. However the cells responsible for the formation of most of the 40 or so proteins are unknown: it is unlikely that the "mucin" cells produce any specific protein, but recent evidence suggests that avidin is formed in the epithelial goblet-cells and ovalbumin and lysozyme may be produced by the tubular glands.

The general processes involved in protein synthesis are well-understood: amino acids and ATP are activated to form a complex which reacts with *transfer* RNA; this in turn combines with the complex ribosomal *messenger* RNA (mRNA) formed under the influence of DNA; the amino acids are then incorporated into the peptide chain according to the coding supplied by the mRNA. Most of the slender evidence suggests that similar processes occur in the cells of the chicken oviduct. Estrogens are involved for their administration increases total oviductal proteins and specifically lysozyme, ovalbumin and avidin. In contrast, progesterone may be involved in the formation of only one protein, avidin, though circumstantial evidence suggests that it may have other actions also.

Release of the formed protein occurs as the developing egg passes down the magnum; on a priori grounds there are three possibilities: (*a*) the developing egg by mechanical or other means causes release of the albumen; (*b*) hormonal mechanisms synchronize release; or (*c*) some neural coordinating mechanism is involved. At present (*a*) seems most likely but the others cannot be excluded.

Isthmus. The membrane-secreting region of the oviduct is placed between the magnum and the shell gland and is narrower with thinner walls than either (Plate 16). Its structure is similar to the magnum (Aitken, 1971) although the

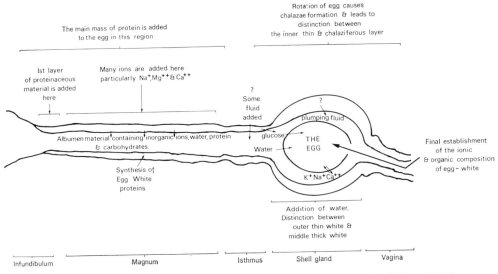

FIG. 15–6.   Diagrammatic representation of oviductal function. (*From Gilbert, 1971b.*)

glands appear to contain material of a keratinous nature.

Little is known of the mechanisms involved in the formation of the protein or its release and deposition on the egg. However, partly-membraned eggs can be obtained and it has been observed that while the albuminous egg is entering the isthmus a membrane is deposited on those parts in contact with the glandular tissue. In the chicken, and probably other species, the typical shape of the egg is produced by the membranes and not the calcified shell which is laid down on an already formed shape (Plate 17). At this stage the membranes are loosely applied to the egg which has about 50% of its final mass.

Shell Gland. Reference to the "uterus" of birds should be avoided for this implies a homology with the mammalian structure which is not justified; moreover the functions are vastly different. The shell gland is characterized by a pouch-like appearance after a short "neck" and extensive muscularization (Plate 16). Microscopically there are extensive mucosal folds, studded with glandular cells.

The egg remains in the shell gland for approximately 20 hours (Fig. 15–5). During the first six hours or so, a watery fluid passes through the membranes ("plumping") resulting in a two-fold increase in the mass of egg white (Fig. 15–6); although this may continue throughout the time the egg stays in the shell gland; thereafter, the main process is calcification. During its stay in the shell gland rotation of the egg around the polar axis leads to the formation of the chalazae, started in the infundibulum, and the stratification of the albumen.

Shell calcification probably starts in the isthmus where small projections from the membrane, the mammillary cores, are formed (Simkiss and Taylor, 1971). If the original "seeding" of the calcium crystals occurs in this region, then the shell gland can be regarded simply as a region of growth of the crystals at a constant rate of mineralization (about 300 mg calcium/hour). The oviduct does not store calcium and about 20% of the calcium in the blood is removed as it passes through the shell gland. The cells responsible for transferring calcium into the lumen may be ciliated epithelial cells.

Once thought to be a function of the vagina, the final task of the shell gland is the formation of the cuticle and pigmentation; the latter consists of porphyrins deposited shortly before laying.

The Vagina. The vagina acts as a passage from the shell gland to the cloaca and plays no part in the formation of the structure of the egg (Plate 17). But it does have an important role to play in the processes involved in fertilization for it contains the "sperm-storage glands" typical of birds. It is a relatively short, "S"-shaped tube closely bound by connective tissue to the shell gland and similar in structure to other regions of the oviduct.

3. TRANSPORT OF THE EGG THROUGH THE OVIDUCT AND OVIPOSITION

The developing egg spends about 15 minutes in the infundibulum (Table 15–3). In the magnum its speed averages about 2 mm/minute and hence it takes about two to three hours to traverse this region. After entering the isthmus the egg may be held for a short while before proceeding on its way; it takes about 1–1½ hours to pass along the isthmus. Most (about 20 hours) of its time in the oviduct (26 hours or so) is spent in the shell gland. Passage through the vagina takes a few seconds usually. The rate of transport is not uniform throughout the oviduct so each region must have its own, unknown, mechanisms for coordination and control.

It is not known precisely how oviposition is initiated but both hormonal and neural mechanisms are involved: it may be related to the ovulation of the ovum 24 hours previously or events associated with it. Whatever the initial stimulus, contraction of the shell gland musculature forces the egg into the

Table 15-3. Number of Hours Spent by the Developing Egg in the Various Regions of the Oviduct

Species	Total	Infundibulum	Magnum	Isthmus	Shell Gland
Chicken	25	0.25–0.5	2–3	1.25	18–20
Turkey	26	0.25–0.5	2.5–3	1–1.5	22–24
Quail	25	0.25–0.5	2–2.5	1.5–2	19–20

vaginal region. This may be controlled neurally, for stimulation of the central nervous system can affect it but hormones are involved: there is a depletion of the oxytoxic hormones (mainly arginine vasotocin in birds) in the pituitary and an increase in their level in the blood; blood oxytocinase decreases in activity; the shell gland musculature becomes increasingly sensitive to these hormones. Steroid hormones may play a similar role to those in mammals (progesterone acts to inhibit muscular activity), though suggestions have been made that estrogen inhibits avian shell gland activity; this has yet to be proved.

With the contractions of the shell gland musculature there occurs relaxation of the uterovaginal "sphincter," possibly analogous to the cervix of mammals. As this occurs the egg is pushed further into the vagina, and the general distension of the vaginal wall brings into operation the "bearing-down" reflex; this involves changes in respiration and stance, and contraction of the abdominal body muscles.

II. THE MALE

A. Semen: Its Physical and Chemical Properties

An ejaculate of a cock varies from 0.2 to 0.5 ml and has an average density of about 3×10^6 sperm/mm.

1. MORPHOLOGY OF THE SPERMATOZOON

Sperm of avian species differ from those of domesticated mammals in being smaller, in having long and filamentous heads and in having no kinoplasmic droplet. An acrosome cap, head-, middle-and principal-piece of the tail are present (Lake, 1971).

Although differences exist between domestic birds, the shape and size of spermatozoa are reasonably consistent (Plate 16). In the fowl the headpiece, curved and 12–13 μm long, is surmounted by the acrosome cap (2 μm long). The tail midpiece is about 4 μm long and the remainder of the sperm's length of 100 μm is made up of the principal-piece of the tail. At its widest part the sperm measures about 0.5 μm.

The spermatozoon is surrounded by the cytoplasmic membrane. The acrosome is simple with an inner spine surrounded by a conical-shaped cap. The head consists essentially of only the nuclear material of the gamete. The midpiece contains the cylindric centriole surrounded by a sheath of about 30 curved platelike mitochondria; the latter are dissimilar to the elongated curved cylinders of mammals and fewer in number. The typical outermost fibers of the mammalian posterior midpiece and the principal-piece of the tail appear to be missing. Throughout the tail runs the ring of nine doublet fibers surrounding two central ones.

2. CHEMICAL COMPOSITION OF THE SPERMATOZOON AND SEMINAL PLASMA

Complex lipids and glycoproteins can be demonstrated around the sperm and their presence in the acrosomal cap may be related to sperm penetration of the female gamete. The tail contains phospholipid which may be metabolized as one source of energy.

The chemical composition of seminal plasma has been studied and between

50 and 60 constituents are known (Lake, 1971). However, it is still not possible to distinguish those substances which are essential for spermatozoon function from those which are present as accidents of the reproductive system. Seminal plasma is essentially an aqueous solution of salts with some amino acids. When uncontaminated with "transparent fluid," fowl seminal plasma differs considerably from that of the mammal: it is almost completely lacking in fructose, citrate, ergothioneine, inositol, phosphoryl choline and glycerophosphoryl choline. Another interesting feature is the low chloride content, for glutamate is the chief anion.

B. Formation of Semen

1. The Testis and Spermatogenesis

The anatomy of the reproductive organs of the bird differs in several important respects from that of mammals (Fig. 15–7 and Plate 16) (Lake, 1971). The avian testes are situated in the abdomen, not in a scrotum, and the ductus epididymidis is short by com-

parison. It is not known whether or not these differences are related to the peculiar reproductive phenomena in birds, where sperm can survive for many days in the female oviduct.

In the avian male, unlike the female, both gonads and ducts develop. The testes together weigh between 14 and 60 gm in the cock, depending on the breed, and are suspended from the dorsal body wall just posterior to the lung and ventral to the kidney. Blood is supplied via the anterior renal artery and a variable testicular artery but the arrangement of blood vessels within the testis is much simpler than in mammals and there is no pampiniform plexus.

In contrast to the arrangement in the mammals, the seminiferous tubules are not grouped into evident lobules surrounded by connective tissue, but branch and anastomose freely within the tunica albuginea; in the mature cock branches of the latter penetrate between the tubules to act as a supporting skeleton. Interstitial tissue is negligible, but it contains the androgen-secreting Leydig cells. The

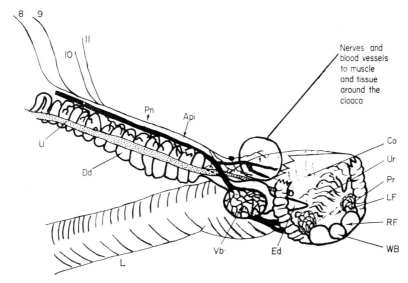

FIG. 15–7. Diagrammatic representation of the pelvic portion of the right male reproductive organs, including the distal part of the ductus deferens and the erectile phallic structures in the cloaca. *Dd*, ductus deferens; *Api*, artery, pudenda interna; *L*, large intestine; *Pr*, proctodaeum; *Ur*, urodaeum; *Co*, coprodaeum; *LF*, lymph fold; *WB*, white body; *RF*, round fold; *Vb*, vascular body (gefassreicher Korper); *U*, ureter; *Pn*, pelvic nerve with contributions from lumbosacral nerves, 8,9,10 and 11; *Ed*, ejaculatory duct (papilla). (*From Lake, 1971.*)

seminiferous tubules of immature males are lined by a single-celled layer of Sertoli cells and stem spermatogonia, while mature males have irregularly-shaped tubules lined by a multi-layered germinal epithelium. The spermatogonia give rise to the primary spermatocytes, secondary spermatocytes and spermatids: the latter eventually metamorphose into spermatozoa.

The time course for maturation of the sperm in the testis depends on the avian species. In the White Plymouth Rock, spermatogonia multiply at about the fifth week after hatching and primary spermatocytes appear at about the sixth week. At about the 10th week these multiply rapidly, secondary spermatocytes appear and the tubules increase in size. Spermatids first appear soon afterward and continued development occurs in the tubules until the 20th week. Thereafter the testis appears capable of producing spermatozoa in large quantities.

2. The Duct System

The fowl lacks the characteristically mammalian coiled and subdivided ductus epididymidis. Moreover, spermatozoa are stored mainly in the ductus deferens and there are no accessory reproductive organs, namely seminal vesicles, prostate, Cowper's glands and urethral glands.

Ductules corresponding with the rete testis and ductuli efferentes are embedded within the connective tissue attachment of the testis to the body wall. The seminiferous tubules unite with those of the rete testis, and these with the ductuli efferentes. Thereafter the ductuli efferentes connect in several places with the short ductus epididymidis which opens into the ductus deferens, an extensive convoluted tube running the whole length of the abdomen on the surface of the kidney (Fig. 15–7). At the distal end the ductus enlarges before passing through an erectile papilla extending into the cloaca; there is no comparable structure to the ampulla of mammals (Fig. 15–7).

The male has no penis, but an erectile phallus (called a penis in waterfowl) is present, and it is believed that this makes contact with the everted vagina during copulation (Plate 16). Erection of the phallus is caused by it becoming engorged with blood, as are related cloacal structures (lymph folds). Presumably erection is controlled by nerves.

C. Reproductive Function in the Male

Though evidence is lacking, it is generally held that FSH controls tubular growth and differentiation and that LH affects the interstitial Leydig cells which produce the steroid hormones.

Other hormones—prolactin, TSH and ACTH—have been implicated in controlling mechanisms, but clear evidence for their normal role has yet to be found. Androgens may not have any effect on spermatogenesis, except perhaps the transformation of secondary-spermatocytes to spermatids. Their main action is a negative feedback on gonadotropin output by the pituitary.

III. FERTILIZATION AND FERTILITY

"Fertilization," "fertility" and "fecundity" have been used ambiguously though "fertilization," the term used for the penetration of the female gamete by the male and the fusion of the gametic pronuclei, is seldom now used incorrectly for insemination. On the other hand, "fertility" has been used to mean "producing many offspring." This had led to confusion because "poor fertility" has been used of flocks in which both fertilization and embryonic mortality are high and of those where true fertilization is low. In the industry "fertility" has yet another meaning. In the present text its use is restricted to its association with fertilization. Fecundity, "producing abundantly," is used to include all aspects of fertility, embryogenesis and hatchability which affect the number of offspring produced.

A. Mating and Insemination

1. Natural Mating

The introduction of semen into the female, a vital stage, is associated with complex courtship behavior in birds (Hafez, 1962; Wood-Gush, 1971).

A diurnal rhythm in mating frequency exists in males and is correlated with semen production. But libido in the male is not necessarily correlated with high fertility and there is a tendency for the most frequent copulators to produce many aspermic ejaculates. Embryonic mortality may increase in flocks served by highly active roosters.

Best fertility is obtained with a ratio of males to females that prevents the formation of a "peck order" among them. Particularly troublesome is the "dominant male" (associated with aggression); he may not be sexually the most active but may prevent other males from mating successfully. Older males introduced into a society may not be able to express their full reproductive potential.

Cocks tend to prefer mating with their own breed or strain; possibly this is related to early learning of visual and behavioral cues, for contact in early life with other strains enhances future mating. Hence heterosexual rearing is important.

Libido of the female is affected by her rearing environment—whether or not males were present—and by her social ranking: receptivity (crouching) varies from hen to hen and this affects the chance of successful mating because the female probably determines whether or not the sexual advances of the male will lead to copulation. Following persistent courtship the female may crouch and allow the male to mount.

2. Artificial Insemination

Factors to be considered in artificial insemination include: (a) laying intensity of female; (b) age of hen and/or advancing season; (c) position of egg in the oviduct; (d) placement of semen in the oviduct; (e) number of spermatozoa inseminated; (f) frequency of insemination; (g) quality of semen due to inherent factors and to its collection; and (h) dilution and storage of semen (cf. Lake, 1967).

B. Survival of the Spermatozoa in the Oviduct

The ejaculated semen is deposited at the entrance to the oviduct and contractions of the oviduct aid in moving the semen further into the vagina. Spermatozoa survive within the avian female for long periods (Lake, 1967): up to 32 days in the chicken and 70 days in the turkey, though the commonly known "fertile period" is usually shorter. Individuals and breeds vary considerably.

This longevity of spermatozoa is associated with at least one specialized region of the oviduct. At the uterovaginal junction, specialized tubular glands, the "sperm host glands," are located (Bobr et al., 1964). At present the factors concerned in survival of the sperm within the glands are unknown and it is common experience that some hens fail to maintain sperm internally.

There is still controversy whether dead sperm enter the gland or whether active movement by the sperm is necessary. If the latter, some chemotactic substance may be involved. Whatever the cause, sperm rapidly enter the glands and disappear from the vaginal lumen. The glands may provide some barrier to the passage of defective sperm since more embryonic deaths result when insemination is intramagnal. Similar glands occur in the infundibular region of the oviduct and, following intramagnal insemination, sperm can be found there in large numbers; however this is contrary to the findings following normal vaginal insemination, and the role of the infundibular glands has yet to be elucidated.

Antibodies to sperm can be produced following the intravenous injection of spermatozoa and that leads to a decrease in fertility; however, present evidence suggests that immune responses play little part in affecting fertility under natural conditions. As fertilization occurs in the infundibular region another requirement is the release of sufficient spermatozoa from the sperm host glands and their transport along the oviduct. Small, discrete "packets" of sperm may be released from the glands at about the time of ovulation, for sperm can easily be collected from the oviducal lumen at this time. On the other hand, recent evidence indicates that release may be a continual process and that sperm are present, though difficult to recover, at all times.

Oviduct motility may be important for sperm to ascend the oviduct and waves of muscular contraction may pass along the oviduct in a direction opposite to that for the transport of the egg. Also secretions of the oviduct may aid transport since extracts of the magnum increased respiration and sustained motility of spermatozoa better than extracts from other regions.

C. Fertilization

Only one male nucleus fuses with the female pronucleus but about 150 spermatozoa are embedded in the vitelline membrane in a way similar to that in which a large number of sperm are present in the mammalian zona pellucida. It is not certain that this number is essential, although a reduction in numbers of spermatozoa at the anterior end of the oviduct leads to reduced fecundity. Alternatively an excessive number of spermatozoa at the infundibulum may lead to early disturbances of embryonic development: perhaps the vitelline membrane has no mechanism to prevent polyspermy, as it has in the mammal, and the rapid coating of the ovum with a thick layer of albumen may have made this development unnecessary under normal conditions. In fact polyspermy may be normal though it has not been established conclusively that it is so. Capacitation is of considerably less importance than in mammals.

Lysine in the acrosomal cap of mammalian sperm is involved in the dispersal of the corona radiata and penetration of the ovum. Lake (1971) suggested that the lipoglycoprotein associated with the acrosome of cock sperm may aid penetration and others have demonstrated the presence of an enzyme capable of initiating rupture of the vitelline membrane.

IV. DEVELOPMENT OF THE EMBRYO

A. Embryogenesis

It is not proposed to consider embryogenesis in detail, for the chick embryo has been the subject of study for many years (Romanoff, 1960). The maturation division and extrusion of the second polar body occurs after penetration by the sperm. In both chickens and turkeys the first cleavage division occurs in the isthmus (i.e. about five hours after ovulation). The second cleavage occurs 20

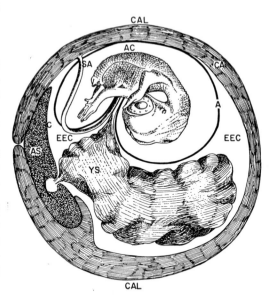

Fig. 15–8. A semi-diagrammatic presentation of the in situ relationship between the 10-day-old chicken embryo and the extra-embryonic membranes. *A*, Amnion; *AC*, amniotic cavity; *AS*, albumen "sac"; *C*, chorion; *CA*, allantoic cavity; *CAL*, chorioallantois; *EEC*, extra-embryonic cavity; *SA*, allantoic stalk; *YS*, yolk sac. (*Adapted from Needham, 1942. Biochemistry and Morphogenesis. Cambridge, courtesy of Cambridge University Press.*)

minutes later, and by the time the egg has been in the shell gland for four to five hours it has reached the 256-cell stage. Thereafter development continues to the gastrula stage at about the time of laying.

On incubation the embryo appears within the transparent area pellucida. As in other vertebrates, the first major development after 24 hours is in the brain and neural tissues. By the time the embryo is 48 hours old, the heart, blood system and rudimentary gut have been established.

Lacking a placenta, the avian embryo has to depend on the extra-embryonic membranes for the necessary metabolic functions associated with nutrition, respiration and excretion. By the end of the third day the allantois, chorion and amnion start to form: the allantois serves initially as a storage organ for nitrogenous wastes but, after joining with the chorion, it forms the respiratory organ that serves the embryo until shortly before hatching (Fig. 15–8).

B. Factors Affecting Embryogenesis and Hatching

1. Biologic Factors

Hatchability is a heritable trait, to which both the male and female contribute. Certain embryonic deformities are known to be associated with certain genes ("lethals"). The position of the egg in a sequence is not related to hatching, but eggs from one-egg cycles hatch worse than, and those from seven-egg cycles hatch better than those from cycles of other sizes.

High embryonic mortality occurs with spermatozoa that have resided in the oviduct for a long period; often this is associated also with low fertilizing capacity of the sperm, but it is not known how the sperm are damaged. However, where it is likely that abnormal sperm have succeeded in fertilizing eggs, more embryonic mortality occurs (Ogasawara et al., 1966).

2. Environmental Factors

Hatchability reflects the level of embryonic mortality. Though mortality occurs over the whole period of incubation, there are two peaks at two to three days and 18 days in the chicken, and four to five days and 25 days in turkeys. Similar peaks have been found in other domestic species. Why this should be so is not apparent, though presumably the peaks indicate periods of severe difficulty for the embryo; environmental conditions are likely to be critical at these times and variations, unimportant at other times, may be sufficient to prevent the embryo's survival.

Temperature. The temperature that produces the best hatch from a group of eggs falls within the range 37°–38° C but some eggs hatch at temperatures of 35° C and 40° C. Though important during the early stages, the actual temperature is less critical after the 16th day and the maintenance of a constant embryonic temperature may be more important than the actual level. Short periods of increased or decreased temperature may lead to embryonic defects, depending on embryonic age and size, the deviation in temperature from the normal and duration of exposure. Younger embryos tend to be more resistant to temperatures below the optimum and older ones more resistant to those above the optimum.

Relative Humidity. Environmental humidity affects water loss from the egg but wide fluctuations in relative humidity are tolerated (between 40% and 70%). During incubation the chicken egg loses about 25% of its water content and abnormal losses, either high or low, lead to decreased hatchability. Excessive loss may lead to drying of the chorioallantoic complex preventing adequate gaseous exchange. Humidity and temperature interact, but details are lacking.

Gaseous Concentration. There is a continuous relationship between hatchability and oxygen partial pressure; hatchability falls with decreasing oxygen concentration below 15% and increasing concentration above 40%. Above 1% carbon dioxide, hatchability decreases and the embryo is more sensitive to increases in carbon dioxide than decreases in oxygen. A continuous relationship exists between carbon dioxide concentration and hatchability down to 0.2%.

Turning. It is essential for the avian embryo to have free movement within the amnionic cavity. Unless correct turning is carried out the embryo tends to adhere in a fixed spot within the membranes resulting in abnormal development and death. In artificial incubators complex interactions occur between frequency of rotation, axis of rotation, angle of rotation and critical orientation of the egg. Whether or not these occur naturally cannot be stated because it is not known how closely incubation turning mimics that by the hen.

Broodiness. In modern strains of chicken broodiness plays little part in reproductive processes. However, in some commercially important species, such as the game birds, broodiness is still important. What initiates broodiness is uncertain, although sometimes it is related to the presence of "sufficient" eggs

in the nest or nest site to trigger the hormonal responses; thereafter, gonadotropin output from the pituitary decreases and increased prolactin production brings about the behavioral responses associated with sitting on the nest and incubation of the eggs.

V. REPRODUCTIVE PERFORMANCE IN THE FEMALE

A. Reproductive Cycles

In each estrous cycle of the mammal all mature follicles ovulate simultaneously. If fertilization takes place implantation occurs and further follicular development is suppressed by progesterone from the corpus luteum; if fertilization does not occur, the estrous cycle runs its course followed at some subsequent time by a further cycle, and so on. In contrast to this, a daily sequence of single ovulations typifies the bird, each followed by the completion of an egg and daily oviposition. This sequence is terminated by events about which we know little, though it may be the tactile stimulus of the eggs in the nest: in some species if eggs are removed from the nest further ovipositions occur until the required number remain to stimulate suspension of ovarian activity.

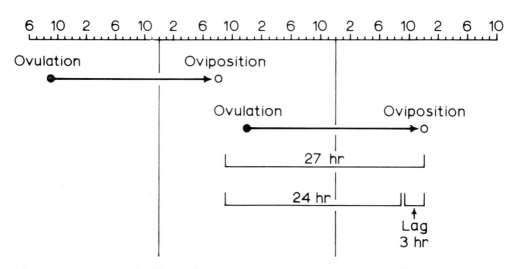

Fig. 15–9.   The relationships between successive ovipositions and lag.
(*From Gilbert and Wood-Gush, 1971.*)

Thereafter incubation behavior (broodiness) occurs; the ovary and oviduct regress to the juvenile condition, a situation presumably analogous to the anestrous condition of the seasonally breeding mammal.

Domestic species, particularly the chicken, may produce several hundred eggs during one laying season; thereafter ovarian activity may continue for many months although at a reduced rate. But egg laying is not strictly daily: a sequence of one or more successive daily eggs is interspersed with usually one day when no egg is laid (pause day). It is difficult therefore to equate the laying patterns of domestic and wild species: to avoid confusion the term "clutch" should be restricted to the sequence of eggs of wild birds, usually of a fixed number within small limits; "sequence" or "cycle" are terms better used for those of domestic species.

Typically hens only oviposit during the daylight hours. The first egg of a cycle is laid usually between 6 and 8 A.M. and each successive egg is laid later in the day than the previous one (Fig. 15–9). This difference in time from day to day is termed "lag" (Fraps, 1961); it depends on the length of the sequence and the position in the sequence of the two eggs concerned: in long sequences lag tends toward zero but may become negative, and lag between the first two members of a cycle tends to be greater than that between others.

The terminal egg is laid toward late afternoon; the next egg follows about 36 hours afterward, i.e. it is the first egg of a sequence laid early in the morning.

Since the time spent in the oviduct is relatively constant for all eggs, it follows that LH release and ovulation must occur successively later each day. It is known that LH release is triggered by the change from light to dark on the night before the first ovulation of a sequence and that this sets the pattern for the following daily LH release: but why is the inherent rhythm somewhat greater than 24 hours when a bird is subjected to a normal solar 24-hour cycle?

The cycle can be terminated in one of several ways:

(a) Ovulation does not occur even though a mature follicle is present which can be ovulated in response to hormonal injection and the pituitary is capable of releasing LH. For an unknown reason LH release does not occur. The follicle ovulates when the next release of LH occurs. During this period more yolk material may be laid down. (b) It may be terminated by the absence of a mature follicle; within the follicular hierarchy "gaps" occur (Gilbert and Wood-Gush, 1971). (c) Occasionally follicles become atretic and are unable to ovulate.

To explain the rhythmic release of LH, Fraps (1961) postulated its release in response to an excitation hormone acting on a neural threshold which varied cyclically. But LH may be released on more than one occasion whereas progesterone levels reach a peak at one time only: whether this peak is in response to or the cause of LH release is uncertain. Future work may show that estrogens act in a way similar to that in mammals.

B. Total Egg Production

Much of the present discussion deals specifically with the chicken though in general terms it is applicable to other avian species (Table 15–4). After the first egg, production rises sharply and a peak of 90% or more is maintained for several weeks; thereafter a decline takes

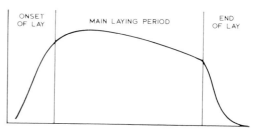

Fig. 15–10. The three periods of the chicken's laying season.

Table 15-4.   Comparative Reproductive Performance* of Some Birds of Commercial Importance

Species	Incubation Period (Days)	Age at Sexual Maturity	Egg Weight (gm)	Number of Eggs/Year	Fertility (%)	Hatchability of Fertile Eggs (%)
Chicken (*Gallus gallus*)	21	5–6	58	230–270	90	90
Turkey (*Meleagris gallopavo*)	28	7–8	85	90	80–85	80
Duck (*Anas platyrhychos*)	27–28	6–7	60	180	95	70
Goose (*Anser anser*)						
Small type	30	9–10	135	60	70	80
Large type	33	10–12	215	50	65	75
Pheasant (*Phasianus colchicus*)	24–26	10–12	30	60	95	85
Guinea fowl (*Numida meleagris*)	27–28	10–12	40	70	90	95
Quail (*Coturnix*)	15–16	1.5–2	10	300	90	75–85

* Only general figures are given since values are affected by breed, location and nutrition. In particular the values given in the last three columns are very dependent on management practices and it is likely that, biologically, fertility and hatchability will be the same for all species.

place and after about one year production falls to 40% (Fig. 15–10).

Since there is an early period of growth and development during which reproduction does not occur and since the hen's reproductive phase lasts for a limited period, the number of eggs which can be produced by a hen is dependent on when her sexual maturity starts, when it ends and her rate of laying during this period.

1. Onset of Sexual Maturity

Reproductive potential of any animal depends on its genetic makeup, which can be altered by selective breeding. In poultry, breeders have concentrated on selecting for early sexual maturity. The most important factor is photoperiod (Fig. 15–11); food and temperature have little effect on sexual maturity.

The hypothalamus-pituitary axis is regulated by day length. Light strength, above a very low level, and wavelength are unimportant. Hence there is a direct correlation between day length during the rearing period and the age at onset of sexual maturity; moreover of more importance is a steadily increasing day length during this period (Morris, 1967).

At this time a general increase in output of the gonadotropins from the pituitary stimulates ovarian production of follicles and steroid hormones. These bring about changes in the oviduct and the secondary sexual characters. Many problems arise with egg quality during this early period: among them are the

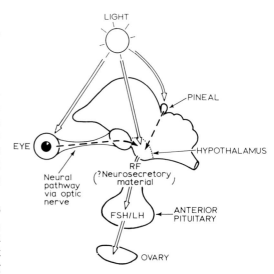

Fig. 15–11.   Light may affect reproductive activity through at least three possible pathways. Though their relative importance is not entirely known, it seems most likely that the main one is through the eye. In reponse to light the hypothalamus produces the releasing factors (*RF*) which in turn regulate pituitary formation of the gonadotropins. (*From Gilbert, 1971c.*)

production of defective eggs or soft-shelled eggs and 40% of the ova produced may fail to enter the oviduct.

2. MAIN PERIOD OF LAY

After the first period, hens typically lay regular sequences, the length of the sequence being a characteristic of the hen. This period lasts about one year; though eggs may be produced longer, production levels are low (and uneconomical) and egg quality deteriorates. Commercially it is customary to kill birds at this time or to induce a cessation of lay for several months followed by a shorter second year's production when egg quality and performance are improved.

Total production depends on the length of the period but rate of lay is also important since a hen laying at 60% for nine months produces the same number as those laying at 90% for six months. Rate of lay depends on the length of the sequence, e.g. a hen laying three-egg sequences can only attain a maximum of 75% production whereas one having sequences of 24 eggs can reach 96%.

Other factors are important (Gilbert, 1972). There appears to be an optimum egg size for production; large and small eggs are associated with poorer production figures. Since one important factor controlling egg size is the size of the yolk, yolk transport is related to total production. The best producers tend to have more developing follicles in the ovary than poor producers.

Calcium is involved not only in shell formation but also in regulating total production. This is not surprising: the turnover of calcium in the formation of one shell is equivalent to 10% of the total body calcium and few animals have such a consistent drain on their calcium resources as the bird (Simkiss and Taylor, 1971). Also, mechanisms exist to protect the skeletal tissues; surprisingly, however, these do not reside within the shell gland and while eggs are produced the shell gland forms a shell. The protective mechanisms appear to be associated with control of the production of ova through the pituitary gonadotropin output: hens on a calcium-deficient diet will produce eggs if given gonadotropin but the birds die with severe skeletal defects. Hens laying eggs without shells ovulate at a higher rate than normal hens and will do so even when on a severely calcium-deficient diet.

Physical environment, nutrition and disease affect reproductive performance but little is known of the way in which they affect the bird's physiology (Gilbert and Wood-Gush, 1971). Below about 10 hours daylight and/or 10 lux at the food trough reproductive performance is curtailed. Temperatures between 13° C and 21° C are recommended as optimum for egg production, though higher temperatures may lead to increased production figures. Altitude, gaseous environment, humidity and noise may also affect egg production.

Nutritional deficiencies affect egg production, but excessive amounts of certain substances may have deleterious effects (Scott et al., 1969). The increasing practice of feeding artefactual food stuffs, such as drugs, may be beneficial in certain circumstances but a balance must be reached between therapeutic and toxic effects. Disease causes decreased egg production and an increase in defective eggs.

3. THE END OF THE LAYING SEASON

Although the pituitary is still capable of producing gonadotropins and the circulating plasma levels of these are high, reproductive efficiency decreases. This may indicate a change in relative hormone levels which adversely affects reproductive physiology. On the other hand, the ovary appears to become more refractive to stimulation by the gonadotropins. The deterioration in shell quality is, at least partly, a result of the steady increase in egg size with age. Apparently the hen secretes only a certain quantity of shell on each egg irrespective of its size and as egg size increases overall shell thickness decreases.

VI. REPRODUCTIVE PERFORMANCE IN THE MALE

Photoperiod is a potent stimulator of sexual maturation (Siegel et al., 1969): cocks reared on about 13 hours day length produced semen earlier than those on day lengths less than eight hours. However, increasing the daily light period after 20 weeks of age was more effective than rearing on a constant day length. The color of light is important and red seems to be the most efficient. Often closely associated with light in the wild state, temperature is known to affect reproductive performance. In general low ambient temperatures (8° C) retard sexual development of males, but may give optimum semen production when males are active.

Nutritional factors may affect semen production; males are usually expected to consume rations designed for hens, but if an adequate quantity of food is provided, it is likely that most specific requirements will be obtained. However, dietary linoleic acid is important and deficiency decreases the fertilizing capacity of sperm. Semen production is inherited: heritable traits include volume of semen, sperm motility, sperm concentration, sperm vigor and the number per ejaculate. However fertility is not necessarily correlated with any of these and enormous variability is found both within and between different genetic lines.

REFERENCES

Aitken, R. N. C. (1971). "The Oviduct." In *Physiology and Biochemistry of the Domestic Fowl*. D. J. Bell and B. M. Freeman (eds.), London, Academic Press, pp. 1237–1289.

Bellairs, R. (1964). "Biological Aspects of the Yolk of the Hen's Egg." In *Advances in Morphogenesis*. M. Abercrombie and J. Brachet (eds.), New York, Academic Press, pp. 217–272.

Bobr, L. W., Lorenz, F. W. and Ogasawara, F. X. (1964). Distribution of spermatozoa in the oviduct and fertility in domestic birds. I. Residence sites of spermatozoa in fowl oviducts. *J. Reprod. Fert.* 8, 39–47.

Carter, T. C. (1970). The hen's egg: Some factors affecting deformation in statically loaded shells. *Brit. Poult. Sci. 11*, 15–38.

Cunningham, F. J. and Furr, B. J. A. (1972). "Plasma Levels of Luteinising Hormone and Progesterone During the Ovulatory Cycle of the Hen." In *Egg Formation and Production*. B. M. Freeman and P. E. Lake (eds.), Edinburgh, British Poultry Science, pp. 51–64.

Fraps, R. M. (1961). "Ovulation in the Domestic Fowl." In *Control of Ovulation*. C. A. Villee (ed.), London, Pergamon Press, pp. 133–162.

Fraps, R. M., Fevold, H. L. and Neher, B. H. (1947). Ovulatory response of the hen to presumptive luteinizing and other fractions from fowl anterior pituitary tissue. *Anat. Rec. 99*, 571–572.

Gilbert, A. B. (1971a). "The Egg: Its Physical and Chemical Aspects." In *Physiology and Biochemistry of the Domestic Fowl*. D. J. Bell and B. M. Freeman (eds.), London, Academic Press, pp. 1379–1399.

Gilbert, A. B. (1971b). "Egg Albumen and Its Formation." In *Physiology and Biochemistry of the Domestic Fowl*. D. J. Bell and B. M. Freeman (eds.), New York, Academic Press, pp. 1291–1329.

Gilbert, A. B. (1971c). "The Ovary." In *Physiology and Biochemistry of the Domestic Fowl*. D. J. Bell and B. M. Freeman (eds.), London, Academic Press, pp. 1163–1208.

Gilbert, A. B. (1971d). "Control of Ovulation." In *Physiology and Biochemistry of the Domestic Fowl*. D. J. Bell and B. M. Freeman (eds.), London, Academic Press, pp. 1225–1235.

Gilbert, A. B. (1971e). "The Endocrine Ovary in Reproduction." In *Physiology and Biochemistry of the Domestic Fowl*. D. J. Bell and B. M. Freeman (eds.), London, Academic Press, pp. 1449–1468.

Gilbert, A. B. (1971f). "The Female Reproductive Effort." In *Physiology and Biochemistry of the Domestic Fowl*. D. J. Bell and B. M. Freeman (eds.), London, Academic Press, pp. 1153–1162.

Gilbert, A. B. (1972). "The Activity of the Ovary in Relation to Egg Production." In *Egg Formation and Production*. B. M. Freeman and P. E. Lake (eds.), Edinburgh, British Poultry Science, pp. 3–21.

Gilbert, A. B. and Wood-Gush, D. G. M. (1971). "Ovulatory and Ovipository Cycles." In *Physiology and Biochemistry of the Domestic Fowl*. D. J. Bell and B. M. Freeman (eds.), London, Academic Press, pp. 1353–1378.

Hafez, E. S. E. (1962). *The Behaviour of Domestic Animals*. London, Balliere, Tindall and Cox.

Lake, P. E. (1967). "The Maintenance of Spermatozoa in the Oviduct of the Domestic Fowl." In *Reproduction in the Female Mammal*. G. E. Lamming and E. C. Amoroso (eds.), London, Butterworths, pp. 254–266.

Lake, P. E. (1971). "The Male in Reproduction." In *Physiology and Biochemistry of the Domestic Fowl*. D. J. Bell and B. M. Freeman (eds.), London, Academic Press, pp. 1411–1447.

McIndoe, W. M. (1971). "Yolk Synthesis." In *Physiology and Biochemistry of the Domestic Fowl*. D. J. Bell and B. M. Freeman (eds.), London, Academic Press, pp. 1209–1223.

Marza, U. D. and Marza, R. V. (1935). The formation of the hen's egg. 1-IV. *Q. J. Microsc. Sci. 78*, 134–189.

Morris, T. R. (1967). "Light Requirements of the Fowl." In *Environmental Control of Poultry Production*. T. C. Carter (ed.), Edinburgh, Oliver & Boyd, pp. 15–39.

Ogasawara, F. X., Lorenz, F. W. and Bobr, L. W. (1966). Distribution of spermatozoa in the oviduct and fertility in domestic birds. III. Intra-uterine insemination of semen from low-fecundity cocks. *J. Reprod. Fert. 11*, 33–41.

Ralph, C. L. and Fraps, R. M. (1960). Induction of ovulation in the hen by injection of progesterone into the brain. *Endocrinology 66*, 269–272.

Romanoff, A. L. (1960). *The Avian Embryo*. New York, Macmillan.

Scott, M. L., Nesheim, M. C. and Young, R. J. (1969). *Nutrition of the Chicken*. Ithaca, N.Y., Scott.

Siegel, H. S., Siegel, P. B. and Beane, W. L. (1969). Semen characteristics and fertility of meat-type chickens given increasing daily photoperiods. *Poult. Sci. 48*, 1009–1013.

Simkiss, K. and Taylor, T. G. (1971). "Shell Formation." In *Physiology and Biochemistry of the Domestic Fowl*. D. J. Bell and B. M. Freeman (eds.), London, Academic Press, pp. 1331–1343.

Wood-Gush, D. G. M. (1971). *The Behaviour of the Domestic Fowl*. London, Heinemann.

IV. Cytogenetics of Reproduction

Chapter 16

Cytogenetics of Mammalian Reproduction

H. Kanagawa and R. A. McFeely

Cytogenetics is a dual science, originally compounded from genetics and cytology, and now reinforced by discoveries arising out of cell biology, electron microscopy, tissue culture and biochemistry. It is a discipline devoted to the study of chromosomes and genes which are concerned with heredity. Essentially, it encompasses the origin and composition of chromosomes, their behavior during mitosis and meiosis, and their relation to the action and recombination of genes. Because of advances in karyologic techniques, a remarkable renewal of interest in studies of mammalian chromosome has developed. To demonstrate chromosomes a large number of dividing cells is required. Tissue culture techniques have provided a source for these cells in the laboratory.

In the older literature, the chromosome number of farm animals, as established by sectioning methods, varied from one investigator to another. After the tissue culture technique was introduced for chromosome study, the correct number of chromosomes for horses (1959), sheep (1959), cattle (1962), pigs (1962) and goats (1964) were confirmed by several investigators (Hsu and Benirschke, 1969).

I. THE CELL

Animal cells contain cytoplasm, membrane and a nucleus (except erythrocytes) within the cell. Between the nucleus and the membrane is the cytoplasm, which contains certain components necessary for the function of that particular cell. Within the nucleus are the chromosomes. In somatic cells, the chromosomes occur in pairs; without exception each member of a pair is similar in size, shape and proportion to its mate. A pair of such chromosomes is known as homologous chromosomes, and the number of pairs, shape and organization of homologous chromosomes is constant in normal individuals within a species. One pair of chromosomes in each cell is known as the sex chromosome, since they are related to the sex of the individual. In mammals sex chromosomes are referred to as the X and Y chromosomes. In farm animals, the XX sex is female, whereas the XY sex is male. The XX and XY sexes are also called "homo-" and "heterogametic," respectively. All chromosomes other than the sex chromosomes are known as autosomes. One member of each pair is descended from the sperm and the other from the egg. The total number of chromosomes in a cell is called the diploid number (symbol 2X or 2N). In the gametes where only one of each pair of chromosomes is found, the number of chromosomes is referred to as the haploid number (one chromosome set, 1X or 1N).

A. Cell Division

Two kinds of cell division are known, mitosis and meiosis. Mitosis refers to the kind of cell division providing for the production of daughter nuclei, both of which possess two complete sets of chromosomes and are genetically identical to one another and to the mother nucleus. Meiosis refers to a type of cell division in the sex cells, the sperm and egg, in which the chromosome number is reduced by one half from the zygotic chromosome number (diploid, 2N) to the gametic number (haploid, 1N). The reduction is necessary so that the union of the sperm and egg at fertilization will restore the diploid number of chromosomes in the somatic cells.

Mitosis has several different phases: the prophase, prometaphase, metaphase, anaphase and telophase. The prophase is the beginning of mitosis. In the early prophase, the chromosomes are long threads consisting of two chromatids each connected by a nonstaining area known as the centromere. During prometaphase, the spindle fibers form and the centromere becomes attached to a spindle fiber. During the metaphase, the chromosomes line up on the equator of the cell, and by means of the spindle are attached to the centrioles which appear at each pole. Anaphase is characterized by each centromere splitting longitudinally. The two sister chromatids of each chromosome separate by their movement toward opposite poles of the spindle. In the telophase, the nucleus is reformed, resulting in two new daughter cells which possess the same chromosome complement as the original cell. The period of time between the telophase of one cell division and the prophase of the next is known as the interphase.

Meiosis involves two successive meiotic divisions: first meiosis and second meiosis. The first meiosis is subdivided into

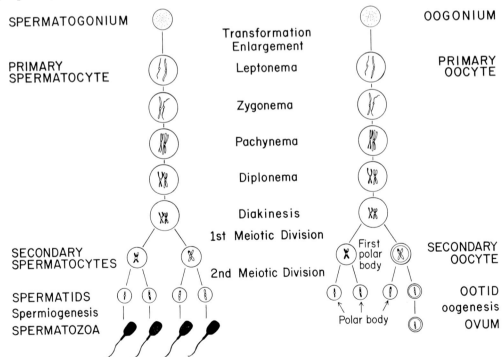

Fig. 16–1. Schematic illustration of meiosis in an organism with one pair of chromosomes. The stages which can be identified during maturation of the gametes in the male (spermatogenesis) and in the female (oogenesis) are listed to the left and right of the diagram, respectively. (*Redrawn from Crispens, Jr., 1971. Essentials of Medical Genetics. New York, Harper & Row.*)

five stages: leptonema, zygonema, pachynema, diplonema and diakinesis. The two homologous chromosomes are paired (synapsis) at zyotene, giving the appearance of four chromatids at pachytene. These synapsed pairs are known as tetrads. During pachytene, the homologous chromosomes may exchange parts reciprocally, which is called chiasmata and each chiasma is understood to represent the physical result of crossing over which provides a means for genetic recombination. After diakinesis, the spindle fibers then appear, and the synapsed chromosomes become attached to the spindle fibers at the centromeres at metaphase. In the first anaphase of the first meiotic division, each tetrad separates into two dyads (the two chromatids connected at the centromere). In the first telophase, the nucleus and cytoplasm reform, resulting in the production of two new cells, each one of which contains only one chromosome of each original homologous chromosome pair. Thus, it is reductional division. The second meiotic division is mechanically similar to mitosis. In the prophase of the second meiotic division, each chromosome in the nucleus is present in the haploid state and appears as a pair of chromatids connected at the centromere. After aligning themselves at the equator at metaphase, each of the two chromatids goes to opposite poles of the cell at anaphase. In the telophase, the nucleus reforms, the cytoplasm divides, and two cells are formed, each having the haploid number of chromosomes. The process of meiosis in the male gonads produces four spermatozoa from each spermatogonium. Each oocyte, in meiosis, produces an ovum and three polar bodies due to uneven distribution of the cytoplasm (Fig. 16–1).

II. THE CHROMOSOME

The genes, which are the units of heredity, are located on the chromosome. The gene is a portion of DNA (deoxyribonucleic acid) molecule. The

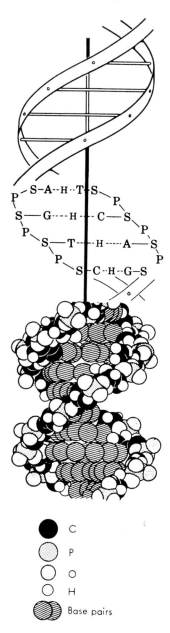

C
P
O
H
Base pairs

Fig. 16–2.   The helix of DNA, with three different ways of representing the molecular arrangement. General picture of the double helix, with the phosphate-sugar combinations making up the outside spirals and the base pairs the cross-bars (*top*). A somewhat more detailed representation: *P*, phosphate; *S*, sugar; *A*, adenine; *T*, thymine; *G*, guanine; *C*, cytosine; and *H*, hydrogen (*middle*). Detailed structure showing how the space is filled with atoms: *C*, carbon; *O*, oxygen; *H*, hydrogen; *P*, phosphorus; and the base pairs (*bottom*). (*From Swanson, 1965. The Cell. New York, Prentice-Hall.*)

DNA molecule may be described as the backbone of the chromosome. The DNA molecule resembles a long, twisted ladder in which the two strands are joined by rungs. Each strand is called a polymer because it is composed of many repeated units, called nucleotides. A nucleotide is composed of a nitrogenous base linked to a sugar, which is in turn linked to a phosphoric acid molecule, containing four bases, adenine, guanine, thymine and cytosine, in an enormous variety of sequences. These make up the genes which direct and control the chemical processes of each cell. The genes exert their effect by means of messenger and transfer RNA (ribonucleic acid), another nucleoprotein (Fig. 16–2).

A. Karyotypes

The analysis of the chromosome complement is routinely performed on somatic cells which are in metaphase. Such a cytologic analysis is termed a karyotype and diagrammatic representation of a karyotype based on the morphologic features of the chromosomes is called an idiogram. The metaphase chromosomes have been classified into three main types, based on the position of the centromere and relative lengths of the long and short arms (Fig. 16–3).

Metacentric. Chromosomes the centromere of which is localized in a roughly median position; the two arms have equal length.

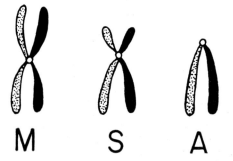

FIG. 16–3. Diagrammatic illustration of the different types of chromosomes based on the relative length of the long and short arms. *M*, metacentric; *S*, submetacentric; and *A*, acrocentric.

Submetacentric. A centromere closer to the center than end, yet not in the center; the ratio of the short arm to the long arm is less than 1:2.

Acrocentric. Chromosomes where the centromere is very close to one end so that one chromosome arm is small or minute and the other very much longer.

The normal karyotypes of farm animals have been well-established (Table 16–1 and Fig. 16–4). The identity of the chromosomes is based upon their size and the location of the centromere. Identification of the sex chromosomes is facilitated in the male by the fact that they are the only two which do not have a homologous counterpart. The *X* chromosome is usually larger than the *Y* chromosome in mammals. Sometimes identification of the *X* chromosome by morphology alone is unreliable and special techniques like thymidine autoradiography are employed to identify the *X* chromosome in the female.

Knowledge of the karyotype has been of interest to those who study evolution and speciation. Animals which have evolved from a common ancestor might be expected to have similar karyotypes. Such is not always the case. In the process of evolution, rearrangement of chromosomes has occurred. Classically, it has been postulated that biarmed chromosomes have resulted from the fusion of two acrocentric chromosomes (Robertsonian fusion). Thus, if one regards each arm of a metacentric or submetacentric chromosome to be equivalent to a single acrocentric, the total number of arms (fundamental number) becomes a guide in evaluating phylogenetic relationships (Matthey, 1945). The mechanisms by which a metacentric chromosome splits into two acrocentric chromosomes or acrocentrics join to form a metacentric chromosome are not understood (Hsu and Mead, 1969).

The chromosome morphology of about 50 species of *Bovidae* has been investigated. Although the diploid number varies from 30 to 60 among these species,

Table 16-1. The Autosomes and Sex Chromosomes of Farm Animals

Species	Chromosome Number 2N	Autosomes		Sex Chromosomes	
		Metacentrics	Acrocentrics	X	Y
Horse	64	26	36	M*	SA‡
Cattle	60	0	58	M	SM§
Goat	60	0	58	A†	SM
Sheep	54	6	46	A	SM
Pig	38	24	12	M	SM

* Metacentric or submetacentric
† Acrocentric
‡ Small acrocentric
§ Small metacentric

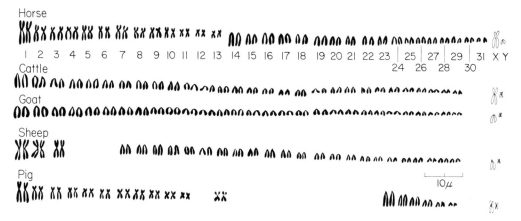

FIG. 16–4. Species differences in the morphology of chromosomes.

the fundamental number varies from 58 to 62 with only a few exceptions (Appendix I). This indicates an almost exclusive use of the Robertsonian fusion mechanism of karyotype evolution in this group of species (Wurster and Benirschke, 1968). This is in striking contrast to its variability in *Equidae* and suggests more complex rearrangement of chromosome sets during the evolution of this family. The diploid number of *Equidae* varies from 32 to 66 and the fundamental number is not constant and varies from 60 to 104 (Benirschke, 1966). However, the fundamental number of 60 is very constant in *Caprinae* while the diploid number of *Caprinae* varies from 42 to 60 (Heck et al., 1968; Appendix I).

B. Techniques of Chromosome Analysis

For the display of chromosomes a source of rapidly dividing cells is required. These can be obtained directly from tissues such as the bone marrow, testes or certain tumors. Other types of cells can be induced to grow in vitro using standard tissue culture techniques. Tissues such as gonad, lung, spleen, kidney and skin are commonly cultured. Specimens from young, healthy animals tend to yield more rapid growth and are less troublesome to get started than tissue from old or debilitated animals. Mononuclear leucocytes when stimulated with a mitogen, e.g., phytohemagglutinin, pokeweed mitogen, produce satisfactory growth in 48–72 hours (Fig. 16–5). All tissues to be cultured must be free of microbial contamination. With proper care many tissues will survive in growth medium for some time and it is possible to ship specimens long distances by air.

Whether cells are used directly or after in vitro culture, the drug colchicine, which is thought to interfere with the spindle, is added to the cell suspension. This interrupts mitosis at metaphase. After a short incubation period the cells are placed in a hypotonic solution which causes swelling of the cells facilitating spread of the chromosomes and then the cells are fixed in acetic acid and alcohol. The cell suspension is placed on a glass slide, air dried and stained (e.g. with Giemsa or Orcein). The preparations are examined under a light microscope.

For analysis, chromosome spreads are photographed and the chromosomes are counted and measured. To construct an

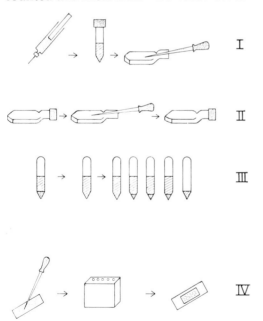

Fig. 16–5. Blood culture technique. I, Mixtures of heparinized blood (1 ml), phytohemagglutinin (0.2 ml) and medium (3 ml). II, Incubation (72 hours at 37°C) and colchicine treatment (4 hours). III, Hypotonic treatment (10 minutes) and Carnoy fixation (5 times) IV; Air dry and stain (*Kanagawa et al., 1965. Jap. J. Vet. Res. 13, 43.*)

idiogram it is necessary to cut out the chromosomes and match them up. While the techniques for the display of chromosomes are not difficult, analysis is tedious and time consuming. There are many new methods utilizing differential staining which promise to make identification of individual chromosomes more precise. It is hoped that subtle structural rearrangements of the chromosomes will be detected more readily.

C. Chromosomal Aberrations

Chromosome aberrations can be divided into two basic groups. The first group consists of individuals with somatic cells that contain other than the diploid number of chromosomes. This can be subdivided into a group with exact multiples of the haploid number of chromosomes (other than the diploid number) which are called polyploids. The other subgroup consists of individuals with chromosome numbers which are not exact multiples of the haploid set, and are referred to as aneuploids. Monosomy is a term used to describe the situation where only one member of a homologous pair is present. Trisomy refers to the addition of a similar chromosome to a pair.

The second major group is composed of individuals in which some structural rearrangement of one or more of the chromosomes has occurred. Various combinations of abnormalities can occur in an individual. If the various aberrations have originated in the same individual it is described as mosaicism. If the mixture has resulted from a fusion or an exchange of cells between two separate individuals the animal is called a chimera (McFeely, 1969).

1. POLYPLOIDY

Polyploidy can result from the fertilization of an egg by more than one sperm. Failure of normal cell division can also produce this condition. True polyploidy in animals appears to be lethal because it has not been reported in antenatal individuals although it is described in embryos from several species. In the pig the incidence of polyspermy was shown to increase with aging of the gametes due to delayed fertilization (Hunter, 1967). Apparently even a few hours of aging of the ovum can increase the incidence of polyploidy due to polyspermy. In most species ovulation is closely related to the end of the period of sexual receptivity thereby minimizing the chance for polyploidy as a result of delayed fertilization.

2. Aneuploidy

Aneuploidy usually results from an abnormal cell division in which the chromatids do not separate normally and both daughter chromosomes are carried into one of the daughter cells. This is referred to as nondisjunction and produces a trisomy in one cell and a monosomy in the other. Aneuploidy in a strain of European wild pigs which ap-

parently arose spontaneously and without detriment to the affected animals has been described (McFee et al., 1966). Several other cases of developmental anomalies associated with autosomal aneuploidy have been reported (Fig. 16–6). Aneuploidy of the sex chromosomes has also been described in a variety of species. These are associated with developmental anomalies of the reproductive system (see Chapter 20).

3. Structural Rearrangement

Chromosomes in which apparent structural abnormalities are seen, may be brought about by the processes of deletion, inversion, translocation, isochromosome formation, duplication and centric fusion. Deletions are structural changes resulting in breakage of a chromosome and a section of the genetic material. Deletions can be lethal when a portion

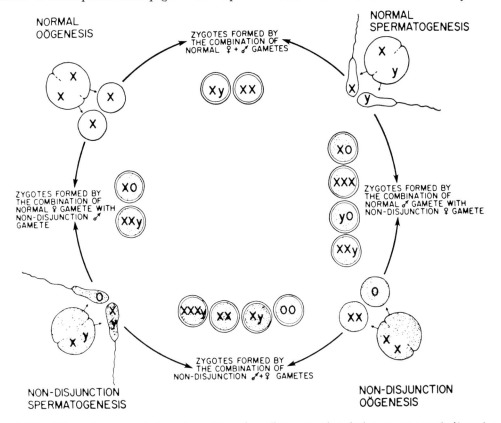

Fig. 16–6. Schematic representation of nondisjunction effects occurring during gametogenesis (*bottom*) on the sex chromosome complement found in the zygote (*Moore et al., 1964. Int. J. Fert. 9, 469.*)

of a chromosome which normally carries genes that perform a function vital to the life of an individual are lost.

When a portion of a chromosome has been rearranged so that the genes occur on the chromosome in a new or inverse order from their original sequence, it is called an inversion. In a mare, a suspected pericentric inversion of one X chromosome led to developmental abnormalities. A similar case has been described in a cow. Translocation is the process involving simultaneous breakage of the arms of two nonhomologous chromosomes with exchange of the broken segments forming two new chromosomes. A translocation between two autosomes was associated with reduced fertility in a boar. His son, with a similar abnormality, also had reduced fertility (Henricson and Backstrom, 1964). Isochromosomes arise from transverse splitting of a centrometric region of the chromosome at the beginning of anaphase. As a result, one daughter cell received a metacentric chromosome composed of two identical short arms and the other a metacentric chromosome made of two identical long arms (Fig. 16–7). Duplication occurs when a portion of a chromosome at-

taches itself to its homologous chromosome. This is probably the result when homologous chromosomes which synapse in meiosis do not completely separate at the reductional division. Centric fusion occurs when two acrocentric chromosomes combine to form a metacentric or submetacentric chromosome. Aneuploidy also results from this structural rearrangement. An extensive study in Swedish cattle has demonstrated the presence of a population of cows with a centric fusion of two autosomes (Gustavsson and Rockborn, 1964). A similar abnormality has been described in the U.S. and Canada. Some evidence suggests it may be associated with some infertility.

Structural rearrangement of chromosomes usually involves at least one break in the integrity of that chromosome. Breaks tend to repair by fusion of the two broken ends. Little or no lasting damage is apparent if the healing takes place between the two segments that were originally together. New genetic combinations result. If fusion involves parts of different chromosomes it is obvious that the more breaks that occur within a cell, the greater will be the chance for struc-

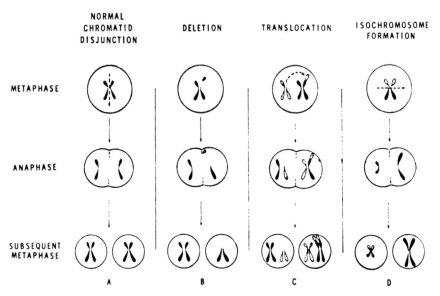

FIG. 16–7.   A representation of some of the mechanisms which may lead to abnormal morphology of the sex chromosomes. (*Moore et al., 1964. Int. J. Fert. 9, 469.*)

tural rearrangement. Certain types of aberrations appear to interfere with the process of cell division and are lethal, while others do not seem to be detrimental. A number of agents have the ability to fragment chromosomes, e.g., x rays, some viruses and certain drugs.

4. CHIMERISM AND MOSAICISM

A variety of developmental abnormalities associated with a mixture of chromosome aberrations has been reported (Benirschke, 1969). The best known example of chimerism is the freemartin (see Chapter 20).

The effect that any of the chromosome aberrations may have depends in part upon the time at which the aberration takes place. It can arise at any time during the life cycle of an animal. If the cells carrying the anomaly are not selected against, all of their progeny can be expected to show the same change. However, it is quite possible that certain changes are deleterious to the affected cells causing death of the cell. It is apparent that the earlier an aberration occurs in the life cycle of an individual, the more widely the defect will be distributed. The result of a meiotic error is expected in all cells of the embryo. Likewise, errors at the time of fertilization will be found universally in the conceptus. Chromosome abnormalities that occur early in the life cycle of an individual can be expected to have a greater effect upon the development of the individual because of the larger number of cells that will be involved. An aberration occurring late in the development can be expected to have far less influence upon the individual because relatively few cells will be descended from the abnormal one.

The loss or addition of small amounts of chromatin material is generally deleterious to normal growth and development; and when larger amounts of genetic material are either lost or gained the resulting genome may be lethal. Therefore, major abnormalities in the normal karyotype may exert their in-

fluence very early in the development of an individual and contribute to early death of an embryo.

D. Hybrids

From time to time attempts have been made to combine the desirable traits of closely related species. For the most part hybridization experiments failed because of several barriers to reproduction. A variety of isolating mechanisms operates under natural conditions to prevent sexual hybridization between species. Of these, geographic, ethologic, mechanical and morphologic barriers can be overcome by artificial insemination. With the new knowledge of cytogenetics attention has been given to what happens to the chromosomes when animals with different karyotypes are mated. It is interesting that some interspecies hybrids are fertile while others are only partially fertile and many are sterile.

Cell biologists have performed some fascinating experiments involving the in vitro hybridization of cells from different species. Chromosome analysis has divulged some very interesting karyotypic changes which may serve in the future as a basis for taxonomic reevaluation. Cellular material from several different species has been combined to form viable polyploid somatic cell hybrids in tissue culture. Somatic cells of different species have been crossed to form somatic hybrids. These laboratory studies are considerably more simple than breeding studies and should facilitate the study of genetics and provide insights into differentiation.

1. HORSE × DONKEY

The ass ($2n = 62$) when mated with a horse ($2n = 64$) produces a viable hybrid (mule or hinny) with a chromosome number of 63 (Appendix II). The karyotype consists of a donkey haploid set and a horse haploid set (Fig. 16–8). The female mule has a virtual absence of germ cells from the mature ovary although most exhibit regular estrus, produce follicles and normal corpora lutea. The males show testicular hypoplasia. Although meiosis appears to start normally the primary spermatocytes degenerate, probably as a result of failure of synapsis. The Leydig cells function normally and aggressive sexual behavior is exhibited. Reports of fertile mules have not been accompanied by chromosome analysis. Where this has been done the animal has been either a donkey or a horse. Other equine hybrids, e.g., zebra × horse, have chromosome numbers intermediate to the diploid numbers of the parents and are similar to mules (Benirschke, 1966).

2. Cattle × Bison

The chromosome complements of domestic cattle and the American bison were compared with those of their hybrid, the cattle. The diploid numbers of all three were 60 and the autosomes and the sex chromosomes were morphologically similar in cattle and American bison (Appendix II). There is no apparent difference in the chromosome makeup of European bison and domestic cattle. These hybrids produce more meat which is low in connective tissue content, have strong resistance to disease and are especially winter-hardy. These advantageous traits have prompted hybridiza-

tion experiments with a view to producing a line of cattle. However, it has failed on a commercial basis, mainly because of the male sterility in F1 hybrids and the poor viability and low male fertility in the backcross progenies. The abortive first meiotic division in male hybrids has been attributed to structural inequalities between their parental chromosomes. However, convincing evidence is scarce (Basrur, 1969).

3. Goat × Sheep

Several reports concerning goat × sheep hybridization have appeared in the past few years. Successful development of hybrids in goat (female) × sheep (male) has been achieved for about the first two months of pregnancy (Chang et al., 1969). The diploid chromosome number of sheep is 54, of goats, 60, and of their embryonic hybrids, 57, as expected (Appendix II).

Death of the hybrid fetus is not likely to be the direct result of gross chromosomal abnormalities. Hemolytic antibodies against paternal red blood cells in the serum of female goats after the time of cotyledonary attachment of hybrid fetuses was observed (Alexander et al., 1967). Whether this occurs during normal pregnancy is not known, but it was considered that the conditions of fetal

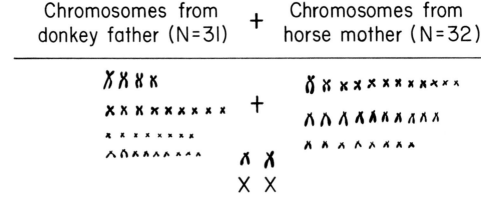

Fig. 16–8.   The karyotype of female mule consists of a donkey haploid set (N = 31) and a horse haploid set (N = 32).

death were consistent with hemolytic disease developing as a result of an imperfect placental barrier. In reciprocal cross, sheep (male) × goat (female), incidence of fertilization is very low.

The chromosome number in goats is 60 and 54 in sheep. However, all chromosomes in goats are acrocentric and three pairs are metacentric in sheep. If we could separate these three pairs of metacentrics at the centromere, the chromosome number in sheep would be 60 and all acrocentrics. It provides an interesting case to speculate on Robertsonian evolution.

E. Clinical Application

The modern techniques of cytogenetics have provided scientists with a valuable tool for the investigation of many types of chromosome disorders. They have only quite recently been used by researchers interested in increasing productivity of farm animals. Two general areas are of particular importance in reproduction in farm animals: (a) developmental abnormalities resulting from chromosomal aberrations which cause reduced fertility or failure to thrive; and (b) chromosomal abnormalities which are lethal to the embryo or fetus.

Sex chromosome anomalies are reviewed in Chapter 20. Chromosomal abnormalities have been found in about one out of every five spontaneous abortions in man. Similar chromosomal aberrations may account for a substantial portion of embryonic loss in farm animals. The effect of aging of gametes on chromosome abnormalities of the conceptus takes on particular significance in species where artificial insemination is used extensively. Timing of insemination is under the direct control of man and is not dependent upon the sexual receptivity of the animal, thereby increasing the possibility of delays in fertilization.

The technique of amniocentesis which is becoming widely used in man for determining the karyotype of the fetus has been used successfully in the cow as early as 40 days of gestation.

III. THE SEX CHROMATIN

In most female mammals a well-developed DNA-positive chromocenter, the sex chromatin body, is observed in the nucleus of most cells from certain tissues. Sex chromatin represents a heterochromatic X chromosome which is genetically inactive. Characteristically, it lies against the inner surface of the nuclear membrane or adjacent to the nucleolus. It has been established that the maximum number of sex chromatin bodies in a given interphase nucleus is one less than the number of X chromosomes. Thus, XY, or XO individuals are

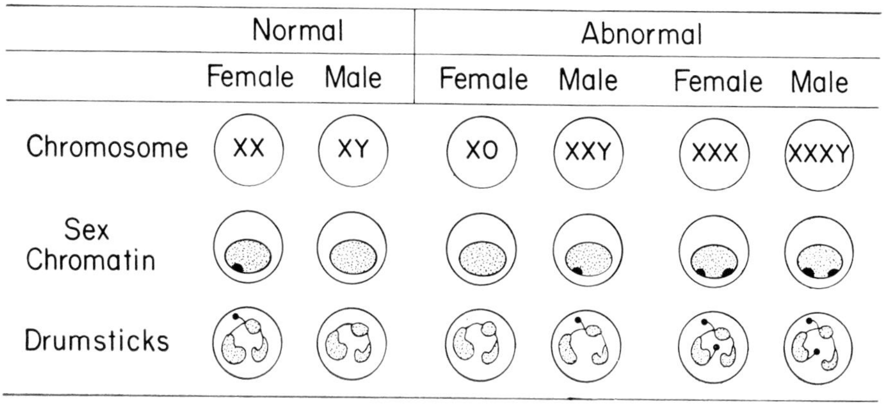

	Normal		Abnormal			
	Female	Male	Female	Male	Female	Male
Chromosome	XX	XY	XO	XXY	XXX	XXXY
Sex Chromatin						
Drumsticks						

FIG. 16–9.   Sex chromatin and drumstick. The maximum number of sex chromatins and of drumsticks is one less than the number of X chromosomes in the cell.

sex chromatin-negative, whereas those who are XX or XXY have one sex chromatin body per nucleus (Fig. 16–9).

In farm mammals, sexual dimorphism, as indicated by presence of the sex chromatin, is detectable mainly in neurones of the brain and spinal cord. In fetal cells obtained from amnionic fluid, sex chromatin was detected in cattle, sheep and pigs. This technique may be a useful research tool in studies of prenatal animals where the genetic sex must be known quickly and without damage to the fetuses.

A mass of chromatin attached by a threadlike stalk to one of the nuclear lobes is seen in a small percentage of polymorphonuclear leukocytes in normal females, but not in normal males. Frequency of neutrophil drumsticks in blood smears from normal female horses is 5.5%. Again, there is evidence to indicate that the maximum number of nuclear appendages is one less than the number of X chromosomes.

Sex chromatin studies are quicker and cheaper than chromosome analysis in determining genetic sex. However it is not possible to detect sex chromosome chimerism in females with this technique.

REFERENCES

Alexander, G., Williams, D. and Bailey, L. (1967). Natural immunization in pregnant goat: Against red blood cells of their sheep $\times$ goat hybrid foetuses. *Austr. J. Biol. Sci. 20*, 1217–1227.

Basrur, P. K. (1969). "Hybrid Sterility." In *Comparative Mammalian Cytogenetics*. K. Benirschke (ed.), New York, Springer-Verlag, pp. 107–131.

Benirschke, K. (1966). "Sterility and Fertility of Interspecific Mammalian Hybrids." In *Comparative Aspects of Reproductive Failure*. K. Benirschke (ed.), New York, Springer-Verlag, pp. 218–234.

Benirschke, K. (1969). Spontaneous chimerism in mammals, a critical review. *Current Topics in Path. 51*, 1–57.

Chang, M. C., Pickworth, S. I. and McGaughey, R. W. (1969). "Experimental Hybridization and Chromosomes of Hybrids." In *Comparative Mammalian Cytogenetics*. K. Benirschke (eds.), New York, Springer-Verlag, pp. 132–145.

Gustavsson, I. and Rockborn, G. (1964). Chromosome abnormality in three cases of lymphatic leupaemia in cattle. *Nature (Lond.) 203*, 990–991.

Heck, H., Wurster, D. and Benirschke, K. (1968). Chromosome study of members of the subfamilies *Caprinae* and *Bovinae*, family *Bovidae;* the *Musk Ox, Ibex, Aoudad, Congo buffalo* and *Gaur. Sonderdrunck ans Z. F. Saugetierkunde 33*, 172–179.

Henricson, B. and Backstrom, L. (1964). Translocation heterozygosity in a boar. *Hereditas 52*, 166.

Hsu, T. C. and Benirschke, K. (1969). *An Atlas of Mammalian Chromosomes*. New York, Springer-Verlag.

Hsu, T. C. and Mead, R. A. (1969). "Mechanisms of Chromosomal Changes in Mammalian Speciation." In *Comparative Mammalian Cytogenetics*. K. Benirschke (ed.), New York, Springer-Verlag, pp. 8–17.

Hunter, R. H. F. (1967). The effects of delayed insemination on fertilization and early cleavage in the pig. *J. Reprod. Fert. 13*, 133.

McFee, A. F., Bonner, M. W. and Rory, J. M. (1966). Variation in chromosome number among European wild pigs. *Cytogenetics 5*, 75–81.

McFeely, R. A. (1969). "Aneuploidy, Polyploidy and Structural Rearrangement of Chromosomes in Mammals Other Than Man." In *Comparative Mammalian Cytogenetics*. K. Benirschke (ed.), New York, Springer-Verlag, pp. 434–444.

Matthey, R. (1945). L'evolution de la formule chromosomiale chez les vertebres. *Experientia 1*, 50–78.

Wurster, D. H. and Benirschke, K. (1968). Chromosome studies in the superfamily *Bovoidea. Chromosoma 25*, 152–171.

Chapter 17

Cytogenetics of Avian Reproduction

P. K. Basrur

Birds belong to the class *Avis* which is an offshoot of *Reptilia*. This interesting class antedates placental mammals on the scene of life and manifests many of the reptilian traits and some of the characteristics of mammals. Birds, like mammals, belong to the warmblooded group of vertebrates, but they also exhibit several cytogenetic and reproductive characteristics which place them closer to the most popular group of reptiles belonging to the order *Squamata*. Although the essential aspects of reproduction are similar in mammals and birds, the latter exhibit several points of departure from mammals. Some of these include the morphologic features of their chromosomes, their cytologic sex-determining mechanisms, the mode of expression of the genes carried on the sex chromosomes, the pattern of gonadal induction and sex differentiation, and their response to environmental changes. Birds exhibit varying degrees of sexual dimorphism in anatomic and external features and thus offer interesting opportunities for studying sex differentiation and the causative mechanisms of intersexuality. Some of these features will be dealt with in the following pages with emphasis on the points of departure of birds from mammals in cytogenetic makeup, and mechanisms of determination and differentiation of sex.

I. MITOTIC CHROMOSOMES

Avian chromosomes vary greatly in length and fall under two morphologic categories: "macrochromosomes" which are large enough to be classified into groups according to their centromere position and total length, and "microchromosomes" which are very small and are difficult to categorize into conventional groups. The minute size, and the apparent inconsistency of distribution of the microchromosomes between metaphase plates of a given species once led to the suggestion (Newcomer, 1957) that they are not true chromosomes but accessory supernumeraries (chromosomoids). It was hypothesized that the microchromosomes may form part of the nucleic acid pool for chromosome replication and that the "true" avian chromosomes are the 11 or 12 large macrochromosomes. However, more recent observations on the chromosomal architecture of the microchromosomes and their meiotic behavior unequivocally confirmed the chromosomal nature of the microchromosomes (Krishan and Shoffner, 1966; Bhatnagar, 1968).

In the many species examined including duck, white Chinese geese, domestic chicken and the barn owl the decrease in chromosome length is so gradual that the smaller macrochromosomes and the

Table 17-1. Chromosome Number of Some of the Domesticated and Wild Species of Birds

Species	Diploid Number
Parakeet (*Melopsittacus undulatus*)	60
Chicken (*Gallus domestica*)	78
Japanese quail (*Coturnix coturnix japonica*)	78
Ring necked pheasant (*Phasianus colchicus*)	78
Canary (*Serinus canarius*)	80
Chinese geese (*Cignopsis cignoid var. domestica*)	80
Pigeon (*Columba livia domestica*)	80
Turkey (*Meleagris gallopavo*)	80
Duck (*Anas platyrhynea domestica*)	82
Great horned owl (*Bubo var. virginianus*)	82

larger microchromosomes overlap in length. However, the first 9–10 pairs of large elements are conventionally regarded as macrochromosomes and the balance as microchromosomes. The macrochromosomes include biarmed (submetacentric and metacentric) elements whereas microchromosomes appear to be singlearmed (acrocentric) elements (Ford and Wollam, 1964; Ohno, 1967).

The diploid number in a majority of birds ranges from 78 to 82 (Table 17–1), the exact number of chromosomes in many species often being a matter of dispute. In the Australian parakeet (budgerigar), the diploid number is 60 with 12 pairs of macrochromosomes and 18 pairs of microchromosomes. In the domestic chicken, the diploid number is 78 including nine pairs of macrochromosomes and 30 pairs of microchromosomes (Plate 18). The microchromosomes in domestic chicken, and in a majority of birds examined so far, amount to 22–30% of the genome (Ohno, 1967).

II. SEX-DETERMINING MECHANISMS

A. Sex Chromosomes

Studies based on sex-linked inheritance have shown that the female is the heterogametic sex in birds. The Z chromosomes of the homogametic male were morphologically identified by investigators during the early part of this century. In a majority of avian species the Z chromosomes rank as the fourth or fifth largest pair in the male karyotype. One of the notable features of the avian Z chromosome is that it exhibits a remarkable consistency in morphologic features and total DNA content. Thus, in many species belonging to diverse orders, the Z chromosome is biarmed, is easily distinguishable from the autosomes, ranges from 2.3μ to 3μ in length and constitutes approximately 10% of their total genome (Ohno, 1967).

The cytologic sex-determining mechanisms in birds was originally considered to be of the $ZZ\male:ZO\female$ type; however, the demonstration of a submetacentric W chromosome for the first time in domestic chicken and subsequently in females of budgerigar, canary, pigeon, the great horned owl and the barn owl, indicates that a $ZZ\male:ZW\female$ type sex-determining mechanism prevails in birds. The failure to detect a W chromosome in some of the species may be attributable to its morphologic similarity to the microchromosomes. In domestic fowl, the W chromosome is a submetacentric (Plate 18, *A*) ranking in

---

PLATE 18

A, Metaphase plates and the macrochromosomes of a male and a female domestic chicken. × 2500.
B, Domestic rooster (*left*), pheasant hen (*right*) and their hybrid (*middle*).
C, Testes of a mature rooster (CH♂), pheasant cock (PH♂) and a male hybrid (HYo) of comparable age.
D and *E*, Gonads of mature hybrids: male (*D*) and female (*E*). × 100.
F and *G*, Histologic sections of hybrid gonads: male (*F*) and female (*G*).
(*Courtesy of Dr. M. K. Bhatnagar, Department of Biomedical Sciences, University of Guelph, Guelph, Ontario, Canada.*)

PLATE 18

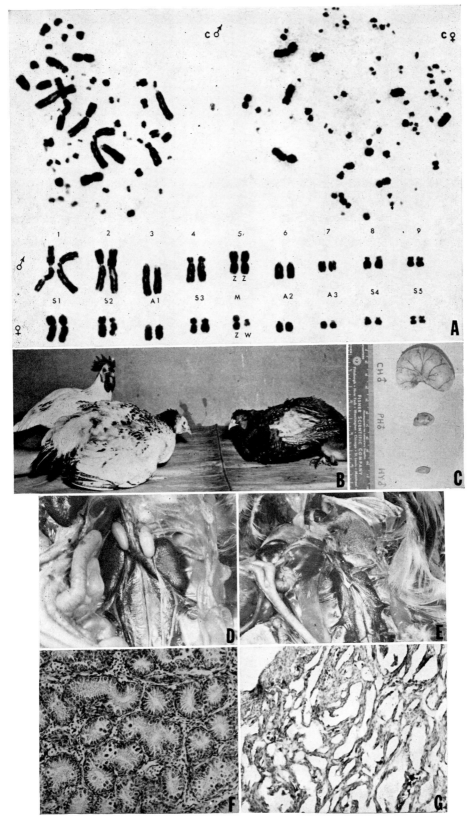

Legend on opposite page.

size between the 9th and 10th chromosomes (Krishan and Shoffner, 1966).

B. Sex Chromatin

A sex-associated dimorphism in the chromatin pattern of interphase nuclei similar to that noted in the homogametic mammalian female, has not been unequivocally demonstrated in birds. The avian male, by virtue of its cytogenetic analogy to the female mammal would be considered the likely candidate to exhibit the genetically inactivated and condensed Z chromosome. However, many investigators were unable to detect the sex chromatin in the domestic rooster. On the other hand, several investigators claim to have observed a sex chromatin in the interphase nuclei, and a positively heteropyknotic chromosome in the early mitotic stage of the domestic hen (Moore and Hay, 1961; Ohno, 1967).

The difference in the sex chromosomes of the homogametic sex of mammals and birds is emphasized by the evidence emanating from genetic studies. The gene for barring (B) which causes white bands lacking melanin across the feathers in Plymouth Rock breed chicken, the gene cinnamon (Cn) which replaces black melanin with brown pigments in budgerigar and the gene faded (B^{of}) which causes light plumage in the domestic pigeon are some of the examples of sex linked neomorphs. Males homozygous for these genes exhibit more extreme phenotypes as compared to the heterozygous males or the hemizygous females. It would appear that these Z-born genes are fully expressed in homozygous males and that a dosage compensating mechanism similar to that facilitated by the inactivation of one of the X chromosomes of female mammals, is not operative in birds (Cock, 1964).

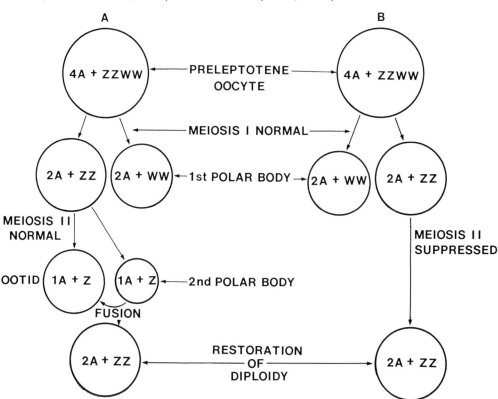

Fig. 17–1. Schematic representation of meiotic processes leading to diploid parthenogenesis. *A*, Diploidy through polar body fusion with ootid. *B*, Diploidy through suppression of second polar body formation.

The positively heteropyknotic element noted in the early mitotic stages of domestic hen, the sex chromatin body and the late labeling element detected in the females of certain avian species (Moore and Hay, 1961) probably represent the W chromosome. Since at present there is no genetic evidence of the W-born genes, it is difficult to ascertain whether or not the heterochromatinization and late labeling tendencies noted in avian W chromosome also reflect its genetic inactivation in somatic cells. The two Z chromosomes of the avian male do not exhibit asynchrony of DNA synthesis or genetic inactivation in somatic cells.

C. Parthenogenesis

Parthenogenesis (birth from virgins) occurs spontaneously in turkeys and in certain strains of chickens (Olsen, 1966) although it is a relatively rare phenomenon in vertebrates. The parthenogenetic progeny in turkeys and domestic fowl are exclusively males and are of diploid makeup, capable of siring offspring. Based on homograft response studies indicating that the parthenogenetic males are not entirely homozygous, the origin of parthenogenesis in these birds is believed to be through a faulty second meiotic division in the female. The fault is presumed to involve either a fusion of the second polar body with the ootid or the failure of cytokinesis and second polar body formation (Fig. 17–1). Either of these processes will restore diploidy in the parthenogenetic egg and will facilitate the occurrence of all male offspring with a $2A + ZZ$ genetic makeup. If the failure of second polar body formation occurs in a meiotic cell of $2A + WW$ constitution, it will be nonviable, since it lacks the Z chromosome.

III. SEX DIFFERENTIATION

A. Gonadal Induction

In birds, as in mammals, the primordial germ cells originate from the extraembryonic yolk sac endoderm. They are distinguishable from the rest of the cells as early as at the first somite stage. They are large round cells containing vesicular nucleus and a cytoplasm with elaborate Golgi complex, an abundance of glycogen, lipid droplets and pinocytotic vesicles (Dubois and Croisille, 1970). These cells migrate to the genital ridge via the vascular system. The germ cells may penetrate the vascular network as well as a variety of organs including skin, liver, lungs and the mesonephros. The primordial germ cells that enter the blood vessels are governed by the blood flow and the chemotactic attraction exerted by the gonadal ridge cells. Studies using radioactive amino acids have shown that the gonadal ridge cells (coelomic epithelial cells) are secretory in nature and that the secretions include two types of proteins: one with a slow turnover rate and probably belonging to the structural proteins, and the other with a rapid turnover rate and excreted by a merocrine type process through the cytoplasmic protrusions on the coelomic epithelial cells. The arrival of the primordial germ cells at the genital ridge is thus conditioned by the secretory property of the coelomic epithelium and the property of the primordial germ cells which pinocytotically pick up the protein.

In female avian embryos, the distribution of primordial germ cells in the left gonadal ridge is greater than that in the right gonadal ridge, the cell count ratio varying according to the stage of development: until the third day of incubation the distribution of primordial germ cells is equal in both gonadal ridges, at the end of the fourth day the ratio of cell counts in the left to the right gonads is 2.5:1, and after five days it is approximately 5:1. This type of asymmetry in germ cell distribution has been demonstrated in the English sparrow, redwing black bird, duck and domestic chicken, all of which, as a rule, exhibit no right gonad in adult females. On the other hand, in some species including hawks, falcons and vultures in which well-developed bilateral ovaries are present in the adults, the ratio of primordial germ cells in the right and left gonadal ridges of the developing embryo approximates one. It would appear that the presence of primordial germ cells is essential for the development of a normal gonad and that the absence of the right gonad in most female birds is attributable to the poor distribution of primordial germ cells in the right gonadal ridge. The unequal distribution of primordial germ cells is not attributable to an asymmetry in their initial migration pattern, orientation of the embryo or to an asymmetry in blood supply in favor of the left gonadal ridge. It is believed that a selective chemotaxis due to the nature of the secretory process of the female coelomic epithelium, may be responsible for the preferential homing tendency of the primordial germ cells in the left gonad. The migratory trait of the germ cells to the germinal epithelium has been demonstrated to be sex-dependent in that the male germ cells retain this property for up to 13 days, whereas the female germ cells lose their ability to migrate after the eighth day of incubation.

The spermatogonia are distinguishable from oogonia by about the eighth day of incubation. In spermatogonia the nucleus is vesicular, the mitochondria are clear with very few cristae and the pinocytotic vesicles and lipid droplets are persistent in the cytoplasm. In oogonia, the mitochondria are dense and elongated and are grouped at the pole where Balbiani's vitelline body is organized. The cytoplasmic lipid droplets begin to disappear from the cortical oogonia by about the eighth day of incubation, whereas those in the medullary region maintain the characteristics of primary gonocytes including the vesicular nature of nucleus, the small mitochondria and the abundance of lipid droplets.

B. Gonaduct Differentiation

In birds, both oviducts develop in the embryo, but about the 10th day of incubation, the right oviduct starts to regress even in those species in which the right ovary is fairly well-developed in the adults. The Wolffian ducts develop into vas deferens in males but remain as rudimentary organs in females. The presence of rudimentary Wolffian ducts allows for the differentiation of genetic females in the male direction under the influence of hormones or other environmental changes. The intersexual modification is thus generally more prevalent in genetic females in the avian species.

IV. SEX REVERSAL

The assumption of male characteristics by birds which start their life as functional females is noted frequently. Crew (1923) described the case of a domestic hen which underwent a fully functional sex reversal. In Crew's bird, which had a tubercular infection, the ovary was represented by a large tumor and the paired testes had the normal testicular architecture. Most of the birds reported to have undergone sex reversal since then, had an ovo-testis in the left side and a testis on the right side suggesting that they may represent true hermaphroditism instead of sex reversal. True hermaphroditism, described previously as gynandromorphism or mosaicism, occurs in chaffinch, finch, canary, pheasant and domestic hen (Taber, 1964). In all these instances the left side of the bird showed the female plumage and contained an ovary or ovo-testis, while the right side, exhibiting male plumage, carried a testis. Some of these birds had both mature oocytes and mature spermatozoa in their gonadal tissues and may be the results of chromosomal nondisjunction at or after the first cleavage in genetic females. It is conceivable that these true hermaphrodites, and those reported as "sex reversed" genetic females are triploids similar to the Rhode Island Red "hen" described by Ohno et al. (1963). This hen also had an ovo-testis and an oviduct on the left, and a rudimentary gonad and genital tract on the right. Triploidy is a common cause of embryonic mortality in domestic fowl. However, the occur-

Table 17-2. Pattern of Sex Differentiation in Birds
Subjected to Various Treatments

Bird	Genetic Sex	Treatment	Gonadal Differentiation
		Gonadal Graft	
Domestic fowl	ZW	Ovary on undifferentiated testis	Regression of right testis: Left gonad—ovo-testis
Domestic fowl	ZW	Regressing right ovary on undifferentiated testis	Left gonad—ovo-testis
		Hormone Administration	
Domestic fowl	ZZ	Estrogen	Left gonad—ovo-testis
Duck	ZZ	Estrogen	Left gonad—ovo-testis
Herring full	ZZ	Estrogen	Persistence of cortex
Domestic fowl	ZZ	Androgen	Left gonad—ovo-testis

rence of triploidy in the adult is rare and the Rhode Island Red hen mentioned before is the first case of a triploid, warm-blooded vertebrate surviving to adulthood. Chick embryos noted to be triploid on detailed cytogenetic analysis have been assumed to result from fertilization of female gamete of $1A + W$ constitution by two spermatozoa, each of $1A + Z$ makeup. The resulting bird $(3A + ZZW)$ could be a hermaphrodite with intersexual modification of the reproductive tract and, as in some instances, with mosaic plumage pattern.

V. AVIAN FREEMARTINS AND INTERSEXES

Conditions similar to the bovine freemartins have been reported in ducks and domestic chicken when two embryos of disparate sex developed in double yolked eggs (Taber, 1964). Unlike bovine freemartinism, the male "twin" is more drastically affected in birds. The female is affected only slightly in avian twins sharing the anastomosed allantoic blood vessels with a male. The genetic female is relatively unaffected although the ovary is smaller than normal and a masculine syrinx and genital tubercle develop whereas the genetic males carry extensive amounts of cortex on the left gonad and the Müllerian duct regression in these

males is inhibited. The feminization of the male embryo developing in the same egg with a female, may be related to the amount of estrogen deposited in the yolks by the hen. When two yolks are present, the amount of estrogen to which the male embryo is exposed is considered to be twice as much as the normal and, as such, sufficient to cause feminization of the male.

Unlike in mammals, in avian species, the fetal testes are affected by the ovary. Retardation of the male gonads can also be brought about by the administration of physiologic doses of estrogens or the steroid-free extract of the ovary, to the developing avian fetus. If an ovariectomy is performed before the pullet is 30 days old, the right gonad may develop into a testis or ovo-testis while similar operations after 30 days produce ovo-testis or ovaries. This difference in the pattern of gonadal differentiation dependent on the time of ovariectomy may be explained in terms of the effect of estrogen on avian gonadal tissue (Table 17–2). Estrogens may have a strong and direct inhibitory effect on the gonadal medullary tissue. The presence of a developing ovary for over 30 days is sufficient to effectively destroy the medullary tissue of the right gonad so that the predominating gonadal tissue after 30 days is the cortex.

VI. HYBRIDIZATION

There were many attempts in the past of experimental hybridization with the hope of combining the desirable traits of different avian species. These efforts and similar efforts with mammalian species have generally been thwarted by partial or total sterility in the hybrids. However, unlike in mammals where the male exhibits sterility and/or malformation, in birds it is the female hybrids which are generally at a developmental or reproductive disadvantage. The cytogenetic basis of this difference between mammals and birds may be inherent in the avian sex-determining mechanism which confers heterogamy in females.

Hybridization of domestic fowl with a variety of wild and domestic birds has been attempted with varying degrees of success. Intergeneric hybridization of domestic fowl with ring necked pheasants generally results in reduced hatchability and a strikingly reduced incidence of female hybrids (Plate 18), whereas domestice fowl-Japanese quail hybridization produces exclusively male offspring. The gonadal development in male chicken-pheasant hybrids proceeds in a manner similar to that noted in roosters up to five months, after which the hybrid male starts showing signs of testicular degeneration. The male hybrids do not crow or exhibit male type wattles and head furnishings whereas the few females that survive to reach adulthood exhibit no ovary and reproductory tract or just a streak gonad on the left side (Plate 18). The testosterone level in male chicken-pheasant hybrids is much lower than that in males of the parental species, whereas the female hybrids generally show plasma testosterone levels much higher than that of the females of the parental species. This observation is more striking since the gonads are generally absent in the adult female hybrids.

Interspecific hybridization between divergent species of birds can cause intersexual modification in females (Taber, 1964). The intersexuality of female hybrids was originally believed to be due to the aberrant chromosome constitution of the hybrids. However, male and female hybrids resulting from chicken-pheasant crosses generally exhibit a normal haploid set each of the chicken and pheasant chromosomes even though the female hybrid tends to be an "intersex." Chromosomal aberration per se may not be the cause of intersexuality in genetic females since the females of our chicken-pheasant hybrids are no more chromosomally aberrant than the males (Basrur and Yamashiro, 1972). A set of co-adapted stabilizing genes carried by the Z chromosome is probably essential for the expression of the female determining genes carried on the W chromosome. Alternatively, the inherent weakness of the avian hybrid females may be attributable to the absence of a Z chromosome belonging to the same parental species from which the W chromosome was derived and the consequent unmasking of the W-born genes with undesirable effects. The higher mortality rate of avian female hybrids may thus be causally related to the lack of protection from the co-adapted genes generally present in the sex chromosome complement of a normal diploid bird.

VII. SUMMARY

Birds represent a unique class of vertebrates which combine many characteristics of reptiles and mammals. Avian karyotype is made up of macro- and microchromosomes which have no counterparts in a mammalian karyotype. The female is the heterogametic sex in birds while in mammals the male is the heterogametic sex. Functional sex reversal, parthenogenesis and total triploidy are common occurrences in birds whereas these phenomena are hitherto unheard of in mammals. In birds, unlike in mammals, the gonadal induction and the development of the reproductive system follow an asymmetric pattern in genetic females, and unlike the bovine freemartin condition, it is the genetic male among avian "co-twins" that undergoes gonadal changes. The avian female maintains the vestigeal right gonad and the Wolffian ducts and thus retains the potential for sex reversal in response to hormonal or other environmental changes.

REFERENCES

Basrur, P. K. and Yamashiro, Y. (1972). Chromosomes of chicken-pheasant hybrids. *Ann. Génét. Sél. Anim. 4*, 495–503.

Bhatnagar, M. K. (1968). *Cytogenetic Studies on Some Avian Species.* Ph. D. Thesis, University of Guelph, Ontario, Canada.

Cock, A. G. (1964). Dosage compensation and sex chromatin in non mammals. *Genet Res. 5*, 354–365.

Crew, F. A. E. (1923). Studies in intersexuality. II Sex reversal in the fowl. *Proc. Roy. Soc. (B) 95*, 256–278.

Dubois, R. and Croisille, Y. (1970). Germ cell line and sexual differentiation in birds. *Phil. Trans. Roy. Soc. Lond. (B) 259*, 73–79.

Ford, C. R. and Wollam, D. H. M. (1964). Testicular chromosomes of *Gallus domesticus. Chromosoma 15*, 568–574.

Krishan, A. and Shoffner, R. N. (1966). Sex chromosomes in the domestic fowl, *Gallus domesticus*, turkey, *Meleagris gallopavo* and Chinese pheasant, *Phasianus colchicus. Cytogenetics 5*, 53–63.

Moore, K. L. and Hay, J. C. (1961). Sexual dimorphism in intermitotic nuclei of birds. *Anat. Rec. 139*, 315–323.

Newcomer, E. H. (1957). The mitotic chromosomes of the domestic chicken. *J. Hered. 48*, 227–234.

Ohno, S. (1967). "Apparent Absence of Dosage Compensation for Z-linked Genes of Avian Species." In *Sex Chromosomes and Sex-Linked Genes.* Berlin, Heidelberg, New York, Springer-Verlag, p. 144.

Ohno, S., Kittrell, W. A., Christian, L. C., Stenius, C. and Witt, G. A. (1963). An adult triploid chicken (*Gallus domesticus*) with a left ovotestis. *Cytogenetics 2*, 42–69.

Olsen, M. W. (1966). Frequency of parthenogenesis in chicken eggs. *J. Hered. 57*, 23–25.

Taber, E. (1964). "Intersexuality in Birds." In *Intersexuality in Vertebrates Including Man.* C. N. Armstrong and A. J. Marshall (eds.), New York, Academic Press, 285–310.

V. Reproductive Failure

Chapter 18

Reproductive Failure in Females

E. S. E. Hafez and M. R. Jainudeen

The reproductive process consists of a chain of well-synchronized events extending from estrus and ovulation through fertilization, implantation and pregnancy, and terminating in parturition and lactation. Sterility is some permanent factor preventing procreation, and infertility or temporary sterility is the inability to produce viable young within a stipulated time characteristic for each species. In this chapter we shall examine the phases of the reproductive process which are most vulnerable and show how hormonal imbalances or adverse environmental, genetic and hereditary factors exert their influences (Fig. 18–1). An important factor—reproductive infections—is discussed in a subsequent chapter.

I. OVARIAN DYSFUNCTION

The two main functions of the ovary, the production of ova and secretion of ovarian hormones, are intimately related and are directed toward successful reproduction.

A. Anestrus

Anestrus (absence of estrus) is not a disease but a sign of a variety of conditions (Table 18–1). Although anestrus is observed during certain physiologic states, e.g., before puberty, during pregnancy and lactation and in seasonal breeders, it is most often a sign of temporary or permanent absence of ovarian function caused by seasonal changes in

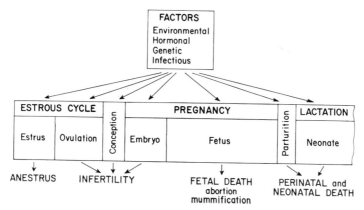

Fig. 18–1. Schematic representation of factors that adversely affect the reproductive process. Common types of reproductive failure are anestrus, infertility, fetal, perinatal and neonatal deaths. Note that infertility may result from fertilization failure or embryonic mortality.

Table 18-1. Abnormalities of Estrus

Species	Abnormality	Causes	Physiologic Mechanisms
Cattle	Anestrus	Pyometra, mummification	Maintenance of corpus luteum
		Lactation	Suckling stimulus inhibits gonadotropin release
		Cystic ovaries	Deficiency of LH
		Ovarian hypoplasia and freemartin	Failure to produce ovarian estrogens
		Nutritional and vitamin deficiencies	Gonadotropin production by anterior pituitary
	Subestrus, silent estrus (quiet ovulation)	High lactation	Endocrine imbalance
	Nymphomania	Cystic ovaries	
Sheep	Anestrus	Season, lactation	Effect of photoperiod on gonadotropin secretion
Swine	Anestrus	Lactation	As for cattle
Horse	Anestrus	Season	As for sheep
	Prolonged estrus	Early in breeding season; old age; undernutrition	
	Split estrus		

the physical environment, nutritional deficiencies, lactational stress and aging. Certain pathologic conditions of the ovaries or the uterus will also suppress estrus.

1. Seasonal Anestrus

During seasonal anestrus there are no cyclic changes in the ovaries and reproductive tract. The extent of seasonal anestrus varies with the species, breed and physical environment, and is more pronounced in sheep and horses than in cattle, pigs and most laboratory mammals.

2. Anestrus During Lactation

In several species ovulation and related reproductive activity are suppressed for a variable period after parturition and during lactation. The incidence and duration of anestrus varies greatly between different species and breeds, and is also influenced by the season of parturition, level of milk production, number of young being nursed and the degree of the postpartum involution of the uterus. For example, during periods of high temperatures and on poor diets, Brahman cows nursing calves are particularly subject to anestrus. The dura-

tion of anestrus in cows nursing calves is longer than in similar cows milked twice daily; this suggests that nursing or frequency of milk removal may influence the pituitary gonadotropic activity. Injections of progesterone in combination with estrogen will reduce the anestrus period in these cows. In sheep, lactational anestrus lasts five to seven weeks. Some ewes suckling lambs will come into estrus, but most ewes show estrus about two weeks after the lambs are weaned. Most foaling mares come into estrus within 5–15 days after foaling but some nervous lactating mares may experience lactational anestrus due to some psychologic disturbance rather than to the stress of lactation.

Physiologic interactions between lactation and depression of ovarian function have not been fully established but may be related to pituitary dysfunction associated with lactation. During intense lactation, prolactin function is maximal, and prolactin inhibiting factor (PIF) function is minimal. Any factor which limits PIF secretion will also suppress the secretion of luteinizing hormone releasing factor (LRF) which would in turn inhibit the production and secretion of

LH. Thus final follicular maturation, estrus and ovulation cannot occur (Symington, 1969). The duration of anestrus is closely related to length and intensity of lactation, and ovarian cysts are common during abnormally high lactation.

3. Anestrus Due to Aging

Animals under wild conditions presumably die or are killed before reaching a stage when their reproductivity declines due to natural aging. Farm animals with the exception of the horse are rarely maintained into old age for economic reasons, and even more rarely given the opportunity to breed late in life. In rodents the incidence of irregular or anovulatory cycles rises steadily with age, and even when they are no longer fertile, mating frequently results in pseudopregnancy. The cessation of menstrual cycles (menopause) long before the end of life itself appears to be peculiar to man. In farm animals there is no evidence indicating a cessation of reproductive activity or irregularities of the estrous cycle. Regardless of the mechanism involved, anestrus due to aging probably alters the functional relationship of the hypothalamus-pituitary-ovarian axis thereby leading to a decrease in gonadotropin secretion or a change in the ovarian response to these hormones.

4. Nutritional Deficiencies

Energy level has a marked effect on ovarian activity. Inadequate nutrition suppresses estrus in young growing females more than it does in adults. Low levels of energy lead to ovarian inactivity and anestrus in beef cows suckling calves. Plane of nutrition affects pituitary size rather than pituitary concentrations of FSH and LH. However, the interaction between nutrient intake and hypothalamic release factor function is obscure.

Deficiencies of minerals or vitamins cause anestrus. Phosphorus deficiency in range cattle and sheep causes ovarian dysfunction which in turn leads to delayed puberty, depressed signs of estrus

and eventually cessation of estrus. Gilts or cows fed a manganese-deficient diet experience ovarian disturbance ranging from weak signs of estrus to anestrus. Vitamin A or E deficiencies may cause irregular estrous cycles or anestrus.

5. Abnormalities of the Ovary or Uterus

Ovarian hypoplasia occurs in Swedish mountain cattle. Affected animals have infantile reproductive tracts and never exhibit estrus. The morphology of the ovary differs from that of seasonal anestrus. Follicles of varying diameter up to the preovulatory size commonly present in the ovaries of anestrous animals are absent in ovarian hypoplasia. There is a marked tendency for ovarian hypoplasia to be associated with white coat color, being inherited as an autosomal recessive.

Freemartins or heifers born co-twin to bulls have poorly developed ovaries and fail to show estrus.

Uterine distension in cattle due to pathologic conditions, e.g., pyometra, mucometra, fetal mummification or maceration, is associated with a retention of the corpus luteum and therefore to a suppression of the estrous cycle. Persistence of the corpus luteum may be attributed to an absence or interference with pathways of the uterine luteolytic mechanism (Ginther, 1968).

B. Atypical Estrus

Short estrus, prolonged estrus, "split" estrus, nymphomania and "silent" estrus are not uncommon in farm animals (Table 18–1). Estrus may be of short duration and without well-marked signs. It may be undetected in young animals without a teaser male or if it occurs during the night particularly in cattle. Prolonged estrus, lasting from 10 to 40 days, characterizes the transition from seasonal anestrus to the sexual season in heavy draft horses. *"Split"* estrus or behavioral estrus interrupted by one or two days of sexual nonreceptivity is observed in mares, especially at the start of the breeding season.

Nymphomania is more frequent in dairy cattle than in beef cattle and horses. In cattle, nymphomania is one of the signs of cystic ovaries. Nymphomaniac cows show intense estrous behavior persistently or at frequent but irregular intervals, depressed milk production, a frequent copious discharge of clear mucus from the vulva, edema and relaxation of the sacrosciatic ligaments and a raised tail head. Nymphomaniac mares are excitable, vicious and intractable. They will not tolerate the approach of another horse nor will they stand for mating. Ovarian cysts are not associated with this condition in the horse.

"Silent" estrus (quiet ovulation) or the occurrence of ovulation without overt estrus is seen between regular estrous cycles in all farm animals, particularly young animals and those on a submaintenance ration. It is suspected when the interval between two consecutive estrous periods is double or triple the normal length. A high incidence of silent estrus occurs in sheep during the first estrous cycle of the breeding season apparently related to the absence of a corpus luteum from a previous cycle, and at the end of the breeding season probably due to an estrogen deficiency. Several silent estruses occur in beef cows and ewes which suckle young, and dairy cows milked three times daily.

C. Ovulatory Failure

Ovulatory failure may be due to a failure of the follicle to ovulate during a normal cycle or due to cystic ovaries.

Anovulatory estrus is more common in swine and horses than in cattle and sheep. The animal shows normal behavioral estrus and the ovarian follicle reaches preovulatory size but does not rupture. Anovulatory follicles become partly luteinized and then regress during the estrous cycle, as does a normal corpus luteum.

Cystic ovaries, common in dairy cattle

and swine, are rarely encountered in beef cattle or in other species. In dairy cattle they are most often seen in high producers during the first few months of lactation. Affected animals are either nymphomaniacs or anestrous. One or both ovaries contain multiple small cysts or one or more large cysts. These are either follicular or luteal cysts. Follicular cysts undergo cyclic changes, i.e., they alternately grow and regress but fail to ovulate. Luteal cysts contain a thin rim of luteal tissue, also fail to ovulate, but persist for a prolonged period. The cyst fluid has a very high concentration of progesterone and is low in estrogen (Short, 1962), but there is no relationship between these hormonal concentrations in the cyst fluid and behavioral signs (nymphomania or anestrus).

In swine, cystic ovaries are an important cause of reproductive failure. Large multiple luteinized follicles are more common than small multiple cysts and contain progesterone. Estrous cycles are

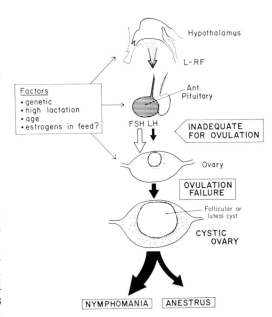

Fig. 18–2. Endocrine sequence, types of cysts and behavioral manifestations associated with cystic ovaries in the cow. Inadequate secretion of LH results in ovulatory failure and formation of follicular or luteal cysts. Note that affected cows are either nymphomaniac or in anestrus.

irregular with prolonged periods between cycles. Signs of estrus are very pronounced but nymphomania does not occur. It is not certain whether cystic ovaries in cattle and swine result from a failure of the ovulatory mechanism, from an adrenal cortex hyperfunction or from a disturbance in the hypothalamo-pituitary mechanism which depresses LH secretion (Fig. 18–2). Some success in treatment of cystic ovaries with high doses of LH points to LH deficiency. Daughters of affected cows show a higher incidence of cystic ovaries than do daughters of normal dams; this suggests a hereditary basis for this condition.

II. DISORDERS OF FERTILIZATION

A. Fertilization Failure

Fertilization failure may be the result of death of the egg before sperm entry, structural and functional abnormality in the egg or sperm, physical barriers in the female genital tract preventing gamete transport to the site of fertilization, or due to ovulatory failure and cystic ovaries discussed previously.

1. Abnormal Eggs

Spontaneous fragmentation and cytoplasmic division occur in both ovarian and oviductal eggs, often resembling cleavage. Fragmentation of eggs is very frequent in immature and superovulated animals. The frequency of fragmentation also increases as a result of delayed insemination or radiomimetic treatment of sperm before fertilization.

Several types of morphologic and functional abnormalities have been observed in the unfertilized egg, e.g., giant egg, oval-shaped egg, lentil-shaped egg, and ruptured zona pellucida (Hafez, 1961). Failure to undergo fertilization and normal embryonic development may be due to inherent abnormalities of the egg or to environmental factors. For example, fertilization is lower in animals exposed to elevated ambient temperature prior to breeding. In sheep, some of the conception failures at the beginning of the breeding season are associated with a high incidence of abnormal ova.

2. Abnormal Sperm

Semen usually contains a varying percentage of morphologically abnormal sperm. The physiologic significance of abnormal sperm in relation to fertilization failure has not been studied in animals other than cattle. The sperm of infertile bulls frequently contain low levels of DNA, and the ratio of DNA to arginine in the nucleus is highly variable.

Sperm and seminal plasma possess antigenic properties. Also the blood sera of cattle, sheep and swine contain antibodies capable of agglutinating sperm. Although this antigenic incompatibility could lead to sperm rejection and fertilization failure, there is at present no convincing evidence that antibodies pass from serum into the lumen of the uterus or oviduct to agglutinate sperm. The interaction between sperm and the egg prior to fertilization is also influenced by an antigen-antibody mechanism, a reaction which is highly specific for rejection or acceptance.

3. Structural Barriers to Fertilization

Congenital or acquired defects of the female genital tract interfere with transport of the sperm and/or the ovum to the site of fertilization (Table 18–2). Congenital defects are the result of arrested development of the different segments of the Müllerian duct (oviduct, uterus and cervix) or of an incomplete fusion of these ducts caudally. Acquired defects are caused by trauma or infection, particularly at time of parturition.

Several types of anomalies of development affect the reproductive tract of all species expressed, albeit in different degrees. Animals with anatomic defects have cycles of normal length and normal estrus, and their condition cannot be detected clinically. Anatomic abnormalities are very common in swine and account for about half of the total cases of reproductive failure.

Table 18-2.   Structural and Functional Causes of Fertilization Failure

Cause	Abnormality	Affected Species	Mechanism Interfered With
Structural Obstructions			
Congenital	Mesonephric cysts Uterus unicornis Double cervix	More common in swine, sheep and cattle than in horses	Sperm transport
Acquired	Tubal adhesions Hydrosalpinx Occluded uterine horns	All species, sheep and swine particularly	Egg pick-up, fertilization Egg transport
Functional			
Hormonal	Cystic ovaries Abnormal cervical and uterine secretions	Cattle and swine Cattle; sheep on estrogenic pastures	Ovulation Gamete transport
Management	Delayed insemination	In all species, horses and swine particularly	Death of egg
	Insemination too early	Cattle	Death of sperm

Common anatomic abnormalities are adhesions of the infundibulum to the ovary or uterine horns, which interfere with the pick up of the egg and the mechanical obstruction of one part of the reproductive duct system. Bilateral or unilateral missing segments of the reproductive tract also cause anatomic sterility. Other congenital abnormalities, such as a double cervix, may interfere with reproduction but do not always cause infertility. Abnormal shape or position of the cervix, a narrow cervical canal preventing the transport of sperm to the oviducts, or a torn cervix may also cause reproductive failure.

Small unilateral or bilateral cysts found in the mesosalpinx or on the oviducts do not interfere with fertility, however large cysts may occlude the lumen of the oviduct and prevent egg transport. The origin of these cysts is obscure.

A classic congenital anomaly associated with the gene for white coat color is the "white heifer disease" in cattle, in which the prenatal development of the Müllerian ducts is arrested, and the vaginal canal is obstructed by the presence of an abnormally developed hymen. This syndrome is expressed in different degrees depending upon the time when prenatal development is arrested. The degree and area of hypoplasia differ so that varying anomalies of the oviducts, uterus, cervix and vagina are formed. This congenital anomaly can be differentiated from the freemartin syndrome by the presence of normal ovaries and a normal structural vulva and labia in the former.

B. Atypical Fertilization

The complex process of fertilization is subject to several possible aberrations, namely polyspermy, monospermic fertilization (an egg containing two female pronuclei), failure of pronucleus formation and gynogenesis or androgenesis. Atypical fertilization may occur spontaneously as a result of aging of the gametes or elevation of environmental temperature. It has also been induced experimentally by x rays or the administration of certain toxic substances.

The aging of the ovum is a gradual one, in which various functions are successively lost (Table 18–3). An early effect of egg aging is that the resulting embryo is not viable and is resorbed before birth. Further aging leads to abnormalities in fertilization involving particularly the pronuclei. The biophysical and biochemical reactions associated with sperm entry into the egg become slower, a condition leading to increased polyspermy or the entry of more than one sperm.

Table 18-3. Aging of Gametes and Atypical Fertilization

Gamete	Mechanism	Abnormality
Sperm	Reduction or loss of DNA	Reduced viability of embryo
Egg	Incomplete maturation with failure to release the second polar body	Triploid embryo
	Inhibition of the block to polyspermy	Triploid or hetero-ploid embryo

Polyspermy occurs in several species of laboratory and farm animals. In swine, a delay in copulation or progesterone injections 24–36 hours before ovulation leads to some eggs with more than two pronuclei (Day and Polge, 1968). It is not clear whether these potential triploid embryos are caused by failure of extrusion of the second polar body or to polyspermy resulting from a failure of the block to polyspermy during ovum aging. The incidence of polyspermy increases when mating or insemination is delayed, resulting in triploid embryos which do not survive. This means that in horse and swine with a relatively long estrus, timing of breeding in relation to ovulation is critical for normal fertilization and embryonic survival. Similarly, polyspermy could account for some of the returns to service by delayed insemination of cattle. The frequency of polyspermy also seems to be controlled by genetic factors. In rats, hyperthermia induces a higher incidence of dispermic eggs in the Sherman and Long-Evans strains than in the Wistar CF strain.

III. PRENATAL MORTALITY

Many interacting factors determine prenatal development such as, nutritional and genetic factors which control the size and viability of the offspring, and uterine infections which interfere with implantation, placentation or prenatal survival. Evidence is also accumulating in farm animals that incompatibility of the fetus and dam may lead to prenatal mortality.

A small percentage of prenatal loss is involved in the normal reproductive process and may be regarded as unavoidable. Prenatal mortality, responsible for approximately a third of all pregnancies, can be divided into *embryonic* and *fetal* mortality.

A. Embryonic Mortality

Approximately 25–40% of embryos in cattle, sheep and swine are lost around the time of implantation. Estimates of embryonic deaths are based on comparisons between the number of eggs fertilized and the embryos that failed to survive. This data can be obtained by slaughtering groups of animals at two or more intervals after mating or insemination. One group is slaughtered at three days after service and eggs are recovered by flushing the oviducts. The number of cleaved eggs expressed as a percentage of the total number of ovulations provides an estimate of the fertilization rate. A second group of animals is slaughtered at the end of the first month of gestation and the number of viable embryos are then counted. The embryonic death rate is the number of nonviable embryos expressed as a percentage of the total number of fertilized ova.

Embryonic mortality occurs most commonly before or immediately following implantation, resulting in complete resorption of the conceptus (Table 18–4). It is also seen in large litters of swine and during multiple pregnancies in cattle and sheep (Hafez et al., 1965). The time of mortality affects the returns to service in two ways. In the first form, the fertilized egg develops to the morula or early blastocyst stage but degenerates before the middle of the estrous cycle. The corpus luteum regresses as in a normal cycle and the animal returns to estrus. In the second form, the blastocysts degenerate after midcycle but prior to, or immediately following, implantation. The regression of the corpus luteum is thus de-

Table 18-4. Causes of Embryonic Mortality

| Species | Period of Maximum Mortality | | Possible Causes |
	(Days of Gestation)	(Stage of Development)	
Cattle	16–25	Rapid growth and differentiation of embryo and extraembryonic membranes	Progesterone deficiency; inbreeding; multiple pregnancy; blood group homozygosity; J-antigen in sera
Sheep	14–21	Transition from yolk sac to allantoic placentation	Estrogenic pastures; inbreeding; increasing maternal age; hemoglobin types; overfeeding; multiple pregnancy; high environmental temperature
Swine	16–25	Spacing of embryos, transuterine migration	Inbreeding; chromosomal aberrations; overcrowding; overfeeding; increasing maternal age; high environmental temperature; transferrins
Horses	30–60	Corpus luteum of pregnancy regresses and accessory corpora lutea is formed	Lactation; twinning

layed for a period which is longer than the length of one estrous cycle.

1. Causes

Prenatal mortality can be due to maternal factors, embryonic factors or to fetal-maternal interactions. Maternal failure tends to affect an entire litter resulting in complete loss of pregnancy. Embryonic failure seems to affect embryos individually causing only a partial loss of pregnancy. In other cases the maternal environment may be insufficient, allowing the support of only a few strong embryos. Heredity, nutrition and age of the dam, overcrowding in utero, hormonal imbalance and thermal stress all contribute to embryonic mortality.

Endocrine Factors. The transport of the fertilized egg through the oviduct to the uterus is governed by a gradual fall of estrogen level after estrus and an increased production of progesterone as the corpus luteum replaces the ruptured follicle. For about $2\frac{1}{2}$ days after estrus "tube locking" exists at the ampullary-isthmic junction, preventing the embryo and the oviductal fluids from entering the uterus. Accelerated or delayed transport of the egg, as a result of estrogen-progesterone imbalance, leads to preimplantation death.

Regression of the cyclic corpus luteum in the nonpregnant animal is due to a continuous luteolytic effect from the uterus. The presence of a normal embryo in the uterus prevents this action, and thus the corpus luteum does not regress (Moor and Rowson, 1966). An abnormally undersized conceptus might not be able to counteract this luteolytic effect with consequent regression of the corpus luteum and termination of pregnancy. Retarded growth of the implanted embryo may be associated with the lack of development of binucleate cells appearing in the ruminant trophoblast at approximately the time of normally occurring corpus luteum regression. These cells may have either a luteotropic or an adhesive function.

A critical period of embryonic survival is the late blastocyst stage. Normally, the developing corpus luteum secretes progesterone which acts on the female tract in close synchrony with the development of the embryos. Thus failure of blastocyst implantation may result from delayed progestational changes in the endometrium at the appropriate time. For example, the pregnancy rate was increased in normal cows by the injection of 100 mg of progesterone one week after breeding.

Lactation. During lactation embryonic mortality occurs in cattle, sheep and horses but not in swine and is characterized by prolonged estrous cycles after breeding. The detrimental effects of lactation on embryonic development is not clear but probably involves a hormonal imbalance leading to an unfavorable uterine environment.

Chromosomal Aberrations. Cytogenetic studies have demonstrated a relationship between chromosomal anomalies and embryonic mortality, particularly in swine. In addition, heteroploidy produced by delaying mating for 36 hours causes embryonic mortality.

Heredity. The frequency and repeatability of embryonic loss is partly determined by the genotype of the sire and dam and the breeding system. In cattle, embryonic mortality is higher in inbred than in outbred lines. In swine and sheep, inbreeding of the dam contributes more to a reduction in litter size than does inbreeding of the embryo. The decrease most likely stems from the combined decrease in the maladjustment of the uterine environment and the number of genetically defective embryos.

Nutrition of the Dam. Caloric intake and specific nutritional deficiencies affect ovulation rate and fertilization rate, as well as cause prenatal death. In swine, high caloric intake or continuous unlimited feeding increases ovulation, thereby increasing the incidence of embryonic mortality prior to implantation. However, following implantation, fetal death is decreased by unlimited feeding. In sheep, full feeding before breeding also increases ovulation as well as embryonic death. The effect of caloric intake on prenatal death in cattle is at present controversial, although hypoglycemia induced by insulin reduced conception rates in lactating cows probably due to embryonic mortality.

Reproductive failure occurs in sheep and cattle grazing on plants which contain compounds with estrogenic activity. It may be due to aberrations in the estrous cycle, abnormal transport of eggs or sperm, or to implantation failure.

The estrogenic activity is due to plant isoflavones and related substances with hydroxyl groups (Fig. 18–3). Most of these compounds have a relatively low activity but are present in considerable amounts. Substances isolated from forages having estrogenic activity on mice are genistein, biochanin-A, daidzein, and coumestrol. The first three are isoflavones; the latter a benzofuranocoumarin. An amino acid (mimosine) extracted from the pasture legume (*Leucaena leucocephala*) caused both a lowered ovarian response to gonadotropins and an increase in embryonic death.

Compounds that have estrogenic activity can be found in several common plant materials, e.g., barley grain (*Hordeum vulgare*), oat grain (*Avena sativa*), the fruits of the apple (*Pyrus malus*) and cherry (*Prunus avium*), the tuber of the potato (*Solanum tuberosum*), and Bengal gram (*Cicer arietinum*). There are probably a number of plant materials with estrogenic activity still awaiting investigation.

Age of Dam. A higher incidence of embryonic mortality is observed in gilts and in sows after the fifth gestation. In the ewe it is highest beyond six years. There is no evidence of age effects on embryonic death in cattle. Embryonic loss in old mares may be associated with uterine atony.

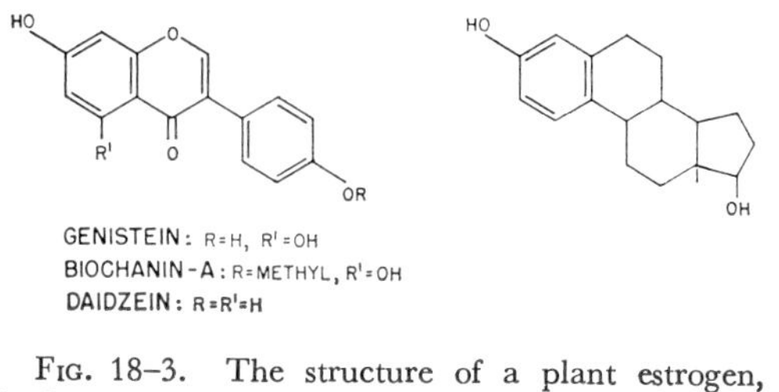

ISOFLAVONES   ESTRADIOL

GENISTEIN: R=H, R'=OH
BIOCHANIN-A: R=METHYL, R'=OH
DAIDZEIN: R=R'=H

Fig. 18–3. The structure of a plant estrogen, isoflavones, and of estradiol. Note the similarity in hydroxyl groups. Isoflavones are found in many forage species of the family *Leguminosae;* they have low estrogenic activity but are found in considerable amounts.

Overcrowding in Utero. As pregnancy advances, the embryo becomes increasingly dependent upon the placenta for its survival. Since the degree of placental development is primarily influenced by the availability of space and vascular supply within the uterus, increasing the number of implantations decreases the vascular supply to each site and restricts placental development. This results in a high embryonic and fetal death rate (Plate 19). This probably explains the higher incidence of embryonic mortality in cattle and sheep following twin rather than single ovulation, especially when both embryos are in the same uterine horn. Thus some uterine factors effectively limit litter size to a level characteristic of the species.

In multiple ovulated cattle and sheep the number of embryos surviving is reduced to a fairly constant number ($2\frac{1}{2}$ to 3 embryos per female) within the first three or four weeks of pregnancy, which implies that embryonic loss increases as the number of ova shed increases. This has been shown by Moore and his associates who transferred different numbers of sheep embryos to foster mothers. Embryonic mortality increased as the numbers increased. Most of the deaths due to crowding occur during the early stages of attachment, at about the 14th day. Mortality does not seem to be due to a deficiency of progesterone. Overcrowded uteri in rodents also have a high percentage of fused placentae leading invariably to embryonic or fetal loss. Transuterine migration of the embryos is of special physiologic importance in certain species for equal distribution of conceptuses, for example, a high incidence of embryonic mortality occurs in pigs if transuterine migration is prevented.

Thermal Stress. Embryonic mortality increases in a number of species following exposure of the mother to elevated ambient temperature, especially in tropical areas. In early stages of development the embryo is directly affected by increased maternal body temperature accompanied by thermal stress. The pig embryo is most susceptible to heat stress during the first two weeks of gestation (Edwards et al., 1968). When unacclimatized ewes are exposed to continuous high ambient temperatures during early pregnancy, approximately 75% of the embryos are lost at a stage of pregnancy which did not interfere with subsequent returns to service after one estrous cycle of normal length (Thwaites, 1967). A lower percentage of embryonic deaths is, however, associated with longer estrous cycles. But in similar ewes exposed to a diurnally variable heat stress of an eight-hour "day" and a 16-hour "night," the embryonic mortality was only 35%. Comparable diurnal variations in environmental temperatures under natural conditions occur only during the hottest periods of the summer (Thwaites, 1969).

The effects of thermal stress on the early embryo are not apparent until later stages of its development. Alliston et al. (1965) cultured in vitro one-cell fertilized rabbit eggs at elevated (40° C) and normal (38° C) body temperature for six hours prior to transferring the eggs into synchronous pseudo-pregnant females. The higher temperature increased the postimplantation embryonic mortality; in other words, the thermal stress was not apparent for nine days of development. Similarly, fertilized eggs of sheep and cattle when subjected to high temperatures either in vitro or in vivo are damaged but continue to develop only to die during the critical stages of implantation (Ulberg and Burfening, 1967). This may explain prolonged estrous cycles in cows failing to conceive during hot weather.

Semen. A part of the embryonic mortality can be attributed to the semen. Infertile matings by highly fertile bulls are primarily due to embryonic mortality while those of bulls with low fertility are due to fertilization failure and embryonic deaths. These fertility differences between bulls in artificial in-

PLATE 19

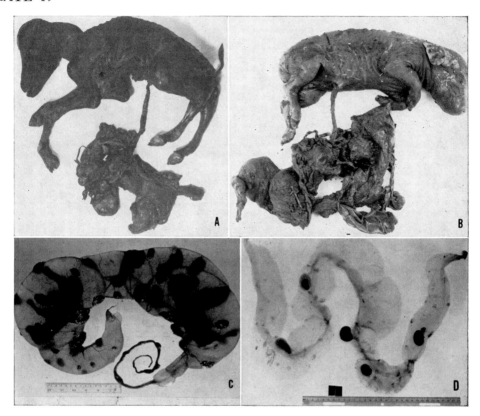

A, "Hematic" mummification of a 7-month-old bovine fetus, removed by Cesarean section from a Guernsey cow approximately 1 month after expected date of calving. (*Photograph by S. J. Roberts.*)

B, "Papyraceous" mummification of a pig fetus expelled at farrowing together with viable young. The fetus was surrounded by fetal envelopes which resemble parchment; note torsion of the umbilical cord.

C, Conceptus of beef cattle at 105 days of pregnancy showing two viable fetuses and five degeneration remnants of placentae and of the chorioallantois. The cow was previously treated with pregnant mare serum and human chorionic gonadotropins to induce multiple pregnancy.

D, Conceptus of beef cattle at 105 days of gestation showing degeneration of the placenta and five fetuses. The cow was previously treated with gonadotropins in order to induce multiple pregnancy.

PLATE 20

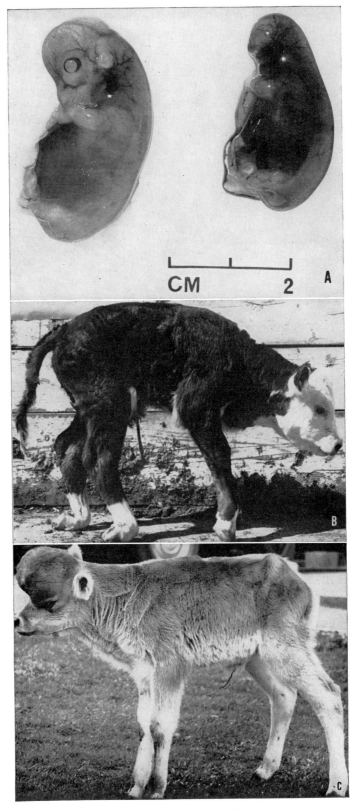

CM 2 A

B

C

(Legend on opposite page.)

semination is attributable to genetic factors caused by spermatozoa which are not revealed in routine tests of semen quality. In swine, semen stored for three days before insemination produced zygotes much more susceptible to early embryonic mortality (Dziuk and Henshaw, 1958) presumably due to the reduced DNA in aged spermatozoa.

Incompatibility. Immunologic incompatibilities may block fertilization (prezygotic selection), or cause embryonic, fetal or neonatal death. Blood group homozygosity is associated with an increase in embryonic death in cattle. The fertility of the ewe may be influenced by the combination of ram and ewe hemoglobin types varying, however, from year to year. There was no evidence that rams of hemoglobin type *AB* and *B*, or ewes of hemoglobin type *A*, *AB* or *B* showed differences in fertility associated with their hemoglobin type.

Electrophoretic differences in the transferrin (β-globulin) in sera are genetically controlled; this has been expounded in terms of biochemical polymorphism (Ogden, 1961). The transferrin locus can influence reproductive performance. In cattle, this influence is exerted by reducing fertility and by increasing embryonic mortality in certain crosses of β-globulin genotypes. In swine reproductive failure in the *BB* × *AB* transferrin matings is due to a higher embryonic mortality rather than to fertilization failure.

Cattle sera can be classified quantitatively into *J*-antigen positive and *J*-antigen negative types, a phenomenon associated with reproductive failure. As *J*-antigen negative cows can produce

J-antibody, Jamieson postulated that cows capable of producing *J*-antibody might show a reproductive failure when inseminated with semen from a *J*-antigen bull or caused to conceive *J*-antigen embryos. Blood systems closely related serologically to *J*-antigen are known in a variety of domestic animals (Stone, 1964).

2. Repeat-Breeders

"Repeat-breeder" females return to service repeatedly after being bred to a fertile male. These females are not sterile because, if bred repeatedly, many of them will ultimately conceive. Fertilization failure and early embryonic mortality are two major factors in reproductive failure (Fig. 18–4).

The nature of the reproductive failure can be determined by the interval between insemination and return to estrus. An interval equal to a normal estrous cycle length suggests failure of fertilization particularly in animals with normal ovarian function. Poor sperm transport and/or survival, abnormal ova, adverse tubal environment or genital tract abnormalities are factors which could prevent fertilization. Depending on the stage at which embryonic deaths occur this interval will either be normal as in fertilization failure or may be prolonged.

Fertilization failure more common in sheep and swine than cattle is due to genital tract abnormalities (Fig. 18–4). The cause(s) for loss of approximately 50% of the embryos within three weeks of pregnancy in repeat-breeder cows are obscure although infections, hormonal imbalances, hereditary and management factors have been suspected.

Explanation of Plate 20.

A, Facelessness in 35-day-sheep fetus (*right*) and a control fetus (*left*). The anomaly was manifested in the prenatal stage; this anomaly may result in fetal death at later stages of gestation.

B, Deformed Hereford calf born to a manganese-deficient dam receiving 16 ppm manganese; note enlarged joints and twisted legs (*Photograph by I. A. Dyer*).

C, Hydrocephalic calf with collection of fluid in the cerebral ventricles; the pressure causes an enlargement of the cranial vault. The syndrome is associated with asymmetry (wry face) and muscular incoordination. The abnormality was manifested in the prenatal life and the young was born alive but it died shortly after birth.

B. Abortion

Abortion refers to pregnancies which terminate with the expulsion from the uterus of a fetus of recognizable size before term. Fetal death is not an essential prelude to abortion. In cattle abortion may be regarded as the termination of pregnancy before term (260 days).

Abortions may be *spontaneous* or *induced*, infectious or noninfectious. Spontaneous abortion is more prevalent in cattle, particularly dairy cattle, than in sheep and horses. Noninfectious causes of spontaneous abortion due to genetic or chromosomal, hormonal and nutritional factors will be discussed here (Table 18–5). It may also occur in animals bred immediately after puberty or immediately after parturition. Mares seem to be endocrinologically susceptible to abortion between the 5th and the 10th months of pregnancy.

Chromosome abnormalities of the fetus are frequently associated with spontaneous abortions in man (Carr, 1965) but their importance in abortion of farm animals is unknown. Abortion due to hereditary factors results from abnormal development of some vital organ or general lowered viability of the fetus. Habitual abortions at 3 to $4\frac{1}{2}$ months of pregnancy in Angora goats is due to a hereditary defect of the anterior pituitary gland, leading to a deficiency of luteotropic hormone secretion required to maintain the corpus luteum of pregnancy. Abortions are occasionally induced with high doses of estrogens or

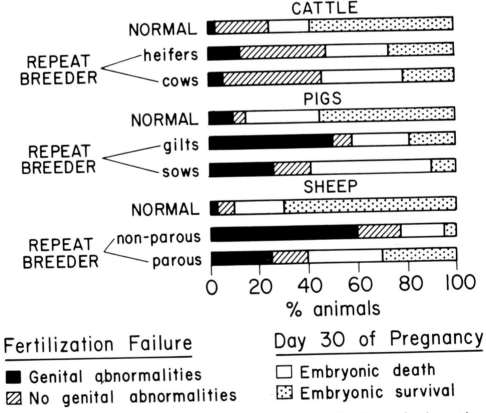

Fig. 18–4. Diagrammatic illustration of the causes of reproductive failure in normal and repeat-breeder animals. Repeat-breeding results from fertilization failure or early embryonic mortality. A high incidence of genital abnormalities causing fertilization failure is more common in sheep and swine than in cattle. (*From the literature.*)

Table 18-5. Noninfectious Causes of Abortion in Farm Animals

Causes	Cow	Mare	Sow	Ewe or Goat
Chemicals, drugs and poisonous plants	Nitrates Chlorinated naphthalenes Arsenic Perennial broomweeds Pine needles	None Phenothiazine (?)	Dicoumarin Aflatoxin Wood preservatives Creosote Pentachlorophenols	Anthelmintics Phenothiazine Carbon tetrachloride Lead, nitrate, locoweeds, lupines, sweet clover, onion grass, veratrum
Hormonal	High doses of estrogens or glucocorticoids Progesterone deficiency	High doses of estrogens or cortisone (?)	High doses of estrogens or glucocorticoids (?)	High doses of estrogens, cortisol or ACTH Progesterone deficiency
Nutritional	Starvation, malnutrition Deficiencies of vitamin A or iodine	Reduced energy intake	Deficiencies of vitamin A, iron and calcium	Lack of TDN or energy, deficiencies of vitamin A, copper, iodine and selenium
Genetic or chromosomal	Fetal anomalies	Fetal anomalies	Congenital or genetic lethal defects	Lethal genetic defects
Physical	Douching or insemination of pregnant uterus, stress (transport, fever, surgery)	Manual dilatation of cervix, natural service during pregnancy (?), rectal palpation of the very young blastodermic vesicle	Stress (transportation, fighting, injury)	Severe physical stress
Miscellaneous	Twinning, allergies, anaphylaxis	Twinning	Poor management	Twinning (?)

(Adapted from Roberts, 1971. *Veterinary Obstetrics and Genital Diseases*, published by the author, Ithaca, N.Y.)

glucocorticoids particularly in young females bred at an early age and in meat-producing animals.

C. Fetal Mummification

Fetal mummification is characterized by fetal death, failure of abortion, resorption of placental fluids, dehydration of the fetus and its membranes, and involution of the uterus. It is more common in cattle and swine than in sheep and horses.

Two types of mummification are known, the "hematic" type common to cattle and the "papyraceous" type of swine (Plate 19). In cattle, the gravid uterus becomes filled with gummy reddish brown material and there is a mas-sive intercotyledonary hemorrhage that causes the maternal and fetal cotyledons to separate. In some cases an "unsuccessful abortion" occurs; the fetus then undergoes gradual autolysis and maceration until only a compact mass of fetal bones is left. This latter form may also be associated with chronic mucopurulent vaginal discharges in cattle.

The syndrome occurs mainly from the fifth to seventh months of gestation in all breeds of cattle. It does not seem to exert any adverse influence on subsequent fertility, since affected cows conceive normally in the subsequent breeding period. Occasionally, bovine mummified fetuses are aborted spontaneously, but in most cases they are carried many

Table 18-6. Neonatal Mortality Due to Nutritional Deficiencies

Disease	Species	Cause	Description
"White muscle" disease	Lambs and calves	Selenium deficiency	Acute and chronic forms, degeneration of cardiac and skeletal muscles, death
Hypomagnesemia	Milkfed calves	Decrease in serum Mg^{++}	Irritability, nervousness and tetany
Piglet anemia	Baby pigs during first week	Iron deficiency	Low hemoglobin levels in blood, weakness and prostration, inability to nurse
Goiter	Foals and lambs	Iodine deficiency	Enlargement of thyroids
Enzootic ataxia	Lambs	Copper deficiency	Locomotor incoordination leading to paralysis
Neonatal hypoglycemia	Newborn piglets	Low blood glucose	Loss of appetite, coma

months beyond the gestation period. Mummification is suspected when a cow fails to calve on the expected date. Rectal palpation reveals the presence of a palpable corpus luteum, absence of the enlarged whirring uterine artery of pregnancy, the absence of fetal fluids and cotyledons, and a dry, mummified fetus.

Twin-bearing ewes may abort a mummified fetus during late gestation and maintain the other lamb to full-term; or they may deliver a mummified fetus attached to the placenta of a viable offspring.

Swine embryos which degenerate in the first six weeks of gestation are completely resorbed. The fetuses which degenerate during midgestation are usually expelled as mummified fetuses along with normal piglets. Those degenerating during late gestation may be expelled as stillborn fetuses. Mummified fetuses are more prevalent in large than in small litters and in some breeds than in others.

Mummification may be due to interference with the fetal blood supply, deficiency in placentation, anomalies in the umbilical cord of the fetus or infection in the gravid uterus. The syndrome may also be genetically inherited since it occurs in numerous pedigreed cows and in consecutive generations of cows bred

to unrelated males. A high incidence of fetal mummification in the Jersey and Guernsey breeds would also tend to support a hereditary influence. What causes the uterus to expel its contents is not known. Probably substances from the mummified fetus or its membranes inhibit the uterine luteolytic mechanism causing persistence of the corpus luteum.

IV. PERINATAL AND NEONATAL MORTALITY

Perinatal mortality refers to death of the offspring shortly before, during or up to 24 hours after parturition at normal term. Nutrition and age of the dam and hereditary factors appear to be the major contributing factors. In cattle, the incidence of perinatal mortality ranges between 5 and 15% of all births, with a high incidence in primiparous animals, in male fetuses and in calves sired by Friesian or Hereford bulls. In swine, one to two piglets in approximately one third of all litters are dead at birth with advancing parity, in extremes of litter size and in litters in which the gestation period is less than 110 days (Randall and Penny, 1970). In sheep, most losses between implantation and weaning occur during the perinatal period as a result of starvation of the neonate and of dystocia, among lambs born to maiden ewes and ewes on poor pasture or as a result of suffering from "clover disease."

Neonatal mortality—death of the neonate during the first few weeks of life—is related to heredity, environmental factors, nutrition and infections (Plate 20). Several nutritional deficiencies may contribute to neonatal mortality (Table 18–6). *White muscle disease*, a myopathy of lambs and calves results from a deficiency or metabolic disturbance of selenium. The hyperacute type, usually due to myocardial damage in younger animals, leads to death in a few hours. The subacute type, usually due to skeletal-muscle damage in older animals, causes muscular weakness, difficulty in nursing and death in one to two weeks. *Hypomagnesemia*, a decrease in blood magnesium levels, is characterized in milkfed calves by irritability, nervousness and, in severe cases, by tetany.

Hemolytic icterus or *neonatal isoerythrolysis* occurs in the neonate of the horse and swine. The disease, rarely observed in foals from a primiparous mare, is due to passage of fetal erythrocyte antigens into the maternal circulation. The production of maternal antibodies becomes progressively elevated until in the third or fourth parturition these antibodies are transmitted through the colostrum to affect the foal during the first two days of life. The disease in foals, apparently not due to an Rh-like factor as in man, can be controlled by the use of a foster mother or frozen colostrum from another mare. Hemolytic icterus, affecting a few piglets in a litter one to seven days after farrowing, results from vaccination of pregnant sows against hog cholera with swine tissue vaccines.

Neonatal mortality may also be a result of long labor, poor maternal nutrition, weakness of the mother or the young, bacterial infection through the umbilical cord of the young, poor maternal behavior, or delayed onset of lactation. Exposure of the newborn pig to low environmental temperature leads to hypothermia, hypoglycemia and death. Heat prostration and some deaths occur in newborn lambs exposed to high environmental temperature. Another source of danger to the neonate is mammalian or avian predators such as the feral pig (*Suis scrofa*), fox (*Vulpes vulpes*), dingo (*Canis dingo*), raven (*Corvus coronoides*), wedge-tailed eagle (*Aquila audax*) and sea eagle (*Haliaetus leucogaster*).

V. COMPLICATIONS OF PREGNANCY, PARTURITION AND LACTATION

A. Dystocia

Dystocia, difficult or obstructed parturition, may be due to fetal or maternal causes. Fetal dystocia (Fig. 18–5) results from abnormalities in the presentation or position of the fetus, and from postural irregularities of its head or limbs; it may be due to relatively or absolutely oversized fetus, and to fetal monstrosities. Fetal dystocia is common in certain breeds of dairy cattle, in cattle and sheep with multiple pregnancies, and in sows with small litters. It may also result when breeds of small stature are crossbred to large-sized males. An excessively large fetus in relation to the size of the birth canal of the dam presents difficulties even though the presentation is normal. Deviations of the head and flexion of the various joints in anterior presentation and flexion of both hind limbs (breech) in posterior presentation may obstruct the progress of the fetus.

During multiple birth, dystocia takes one of three forms: (*a*) both fetuses are simultaneously presented and become impacted at the maternal pelvis (Fig. 18–6); (*b*) one fetus only is presented but cannot be born because of defective posture, position or presentation; the lack of extension of limbs or head being due to insufficient uterine space; and (*c*) uterine inertia (Arthur, 1964).

Maternal dystocia is more frequent in dairy cattle and sheep than horses and swine. It occurs frequently in primiparous animals and in animals with multiple young. Absence of uterine contractions or inertia may be primary or second-

13

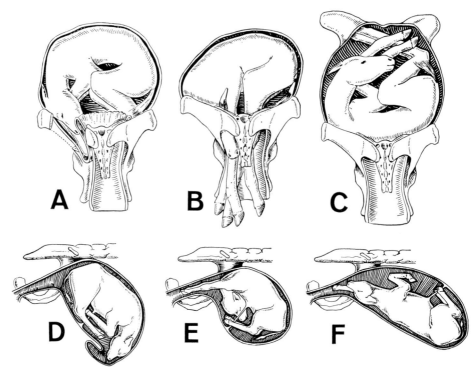

FIG. 18–5. Malpresentation in horses (*A, B, C*) and cattle (*D, E, F*). *A,* Ventro-transverse presentation with ventral displacement of the uterus. *B,* Ventro-transverse presentation; uterine body gestation. (*A, B* and *C from Arthur, 1964. Wright's Veterinary Obstetrics. London, Bailliere, Tindall & Cox. D, E, and F redrawn from Diseases of Cattle. U.S.D.A., Special Report, 1942.*)

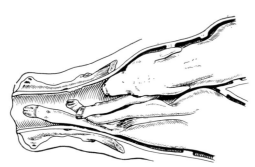

FIG. 18–6. Simultaneous presentation of cattle twins: one twin in anterior presentation, dorsal position, shoulder-flexure posture; the other twin in posterior presentation, dorsal position and extended posture. (*From Arthur, 1964. Wright's Veterinary Obstetrics. London, Bailliere, Tindall & Cox.*)

ary. Primary uterine inertia due to excessive stretching is common in multiple pregnancy in cattle and in large litters in swine. Secondary uterine inertia is due to exhaustion of the uterine muscle secondary to an obstructive dystocia.

Anomalies of the soft parts of the passages or the bony pelvis are occasional causes of dystocia. One group of anomalies causing a narrowing of the birth canal includes abnormalities or fractures of the pelvis, and stenosis or obstruction of the cervix, vagina or vulva. Another group of abnormalities preventing entry of the fetus into the birth passages includes failure of the cervix to dilate or torsion of the uterus.

B. Metabolic Disorders of Late Gestation and Parturition

Metabolic disorders associated with reproductive failure fall into two groups. Neuromuscular disorders are associated with disturbances in the metabolism of calcium, magnesium, and phosphorus. These syndromes are chiefly observed prior to and at the time of parturition, e.g., milk fever in cattle, lactation tetany

Table 18-7. Complications of Pregnancy, Parturition and Lactation

Syndrome	Species	Causes	Description
Milk fever	Cattle	Hypocalcemia	At parturition, recumbency, circulatory collapse, loss of consciousness and death
Grass tetany	Cattle	Hypomagnesemia	Late gestation or lactation, convulsions and tetany
Ketosis	Cattle	Hypoglycemia	Disturbance of carbohydrate metabolism during first month after calving; ketones in blood and urine; rapid loss of weight, drop in milk
Pregnancy toxemia	Sheep	Hypoglycemia	During late gestation in ewes carrying twins, disturbance of carbohydrate metabolism
Retained placenta	Cattle and sheep	Dystocia, infections	Failure to expel placenta after calving
Vaginal prolapse	Cattle and sheep	Excessive relaxation of pelvic ligaments, restricted exercise, twinning	Late gestation
Uterine prolapse	Cattle and sheep	Dystocia, placental retention	Follows parturition
Hydrops of fetal membranes	Cattle and sheep	Fetal anomalies, placental dysfunction, fetal-maternal incompatibility	Excessive accumulation of fluids within the amniotic or allantoic cavity
Twinning	Cattle, sheep and horses	Spontaneous or induced	Abortion
Prolonged gestation	Cattle	Genetic and fetal abnormalities	Fetal death
	Sheep	Ingestion of teratogens in early pregnancy	Fetal death
	Swine	Genetic, in certain inbred lines	Fetal death

in cattle, and lambing sickness. Disturbances in carbohydrate metabolism are most common in sheep and cattle (Table 18–7).

Hypocalcemia or Milk Fever. This is one of the most common disorders associated with parturition in dairy cows, characterized by recumbency, circulatory collapse and loss of consciousness. Milk fever in cattle is due to a sharp fall in plasma calcium and inorganic phosphorus levels. If the disease is not treated with calcium borogluconate, it often terminates in death.

Grass Tetany. This is a metabolic disease of cows grazing on lush pasture in late gestation or lactation and is characterized by hypomagnesemia, convulsions and tetany. It resembles milk fever.

Ketosis in Cattle. Ketosis which occurs one to six weeks after calving is characterized by lack of appetite, rapid loss of weight and milk production, nervous signs, hypoglycemia, ketonemia and ketonuria, and a depletion of the alkali reserves.

Pregnancy Toxemia, "Twin-Lamb" Disease, Ketosis of Pregnancy or Prepartum Paralysis. This syndrome occurs in sheep during the last two weeks of pregnancy and is caused by hypoglycemia which may stem from twin pregnancy, sudden changes in caloric intake or sudden restriction of exercise, such as occurs with heavy snowfall. Ewes overfed in early pregnancy or underfed in late pregnancy are highly susceptible to pregnancy toxemia resulting in their death or loss of lambs due to lack of vigor at birth. Ewes often recover if fetuses are aborted.

C. Retained Placenta

Failure of the placenta to be expelled during the third stage of labor is a common complication in ruminants due primarily to failure of the fetal villi to detach themselves from the maternal crypts. Retention of the placenta beyond 12 hours in cattle is considered pathologic and is associated with abortions due to *Brucella abortus* or *Vibrio fetus*, dystocia, uterine inertia and twinning. It is more common in dairy than beef breeds and adversely affects milk production and fertility due to delayed uterine involution.

Although extensive putrefactive changes occur if the placenta is retained for several days, antibiotic therapy is more effective than manual removal of the placenta.

D. Lactation Failure

Lactation failure following parturition is due to physiologic as well as physical factors. Milk let-down is inhibited by fright or painful conditions of the udder particularly in heifers. Underfed ewes during late pregnancy often fail to produce colostrum during the first few hours of the neonatal period. Colostrum may also be inaccessible to the lamb as a result of damage to the teats, presence of wool and dirt around the udder or abnormal maternal behavior.

A relationship exists between lactation and physiologic stresses associated with high environmental temperatures. The lactation failure in these instances is probably related to a depression in the secretion of the anterior pituitary and other hormones essential for lactation.

In cattle, inflammation of the mammary glands or mastitis results in a complete failure or a reduction in the production of milk depending on whether the inflammation is acute or chronic.

E. Prolapse of the Vagina or Uterus

Prolapse of the vagina due to relaxation of the pelvic ligaments during the latter half of pregnancy is common in multiparous cows and in ewes on restricted exercise, grazing estrogenic pastures or carrying twins. Prolapse of the uterus is more common in cows and ewes than in mares and sows and occurs after a few hours of parturition or may be delayed by several hours. Prolapse of the uterus may result from dystocia or pathologic retention of the placenta or hyperestrogenism in ewes on pastures rich in estrogenic compounds.

F. Hydramnios and Hydrallantois

Hydramnios, the excessive accumulation of amniotic fluid, is less common than hydrallantois, the accumulation of allantoic fluid. Hydramnios, observed more often in cattle than sheep or swine, is associated with certain cranial abnormalities of the fetus. In these defective fetuses, swallowing is impaired causing amniotic fluid to accumulate as gestation progresses (Fig. 8–2). Fetuses in prolonged gestation in the Guernsey and Jersey breeds have hydramnios.

Hydrallantois occurs in cattle especially in a twin pregnancy and is characterized externally by an enormous enlargement of the abdomen after the sixth month of gestation. The syndrome is due to fetal-maternal incompatability and placental dysfunction. The endometrium undergoes degeneration and necrosis, and fetal size is greatly reduced. The functioning caruncles decrease in number, the nongravid horn does not participate in placental formation and there is a compensatory accessory caruncular development in the gravid horn (Arthur, 1969). After delivery, either normally or by

Cesarean operation, the placenta may be retained or metritis may develop as a result of delayed involution of the uterus.

G. Multiple Pregnancy

In cattle, horses, sheep and goats, the frequency of multiple pregnancies is higher than that of multiple births, due to the high incidence of abortion and fetal resorption. In the cow the sequelae of twinning include shortened gestation periods, abortions, stillbirths, dystocia and retained placenta. Economic losses are related to decreased fertility, neonatal mortality, decrease in birth weights of calves, longer calving intervals, and lower butter fat production. Over 90% of the females born co-twin to a male are sterile (freemartins). Neonatal mortality in sheep is greater among twins than among singles. Ewes carrying twins are more susceptible to pregnancy toxemia ("twin lamb" disease). In mares a high percentage of twin fetuses are aborted.

H. Prolonged Gestation

Abnormally long gestations due to fetal abnormalities in cattle, sheep and swine result from genetic and nongenetic factors. Prolonged gestation is most common in dairy breeds: Guernsey, Jersey, Ayrshire, Brown Swiss, Holstein-Friesian, Swedish Red-and-White and other European breeds.

Kennedy and his associates (1957) reported that prolonged gestation in dairy cattle has a hereditary background. If the sires and dams of the fetuses are both heterozygous for the autosomal gene P, one-fourth of the fetuses will be carried beyond normal term by virtue of their having two recessive genes and three-fourths will be delivered at the normal time. Prolongation occurs only if the fetus is homozygous recessive for this gene, and is independent of the number of calves born previously. A cow carrying a fetus with two recessive genes may have a normal course of pregnancy for seven to eight months, but before parturi-

tion, the pelvic ligaments do not relax, the vulva does not become edematous and the mammary glands do not develop.

There are two general types of the syndrome in cattle. In the type seen in Holstein and Ayrshire breeds, the large fetuses have no facial abnormalities and when delivered by surgery are weak, unable to nurse and die in six to eight hours of severe hypoglycemia. In the type seen in the Guernsey and Jersey breeds, fetuses are small, many exhibit facial abnormalities and hydramnios, lack an adenohypophysis and survive in utero for long periods past term but live for only a few minutes when delivered by surgery (Kennedy et al., 1957). Prolonged gestation occurred in an inbred line of swine and in ewes consuming the teratogenic plant *Veratrum californicum*, 14 days after conception (Binns et al., 1965).

REFERENCES

Alliston, C. W., Howarth, B., Jr. and Ulberg, L. C. (1965). Embryonic mortality following culture *in vitro* of one- and two-cell rabbit eggs at elevated temperature. *J. Reprod. Fert. 9*, 337–341.

Arthur, G. H. (1964). *Wright's Veterinary Obstetrics.* London, Bailliere, Tindall & Cox.

Arthur, G. H. (1969). The fetal fluids of domestic animals. *J. Reprod. Fert.* Suppl. *9*, 45–52.

Binns, W., Shupe, J. L., Keeler, R. F. and James, L. F. (1965). Chronologic evaluation of teratogenicity in sheep fed *Veratrum californicum. J. Amer. Vet. Med. Assn. 147*, 839–842.

Carr, D. H. (1965). Chromosome studies in spontaneous abortions. *Obstet. Gynec. 26*, 308–326.

Day, B. N. and Polge, C. (1968). Effects of progesterone on fertilization and egg transport in the pig. *J. Reprod. Fert. 17*, 227–230.

Dziuk, P. J. and Henshaw, G. (1958). Fertility of boar semen artificially inseminated following *in vitro* storage. *J. Anim. Sci. 17*, 554–558.

Edwards, R. L., Omtvedt, I. T., Tuman, E. J., Stephens, D. F. and Mahoney, G. W. A. (1968). Reproductive performance of gilts following heat stress prior to breeding and in early gestation. *J. Anim. Sci. 27*, 1634–1637.

Ginther, O. J. (1968). Utero-ovarian relationships in cattle: Physiologic aspects. *J. Amer. Vet. Med. Assn. 153*, 1656–1664.

Hafez, E. S. E. (1961). Structural and developmental anomalies of rabbit ova. *Int. J. Fert. 6*, 393–407.

Hafez, E. S. E., Jainudeen, M. R. and Lindsay, D. R. (1965). Gonadotropin-induced twinning and related phenomena in beef cattle. *Acta Endocr.* Suppl. *102*, 1–43.

Kennedy, P. C., Kendrick, J. W. and Stormont, C. (1957). Adenohypophyseal aplasia an inherited defect associated with abnormal gestation in Guernsey cattle. *Cornell Vet. 47*, 160–178.

Moor, R. M. and Rowson, L. E. A. (1966). Local maintenance of the corpus luteum in sheep with embryos transferred to various isolated portions of the uterus. *J. Reprod. Fert. 12*, 539–550.

Ogden, A. L. (1961). Biochemical polymorphism in farm animals. *Anim. Breed. Abstr. 29*, 127–138.

Randall, G. C. B. and Penny, R. H. C. (1970). Stillbirths in the pig: An analysis of the breeding records of five herds. *Brit. Vet. J. 126*, 593–603.

Short, R. V. (1962). Steroid concentrations in normal follicular fluid and ovarian cyst fluid from cows. *J. Reprod. Fert. 4*, 27–46.

Stone, W. H. (1964). Significance of immunologic phenomena in animal reproduction. *Proc. 5th Internat. Congr. Anim. Reprod. Artif. Insem.* Trento, Italy, *2*, 89–90.

Symington, R. B. (1969). Factors affecting postpartum fertility in cattle with special emphasis on the hormonal aspects of the problem in ranch cows in South Africa. *Proc. S. Afr. Soc. Anim. Prod.* 29–34.

Thwaites, C. J. (1967). Embryo mortality in the heat stressed ewe. I. The influence of breed. *J. Reprod. Fert. 14*, 5–14

Thwaites, C. J. (1969). II. Application of hot-room results to field conditions. *J. Reprod. Fert. 19*, 255–262.

Ulberg, L. C. and Burfening, P. J. (1967). Embryo death resulting from adverse environment of spermatozoa or ova. *J. Anim. Sci. 26*, 571–577.

Chapter 19

Reproductive Failure in Males

L. C. FAULKNER AND E. J. CARROLL

Reproductive failure may be associated with defective seminal quality, accessory gland abnormality, a failure in mating performance, improper management or a combination of these.

I. DEFECTIVE SEMINAL QUALITY

No single measure or set of measurements has been found to give an infallible estimate of the fertility of semen. Correlations of seminal characteristics and fertility tend to be reasonably high when the samples are drawn at random from a large, unselected population of males, and each ejaculate is used to inseminate a large number of females. Correlations are especially poor when the range of values for the seminal characteristic in the sample is small and when levels of fertility are established from natural matings or from insemination of a small number of females.

Measures of seminal characteristics reflect the functions of the testes, the excurrent ducts, and the accessory sex glands. They also reflect ejaculatory frequency and the skill and care exercised in collecting and handling semen. A careful physical examination is a useful adjunct to seminal analysis in the diagnosis of subnormal seminal quality. Testicular biopsies and blood levels of steroids and gonadotropins will be useful supplements as advances in technology make them available and knowledge makes them meaningful. Some infectious agents produce male infertility without causing any detectable sign.

Variations in size and consistency are the most commonly observed testicular defects (Table 19–1), and reasonably precise techniques for clinical measurements are available. Testicular circumference is positively correlated with other testicular measurements, is highly repeatable and is the simplest measurement to obtain. Testicular circumference can be measured easily with a flexible tape around the greatest diameter of the testes and scrotum. Testicular size is positively correlated to spermatozoan output in young bulls, providing a valuable method for predicting spermatozoan output and subsequent testicular development. It is of less value in bulls more than five years old, in which pathologic alterations frequently obscure the relationship of testicular size to spermatogenic potential (Hahn et al., 1969a). Soft consistency of the testes is often related to poor seminal quality and low fertility. Because manual palpation is too subjective to classify gradations in consistency, a testicular tonometer was developed to provide an objective measurement (Hahn et al., 1969). The tonometer provides a simple, quantitative means of predicting seminal quality and potential fertility in dairy bulls. The usefulness of the instrument

Table 19-1.   Occurrence of Testicular and Penile Defects in 10,940 Bulls

Testicle		Penis	
Reduced size	7.4%	Deviation	1.7%
Soft	7.4%	Fibropapilloma	0.6%
Hypoplasia	1.3%	Persistent frenulum	0.5%
Abnormal shape	0.9%	Laceration	0.2%
Fibrosis	0.4%	Urethral fistula	0.2%
Cryptorchid	0.1%	Balanitis	0.1%

(From Carroll et al., 1963. *J. Amer. Vet. Med. Assn. 142*, 1105.)

has not been adequately tested as a measure of the breeding potential of young beef bulls or males of other species.

Many conditions have been described in which various portions of the male reproductive tract are absent (*aplasia*) or incompletely developed (*hypoplasia*). Males with unilateral deficiencies or occlusions often have normal fertility in limited service. Segmental aplasia of the Wolffian ducts is more common among sons of bulls with unilateral segmental aplasia. The condition is characterized by a total or partial absence of the epididymis on one or both sides, but most often on the right side (Blom and Christensen, 1972). Rectal examination may reveal marked asymmetry in the seminal vesicles or ampullae in such bulls. Cysts, remnants and miscellaneous anomalies of the mesonephric and Müllerian ducts are of little consequence.

Reproductive inefficiency, usually associated with defective seminal quality, constitutes a major reason for disposal of bulls from insemination centers. Semen from bulls with lowered fertility depresses reproductive efficiency through both failure of fertilization and embryonal mortality. Seminal quality in males of polytocous species is related to fertility (percentage of fertile matings) and to fecundity (percentage of eggs fertilized).

One or a few seminal ejaculates collected at irregular intervals may be adequate to detect gross abnormalities, such as azoospermia, a high incidence of morphologically abnormal sperm, or the presence of leukocytes or erythrocytes. Even a 50% decrease in potential daily spermatozoan output may go undetected in a single seminal sample or a series of ejaculates collected at low ejaculatory frequency, because extragonadal sperm reserves may mask the deficiency (Amann and Almquist, 1961). To detect spermatogenic deficiency or unilateral occlusion of the excurrent ducts, seminal samples should be collected frequently with a short, uniform interval between collections. The period of sampling should follow pre-experimental collections to stabilize the extragonadal sperm reserves and the rate of loss of sperm through resorption or into the urine. The period of sampling should be of sufficient duration to accommodate for the duration of the spermatogenic cycle and epididymal transit when studying the effects of agents which damage the early stages of spermatogenesis.

Fertility of mature males declines with advancing age. The magnitude and nature of these changes are best described in dairy bulls. It is likely there are similar changes in other domestic males. Seminal quality and fertility of dairy bulls are highest at three to four years of age, before progeny tests to determine the genetic potential of the bulls have been completed (Hahn et al., 1969b).

A lowered spermatozoan output in aged sires is due in part to difficulties in sexual preparation of older bulls with physical disabilities, but degenerative changes are commonly observed in the testes of older bulls, including focal tubular degeneration, calcified tubules, interstitial fibrosis and Leydig cell hyperplasia. A progressive fibrosis, beginning in the ventral portion of the testis, is associated with senility (McEntee, 1970). The ventral distribution of fibrosis and associated hyaline degeneration of the walls of arterioles suggest that this lesion may be due to vascular degeneration. Testicular degeneration may be caused by a variety of vascular lesions, and inflammation of the testicular artery occurs frequently in the stallion. Migrating strongyle larvae and the equine arteritis virus cause testicular arteritis in stallions, but the cause is not established in many cases. The usual manifestations are localized and are associated with degeneration of seminiferous tubules adjacent to inflamed arteries and arterioles. Vascular and other degenerative testicular lesions are coincident with, but not necessarily attributable to, advanced age.

A. Nutritional Factors

Specific dietary deficiencies, e.g. of zinc and vitamin A, reportedly impair reproductive function, but deficiencies of energy may have been involved since feed intake and gains in body weight were commonly affected adversely. Vitamin E has been highly overrated in reproductive function in domestic animals; however, Dukelow (1967) reviewed the evidence for beneficial effects of feeding wheat germ oil on reproductive processes and suggested that other biologically active factors in wheat germ oil may affect reproductive performance. Supplementation with vitamin A has been recommended for breeding bulls which are in dry lot and fed hay which has been stored for several months.

B. Environmental Factors

Although spermatogenesis is continuous in most domestic males, the level and quality vary with climatic influences. The extent of seasonal variation in spermatogenesis is related to species, breed and latitude. Ambient temperature, humidity, convection currents and solar radiation may exert direct effects on the neuroendocrine system and testes; they also operate indirectly through their influence upon the quality and quantity of forage. Seasonal effects are especially pronounced in rams and stallions, in keeping with the pronounced effects of season on the reproductive cycles of ewes and mares (Fig. 19–1). In general, male reproductive function is depressed during seasons of peak temperatures, especially when accompanied by high humidity.

Local factors which elevate testicular temperature also impair seminal quality. Local elevation in testicular temperature may be associated with vascular changes, and the impaired function may be due to hypoxia, hyperemia or direct thermal damage. Cryptorchidism, ectopic testes, excessive scrotal wool, insufficient scrotal wool, heavy deposits of scrotal fat, orchitis, scrotal dermatitis, scrotal and testicular infections and wounds and febrile diseases may cause testicular degeneration and abnormal maturation or storage of spermatozoa. Cryptorchidism is more common in stallions, billy goats and boars than in bulls. It may be a hereditary defect due to a recessive gene.

Hypothermia may also adversely affect male reproductive function, although the system is more resistant to cold than it is to hyperthermia. Cold did not interfere with testicular development, production of sperm, or seminal quality in boars which were exposed to mean daily temperatures of -15 to $-20°$C for a 15-week experimental period (Swierstra, 1970). On the other hand, bulls which suffered scrotal frostbite as a result of exposure to severe blizzard conditions produced semen of lower quality than bulls without scrotal necrosis in the same herds (Faulkner et al., 1967). Moreover, the incidence of unsatisfactory ejaculates was greater among bulls which were exposed to severe chilling but failed to develop scrotal frostbite than among a sampling of bulls which were not exposed to such extreme cold. Sires exposed to extreme conditions of cold should be protected by dry bedding.

C. Immunologic Factors

Degenerative testicular lesions have been induced in the males of several species by injecting autologous or isologous materials containing spermatozoa. Seminal antigens are normally isolated from the body's immunogenic mechanism by the intact epithelium of the excurrent tract. However, pathologic leakage of seminal products into the tissues could result in the formation of antibodies against seminal components. Autoimmune reactions have not been established as a cause of spontaneous testicular lesions or infertility in any species, but the presence of spermagglutinating antibodies in seminal and blood plasmas has been associated with clinical infertility. The role of autoimmunity in causing spontaneous testicular degeneration in farm animals is difficult to assess, since a variety of agents can induce morphologically identical degenerative lesions.

Seminal antigens might also induce the formation of antibodies associated with

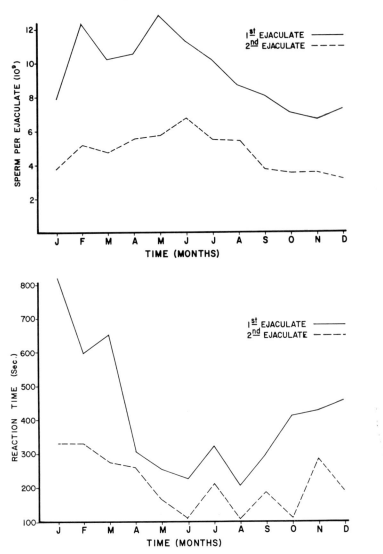

Fig. 19–1. (*Top*). Mean sperm per ejaculation (10^9) of five stallions collected weekly for 52 consecutive weeks. There were highly significant differences ($P < 0.01$) due to season and between first and second ejaculates. (*Bottom*) Mean values for reaction time of five stallions for 52 consecutive weeks. Reaction time was the total time the stallion was in visual contact with the mare until the beginning of copulation. Seasonal differences were highly significant. (*From Pickett, 1970. Proc. 3rd Tech. Conf. A.I. Reprod., Natl. Assoc. Anim. Br. pp. 1–8.*)

infertility in females. Indeed, infertility has been induced in several species of mammalian females, including cattle, by isoimmunization with spermatozoa, semen or testicular material. The seminal antigens which cause experimental infertility in cattle appear to be components of sperm, and the infertility is due to inhibition of fertilization, embryonal or fetal death, or a combination of both factors (Menge, 1969). Antigens in seminal plasma and sperm-coating antigens from the accessory sex glands appear not to cause infertility. The role of isoimmunization of females as a cause of infertility is not defined; even with cattle, in which the response to isoimmunization against sperm is good, the semen must be administered in conjunction with immunologic adjuvants to stimulate the formation of effective antibodies.

Some seminal extenders used in artificial insemination also contain antigens, e.g., egg yolk and milk proteins. Antibodies against egg-yolk antigens have been found in samples of uterine mucus and tissue from cows which had been repeatedly inseminated (Griffin et al., 1971). When cows were inseminated with semen extended in egg-yolk diluent, the fertility of those with uterine titers of antibodies against egg-yolk antigens was lower than that of cows with no uterine titers.

D. Congenital Anomalies

Testicular hypoplasia is a congenital, presumably heritable defect in which the potential for complete development of the spermatogenic epithelium is lacking in one or more epithelial components (Kenney, 1970). Hypoplasias may be characterized by total germinal cell hypoplasia, partial germinal cell hypoplasia, spermatogenic arrest with multipolar spindles, and spermatogenic arrest with sticky chromosomes.

E. Genetic Factors

Despite the generally low heritability estimates of most components of fer-

Table 19-2. **Heritability Estimates for Seminal Characteristics in 534 Linecross Hereford Bulls**

Trait	Heritability
Concentration of spermatozoa	0.28 ± 0.12
Motility	0.23 ± 0.11
Percentage unstained with supravital stain	0.17 ± 0.11
Normal cells	0.24 ± 0.11
Primary abnormalities	0.30 ± 0.12
Secondary abnormalities	—.05 ± 0.01

(Abadia, D., 1972, *Genetics of Semen Traits of Beef Bulls*. Ph.D. Thesis. Colorado State University, Fort Collins, Colorado.)

tility, inbreeding has a detrimental effect on male fertility, while heterosis is associated with increased fertility (Abadia, 1972). The detrimental effect of inbreeding is not expressed uniformly among different lines indicating differences in the frequencies of harmful genes among lines. In some cases, there is little effect. Heterosis is evident in the comparison of all seminal characteristics among linecross, inbred and outbred bulls. The largest amount of heterosis was found for a decrease in primary morphologic abnormalities of spermatozoa. Heritability estimates for seminal characteristics were obtained by paternal half-sib analyses on data from linecross bulls (Table 19-2). All seminal traits, except secondary morphologic abnormalities of spermatozoa, were moderately heritable (17–30%) and should respond to selection, although progress would be slow. In an earlier study on animals from the same herd, physical defects of the genitalia were highly heritable; only bulls with normal reproductive organs should be selected for mating (McNitt et al., 1966).

Specific morphologic abnormalities of spermatozoa appear to be inherited, and new abnormal forms continue to be reported. For example, a specific gametic sterility which is probably regulated by a recessive autosomal gene has been demonstrated in England and

Holland (Donald and Hancock, 1953). Bulls which are homozygous for this gene produce large numbers of abnormal spermatozoa and are completely sterile. The defect, called "knobbed," is an eccentrically placed thickening of the acrosome. A similar defect has been described in boars.

II. TESTICULAR ABNORMALITY

Testicular degeneration is the most common lesion associated with acquired infertility and lowered seminal quality. Acquired hypospermatogenesis, without evidence of other degenerative changes, is probably the most common abnormality in the testicles of bulls culled for reduced reproductive efficiency, but the true incidence is unknown because the testes of culled animals are not routinely examined in a manner that enables a diagnosis (Kenney, 1970). A variety of agents, including hormones, drugs, plant toxins, immune reactions, systemic disease and physical insults to the testes, induce identical degenerative lesions in the seminiferous epithelium, and degenerative processes are not always distinguishable from those of testicular hypoplasia. Changes in seminal characteristics and testicular degeneration have also been associated with adrenal hypertrophy and adrenal tumors (van Rensburg, 1965). Orchitis may cause testicular degeneration and is characterized by the inflammatory response to infectious agents or extravasated spermatozoa, which are the most frequent causes.

The spermatogenic epithelium is sensitive to a variety of insults. Renewal of the seminiferous epithelium is from type A spermatogonia, which appear to be quite resistant to damage and provide for an ability to recover. The damaging effect of heat on the germinal epithelium and subsequent regeneration are known. Type A spermatogonia in the process of mitosis at the time of thermal insult may have their regenerative capacity impaired. Rat testes can withstand repeated applications of heat and still recover spermatogenic function, but the percentage of severely atrophic tubules increases by about 4% with successive heatings (Bowler, 1972).

Primary testicular tumors are common in older animals. Tumors replace functional tissue and may disrupt normal spermatogenesis. Interstitial (Leydig) cell tumors grow slowly and are associated with compressional atrophy in surrounding tubules. Seldom can they be diagnosed by clinical examination. Spermatozoan production and fertility are decreased in bulls with large interstitial cell tumors, but total ejaculatory volume is increased, indicating androgenic stimulation of the accessory glands. Sertoli cell tumors are less common but occur in bulls and stallions. Seminomas have been observed in stallions and rams.

III. ACCESSORY GLAND ABNORMALITIES

The epididymis is important in the maturation and storage of spermatozoa,

A, Chronic bilateral seminal vesiculitis with adhesions between the right seminal vesicle and rectum; a draining sinus tract extended from the vesicle into the lumen of the rectum.

B, Normal pelvic genitalia of the bull. Lobulation of the seminal vesicle is distinct.

C, Bull with subacute seminal vesiculitis. Lobulations are still distinct, but the glands were very firm on rectal palpation.

D, Pelvic genitalia from bull with chronic bilateral seminal vesiculitis. Lobulation in right seminal vesicle is indistinct, and fibrosis is extensive.

E, Deviation of penis caused by persistent penile frenulum in a bull being stimulated with an electroejaculator.

F, Penis after surgical correction of the persistent frenulum shown in E. Most bulls make a satisfactory recovery and are completely healed in 7 to 10 days.

(A, B, C, and D from Ball et al., 1964. Am. J. Vet. Res. 25, 291. E and F from Carroll et al., 1964. J. Amer. Vet. Med. Assn. 144, 747.)

PLATE 21

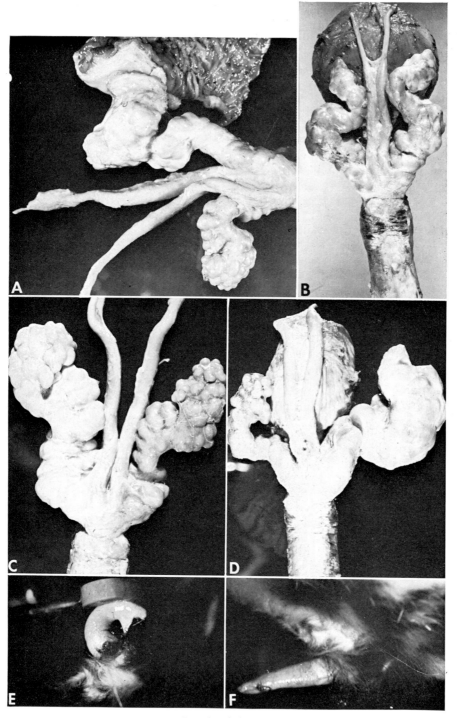

Legend on facing page

and epididymal dysfunctions may adversely affect seminal characteristics and fertility. The most commonly identified epididymal abnormalities are epididymitis, segmental aplasia and vestigial remnants of the mesonephric tubules. Epididymal lesions are capable of interrupting the continuity of the duct, precluding the passage of sperm. Lesions that allow extravasation of spermatozoa into the stromal tissue result in a characteristic, granulomatous response which resembles a tuberculous lesion (McEntee, 1970). Epididymitis is the most common lesion of the reproductive tract in rams. Lesions are most frequent in the tail of the epididymis. *Brucella ovis* is a common cause. Scar tissue in the stroma and epithelial hyperplasia in the duct combine to obstruct the lumen, resulting in seminal stasis. Spermatic granulomas commonly develop as a result of the extravasation of sperm. Most of the extravasations occur proximally to the obstruction in the tail of the epididymis, but some also occur in the tail. Organisms other than *Brucella ovis*, including chlamydia, can initiate this same sequence in the formation of epididymal lesions.

Seminal vesiculitis is the most common inflammatory change diagnosed clinically in the reproductive system of bulls. The incidence is often high in certain groups of beef bulls. Young bulls confined in groups and intensively fed are more often affected than range bulls, and the number of affected bulls varies widely from year to year in the same herd, suggesting that a specific infectious agent may be responsible. Experimental seminal vesiculitis has been achieved with *Brucella abortus* and certain viral and chlamydial agents.

Seminal vesiculitis can be diagnosed by rectal examination, and unless this procedure is used, the condition may be overlooked. Other clinical signs such as leukocytes in semen, arched back, fever, reduced appetite, depressed ruminal function and pain on defecation are described in the literature but may not be observed. Seminal vesiculitis is characterized by increased glandular size and firmness. The extent of enlargement can be quite variable. Either or both glands may be affected, and normal asymmetry should not be confused with pathologic change (Plate 21, A–D). Purulent exudate is often found in the semen of affected bulls, and seminal vesiculitis is the most common cause of leukocytes in bovine semen. Spontaneous recovery occurred in at least half of the cases observed in bulls one to two years old (Carroll et al., 1968). Recovery is associated with systemic antibiotic therapy, but the efficacy of therapy is not known, since spontaneous recovery is common. Older bulls and those cases associated with *Corynebacterium pyogenes* tend to remain chronically affected and some become progressively worse.

IV. INEFFECTIVE MATING PERFORMANCE

Mating performance may be affected adversely by psychic disturbances or physical disabilities. Psychic disturbances are based in the nervous system and cause inappropriate mating behavior. Mating behavior may be affected by environmental factors. Libido may increase in boars maintained at low ambient temperature (Swierstra, 1970). Breed differences exist among rams in response to heat stress, which generally cause diminished sexual activity during and following exposure to high ambient temperature (Lindsay, 1969). Monthly variation has been observed in mating behavior in stallions (Fig. 19–1).

Sexually naive males may be indifferent or respond with inappropriate behavior to sexual stimuli; many of these behavioral deficiencies may be overcome with training. Inhibition of mating behavior is not limited to naive males, however. The incidence of sexual inhibition in yearling rams (35%) is higher than in ram lambs (23%) or mature rams (19%) (Hulet et al., 1964). Severe sexual inhibitions are commonly observed in stallions coming off of tracks, where sometimes heroic methods are employed to discourage sexual behavior.

Physical disabilities may impede or prevent mating performance by causing failure in copulatory behavior; i.e., mounting and intromission. Abnormalities of the penis frequently result in failures to achieve intromission. The most common penile defects in beef bulls are deviations, fibropapillomas and persistence of the penile frenulum (Table 19–1).

Spiral deviation, or "corkscrew penis," has been described as a cause for the inability of bulls to achieve intromission. It may be only a premature occurrence of a normal response (Seidel and Foote, 1969). Normal bulls often show spiral deviation on withdrawal, and the deviation was seen frequently during ejaculation among bulls collected with a transparent artificial vagina. Since the bull's penis dilates very little during erection, spiralling may increase the vaginal stimulus in the female during intromission. Spiral deviations of the normal penis occur frequently in bulls subjected to stimulation with an electroejaculator. A diagnosis of spiral deviation of the penis should be restricted to those bulls which repeatedly fail to achieve intromission; under these circumstances, the incidence and successful surgical correction of this condition would probably be substantially reduced. Downward and lateral deviations are also observed.

Penile fibropapillomas are seen most frequently in young bulls which have been penned together. Young bulls commonly mount penmates and abrade the glans penis during the process; this practice is apparently highly conducive to invasions of the penile epithelium by the virus which causes cutaneous warts and genital fibropapillomas. Housing in individual pens reduces the incidence of this disease. In most cases, surgical removal of penile fibropapillomas is easily accomplished. There may be a regrowth, but the chance of recurrence is decreased when surgical removal is delayed from 10 to 30 days after the initial observation

of the lesions. When the disease exists as a herd problem, prophylactic vaccination with formalin-treated bovine wart tissue vaccine is indicated (Olson et al., 1968).

Persistent penile frenulum causes a marked deviation of the erect penis and prevents normal protrusion. The persistent frenulum extends from the ventral median raphe of the penis to the prepuce and varies from a small, cordlike attachment to a wide band of tissue extending the full length of the median raphe. A higher incidence was found in Beef Shorthorn and Angus bulls. Surgical correction is simple, and most bulls make a satisfactory recovery (Plate 21,E,F). High incidence in certain herds and the differences in breed incidence suggest a genetic basis for the abnormality, and it may be an error to use surgically corrected bulls in purebred herds.

Abnormalities of the bovine prepuce are more prevalent in certain breeds of cattle, and the differences are probably due to the anatomic variations. The prepuce is longer in *Bos indicus* (Zebu) breeds than *Bos taurus* (European) breeds. Of the latter, polled breeds (Angus, Polled Hereford and Shorthorn) have longer prepuces than the horned breeds, and polled bulls do not have preputial retractor muscles (Bellenger, 1971). Chronic preputial prolapse is associated with preputial injuries and infections which may lead to scarring and constriction of the preputial orifice. Affected bulls are reluctant or unable to serve cows. Techniques have been developed for surgical correction of the damaged prepuce and for shortening excessively long prepuces. The predisposition to prolapse is inherited and suggests that selection of afflicted bulls for breeding, made possible by surgical correction, could increase the frequency of this defect (Lagos and Fitzhugh, 1970).

The boar has a preputial diverticulum dorsal to the preputial orifice. Its capacity is 20–30 ml, but in a full-grown boar it can hold more than 100 ml. The contents of the diverticulum have an intense odor and consist mainly of urine, semen and desquamated epithelial cells. Surgical extirpation of the preputial diverticulum in boars used in artificial insemination has been described. This procedure resulted in a substantial reduction of bacteria in semen collected with the artificial vagina, and the intense odor from the preputial fluid disappeared (Aamdal, 1958).

Physical disabilities involving systems other than the reproductive system may interfere with locomotion in seeking estrous females and with copulatory performance. These disabilities include dislocations, arthritides, skeletal fractures, ruptured ligaments, tendons or muscles, foot rot, overgrown hooves, interdigital fibromas, spinal disorders and the spastic syndrome. Abnormalities of the vertebral column and bones and joints of the rear limbs frequently impair the serving ability of older bulls. Ankylosing spondylosis deformans of the thoracolumbar vertebrae may interfere with mobility, and a few bulls have fractured the ankylosed vertebrae and severed the spinal cord. Injuries to and arthritides of the stifle and hock joints are predominant; the knee and fetlock are less commonly involved. These lamenesses have been associated with advancing age, but they may be accelerated or initiated by diets containing excessive amounts of calcium. Bulls kept in confinement are typically fed a ration similar to that of milking cows. Alfalfa hay is generally provided because it is an excellent source of nutrients, but the calcium is high, exceeding the bull's needs for this mineral. Excessive dietary calcium in bulls leads to abnormalities of bone structure (Krook et al., 1971).

Other physical disorders contribute to infertility in more subtle ways, causing general disability or reducing sensory acuities involved in mating behavior, e.g., internal and external parasites, tuberculosis, diseases of the mouth and any condition which impairs the male's ability to perceive olfactory, visual, gustatory or auditory stimuli from estrous females.

V. ERRORS IN MANAGEMENT

Methods of management affect the effective fertility of males, particularly of those used in artificial insemination, e.g., procedures used in storage and handling of semen, the site of semen deposition. Yearling or mature rams were joined singly, or in groups, to groups of ewes which varied from 25 to 100 ewes per ram (Lightfoot and Smith, 1968). Decreasing the number of ewes per ram was associated with more rams serving each estrous ewe, a greater proportion of ewes served during the first two weeks of mating, fewer ewes unserved at the end of a six-week mating period, and higher fertility and fecundity (proportions of ewes lambing and twinning rate). Yearling rams are more fertile when joined with 25 ewes per ram than when joined with 50 ewes per ram, but the fertility of mature rams was similar at both ewe/ram ratios. Mature rams are more fertile than yearlings when joined at 50 ewes/ram but not at 25 ewes/ram.

The number of females assigned to each male depends on nutritional factors, terrain, size of breeding enclosure and tendencies of the herd to disperse or flock. In general, studies on the factors which influence mating behavior in farm animals and clinical parameters associated with high sex drive in males are woefully lacking, especially for animals in natural service.

REFERENCES

Aamdal, J., Hogset, I. and Filseth, O. (1958). Extirpation of the preputial diverticulum of boars used in artificial insemination. *J. Amer. Vet. Med. Assn. 132*, 522–524.

Abadia, D. (1972). *Genetics of Semen Traits in Beef Bulls.* Ph.D. Thesis. Colorado State University, Fort Collins, Colorado.

Amann, R. P. and Almquist, (1961). Reproductive capacity of dairy bulls. V. Detection of testicular deficiencies and requirements for experimentally evaluating testis function from semen characteristics. *J. Dairy Sci. 44*, 2283–2291.

Bellenger, C. R. (1971). A comparison of certain parameters of the penis and prepuce in various breeds of beef cattle. *Res. Vet. Sci. 12*, 299–304.

Blom, E. and Christensen, N. O. (1972). A systematic search for abnormalities in testis-epididymis in pedigree bulls in Denmark. Studies on pathological conditions in the testis, epididymis, and accessory sex glands in the bull. VII. Communications from The State Veterinary Serum Laboratory, Copenhagen, Denmark 514, pp. 1–36.

Bowler, K. (1972). The effect of repeated applications of heat on spermatogenesis in the rat: A histological study. *J. Reprod. Fert. 28*, 325–333.

Carroll, E. J., Aanes, W. A. and Ball, L. (1964). Persistent penile frenulum in bulls. *J. Amer. Vet. Med. Assn. 144*, 747–749.

Carroll, E. J., Ball, L. and Scott, J. A. (1963). Breeding soundness in bulls—a summary of 10,940 examinations. *J. Amer. Vet. Med. Assn. 142*, 1105–1111.

Carroll, E. J., Ball, L. and Young, S. (1968). Seminal vesiculitis in young beef bulls. *J. Amer. Vet. Med. Assn. 152*, 1749–1757.

Donald, H. P. and Hancock, J. L. (1953). Evidence of gene-controlled sterility in bulls. *J. Agric. Sci. 43*, 178–181.

Dukelow, W. R. (1967). Wheat germ oil and reproduction. A review. *Acta Endocr. 56*, Suppl. *121*, 1–15.

Faulkner, L. C., Hopwood, M. L., Masken, J. F., Kingman, H. E. and Stoddard, H. L. (1967). Scrotal frostbite in bulls. *J. Amer. Vet. Med. Assn. 151*, 602–605.

Griffin, J. F. T., Nunn, W. R. and Hartigan, P. J. (1971). An immune response to egg-yolk semen diluent in dairy cows. *J. Reprod. Fert. 25*, 193–199.

Hahn, J., Foote, R. H. and Cranch, E. T. (1969). Tonometer for measuring testicular consistency of bulls to predict semen quality. *J. Anim. Sci. 29*, 483–489.

Hahn, J., Foote, R. H. and Seidel, G. E. (1969a). Testicular growth and related sperm output in dairy bulls. *J. Anim. Sci. 29*, 41–47.

Hahn, J., Foote, R. H. and Seidel, G. E. (1969b). Quality and freezability of semen from growing and aged dairy bulls. *J. Dairy Sci. 52*, 1843–1848.

Hulet, C. V., Blackwell, R. L. and Ercanbrack, S. K. (1964). Observations on sexually inhibited rams. *J. Anim. Sci. 23*, 1095–1097.

Kenney, R. M. (1970). Selected diseases of the testicle and epididymis of the bull. *VI International Conference on Cattle Diseases*, pp. 295–314.

Krook, L., Lutwak, L., McEntee, K., Henrikson, P., Braun, K. and Roberts, S. (1971). Nutritional hypercalcitoninism in bulls. *Cornell Vet. 41*, 625–639.

Lagos, F. and Fitzhugh, H. A. (1970). Factors influencing preputial prolapse in yearling bulls. *J. Anim. Sci. 30*, 949–952.

Lightfoot, R. J. and Smith, J. A. C. (1968). Studies on the number of ewes joined per ram for flock matings under paddock conditions. *Aust. J. Agric. Res. 19*, 1029–1042.

Lindsay, D. R. (1969). Sexual activity and semen production of rams at high temperatures. *J. Reprod. Fert. 18*, 1–8.

McEntee, K. (1970). "The Male Genital System." In *Pathology of Domestic Animals* Vol. I, 2nd ed. K. V. F. Jubb and P. C. Kennedy, New York and London, Academic Press, pp. 355–387.

McNitt, J. I., Stonaker, H. H. and Carroll, E. J. (1966). Breeding soundness in beef bulls. *Proc. Western Section, American Society of Animal Science 17*, 25–30.

Menge, A. C. (1969). Early embryo mortality in heifers isoimmunized with semen and conceptus. *J. Reprod. Fert. 18*, 67–74.

Olson, C., Robl, M. G. and Larson, L. L. (1968). Cutaneous and penile bovine fibropapillomatosis and its control. *J. Amer. Vet. Med. Assn. 153*, 1189–1194.

Pickett, B. W. (1970). Seasonal variation of stallion semen. *Proc. 3rd Tech. Conf. Artif. Insem. Reprod. Natl. Assn. Anim. Breed.* pp. 1–8.

Seidel, G. E. and Foote, R. H. (1969). Motion picture analysis of ejaculation in the bull. *J. Reprod. Fert. 20*, 313–317.

Swierstra, E. E. (1970). The effect of low ambient temperatures on sperm production, epididymal sperm reserves, and semen characteristics of boars. *Biol. Reprod. 2*, 23–28.

van Rensburg, S. J. (1965). Adrenal function and fertility. *J. S. Afr. Vet. Med. Assn. 36*, 491-500.

Chapter 20

Intersexuality

R. A. McFeely and H. Kanagawa

Intersexuality continues to attract the interest of scientists and breeders alike. To the scientist, intersex animals provide clues to understanding the mechanisms of sex determination while the breeder is more concerned with the economics of raising animals destined to be infertile. Despite its current popularity, interest in intersexuality has its origin in antiquity. Mention of ambisexual people appears in the Greek literature as early as the fourth century BC. Ovid tells us the story of Hermaphroditus, the handsome son of Mercury and Venus, who chanced to attract the comely nymph Salmacis. Her passion for him was so great that she convinced the gods to unite them into one individual. Hermaphroditus and other ambisexual individuals have persisted throughout the Graeco-Roman culture in stories, sculpture, paintings and other art forms. Further interest in intersexuality is evident throughout recorded history. Over two centuries ago, John Hunter described an intersex bovine which appears to have been a freemartin.

The discovery of sexual dimorphism in somatic cell nuclei and the development of suitable techniques for the display of chromosomes have renewed interest in intersexuality and some of the developmental defects are beginning to be understood.

I. NORMAL SEXUAL DEVELOPMENT

In the domestic mammals all individuals produce either male or female gametes. Each gamete contains the haploid number of chromosomes composed of one sex chromosome and a set of autosomes. The female mammal is called the homogametic sex because all of the gametes are similar, with one X chromosome. The male, however, has two types of gametes, one type containing an X chromosome and the second type possesses a Y chromosome. At fertilization, the genetic sex of the new individual is determined when the X-bearing ovum is united with sperm carrying either an X chromosome producing a genetic female or a Y chromosome in which case the genetic sex of the zygote will be male. There are a few minor species exceptions to this method of genetic sex determination which are of interest and involve translocations of sex chromosomes to autosomes so that, in fact, the same general principles apply.

There is intriguing evidence to suggest that many rearrangements in chromosomal structure have taken place in the autosomes during the process of speciation. However, in all mammalian species studied, the X chromosome appears to be very similar in DNA content and contains genes controlling the same processes, implying that evolution has not greatly altered this chromosome (Ohno, 1967). A further expansion of this theory suggests that the Y chromosome, which is responsible somehow for determining maleness, has resulted from a deletion of extraneous genes from the original homolog of the X chromosome. Only one gene not directly concerned with sex determination has been proven to be on the Y chromosome.

The second phase of sexual differentiation involves the development of the gonad or gonadogenesis. The genital ridges from which the gonads and reproductive ducts develop are formed very early in organogenesis. Initially the sex of the primitive gonad cannot be distinguished between the sexes. Epithelial and primordial germ cells migrate into the indifferent gonad. If the individual is to be a male, these migratory cells give rise to the seminiferous tubules. If the individual is to be a female, there is additional migration and proliferation of epithelial and germ cells which give rise to the ovarian cortex with the medullary or testicular portion remaining vestigial.

Although the controlling mechanisms of gonadogenesis are not understood it seems logical to assume that the Y chromosome of the somatic cells in the germinal ridge probably mediated through its genes for maleness, causes the medulla to persist. In the absence of a Y chromosome or perhaps as a result of two X chromosomes the cortex predominates to produce an ovary. There are exceptions to this, however, and these theories may require refinement. The chromosome composition of the primordial germ cells is also of importance. It appears that, at least in some mammals, survival of female primordial germ cells is dependent upon both X chromosomes being present and active and that these cells eventually become located in the cortex. They do not survive in a testicular environment.

The development of the gonad is closely associated with the development of the Wolffian or mesonephric duct and the Müllerian or paramesonephric duct. Normally one duct persists while the other regresses. In the male two substances are required for normal development. One hormone which is an androgen causes persistence of the Wolffian system. The second substance which results in regression of the Müllerian system has not been chemically defined. In the absence of the testes the mesonephric ducts regress and the Müllerian system develops into the tubular genital tract. The castrated male fetus develops as a female (Jost, 1970). There is good experimental evidence to show that the control of development of tubular genital organs is largely a local event and is not the result of circulating hormones.

Posteriorly the genital tract is formed by the development of the urogenital sinus and the genital tubercle. Development of the external genitalia is greatly influenced by circulating androgens and a female fetus can be masculinized by male hormones at a critical stage of development.

In normal development the genetic sex, the gonadal sex and the sex of the accessory genital organs are identical and each one is dependent upon events which have occurred in the antecedent stages. It is remarkable, however, that at the indifferent stage of gonadogenesis all individuals have the potential to develop, sometimes almost completely, in the opposite direction of their genetic sex. A variety of factors can distort normal sexual development thereby giving rise to various forms of abnormal genitalia or intersexuality.

II. CLASSIFICATION OF INTERSEXUALITY

Although sex is defined simply as male or female, it is determined at a number of levels: genetic sex, gonadal sex, phenotypic sex, behavioral sex and legal sex. In studies of intersexuality in domestic mammals we are primarily concerned with the first three. Normal individuals have the same sex in all three categories. If one of these levels of sex determination is different from the others, an intersex results. An intersex, therefore, is an animal with congenital anatomic variations which confuse the diagnosis of sex. Such an animal may have some of the reproductive organs of both sexes or be genetically one sex and phenotypically the other.

A true hermaphrodite has both male and female gonads either separately or combined as an ovo-testis (Plate 22). A pseudohermaphrodite has either testes or ovaries but the remainder of the reproductive organs have some of the characteristics of the opposite sex. These animals are classified as male or female depending upon the sex of the gonad. Therefore, a male pseudohermaphrodite has testes but a largely female phenotype.

In this chapter, an attempt is made to classify abnormal sexual development by the level of sex determination where the abnormality occurred. However, as in all other schemes many cases cannot be placed exclusively into a single category and some overlap exists.

III. ABNORMAL SEXUAL DEVELOPMENT

Aberrations in the normal development of the reproductive system can occur at any level of sex determination. If chromosome abnormalities alter the genetic sex, the developmental events which follow will very likely be affected. Altered gonadogenesis does occur in the presence of a normal chromosome picture, thereby raising the question of whether the concerned genes are normal, or if some normal genes were merely de-repressed at the wrong time. It is impossible definitely to categorize all of the developmental defects. The following classification is admittedly arbitrary and will undoubtedly require modification in the future.

A. Aberrations of Genetic Sex

Genetic sex is either XX or XY in the normal animal. Aberrations can result from structural abnormalities in either one or both of the sex chromosomes, by increases or decreases in the number of sex chromosomes, or by the presence of both male and female cells in the same individuals.

Although structural abnormalities have been difficult to detect, new techniques of chromosome identification, i.e. banding, should add to our knowledge in this area in the future. In man and in the mouse, a variety of abnormalities of the X chromosome have been reported, usually associated with some anomaly of development. However, very few cases have been described in domestic mammals.

One case in cattle with scrotal testes epididymides and ducti deferens and a female phenotype associated with an abnormal chromosome, presumed to be an abnormal X chromosome with a pericentric inversion or a translocation of the short arm of the Y chromosome to an autosome, has been reported. It is postulated that the alteration in the structure of the X chromosome influenced the normal development of the genital tract. This animal also possessed a line of normal female cells (McFeely et al., 1967).

Monosomy of the X chromosome (XO) has been reported in man and the mouse and in man this is very often a lethal combination. Perhaps this is why they appear to be rare in farm animals.

Extra sex chromsomes appear to be more compatible with life. The XXY syndrome in man (Klinefelter's syndrome) is associated with testicular pathology and infertility, and in pure cases of XXY sex chromosome constitution in the bull, ram and pig similar pathology seems to prevail with infertility (Bishop, 1972). The XXY was first identified in the rare tricolor male cat.

An equine intersex with four different cell lines with $XO/XX/XY/XXY$ sex chromosome has been reported (Basrur et al., 1969). This interesting case had one abnormal testis and one gonad that had failed to develop. The reproductive tract was mostly male while the external

PLATE 22

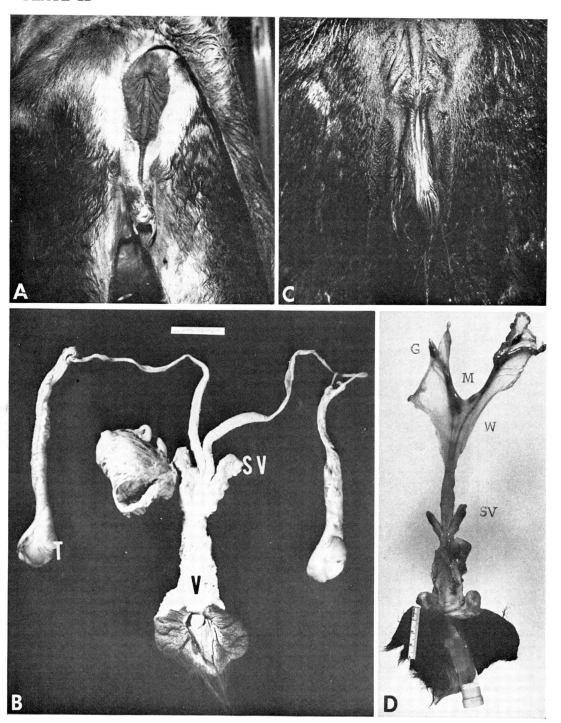

A, External genitalia of bovine male pseudohermaphrodite.

B, Internal genitalia of bovine male pseudohermaphrodite. *T*, testis; *SV*, seminal vesicles; *V*, vagina.

C, External genitalia of bovine freemartin.

D, Internal genitalia of bovine freemartin. *G*, gonad; *M*, Müllerian remnants; *W*, Wolffian remnants; *SV*, seminal vesicle.

genitalia was ambiguous. It was originally postulated that the various cell lines originated from an *XXY* zygote; however, on blood type analysis it was concluded that the intersex was a whole body chimera with *XX* and *XY* cells predominating, and *XXY* and *XO* cells as minor populations derived through nondisjunction of *XY* cells.

Because of the special interest to livestock breeders and the unique etiology, cases with a mixture of male and female cells are discussed separately in the section on freemartinism.

B. Abnormal Gonadal Development

This section deals with the situation where the gonad develops in a manner which is inconsistent with the genetic sex. Obviously in cases of gonadal hypoplasia in both sexes or cryptorchidism in the male there are developmental defects in the gonad but because they do not usually confuse the diagnosis of sex, they are not considered in intersexuality.

True hermaphroditism with various combinations of ovaries, testes and ovotestes have been described in the horse, goat, cow and pig (Biggers and McFeely, 1966). In some cases, this is associated with chimerism of male and female cells. These cases could be readily classified in the previous category and the development of the gonad attributed to the local effect of either male or female cells in the primordial gonad. However, there is no evidence at present to prove or disprove this theory.

Other hermaphrodites with both ovarian and testicular tissue have been described in which only one genetic sex has been demonstrated. While it is impossible to rule out undetected chimerism in these cases there are other possible explanations for this phenomenon. They may have a similar origin to a sizeable number of cases described in the goat, pig and dog which have male gonadal sex but are genetic females (Bishop, 1972). Although the testicles in these animals are not entirely normal, usually because

they are undescended, there is little doubt that they are basically testicular in nature. The accessory sex organs vary considerably in these cases presumably depending upon the hormone production by the embryonic gonad. Several theories have been advanced to explain the contradiction between genetic and gonadal sex. It has been postulated that the *Y* chromosome or a part of it, could be translocated to one of the *X* chromosomes or an autosome and thereby remain undetected (Ferguson-Smith 1966). However, these animals would then have an *XXY* constitution and a different type of abnormality would be expected.

Another hypothesis elaborates on the theory previously mentioned that the *Y* chromosome has developed by successive deletion of the *Y* chromosome. If this is the case, the *X* chromosome may carry genes for maleness which are normally not expressed. In these cases of genetic females with testes, the maleness genes on the *X* chromosome may find a way of expression (McFeely et al., 1967). This theory also has application in other types of intersexuality such as the cow, described previously with a structural abnormality of the sex chromosomes. In fact, the theory can be stretched further to ascribe the function of the *Y* chromosome as somehow permitting expression of maleness genes on the *X* chromosome. This also can help explain the freemartin. However, it is only a theory and there are obvious points which need clarification or modification.

C. Abnormal Development of the Accessory Genital Organs

Aberrations in the accessory genital organs are often the most easily detected as they frequently confuse or distort the phenotypic sex. In these cases the genetic and gonadal sex develop sequentially but there are discrepancies in the development of the derivatives of the Wolffian or Müllerian ducts, or the urogenital sinus and the external genitalia.

1. Male Pseudohermaphroditism

As the development of these structures appears to be dependent upon substances produced by the gonad in the male, male pseudohermaphrodites could result from either a failure of normal production by the testes or by lack of response to these substances by the target organs.

Male pseudohermaphroditism has been well-described in the testicular feminizing syndrome of man (Polani, 1970). In these cases genetic and gonadal sex are male but phenotypic, behavioral and legal sex are female. Apparently the developmental abnormality in these cases is the result of an insensitivity of the target organs to androgens. The testes have been shown to produce testosterone and yet the Wolffian elements do not develop. However, the substance causing regression of the Müllerian elements must also be produced because they are not present in the postnatal individual. These individuals often have quite adequate breast development.

A few cases similar to testicular feminization in man have been described in farm animals, particularly cattle (Bishop, 1972). As there is a suggestion that this syndrome may have a hereditary basis it is interesting to speculate upon the role of some autosomal gene in the development of this syndrome. Further research is required to clarify this abnormality.

2. Female Pseudohermaphroditism

Female pseudohermaphroditism in man occurs in individuals with a deficiency of an enzyme in the metabolic pathway in the adrenal gland causing a block in the production of cortisone and a resultant shunting of the pathway causing an increase in the production of androgens. The adrenogenital syndrome is not well-documented in farm animals.

Increased circulatory androgens from a functional ovarian or adrenal tumor in the dam causing masculinization of a female fetus have been suggested as a possible cause of female pseudohermaphroditism but convincing evidence is lacking. However, animals have been partially masculinized by the administration of drugs with androgenic properties to pregnant females. This has been done both experimentally and clinically when these drugs were used therapeutically.

D. Freemartinism

Freemartinism*which occurs primarily in cattle is the sexual modification of a female twin by in utero exchange of blood from a male fetus; a few cases of "freemartinlike" syndromes have been reported in sheep, goats and pigs (Bishop, 1972). Freemartins are characterized by: (a) internal reproductive organs of both sexes; (b) modified ovary with varying degrees of similarity to male gonads; (c) external genitalia like those of a normal female; and (d) blood chimerism (blood group and sex chromosome, XX/XY).

1. Aberrations of Reproductive Organs

The gonads of the freemartin are intra-abdominal and rarely descend through the inguinal canal. The anatomic structure ranges widely from almost normal ovaries to structures resembling testes with an increase in interstitial cells. Thus, freemartins do not exhibit estrus and are sterile.

The anatomic and histologic structure of the reproductive tract also varies considerably from individual to individual. In general, the Wolffian and Müllerian ducts are poorly developed which is apparent very early in embryonic life. Often a rudimentary epididymis and ductus deferens are observed. The presence of seminal vesicles appears to be a constant finding. The external genitalia,

* The word "freemartin" was first used in print in the 17th century, but it had been in common use for years before that. "Free" may be a contraction of "fallow," used to denote an infertile animal in Scotland; "martin" may come from the Gaelic word for cow, "mart." A sterile cow (farrow mart) was slaughtered on St. Martinmas Day, an important holiday in Scotland and northern England.

with few exceptions, resemble that of the normal female; however, they lead into a much shorter vagina. Frequently the clitoris is enlarged and the tuft of hair at the ventral commissure of the vulva is elongated. The mammary glands are underdeveloped and the teats shorter than those in three- to four-month-old normal heifers.

2. ETIOLOGY OF THE SYNDROME

During multiple pregnancy in cattle, the chorioallantois of adjacent embryos fuse. Even when two chorionic sacs develop in separate uterine horns, they eventually meet and fuse. An early and intimate vascular connection is very likely between fetuses in the same horn, whereas a connection between fetuses in different horns would probably develop later.

Freemartinism has been ascribed by Lillie (1922) to the male twin's hormones that reach the female through vascular anastomoses between the fused placentae. The dominance of the male twin was explained on histologic evidence; namely the appearance of interstitial cells in the fetal testis much earlier than the fetal ovary, suggesting that the testis is active endocrinologically long before the ovary. This "hormonal" theory played a crucial role in establishment of the endocrinology discipline. However, recent knowledge raises questions about the hormone theory; there is no evidence to support the idea that androgens can alter a mammalian ovary into a testis and likewise androgens cannot cause regression of the Müllerian ducts which is a common feature of the freemartin. The external genitalia, a sensitive area for masculinization in the female, remains female.

The fusion of chorioallantois is so intimate that extensive arterial and venous anastomoses occur, and blood-forming cells and other cells are exchanged between the fetuses. A result of this reciprocal exchange between dizygotic twins is identical erythrocyte antigen types in both twins and sex chromosome chimerism XX/XY of leukocytes. Each

twin contains two genetic populations of cells, one corresponding to its own genotype and one produced by the cells acquired from the co-twin.

In view of the recent advances in immunology and cytogenetics, some people feel that the presence of the male cells and, most particularly, the Y chromosomes is the key to the altered development (Herschler and Fechheimer, 1967). However, there is conflicting evidence in this area, as well. In the marmoset monkey, XX/XY chimerism is a common finding and developmental abnormalities of the genital tract do not result (Benirschke and Brownhill, 1962). In experimental chimeras produced by fusing mouse blastocysts, intersexuality is not a common occurrence (Mintz, 1968). Obviously, the mechanisms responsible for freemartinism are still not precisely understood.

The stage in prenatal development and the degree of vascular anastomosis vary with different pregnancies, resulting in varying degrees of expression of the freemartin condition. In about five percent of bovine twin pregnancies the blood vessels of the chorioallantois never fuse and the resulting female is not a chimera and is fertile.

Kanagawa and his associates studied the chromosome constitution of bovine heterosexual twins, triplets, quadruplets and quintuplets by means of blood culture and colchicine treatment of bone marrow cells. All the freemartins manifested sex chromosome chimerism (XX/XY). The chimera ratios were parallel among the freemartins and their co-multiple male (Basrur and Kanagawa, 1969). The degree of morphologic deviation in freemartins does not seem to depend upon the degree of the chimerism.

Occasionally, a single horn animal is demonstrated to be a chimera. It is possible that in certain proportions of the heterosexual twins, one of the twins degenerates during early gestation and the other twin is born as a single with blood chimerism. Fusion of zygotes very early in development is another plausible explanation. A particularly interesting and unique case of a diploid-triploid true hermaphrodite cow was postulated to have arisen as a result of double fertilization and fusion of the fertilized meiotic products (Dunn et al., 1970).

Although the development of the male co-twin appears to be anatomically normal there is increasing evidence that fertility of these animals is reduced (Stafford, 1972).

3. Diagnosis of Freemartinism

The diagnosis of freemartinism is based upon: (*a*) clinical genital abnormalities; (*b*) presence of sex chromatin bodies or drumstick of leukocytes in male co-twin; (*c*) presence of sex chromosome chimerism (*XX/XY*) of hemopoietic cells; (*d*) blood typing; and (*e*) skin grafting which provides indirect evidence for fetal vascular anastomosis.

Clinical diagnostic procedures of rectal palpation and vaginoscopy by a glass speculum are usually used. Freemartins often have only a rudimentary uterus and a small vagina.

Blood group analysis which is used to determine whether twins are monozygotic or dizygotic, is extremely useful for diagnosis of freemartinsim. Mosaicism of erythrocytes permits the technique of "differential hemolysis" to be used. If

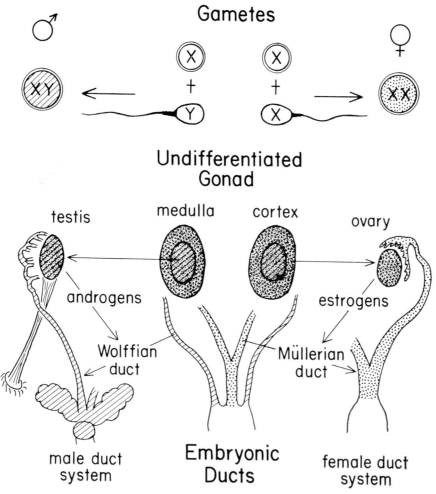

Fig. 20–1. Diagrammatic illustration of sex determination in mammals.

Table 20-1. Intersexuality in Farm Mammals

Syndrome	Sex Chromosomes*	Gonad	Reproductive Tract	External Genitalia
Freemartin	XX/XY	Masculinized ovary	♂♀	♀
True hermaphrodite	XX/XY XX/XY	Ovary and testis or ovo-testis	♂♀	♂♀
Male pseudohermaphrodite	XY XX	Testis	♂	♀
Female pseudohermaphrodite	XX	Ovary	♂♀	♂

* Many other combinations of abnormal chromosome complements are possible.

two animals contain the same proportion of two cell types, then the fraction of cells not lysed by a specific reagent should be the same, whereas if they contain different cells, lysis will be independent (Stone et al., 1964). Thus a female twin possessing the same lysis pattern as her twin brother will be a freemartin.

Fetal blood exchanges result in immunologic tolerance, and therefore dizygotic twin calves of different sex are, with few exceptions, tolerant to grafts of each other's skin. This tolerance is present in monozygotic twins (identical genotype) but is never found in other relationships. Thus it was found that all females that were highly tolerant to male-twin grafts were freemartins (Billingham and Lampkin, 1957).

4. EXPERIMENTAL INDUCTION

Experimentally produced gonadal inversions have been reported in amphibians, birds and opossums treated with gonadal hormones. Several attempts have been made to induce freemartinism in cattle and rodents by injections of steroid hormones in either the pregnant mother or the placenta. To date, all these attempts have been unsuccessful.

5. FREEMARTINISM IN OTHER SPECIES

According to the definition of a freemartin, all cases described for freemartinism in species other than the bovine are not typical of the syndrome, although a few cases of "freemartinlike" syndromes have been reported in goat, sheep and pigs (Biggers and McFeely, 1966).

In the goat the freemartinlike condition has been observed infrequently, although twinning is not rare and intersexes are common. Some intersexes have XX/XY lymphocyte and bone marrow chimerism, and the anatomic findings are somewhat consistent with the freemartin syndrome although not appreciably different from other types of intersexuality in goats.

In sheep, fraternal twins are born more frequently than in cattle, however sex chromosome chimerism occurs in a small percentage of sheep twins. In studying skin grafts exchanged between sheep twins of opposite sex, Moor and Rowson (1958) found that four of five sets showed normal homograft reactions, the skin dying within 10 days—whereas the homografts on the remaining pair behaved as autografts. Furthermore, the female of this pair showed abnormal development of the vulva and clitoris and showed no signs of estrus. She was most likely a chimera.

In the pig, a few cases of a freemartinlike syndrome have been reported. Side-to-side fusion of fetal sacs is common but end-to-end chorioallantoic vascular fusion seems to be rare. Such vascular fusion must occur very early in these rare cases since in later development chorionic overlapping is the rule without parabiosis.

REFERENCES

Basrur, P. K. and Kanagawa, H. (1969). Parallelism in chimeric ratios in members of heterosexual twins in cattle. *Genetics 63*, 419–425.

Basrur, P. K., Kanagawa, H. and Gilman, J. P. W. (1969). An equine intersex with unilateral gonadal agenesis. *Canad. J. Comp. Med. 33*, 297–306.

Benirschke, K. and Brownhill, L. E. (1962). Further observations on marrow chimerism in marmosets. *Cytogenetics I*, 245–257.

Biggers, J. D. and McFeely, R. A. (1966). "Intersexuality in Domestic Mammals." In *Advances in Reproductive Physiology*. Vol. I. A. McClaren (ed.) London, Logos.

Billingham, R. E. and Lampkin, G. H. (1957). Further studies in tissue homotransplantation in cattle. *J. Embryol. Exp. Morph. 5*, 351–367.

Bishop, M. W. H. (1972). Genetically determined abnormalities of the reproductive system. *J. Reprod. Fert*. Suppl. *15*, 51–78.

Dunn, H. O., McEntee, K. and Hansel, W. (1970). Diploid-triploid chimerism in a bovine true hermaphrodite. *Cytogenetics 9*, 245–259.

Ferguson-Smith, M. A. (1966). X-Y chromosomal interchange in the aetiology of true hermaphroditism and of XX Klinefelter's syndrome. *Lancet ii*, 475–476.

Herschler, M. S. and Fechheimer, N. S. (1967) The role of sex chromosome chimerism in altering sexual development in mammals. *Cytogenetics 6*, 204–212.

Jost, A. (1970). Hormonal factors in the sex differentiation of the mammalian foetus. *Phil. Trans. Roy. Soc. Lond. B 259*, 119–130.

Lillie, F. R. (1922). The etiology of the freemartin. *Cornell Vet. 12*, 332–337.

McFeely, R. A., Hare, W. C. D. and Biggers, J. D. (1967). Chromosome studies in 14 cases of intersex in domestic mammals. *Cytogenetics 6*, 242–253.

Mintz, B. (1968). Sex anomalies in allophenic mice. *J. Anim. Sci.* Suppl. I, *27*, 51–60.

Moor, N. W. and Rowson, L. E. A. (1958). Freemartins in sheep. *Nature* (Lond.) *182*, 1754–1755.

Ohno, S. (1967). *Sex Chromosomes and Sex Linked Genes*. Berlin, Heidelberg, New York, Springer-Verlag.

Polani, P. E. (1970). Hormonal and clinical aspects of hermaphroditism and the testicular feminizing syndrome in man. *Phil. Trans. Roy. Soc. Lond. B 259*, 187–204.

Stafford, M. J. (1972). The fertility of bulls born co-twin to heifers. *Vet. Rec.*, 146–148.

Stone, H. W., Friedman, J. and Fregin, A. (1964). Possible somatic cell mating in twin cattle with erythrocyte mosaicism. *Proc. Nat. Acad. Sci. 51*, 1036–1044.

Chapter 21

Reproductive Infections

J. W. KENDRICK AND J. A. HOWARTH

I. VIRAL INFECTIONS

A. Viral Diseases of Cattle

1. GRANULAR VENEREAL DISEASE
   (*Granular Vaginitis*)

This is a widespread condition, observed in cattle, that masquerades under a misnomer. It is, in fact, a hyperplasia of the lymphatic follicles that are normally found in the subepithelial tissues of the vulvar mucosa. These follicles increase in size and protrude above the mucous membrane in nodules 1–2 mm in diameter. In mild cases, there may be only a few nodules in the region of the clitoris. The most severely affected cows will have numerous nodules on the vulvar mucosa, and there may be hyperemia of the mucosa, but swelling, pain and discharge are absent. The relationship of this condition to fertility is not clear. In one survey it was found that the more severely affected animals had approximately a 10% reduction in fertility.

Clinical Signs and Diagnosis. Cattle of all ages are affected, but it is most frequently seen in heifers where it occurs in the most severe form. Granular venereal disease does not interfere with breeding, and one should not withhold an animal from breeding because of the presence of this condition. A similar hyperplasia may occur on the bull's penis. These generally are more hemorrhagic than in the cow, and if they occur in sufficient numbers, the bull may refuse to serve.

The condition is not an infection but, rather, is the nonspecific normal lymphoid reaction to an infectious agent, and perhaps to noninfectious agents. This is most clearly demonstrated in the case of infectious pustular vulvovaginitis. This acute viral disease which causes ulceration of the vulvar mucosa heals in approximately 14 days and shedding of the virus ceases. Following healing, the lymphoid follicles are hyperplastic and there is a moderate to severe "granular venereal disease" that persists for several months. During this time it is not possible to isolate virus or to transmit the condition to other animals. Furthermore, the histologic appearance of the nodules does not indicate an active infection but, rather, a nonspecific lymphoid response. In all probability, other infections of the lower genital tract, including nonspecific and environmental contaminants to which the animal comes in contact during its first years of life, are the cause of this condition.

Treatment. Spontaneous recovery is the rule and, in most cases, treatment is unnecessary. If treatment is used, mild antiseptics in the form of ointments, douches and powders may be used. An exception is the case of bulls in which the condition may prevent service. Mild antiseptics and antibiotics should be used in treatment. In severe or refractory cases, cauterization of the individual lesions with silver nitrate has been used. If this is done, special care must be taken to prevent adhesions, which is accomplished by using antibiotic or antiseptic ointments, and frequently extending the penis during the healing period.

2. BOVINE VIRAL DIARRHEA

This disease is found in all parts of the United States and is widely distributed throughout the world. Four clinical forms of this infection are recognized. The most frequently occurring infection with this virus is subclinical. There are many herds, or groups of herds, in which the infection rate, as determined by the presence of serum antibody, is more than 50% of the animals, and the history of clinical signs referrable to this infection are absent. The acute infection is characterized by a high temperature, nasal discharge, erosions in the oral cavity and digestive tract, and diarrhea. The acute form has been observed as a herd outbreak, in which almost all of the animals were affected, as well as cases in which only an individual animal was affected. In the latter case, in all probability, other animals in the herd were already immune from a previous exposure.

Chronic viral diarrhea is characterized by poor appetite, emaciation and slow rate of growth. The onset may not be accompanied by any of the acute signs but may appear insidiously and be recognized only by the loss of condition in the affected animal. There may be diarrhea. The course of the disease is two to six months, and if death does not occur, those animals that survive represent a continuing economic loss. These cases occur most often in poorly managed herds.

The mucosal disease form presents all of the symptoms of the acute viral diarrhea, except that the lesions progress in severity so that the animals die approximately 14 days after the onset of clinical signs. Mucosal disease usually occurs between 8 and 18 months of age, and differs from the other forms of BVD-MD virus infection because the animal is unable to develop an immunity in mucosal disease.

The etiologic agent is known as the bovine viral diarrhea-mucosal disease (BVD-MD) virus. Spread is by contact among animals housed or pastured together. Infection confers a long lasting immunity that is recognized by the presence of antibody in the serum.

This disease affects the reproductive performance of cattle by causing abortion and birth defects. Contrary to earlier opinion, the incidence of abortion caused by this virus is relatively low. The diagnosis of BVD-MD abortion is difficult, and there is no satisfactory means of confirming a clinical diagnosis. In those cases in which abortion has been ascribed to this virus, such evidence as concurrent appearance of a febrile disease or the appearance of antibody in the serum following abortion, have been taken as circumstantial evidence that BVD-MD virus caused the abortion. Using these criteria, abortion has been diagnosed at all stages of gestation; however, most abortions occur during the first three or four months.

The experimental inoculation of pregnant, susceptible cattle has resulted in abortion only during the first three months of pregnancy. Abortion at this time, when the conceptus is small, often goes unrecognized in the average livestock operation and infertility, rather than abortion, is diagnosed.

Viral infection and fetal damage during the middle third of gestation when body organs are in the developmental stages, results in the birth of calves with defective brain, eye and hair development. Those birth defects referrable to the brain and the eye have occurred most frequently. Brain defects result in inability to stand, walk or nurse. Eye lesions result in blindness and opacity of the cornea. Fetuses may experience BVD-MD infection during the last third of gestation without recognizable sequelae.

All fetuses surviving intrauterine infection are actively immune at the time of parturition.

Control. The prevention of this disease, and thus the prevention of abortion and birth defects, is accomplished by vaccination with a modified live virus vaccine. The best time to administer this vaccine is to replacement animals when approximately one year of age. It creates a lifelong immunity; however, booster vaccinations are often given on an annual basis. Pregnant animals should not be vaccinated (Kendrick, 1971; Kahrs, 1970; Ward et al., 1969).

3. Infectious Bovine Rhinotracheitis-Infectious Pustular Vulvovaginitis Virus Infection (IBR-IPV)

This virus is a herpesvirus that affects cattle and rarely affects other species. Infections are frequently observed in the United States and the distribution is worldwide. In addition to respiratory and genital disease, IBR-IPV virus causes conjunctivitis, encephalitis, acute gastrointestinal disease of newborn calves and abortion. The diseases caused by this virus ordinarily occur singly, i.e., an outbreak of vulvovaginitis usually is not accompanied by the upper respiratory infection, or vice versa. The reason for this unique characteristic is unknown. It is thought by some investigators that different strains of the virus exist; however, these have not been defined. The characteristics of transmission, virulence, immunity and immunization are similar for all diseases. There is no danger of human infection from this bovine strain of herpesvirus.

The virus is a relatively stable virus and can exist in nature for several days. It remains viable at room temperature in a laboratory for more than a week, at 4° C for longer than one month, and can be stored indefinitely at −80° C. It is easily transmitted by contact and application of the virus to any mucous membrane of a susceptible animal results in localized disease in that area, systemic infection and systemic immunity.

Infection, regardless of the type of disease, results in resistance to re-infection and is characterized by the presence of specific viral antibody in the serum. This immunity is lifelong and serves to protect the animal, even at very low levels. Recovered animals rarely shed the virus. The exceptions are the occasional animal which may have a recrudescence of infection at the time of parturition, and the bull which may harbor the virus in the sheath for several months after the acute infection.

Clinical Signs. Infectious pustular vulvovaginitis is characterized by pustule formation in the vulva. The condition is painful and the animal stands with the back arched and tail elevated. Irritation of the vulva is indicated by straining, frequent urination, stamping of the feet, restlessness and switching of the tail. There is a small to moderate amount of tenacious yellow discharge. The vulva is swollen, and in the case of white-skinned animals, a slight reddening can be observed. Small white pustules, approximately 2 mm in diameter, occur on the vulvar mucosa. The pustules may become confluent and form a sheet of fibrinous, necrotic exudate over the affected area. In some cases the lesions extend into the vagina where they occur as a more diffuse necrosis. Examination reveals sheets of tissue peeling from the wall of the vagina.

Recovery is usually uncomplicated and occurs in 7 to 10 days. Within two or three weeks after recovery, the characteristic nodules of granular venereal disease appear in the vulva. Usually, vulvar and vaginal infection do not extend into the uterus. Contamination of insemination pipettes or of semen with the virus will result in infection of the uterus and endometritis.

Breeding should be suspended during the acute phase of the disease; however it may be resumed after this because normal fertility following recovery is to be expected. Endometritis is a cause of infertility and recovery of the uterus results in a return to normal fertility.

Infectious pustular balanoposthitis is the corresponding condition in the bull. The same basic lesions occur but, because of the movement of the penis, the exudate is often scrubbed off and the lesions appear as hemorrhagic ulcers. There may be edema of the penis and prepuce sufficient to cause phimosis or paraphimosis. In severe cases adhesions of the penis and prepuce have occurred. Spontaneous recovery of these lesions also occurs, as in the cow; however, to avoid adhesions and secondary infection, oily solutions of an antibiotic should be placed in the sheath.

IBR-IPV virus is a common cause of abortion in cattle. It occurs following natural infection but has frequently been the result of improper use of the modified live virus vaccine for prevention of this disease. This is a highly effective vaccine and, when properly used, creates a long-lasting immunity with excellent protection against this disease. However, it will cause abortion in approximately 50% of pregnant animals vaccinated during the last half of the gestation period.

The period between infection and abortion is from 21 to 90 days. The aborted fetus is autolyzed. Some cows have a retained placenta; however metritis is usually mild or absent. Recovery is the rule, and except in those cases where secondary infection of the uterus occurs, fertility in the subsequent breeding season is usually normal.

Diagnosis. The disease in mature animals is usually diagnosed by the clinical signs and confirmed by the isolation of the virus from swabbings of the lesions. Serologic examination may give supporting evidence of the infection; however, it often falls short as a diagnostic tool because the presence of antibody in the serum may be the result of infection that took place several months, or even several years, prior to the present condition. Abortion from the IBR-IPV virus is suspected when there is a history of other forms of the disease, a storm of abortions during the last half of pregnancy, or use

of the modified live vaccine during pregnancy. Necropsy of the fetus reveals autolysis which is nonspecific. Diagnosis is confirmed by finding microscopic focal necrosis, which is commonly present in the liver but may occur in many other organs of the body. Frequently the virus can be isolated from the fetus, although in some instances the autolysis precludes this isolation.

Control. The best method of control for this disease is the routine immunization of all potential replacement animals when they are six months to one year of age. Such a regimen selects calves that have lost their maternally-conferred immunity at approximately six months and before breeding age. A single vaccination should be sufficient; however, booster inoculations may be given when the animal is not pregnant (Kendrick et al., 1958; Kendrick and Straub, 1967; McKercher and Wada, 1964).

4. Epizootic Bovine Abortion (EBA)

Epizootic bovine abortion refers to a well-defined clinical condition that occurs principally in the foothill and mountain areas surrounding the Central Valley of California. It occurs most commonly in beef cattle, probably because these are the animals most frequently pastured in the endemic areas. It occurs mostly in heifers, but older animals moved to enzootic areas for the first time may subsequently abort. Usually a cow only aborts once, makes a complete recovery, and fertility is normal in subsequent gestations. The abortion rate may exceed 80% where large numbers of animals are exposed for the first time.

Clinical Signs. Most abortions occur in the last third of pregnancy, and because most abortions are in beef cattle which are on a seasonal breeding program, there is a seasonal incidence of abortion. One of the most striking characteristics of this abortion disease is the freshness of the fetal tissues, indicating that death of the fetus occurs at approximately the time of expulsion from the uterus. This is in marked contrast to most other abortion diseases in which the fetus is badly auto-lyzed before being aborted.

Diagnosis. At the present time the only satisfactory diagnostic criteria are the pathologic lesions of the fetus. There are petechial and ecchymotic hemor-rhages in the conjunctiva and oral mucosa. The muzzle of these fetuses frequently is bright red and dermatitis is present. A subcutaneous edema and a straw-colored pleural and peritoneal ef-fusion is usually present. A swollen nodular liver is a dramatic but irregularly-occurring finding. Petechial hemor-rhages are found throughout the body and the lymph nodes are enlarged and moist.

The most specific lesions are micro-scopic. These consist of a general reticuloendothelial hyperplasia seen most prominently in the liver, lymph nodes and spleen. This lesion may also be present in the brain and heart. A sig-nificant finding is the depletion of lym-phocytes and reticuloendothelial hyper-plasia of the thymus. There may be some focal necrosis and inflammatory changes in the liver and spleen. The lesions described are not necessarily specific for a single etiologic agent, but rather an indication of an infectious agent that causes a chronic disease of the fetus. The mechanism for abortion un-der these circumstances is probably stress, which causes an adrenal cortical hyperplasia, and which, in turn, releases cortical hormones to initiate parturition. The cause of this disease is unknown.

Control. No vaccine or satisfactory control methods are available. Avoiding the foothill and mountain areas will pre-vent the disease; however, this is a vast rangeland and must be used for beef cattle production. Those animals that have experienced the disease once will not abort again, and these should be maintained in the herd. There is no evidence that EBA spreads from cow to cow (Jubb and Kennedy, 1970).

B. Viral Diseases of Sheep

1. Enzootic Abortion of Ewes (EAE)

The disease occurs in many other parts of the world and has been diagnosed in the western United States. The rate of abortion in a newly infected flock may be as high as 30%. In flocks experiencing a reinfection, the rate is less than 5%. Ewes abort during the last month of preg-nancy. Abortions continue until normal lambing time and some fetuses are ex-pelled alive at term but are diseased. Fetuses are usually fresh (not autolyzed) and there are no specific lesions in the fetus. Diagnostic lesions do occur in the placenta. Cotyledons are necrotic and have a dry, dull yellowish or gray color. The allantois-chorion is brownish and thickened and is often described as "leatherlike." The etiologic agent of EAE is a member of the *Chlamydia* group. This name has appeared in the literature only recently. This disease must be differentiated from vibriosis.

Diagnosis and Control. Diagnosis of EAE depends upon finding the elemen-tary bodies of the infectious agent in smears from the cotyledon or from pla-cental exudate.

The disease produces an immunity. A single infection produces lifetime im-munity and the disease can be prevented by vaccination (Marsh, 1965; Parker et al., 1966).

2. Bluetongue

Bluetongue is an arthropod-borne viral disease of sheep, goats, cattle and wild ruminants. The infection of sheep is characterized by elevated temperature, ranging from 104° to 108° F, severe de-pression and anorexia. The name of the disease comes from the cyanotic appear-ance of the mucosa of the mouth and tongue. Erosions occur on the lip, cheek, tongue, in the mouth and in the nostrils. Eventually there is a muco-purulent discharge and encrustations on the lip and muzzle. Lameness, due to laminitis and inflammation of the coro-nary band, may occur. The course of the acute disease is 6–14 days and mortality is 10–40%.

The disease influences reproductive performance primarily by affecting the fetus. Virus of the naturally occurring disease and the modified live virus vaccine used to protect against the disease may cross the placenta and produce fetal disease characterized by abnormal brain development. Although lambs are born at full-term, some are stillborn, others are spastic and lay struggling until death. Others, called "dummy lambs," are unable to stand or may be uncoordinated and fail to nurse. The period of greatest danger from disease or vaccination is between the fourth and eight weeks of pregnancy, but infection at a later date may produce mild but significant birth defects. The disease is spread by a small biting insect, the *Culicoides* gnat. The recovered animal is immune.

Control. A modified live virus vaccine produces a solid immunity which lasts for as long as 30 months. Because of the level of virus that occurs in the blood of vaccinated sheep, insect vectors are able to transmit a severe infection to susceptible sheep and therefore the use of vaccine is contraindicated in nonenzootic areas. Do not vaccinate withing three weeks of breeding or during pregnancy (Marsh, 1965; Osburn et al., 1971).

3. ULCERATIVE DERMATOSIS (*Lip and Leg Ulceration and Venereal Disease*)

Ulcerative dermatosis is a viral infection of sheep, characterized by circumscribed ulceration of the skin of the lips, legs, feet and external genital organs. Transmission under natural conditions occurs when the virus enters through a break in the skin. Venereal transmission occurs and produces lesions of the penis, prepuce and vulva. The condition is characterized by localized ulceration in the areas described. These ulcers become covered with scabs and sometimes may undergo a secondary infection. The genital lesions interfere with breeding. This condition must be differentiated

from a noncontagious balanoposthitis, commonly called "pizzle-rot," which is probably caused by dietary-induced changes in the urine. No vaccine for ulcerative dermatosis is available. Infection produces low-grade immunity and an animal can be reinfected within five months.

C. Viral Diseases of Horses

1. EQUINE RHINOPNEUMONITIS (VIRAL ABORTION, EQUINE HERPESVIRUS I INFECTION)

The equine herpesvirus I infection is primarily a condition of the upper respiratory tract. The horse first experiences exposure to this virus as a foal in the fall near weaning time. Horse farms are very familiar with the mild, upper respiratory infection that affects most of the weanlings each year. The clinical signs include a temperature of 102° to 105° F, serous nasal discharge and congestion of the nasal mucosa. After several days the nasal discharge becomes purulent, and there is a mild to moderate cough. Appetite is affected only in the more severely involved animals. The course of the disease is two to four weeks, and spontaneous recovery is the rule. This epizootic in the foals produces a massive exposure of the mares which are in mid-pregnancy.

Few, if any, horses in the United States escape this infection during the first years of life. Infection confers an immunity, as indicated by the presence of antibody in the serum. However, unlike most viral immunity, soon after infection it begins a relatively rapid decline and, when the titer of antibody in the serum reaches a level of less than 1:100, the horse is again susceptible to respiratory infection by the virus. In this way, individual horses may be repeatedly infected during their lifetime. The rate of abortion in a band of mares depends upon their susceptibility but abortion rates as high as 80% have been reported.

Clinical Signs. Abortion usually occurs after the eighth month of pregnancy and,

14

because of seasonal breeding, most abortions occur between January and April. There are no premonitory signs indicating that abortion is about to occur. The placenta is not retained, and the mare undergoes a prompt recovery without after effects. Fertility in subsequent years is not affected. The fetuses are not decomposed, indicating that death occurs at approximately the time of expulsion from the uterus.

Diagnosis. By far the most frequent cause of epizootics of abortion in mares is the equine herpesvirus I, and this infection should be suspected until eliminated. A positive diagnosis is based on gross and microscopic lesions in the aborted fetus and on virus isolation and identification. Gross examination reveals a straw-colored fluid in the body cavities, and edema of the lung which is present in 80–90% of aborted fetuses. There are small foci of necrosis varying in size from minute up to 5 mm in diameter in the liver in about 50% of aborted fetuses. Hemorrhages and edema may occur in other parts of the body. The diagnosis is confirmed by the presence of intranuclear inclusion bodies in fetal tissues and by viral isolation.

Control. The equine herpesvirus I is highly contagious and probably all susceptible animals are infected at the time of the fall outbreak of upper respiratory infection in the weanlings. It is seldom possible to prevent abortion in a band of mares by isolating the nonaffected from those that have aborted. However, if possible, it should be attempted. Abortion from this infection rarely occurs in successive years, and in the absence of other control methods may not occur in a single band of mares for several years. Immunization is the best method of prevention. Attempts to modify this virus and still retain its immunizing capability have not been successful. However, a hamster-adapted virus is available for use in a "controlled infection" program for the prevention of this disease. This system has been used successfully on farms where the disease has been diagnosed. It should be initiated immediately after an outbreak of abortion when immunity is at a high level. Vaccination of every horse on the farm is done yearly in July and October. In mares without recent exposure to the virus the vaccine may cause abortion. Therefore, this immunization method should be used only following diagnosis of this disease (Doll and Bryans, 1963*a* and *b*).

2. EQUINE VIRAL ARTERITIS (EPIZOOTIC CELLULITIS, PINKEYE, INFLUENZA)

This disease is rarely seen in the United States at this time. In the early part of this century and before, when horses were used for transportation and in Army units, this disease occurred in devastating epizootics. Equine viral arteritis is characterized by elevation of the body temperature to 103°–106° F, severe depression, anorexia, weight loss, colic and diarrhea. There is a severe conjunctivitis with edema. The protrusion of the brick-red mucous membrane around the eye gives this disease the name "pinkeye." There is lacrimation and sometimes purulent discharge. There is congestion of the nasal mucous membranes, increased respiratory rate accompanied by an expiratory lift (heave line). Immunity follows infection. A vaccine has been developed but is not available commercially because of the rarity of this disease. Occurrence of this disease should be reported to state or federal authorities and immediate quarantine effected (Jones, 1969; McCollum, 1969).

This disease causes an abortion rate of approximately 50%. The fetus is expelled without premonitory signs during the latter stages of the disease, or within a few days after recovery. The diagnosis is based upon the close association with the severe clinical disease and the elimination of other causes of abortion, particularly equine herpesvirus I abortion. Fetal lesions in the two diseases differ significantly and assist in obtaining a diagnosis.

Table 21-1. Causes of Sporadic Abortion

Cattle	Sheep	Horses	Swine
Parainfluenza₃ virus	Tick-borne fever	Equine infectious anemia	Foot and mouth disease
Bluetongue virus	Wesselsbron virus	Dourine	Picorna viruses
Malignant catarrhal fever	Rift Valley fever	Piroplasmosis	Influenza
Foot and mouth disease	Nairobi sheep disease		Japanese B virus
Rinderpest	Rinderpest		Hemagglutinating virus
Rift Valley fever	Foot and mouth disease		African swine fever
Tick-borne fever	*Coxiella burnetti* (Q fever)		Pseudorabies
Bovine petechial fever			
Anaplasmosis			
Piroplasmosis			
Trypanosomiasis			
Toxoplasmosis			
Globidiosis			

D. Sporadic Abortion Caused by Viruses

Viral diseases that cause substantial abortion losses have been described. In addition to these viruses, numerous other viruses have been isolated from aborted fetuses. Some of these are foreign diseases that do not occur in the United States, and others are viruses that are only rarely found in the fetus. In this latter category, one must recognize that the fetus is a highly susceptible individual and that many viruses that cause only mild disease in the dam will cause a lethal infection in the fetus. The most important factor in preventing fetal infection with these viruses is the placenta. In most cases this placental barrier is effective; however, occasionally it may fail, and for this reason there are sporadic abortions from viruses that are not generally considered to be abortifacient agents. These viruses and other agents that cause sporadic abortion are listed in Table 21–1.

Several inherent problems make the diagnosis of abortion one of the more challenging aspects of veterinary medicine. Most abortion is caused by a disease of the fetus. Very often this disease causes the fetus to die in utero several days before it is expelled. Autolysis destroys the causative agent and, to a large extent, distorts the lesions. In chronic disease of the fetus the agent may be eliminated by the protective mechanisms of the fetus before it is aborted. It is commonly reported that in approximately 50% of the fetuses submitted for diagnosis the etiology cannot be determined.

II. PROTOZOAN DISEASES

A. Bovine Trichomoniasis

Trichomoniasis is a contagious, venereal disease of cattle characterized by sterility, pyometra and abortion. It is caused by the protozoan parasite *Trichomonas fetus*. Abortion and pyometra are often the first signs of trichomoniasis that are noticed in a herd, but they occur in relatively few animals. Infertility characterized by repeat breeding and irregularly long or short estrous cycles is the most constant symptom and occurs in a high percentage of animals in a recently infected herd. Heifers are usually considered to be most susceptible; however, immunity does not increase with age but is created through exposure. Infection is confined to the genital tract, and with rare exceptions is transmitted only during the breeding act.

Clinical Signs. The introduction of trichomonad infection into a clean herd passes unnoticed for a considerable period of time. Infection of the female results in a mild vaginitis, which is usually not observed. The first recognition of the disease is the occurrence of pyometra, abortion or decreased fertility. The infection is first established in the vagina, and then passes to the uterus within 20 days. It causes an endometritis which results in an inhospitable environment for the developing zygote, or an infection of the zygote which results in death and resorption. Inflammation of the uterus may cause shortening of the estrous cycle. Long estrous cycles occur when the embryo survives beyond 14 days of age and the estrous cycle is interrupted. Later, when the embryo dies and is expelled or resorbed, estrous cycles are again initiated. The pyometra resulting from this infection consists of macerated fetal tissue. It will persist for a long period of time unless recognized and treated. Abortion from trichomoniasis always occurs prior to the fifth month of pregnancy and very often the membranes are expelled intact with the fetus.

Diagnosis. The appearance of the previously-mentioned symptoms, usually following the introduction of new animals into the herd, is suggestive of trichomoniasis or vibriosis. The diagnosis of trichomoniasis is confirmed by finding the organism in at least one animal in the herd. No other tests, such as serologic tests, are satisfactory for the detection of trichomoniasis. The organisms may be found in large numbers in the tissues, stomach contents and placental fluids of aborted fetuses. They are present in the uterus for several days after abortion. Trichomonads are found in large numbers in pyometra fluids, particularly if the cervix has remained sealed and bacterial contamination of the uterus has not occurred. Sometimes, however, bacterial contamination will supersede the trichomonad infection. Trichomonads may be found in aspirated vaginal fluid particularly during the two or three days just prior to the occurrence of estrus. Since the bull is the source of infection and remains infected over a long period of time, he is often selected for a herd diagnosis. Careful technique must be used when the bull is used for diagnosis, because relatively few organisms are present on the penis and in the sheath.

Treatment and Control. Most cows recover spontaneously and individual treatment is unnecessary, except in the case of pyometra or secondary infection associated with abortion. The usual recommendation is 90 days of sexual rest to allow immunity to eliminate the infection. In herds where artificial insemination can be used and there is no danger of animal-to-animal transmission, breeding has not been interrupted and the recovery rate in the female has been satisfactory. Bulls must be treated and those worth less than twice their salvage value should be replaced. The cost of treatment involves not only the treatment itself, but the time and effort necessary to test the bull to be certain that treatment has been successful. The most effective treatment for bulls is the combined use of sodium iodide, Acroflavin, and Bovoflavin ointment.

The best method for controlling trichomoniasis is to use artificial insemination. The herd should be examined to find animals with pyometra and other recognizable genital diseases. These are treated or eliminated. The use of artificial insemination in a herd for two years will, in most instances, eliminate the infection. When replacements are added to the herd they should come from herds with high fertility rates.

B. Toxoplasmosis

Ovine abortion caused by *Toxoplasma gondii* is of major importance in New Zealand. There it is not associated with clinical disease in the ewe. Abortion occurs during the last month of gestation and weak, full-term lambs may be observed. Specific gross lesions are not seen in the fetus, but placental lesions of focal necrosis of the cotyledon up to 2 mm in diameter occur. In some cases the lesions may be found only by microscopic examination. Histologically, foci of necrosis and gliosis are observed in the brain. Diagnosis is confirmed by the recognition of white necrotic areas in the cotyledon, which microscopically include the *Toxoplasma* organism. Impression smears of cotyledons may also reveal *Toxoplasma*. The agent can be isolated by injecting placental or fetal tissue intraperitoneally in mice and observing the organism in peritoneal exudate one to two weeks later (Watson and Beverley, 1971).

III. BACTERIAL DISEASES

A. Vibriosis

Bovine vibriosis is a contagious venereal disease of cattle characterized by infertility and abortion. It is worldwide in distribution and occurs throughout the United States. It is seldom seen in dairy cattle because the semen sources for artificial insemination are essentially free of this agent. On the other hand, natural breeding still predominates in beef cattle, and the disease prevails in many herds. The clinical effects of this infection are limited to the reproductive system and spread is only by breeding (Roberts, 1971).

Clinical Signs. Infertility or delayed conception occurs in almost all infected animals and is economically by far the most damaging aspect of this disease. Some cycles following infection are of normal length. Frequently one or more cycles is abnormally long because conception occurs and the developing embryo delays estrus until it dies and is resorbed. Inflammation of the uterus may cause an abnormally short cycle. Physical signs of the disease are slight in the female and are characterized by endometritis and the transient presence of small amount of pus in the vagina. The disease is self-limiting and recovery occurs in approximately two months in 75% of the cases. Of the remaining 25%, almost all recover in a period of three or four months, although occasionally there are identified carrier animals in which the infection persists through pregnancy and for periods in excess of one year.

The organism inhabits the penis and the prepuce of the bull in a symbiotic relationship that does not cause any physical changes, nor does it affect semen quality and libido. Bulls may be mechanical carriers of the agent at any age, but establishment of the carrier state usually occurs only in bulls four years or older.

While abortion is associated with vibriosis of cattle, it should be emphasized that this occurs in a very small number of cases. Ovine vibriosis is characterized by abortion in late gestation, which may affect up to 80% of the flock.

Diagnosis. The disease is suspected in herds in which there is an increase in the number of services per conception, and in beef herds a prolongation of the breeding and calving seasons. The presence of the disease must be confirmed by one or more laboratory tests. Animals become positive to the vaginal mucus agglutination test within 60 days after infection, and may return to negative within seven months. The best stage of the cycle to collect samples is during diestrus. False positives occur immediately after estrus; false negatives occur during estrus. Because of the variables mentioned, standard procedure for testing is to select infertile cows that were bred at least 60 days prior to the sampling date. Presence of positive titers of several animals is indicative of herd infection. *Vibrio fetus* may be cultured from vaginal and cervical mucus. The sample is collected aseptically in a pipet, refrigerated and taken to a laboratory for culture. Fluorescent antibody is used to identify organisms in washings from the bull's sheath. All diagnostic procedures are subject to considerable variation and accuracy. Therefore, they are used primarily for identifying the presence of infection in the herd. Thereafter, the infected herd is treated as a unit. Abortion in sheep is diagnosed by direct microscopic observation or culture of *Vibrio fetus* from the placenta or the aborted fetus.

Treatment and Prevention. Females with normal genitalia eliminate the disease spontaneously in most cases. Bulls may be treated by systemic administration of streptomycin, local application of streptomycin or a combination of both. The disease may be eliminated from a herd by removing animals with physically abnormal genitalia and using artificial insemination with noninfected semen for a period of two years. A bovine vaccine is available and should be used on females approximately two months prior to the beginning of the breeding season. An ovine vaccine is effective in preventing abortion when used prior to breeding or in early gestation.

B. Listeriosis

This disease primarily affects the nervous system but may be responsible for outbreaks of abortion in cattle and sheep. Abortion occurs in late pregnancy and is accompanied by retained placenta, metritis and death of the dam. Mortality can be prevented by local and systemic antibiotic treatment. The disease is associated with silage feeding and stress. No vaccine is available.

C. Brucellosis

Brucellosis is an infectious disease of animals and man caused by one of the *Brucella* species (*B. abortus*, *B. melitensis*, *B. suis*, *B. ovis*, *B. canis*) and characterized by genital and mammary gland infections in animals and undulant fever in man. In cattle the disease is also known as Bang's disease and contagious abortion.

Brucellosis is a worldwide problem of both public health and economic importance. Loss to the livestock producer occurs from abortion, premature birth, infertility and decreased milk yield.

The usual vehicle of infection for man is a food product of raw milk origin. Contact with the organism in vaginal discharges, fetuses, placentas, urine, manure and carcasses causes a large portion of the human cases. Transmission of brucellosis in animals occurs through the ingestion of feedstuffs contaminated by uterine discharges and milk. Infected male animals may transmit *Brucella* during breeding.

In young animals not yet sexually mature the infection causes no ill effects and is eliminated within a few weeks. In adult, nonpregnant female animals the organism localizes in the mammary gland and later spreads to the uterus when pregnancy occurs. In pregnant animals *Brucella* organisms invade the uterus, placenta and fetus, the final result being the death and premature expulsion of the fetus or abortion.

Clinical Signs. In *B. abortus* infection in cattle, abortion after the fifth month of pregnancy is the most prominent clinical sign. In subsequent pregnancies calving may be normal although second or third abortions may occur with the same cow. Retention of the placenta and uterine infection are common sequelae following abortion. Severe uterine infections may cause sterility. In the bull infection of the testicle may occur and if both testicles are involved sterility may result.

In *B. suis* infection of swine the most common symptom is abortion or birth of weak pigs. Sows that have aborted once will usually farrow normal litters thereafter. A persistent but scanty discharge from the uterus may follow abortion. Sows so affected are either temporarily or permanently sterile, depending on the persistence of the uterine infection. The testicles of boars may be infected with sterility resulting.

The disease caused by infection of sheep with *B. ovis* is characterized by infertility in rams due to epididymitis. Occasionally abortion or the birth of weak lambs occurs. The disease caused by infection of dogs with *B. canis* is characterized by abortion, whelping failure, epididymitis and sterility.

Diagnosis. Laboratory procedures used in the diagnosis of brucellosis include isolation of the organism and tests for the presence of specific antibodies in serum and milk. *Brucella* organisms can be isolated from aborted fetuses, the placenta, the uterine discharges, milk and semen.

The milk ring test for *Brucella* antibodies is the most practical method for locating infected dairy herds and for surveillance of brucellosis free herds. It is performed on samples obtained from milk that is in bulk. In eradication programs, herds that show a positive ring test should be examined by serologic test to identify the infected individuals. The serum agglutination test is the most commonly used procedure for detecting brucellosis in individual bovine animals.

Control. The control of bovine brucellosis is based on hygienic measures, vaccination with strain 19 *B. abortus*, and test and disposal of infected animals. The living attenuated strain 19 vaccine used in the control of bovine brucellosis affords good protection and does not spread from animal to animal when properly used. Heifer calves vaccinated at four to six months of age are resistant to infection for seven years or more and do not react to serologic tests used in eradication programs. Effective vaccination programs can reduce infection rates to a point where procedures based on testing and the elimination of the reactors are possible for eradication.

The control of brucellosis in swine is based on detection and elimination of infected herds. As yet there is no vaccine available for use in swine. The control of *B. ovis* infection in rams can be achieved by the elimination of infected rams and vaccination of immature rams to serve as replacements (Meyer, 1966; Joint FAO/WHO Expert Committee on Brucellosis, 1971).

D. Leptospirosis

Leptospirosis of cattle, horses and swine is caused by a variety of leptospiral serotypes widely distributed throughout the United States and the rest of the world. Both domestic animals and wildlife are sources of leptospiral infection for man. Serotypes responsible for disease in cattle, horses and swine are *L. pomona*, *L. grippotyphosa*, *L. canicola* and *L. icterohemorrhagiae*. In addition, *Leptospira hardjo* has been isolated from cattle.

Only a small portion of the animals infected with *Leptospira* will have obvious sickness, but a high percentage may develop persistent kidney infections and be spreaders of the organism. Cattle recovering from leptospirosis remain kidney carriers for about three months, but swine, wild rodents and skunks may become permanent sources of infection.

Transmission of leptospirosis most commonly occurs when an infected animal contaminates pastures, drinking water and feed with infected urine. Infection may also be spread by aborted fetuses and contaminated semen.

Clinical Signs. Clinical signs of leptospirosis in an animal species are quite similar regardless of the infecting leptospiral serotype. In cattle an acute hemolytic anemia with blood in the urine occurs primarily in younger animals. Abortion occurring during the last third of the pregnancy and a characteristic drop in milk yield (agalactia) are more common clinical signs.

Infection may cause abortion in pigs, sheep, goats and horses. In horses an additional sequela is a disease of the eye called periodic ophthalmia.

Diagnosis. Leptospiral organisms are not readily isolated from animals with clinical signs and this is particularly true with aborted bovine fetuses. Isolation of *Leptospira* from the urine denotes chronic infection and thus does not really assist in making a diagnosis.

The most reliable and most commonly used diagnostic procedures are the tube and plate agglutination tests. Serum samples from animals suspected of having leptospirosis should be tested with each of the several serotypes during the sickness and again when such animals are convalescent.

Control. Streptomycin and tetracyclines are used in the treatment of acutely ill animals and to eliminate the kidney carrier state.

Vaccination of cattle, horses and swine with leptospiral vaccines has been effective in controlling clinical signs of the disease. Because immunity wanes quickly, animals should be vaccinated at least once yearly. Currently, *L. pomona, L. canicola* and *L. icterohemorrhagiae* vaccines are commercially available. Vaccines against other serotypes will become available as the need arises (Szatalowicz et al., 1969; Morter et al., 1969; Hanson et al., 1972).

REFERENCES

Doll, E. R. and Bryans, J. T. (1963a). Epizootiology of equine viral rhinopneumonitis. *J. Amer. Vet. Med. Assn. 142*, 31–37.

Doll, E. R. and Bryans, J. T. (1963b). A planned infection program for immunizing mares against viral rhinopneumonitis. *Cornell Vet. 53*, 249–262.

Hanson, L. E., Tripathy, D. N. and Killinger, A. H. (1972). Current status of leptospirosis immunization in swine and cattle. *J. Amer. Vet. Med. Assn. 161(11)*, 1235–1243.

Joint FAO/WHO Expert Committee on Brucellosis; 5th Report (1971) *WHO Tech. Rep. Ser., No. 464*, 76 pp.

Jones, T. C. (1969). Clinical and pathologic features of equine viral arteritis. *J. Amer. Vet. Med. Assn. 155*, 315–317.

Jubb, K. and Kennedy, P. (1970). *Pathology of Domestic Animals.* 2nd ed. New York, Academic Press.

Kahrs, R. F., Scott, F. W. and deLahunta, A. (1970). Bovine viral diarrhea-mucosal disease, abortion and congenital cerebellar hypoplasia in a dairy herd. *J. Amer. Vet. Med. Assn., 156*, 851–857.

Kendrick, J. W., Gillespie, J. H. and McEntee, K. (1958). Infectious pustular vulvovaginitis of cattle. *Cornell Vet. 48*, 458–495.

Kendrick, J. W. and Straub, O. C. (1967). Infectious bovine rhinotracheitis-infectious pustular vulvovaginitis virus infection in pregnant cows. *Am. J. Vet. Res. 28*, 1269–1282.

Kendrick, J. W. (1971). Bovine viral diarrhea-mucosal disease virus infection in pregnant cows. *Amer. J. Vet. Res. 32*, 533–544.

McCollum, W. H. (1969). Development of a modified virus strain and vaccine for equine virus arteritis. *J. Amer. Vet. Med. Assn. 155*, 318–322.

McKercher, D. G. and Wada, E. M. (1964). The virus of infectious bovine rhinotracheitis as a cause of abortion in cattle. *J. Amer. Vet. Med. Assn. 144*, 136–142.

Marsh, Hadleigh (1965). *Newsom's Sheep Diseases*, 3rd ed. Baltimore, Williams & Wilkins.

Meyer, M. C. (1966). Host-parasite relationships in Brucellosis, I. Reservoirs of infection and interhost transmissibility of the parasite. *U.S. Livestock Sanitary Assoc. Proc. 70*, 129–134.

Morter, R. L., Williams, R. D., Bolte, H. and Freeman, M. J. (1969). Equine leptospirosis. *J. Amer. Vet. Med. Assn. 155(2)*, 436–442.

Osburn, B. I., Silverstein, A. M., Prendergast, R. A., Johnson, R. T. and Parshall, C. J., Jr. (1971). Experimental viral-induced congenital encephalopathies. I. Pathology of hydranencephaly and porencephaly caused by bluetongue vaccine virus. *Lab. Invest. 25*, 197–205.

Parker, H. D., Hawkins, W. W. and Brenner, E. (1966). Epizootiologic studies of ovine virus abortion. *Am. J. Vet. Res. 27*, 869–877.

Roberts, S. J. (1971). *Veterinary Obstetrics and Genital Diseases (Theriogenology).* Ithaca, N.Y., published by the author, distributed by Edwards Bros., Ann Arbor, Mich.

Szatalowicz, F. T., Griffin, T. P. and Stunkard, J. A. (1969). The international dimensions of leptospirosis. *J. Amer. Vet. Med. Assn. 155(12)*, 2122–2132.

Ward, G. M., Roberts, S. J., McEntee, K. and Gillespie, J. H. (1969). A study of experimentally induced bovine viral diarrhea-mucosal disease in pregnant cows and their progeny. *Cornell Vet. 59*, 525–538.

Watson, W. A. and Beverley, J. K. A. (1971). Epizootics of toxoplasmosis causing ovine abortion. *Vet. Rec. 88*, 120–123 and 124 (fig).

VI. Techniques for Improving Reproductive Efficiency

Chapter 22

Artificial Insemination

R. H. Foote

Artificial insemination is one of the most important techniques ever devised for the genetic improvement of animals. Methods have been developed for inseminating cattle, sheep, goats, swine, horses, dogs, poultry, primates and many laboratory animals and insects. The earliest carefully documented use of artificial insemination was in 1780 when Spallanzani, an Italian physiologist, obtained pups by this method. Other scattered reports appeared in the 19th century, but it was not until about 1900 that extensive studies with farm animals began in the U.S.S.R. and shortly thereafter in Japan.

Major advantages of artificial insemination are as follows: (a) genetic improvement; (b) control of venereal diseases; (c) availability of accurate breeding records necessary for good herd management; (d) economical service; and (e) safety through elimination of dangerous males on the farm. It is essential in conjunction with estrus synchronization and has been proposed as a means of sex control through sperm separation.

Because the genetic merit of the dairy bull for milk production can only be determined accurately by progeny testing, it is extremely important that carefully selected young bulls be sampled as soon as possible and rigidly culled after the progeny test. Only a few outstanding

sires need be selected to breed a large population of cows, provided proper procedures of sexual preparation, semen collection and processing are used to harvest and preserve the maximum number of viable sperm from each sire.

Similar types of performance and progeny test programs are important for meat-type farm animals. Artificial insemination facilitates crossbreeding, requiring that only one breed be maintained on the farm. The opportunity for genetic improvement through these testing and breeding programs using artificial insemination is enormous.

When properly done, disadvantages are few. However, it is necessary to have a sufficient number of trained personnel to provide proper service and to provide appropriate arrangements for corraling females for heat detection and insemination, particularly under range conditions. Ram, boar and stallion sperm cannot be preserved as well as bull sperm for long periods.

I. MANAGEMENT OF MALES AND SEMEN COLLECTION

Production of high quality semen depends upon males that have been kept under good conditions. When young males are properly fed and managed, semen can be collected successfully at the following approximate ages: bulls, 12 months; rams, goats and boars, 7–8 months; stallions, 24 months.

A. Physical Condition of Males

Feeding has a marked effect on rate of sexual development. When energy intake is restricted the rate of growth is decreased, testis growth is retarded, age at puberty is increased and sperm output may be reduced. When Holstein bulls were raised on high, medium and low energy intakes semen was first obtained at 39, 46 and 58 weeks of age, respectively. Beef bulls fed liberally appear to require about five weeks longer than Holsteins to reach sexual maturity.

Testicular development and associated sperm output under good conditions for Holstein bulls are shown in Figure 22–1. Testis size increases rapidly as puberty approaches, and by one year of age bulls have reached approximately 50% of their mature potential. There is a high correlation between testis size and sperm-producing potential in different species (Amann, 1970; Swierstra, 1971). Since high sperm production by genetically superior males is of special value in artificial insemination it is worthwhile to measure testis development in sires being considered for use. This can be done easily by measuring scrotal circumference (Foote et al., 1972).

Bulls should be maintained in a thrifty condition. Excessive feeding is wasteful and may produce more awkward and sluggish animals. The same principles apply to rams, boars, stallions and other males. Restricted energy intake of the young animal delays growth and it increases the age at puberty. Prolonged deficiencies of essential nutrients may cause reproductive failure. However, other symptoms of nutrient deficiencies usually appear before reproduction fails. Housing for males should be conveniently located, comfortable for the animals and safe for the handlers. Exercise is presumed to be important for the general health of the animal, but has not been proven to affect fertility. Pens and paddocks provide adequate exercise for bulls, rams, boars and stallions. Many bull studs have electrically driven mechanical exercisers to provide forced exercise for bulls kept in stalls and small pens.

B. Semen Collection

One of the most important parts of an artificial insemination program is the correct collection of semen. This involves the proper scheduling and sexual preparation of males, as well as the use of the proper techniques of semen collection.

1. Mounts and Teasing Procedures

Live mounts such as a teaser female, another male or a castrated male have proven to be the most successful for routine semen collections. Some males can be trained to mount dummies well (Fig. 22–2). Dummies have an advantage over live mounts in stability, and particularly over teaser females with respect to disease control. Boars particularly can be trained to mount a dummy sow. For the more sluggish or discriminating male, estrous females may provide added incentive during the training period. The dummy or live mount used should provide adequate support. It should be placed or restrained in an area with good footing so that the male will not slip and fall. Live mounts should be restrained to minimize lateral as well as

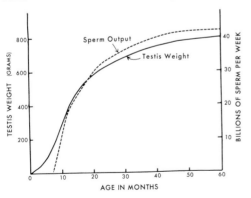

Fig. 22–1. Testicular size and sperm output of growing Holstein bulls ejaculated frequently. (*Adapted from Hahn et al., 1969. J. Anim. Sci. 29,41.*)

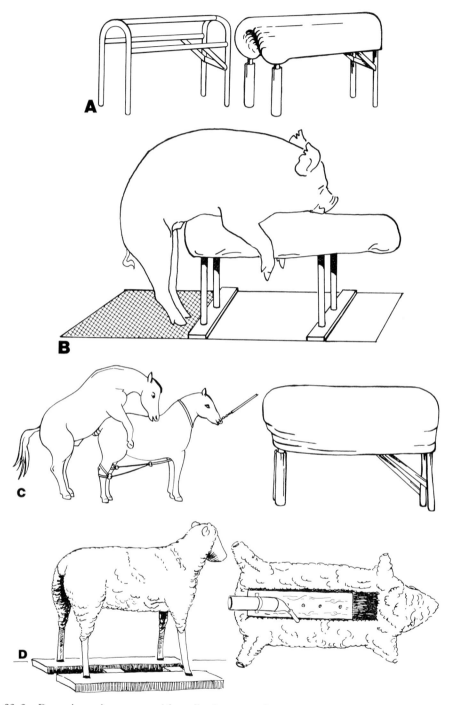

FIG. 22–2. Dummies and mounts used for collecting semen from farm animals.
A, Left to right, dummy cow constructed from steel pipe.   Completed dummy covered and well-padded.
B, Boar mounting padded dummy.   A mat to give secure footing is provided.
C, Left to right, mare hobbled, restrained and tail-bandaged before using as a mount; padded dummy.
D, Left to right, side view of a portable dummy for rams.   Bottom view showing an artificial vagina secured
for semen collection:

Table 22-1. Semen Characteristics and Sperm Output When Normal Mature Farm Animals Are Ejaculated at Different Frequencies

Item	Dairy Cattle	Beef Cattle	Sheep	Swine	Horses
Number of semen collections per week	1–6	1–6	7–25*	2–5	2–6*
Characteristics of average ejaculates†					
Volume (ml)	5–8	3–6	.8–1.2	150–300‡	30–100‡
Sperm concentration (million/ml)	1000–2000	800–1500	2000–3000	200–300	200–400
Total sperm/ejac. (billion)	7–15	5–10	1.6–3.6	30–60	5–10
Total sperm/week (billion)	15–40	10–30	25–40	100–150	15–30
Motile sperm (%)	50–75	40–75	60–80	50–80	40–75
Morphologically normal sperm (%)	70–95	65–90	80–95	70–90	70–90

* One or two days of rest should be provided each week.
† Semen characteristics are markedly influenced by the frequency of collection. The low volumes, concentrations and sperm/ejaculation and the high sperm/week correspond to the high frequency of collection. Size of breed also affects values.
‡ Gel-free volume.

forward movement and still provide easy access for semen collection. Convenient posts should be included for restraining bulls during sexual preparation. Dummies may be constructed to hold the artificial vagina (Fig. 22–2D).

Sexual preparation prior to semen collection improves the quality and quantity of semen obtained (Amann, 1970). False mounting a bull several times may increase the number of sperm collected by as much as 100%. Intensive teasing for 5–10 minutes without false mounting also is effective. Fluids from the accessory glands secreted during this preparatory period may flush out contaminating material from the urethra. Stimuli which are effective in increasing the sexual response of bulls include: (a) change of teaser; (b) change in location of the teaser; (c) bringing a new bull into the collection area; and (d) false mounting. Various combinations should be tried with sluggish males to keep the intensity of the sexual stimuli high. Beef bulls generally exhibit less libido than dairy bulls. They require more ingenuity on the part of the handlers to provide the

proper stimuli for ejaculation into an artificial vagina (Foster et al., 1970).

Methods of sexually preparing boars, rams and stallions have not been studied in as much detail. However, it is clear that exposure of males to various courtship situations before semen collection increases the number of sperm ejaculated.

2. FREQUENCY OF SEMEN COLLECTION

Increasing the frequency of semen collection decreases the number of sperm per collection but increases the number of sperm obtained per week or month. Thus with frequent collection, potentially more females can be inseminated with semen from superior sires. Billions of sperm can be obtained per normal male per week from any of the farm animals when properly managed at the time of semen collection (Table 22–1). Partly because of the large epididymal reserves of sperm, semen can be collected daily for short periods of time with large numbers of sperm per ejaculate. As epididymal reserves are partially depleted, the number of sperm obtained daily will be reduced until an equilibrium is reached with the newly formed sperm leaving the testis each day. An indication of the range in frequency of ejaculation often encountered for several species is shown in Table 22–1.

Under practical conditions artificial breeding organizations often prefer to collect sperm from bulls twice a day two days per week. This harvests most of the sperm that could be obtained by more frequent collections and provides more sperm to process at one time. The latter is convenient when freezing semen. With liquid semen program bulls can be ejaculated three times per week and semen made continuously available from each bull by using it over a period of two to three days (Shannon, 1968). Bulls can be ejaculated daily, if necessary, without reducing fertility. However, the number of sperm per ejaculate is reduced and more stimuli often are required for proper sexual preparation.

The ram can be ejaculated many times a day for several weeks before severely depleting its epididymal reserves of sperm. This is because of the small ejaculates (Table 22–1) and the large epididymal reserves. Rams often mate or are ejaculated several times per day during the breeding season.

Boars and stallions expel large numbers of sperm in each ejaculate and deplete their epididymal reserves more quickly. It is best not to attempt regular collections more often than every other day. If daily ejaculations are required for short periods this should be followed by a rest for two to three days.

3. ARTIFICIAL VAGINA

The best procedure for collecting semen is with the artificial vagina (A.V.). The A.V. is simple in construction and simulates natural conditions. While there are a variety of modifications and different sizes and shapes for different species, the basic construction is illustrated with an artificial vagina for bulls (Fig. 22–3). The unit provides suitable temperature, pressure and lubrication to evoke ejaculation, and a cali-

brated tube is attached to collect the semen. Sanitation and the technical skill of both the semen collector and handler of the male are important. Such skill should lead to obtaining semen of high quality and purity and will minimize the possibility of injury during collection. A separate sterile artificial vagina should be used for collecting each semen sample.

Bull. For the bull, temperature of the A.V. is more important than the pressure exerted on the penis. The latter is usually controlled by the amount of water added but final pressure may be adjusted by pumping in air. The temperature inside the vagina should be near 45° C at the time of collection, although temperatures from 38 to 55° C have been employed (Seidel and Foote, 1969). The proper initial temperature to achieve 45° C at the time of semen collection depends upon the ambient temperature and the interval between preparation of the artificial vagina and collection. An excellent system is to place the assembled vaginas and protective insulated covers several hours before collection in a conveniently located incubator set at 45° C. At the same time the semen collection tube should be maintained near body temperature to prevent sperm damage upon ejaculation due either to overheating or to coldshock.

During collection the artificial vagina should be held parallel and close to the

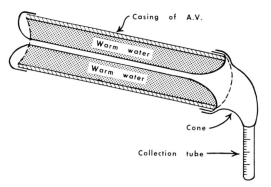

FIG. 22–3. Artificial vagina for bulls shown in longitudinal section to illustrate construction.

cow in a slanted position in line with the expected path of the bull's penis. The penis should be guided into the artificial vagina by grasping the sheath with the flat of the hand immediately behind the preputial orifice (Plate 23). The protruded penis itself should never be grasped, as this may cause either immediate retraction or ejaculation before the penis enters the artificial vagina.

Insertion of the penis is usually best accomplished on the upward movement as soon as the bull mounts. Precise timing produces best results. As the bull thrusts forward for ejaculation, the operator allows the artificial vagina to move forward with the thrust and to be aligned with the penis. Then he gently tilts the artificial vagina down so the semen can run into the collecting tube. The skill of the operator is important in avoiding sharp bending of the penis, which may cause discomfort or even injury to the bull. The experienced operator will alter the procedure slightly according to the sexual behavior of individual males. Rough inner liners and a slight increase in temperature may evoke ejaculation in a bull that otherwise is unresponsive.

Ram. The recommended temperature of the smaller-sized artificial vagina for the ram is identical to that described for the bull. The technique of collection is also similar, but the forward thrust is less vigorous, thus reducing the possibility of injury. Rams often mount quickly so the collector should swiftly coordinate his movements with those of the ram. An A.V. may be attached to a dummy (Fig. 22–2D).

Boar. With boars the collection period must be prolonged because emission of semen, as in natural service, may last for more than five minutes. Pressure is more important for collecting semen from the boar than from the bull. The boar ejaculates when the curled tip of the penis is firmly engaged in the sow's cervix, the artificial vagina, or the operator's hand (Plate 23C). When using the A.V. or the gloved hand, con-

tinuous pressure should be exerted on the coiled distal end of the penis. Once ejaculation has started, the boar usually will remain quiet until ejaculation is completed. Relaxation of pressure on the penis will result in interruption of ejaculation and a return to thrusting by the boar. During the period of thrusting, seminal fluid and gelatinous material may be ejaculated.

The ejaculate from the boar consists of the three fractions: the pre-sperm fraction, the sperm-rich fraction and the post-sperm fraction. The first fraction contains mostly seminal fluids with a high bacterial content and gelatinous pelletlike material from the Cowper's glands which tends to seal the cervix of the sow during mating, preventing loss of semen. This pre-sperm fraction is best discarded before the sperm-rich fraction is collected. The sperm-rich fraction is followed by another seminal fluid fraction which also contains some gelatinous material. The gelatinous material (about 20%) may be discarded at the time of collection and/or removed by collecting the semen through a filter. Usually the boar emits only one sperm-rich fraction at a collection; however, some boars may ejaculate a series of two or three sperm-rich fractions which are interrupted by seminal fluid emission.

Stallion. Prior to ejaculation the penis should be washed with warm soapy water and rinsed with clean water to remove smegma and other debris on the surface of the penis. The mare should be hobbled (Fig. 22–2C). The A.V. for the stallion must be larger than for other farm animals in order to accommodate the stallion's erect vascular-muscular penis. Many types of A.V.s have been developed (Pickett, 1968), and one type is shown in Fig. 22–4. A handle to maintain a firm grip on the A.V. is necessary because of its size and to cope with the vigorous thrusting of a stallion. The A.V. is partially filled with water so as to give an internal temperature of 45° C. Air space connected to an expansion valve is

PLATE 23

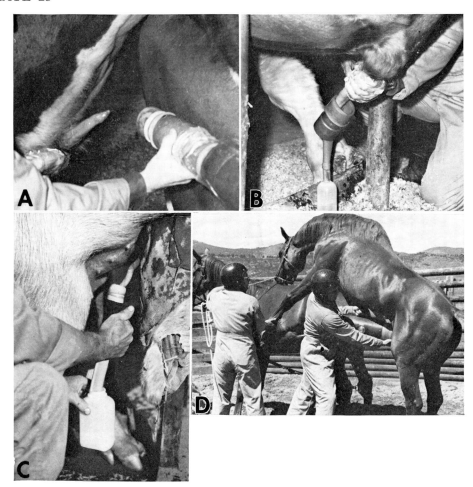

A, Proper technique of collecting bull semen with the artificial vagina.

B, Achieving an erection and semen emission by electroejaculation in the bull. The semen collection tube is protected by a warm water jacket.

C, Collecting boar semen by applying manual pressure to a short artificial vagina.

D, Collecting stallion semen. Note the foreleg of the stallion is held to prevent it from striking the collector. (*Courtesy of B. W. Pickett.*)

PLATE 24

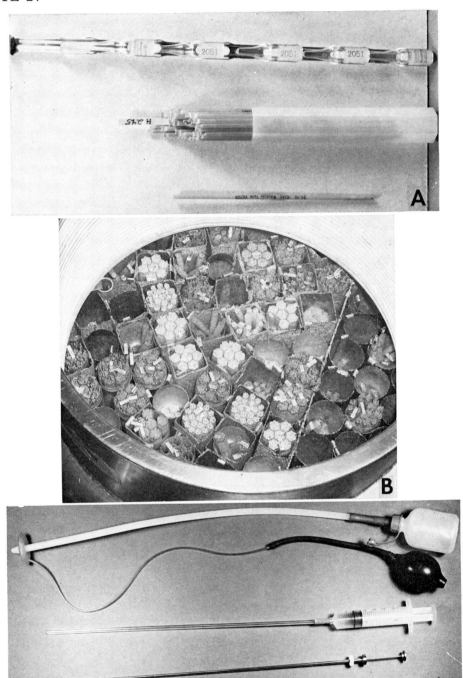

A, A 0.5 ml straw and goblet holding 36 straws in comparison with a 1.0 ml ampule as on a 6-ampule cane.
B, Storage of straws under liquid N₂ at −196° C.
C. Insemination equipment for sows, mares, cows, ewes and bitches (*top to bottom*).

FIG. 22–4. *Top*, artificial vagina (modified Japanese model) for stallions. *Bottom*, enlarged collection bottle is fitted with a filter to remove the gel. (*Adapted from Komarek et al., 1965. J. Reprod. Fert. 10, 337.*)

provided to permit expansion of the liner when the penis engages the A.V. Pressure increases in the A.V. during thrusting. This pressure and friction stimulate the stallion to ejaculate. When pulsations at the base of the penis—characteristic of ejaculation—begin, the end of the A.V. with the attached collection bottle should be lowered sufficiently to allow the semen to flow into the collection vessel (Plate 23D). Ejaculation is completed in about 25 seconds (Pickett, 1968).

4. ELECTRO-EJACULATION

Electrical stimulation is the preferred method for males that refuse to serve the artificial vagina (often under range conditions), or when injuries and infirmities make mounting impossible. This method can be used very successfully in the bull and ram, and may be useful in obtaining a semen sample of reduced volume from the boar. It is not desirable to collect and use semen from males unable to serve because of a probable genetic defect.

Many of the electro-ejaculators on the market run on 110 volts as well as on the 12-volt car ignition system. The newer models are transistorized which makes them more portable. The same basic unit with different probes can be used for bulls, rams and boars.

Bull. A rectal probe with either ring or straight electrodes is used to provide the electrical stimulation. The penis usually erects and the semen is collected without the possibility of contamination in the prepuce. Excess fecal material is removed from the rectum and the lubricated probe is inserted. The low voltage used at first is gradually increased. Repeated rhythmic stimulation is alternated with short rest periods; the voltage is momentarily returned to zero after each increase. Secretion from the accessory sex glands takes place at the lower voltage level and ejaculation at the higher voltages. If the increase is too rapid a pre-erection ejaculation is obtained which may result in contamination by the prepuce.

Bulls tend to stiffen, arch their backs and push forward. There are no apparent ill effects, but a suitable restraining rack with good footing should be used (Plate 23).

Semen samples are usually of larger volume with a lower concentration of sperm, but the fertility and total sperm numbers are the same as samples obtained with an artificial vagina. This is also true of the ram.

Ram. The ram responds exceptionally well to electrical stimulation, and the response is much more rapid than in the bull. Sometimes only three stimulations of two-, five- and eight-volt peaks are required for ejaculation. Semen can be collected with the ram standing or lying on a table. The glans penis should be lightly secured with sterile gauze so that the filiform appendage and urethra are directed into the collection tube prior to ejaculation.

Boar. An electro-stimulator designed for use in the ram can also be used in the boar. Because of the fat insulation, the voltage required to initiate ejaculation may result in discomfort. Anesthesia or

Table 22-2. Composition of Buffers and Extenders for
Frozen Semen as Compiled from the Literature

| | For Frozen Semen ($-196°$ C) | | | | |
| | Ampules, Straws | Pellet Freezing | | | |
Ingredients*	Bull	Bull	Ram	Boar	Stallion†
Sodium citrate dihydrate (gm)	23.2	—	20.0	—	—
Tris (gm)	—	—	—	20	—
Citric acid monohydrate (gm)	—	—	—	10	—
Glucose or fructose (gm)	—	—	—	5	50
Lactose (gm)	—	—	—	40	3
Raffinose (gm)	—	139	99	—	3
Casein (gm)	—	—	—	20	—
Penicillin (units/ml)	1000	1000	1000	1000	1000
Streptomycin (μg/ml)	1000	1000	1000	1000	1000
Polymyxin B (units/ml)	500	500	—	—	—
Glycerol (ml)	70	47	50	—	50
Egg yolk (ml)	200	200	150	—	50
Distilled water to final volume (ml)	1000	1000	1000	1000	1000

* Ingredients usually are dissolved in distilled water and then glycerol, more water and egg yolk are added to bring the final volume to one liter. Two buffers and extenders often are prepared as the glycerol level in the initial medium mixed with sperm usually has a lower glycerol concentration (excepting for boar sperm) than the extender in which sperm are frozen. Antibiotic levels also vary.

† Also frozen in straws. Up to 0.5 gm each of sodium phosphate and sodium-potassium tartrate may be included in the medium (From Nishikawa and Shinomiya, 1972. *Riproduzione Animale e Fecondazione*. Bologna.)

an analgesic can be used to immobilize the boar. Sperm motility, morphology and concentration are good, but the volume usually is small. Thus, the technique is not satisfactory for routine semen collection.

5. Massage Method

Massage via the rectum of the vesicular glands and ampullae of the vas deferens can induce semen flow (Parsonson et al., 1971). Massage of the sigmoid flexure also may be desirable to cause protrusion of the penis. Semen so collected usually has a lower sperm concentration than ejaculates obtained with the artificial vagina.

II. EVALUATING SEMEN

Semen evaluation should be rapid and effective so that the carefully collected sample can be processed to preserve its initial quality and used to its fullest potential, or so that inferior samples can be rejected. There are a number of tests

useful in judging quality. No single test has been developed that is an accurate predictor of the fertility of individual ejaculates, but when several tests are carefully combined, ejaculates can be selected for use which have a higher fertility potential than those discarded. Several of the characteristics of semen usually evaluated are shown in Table 22-1.

A. Appearance and Volume

Semen that has been properly protected to avoid coldshock during collection should arrive at the laboratory slightly below body temperature. It should have a relatively uniform opaque appearance indicative of high sperm concentration. A sample which appears to be translucent contains fewer sperm cells. The sample should be free from hair, dirt and other contaminants. Any semen containing chunks of material (other than the gel in boar and stallion semen) or with a curdy appearance should not be used, as this is indicative of inflammation.

Some bulls consistently produce yellow semen due to the presence of a riboflavin pigment. This is harmless and should not be confused with urine which has a distinctive odor. The tube of semen should be labeled immediately to identify the source of semen properly.

The volume of the sample can be determined from the calibrated collection tube. Values will be more extreme than the averages given previously (Table 22–1). Individual ejaculates of bull semen may range from 2 to 20 ml. In general young animals and those of smaller size within a species will produce smaller volumes of semen. Frequent ejaculation results in lower average volume and when two ejaculates are obtained consecutively the second usually has the lower volume. In addition there is considerable random variation. Small volume is not harmful, but if accompanied by a low sperm concentration it limits the number of sperm available.

B. Motility and Live Cells

The percentage of progressively motile cells should be estimated on a microscope stage incubator at 37°–40° C under high power (400×). The fresh semen should be prepared as a thin film on a microscope slide diluted with enough physiologic saline solution so that individual cells are visible. At 400× the percentage of motile cells, their rate of motility and their gross morphology can be scrutinized. Often television equipment is used to display the sperm cells. The motility can be recorded to the nearest 10% or in units on a scale from 0 to 10. Most semen will not routinely contain more than 75% progressively motile sperm, except dog semen, which normally contains at least 80%. The rate of movement can be estimated microscopically on an arbitrary scale fairly accurately. Experimental instruments have been designed to do this with considerable precision. Circular or reverse motion often are signs of coldshock or of media that is not isotonic with semen. Oscillatory motion is frequently seen in aged semen,

when many spermatozoa have ceased all movement and are presumed to be dead.

The proportion of live cells to dead cells can be estimated by supravital staining with a stain such as nigrosin-eosin. The cells which were alive when the stain was applied remain unstained, and the dead cells stain red with eosin against the dark nigrosin background. The results are highly correlated with the visual estimates of progressively motile cells, but the latter averages are lower than the percentage of unstained sperm.

Photoelectric methods of observing sperm velocity, swimming patterns and survival as sperm pass a phototube have been developed and found to yield values correlated with fertility (van Duijn and Verver, 1970). Instrumentation is sophisticated and not available for general use.

Other techniques used to evaluate motility include impedance change measurements and microscopic estimation of swirling. These criteria are affected by both motility and sperm concentration.

C. Concentration

Accurate determination of the number of sperm per milliliter of semen is extremely important, as this is one of the most variable of semen characteristics used to judge semen quality. Sperm volume and percentage of motile sperm,

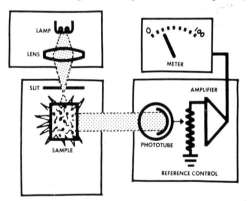

Fig. 22–5. Diagram of a photometer which can be used to accurately determine the concentration of sperm cells in a sample by reflected or transmitted light. The portion of light reaching the phototube is converted to an electrical signal which deflects the needle on the meter scale.

gives the number of motile sperm per ejaculate. This quantity determines how many females can be inseminated, each with the optimum number of sperm.

The simplest way of routinely estimating sperm concentration is to estimate the optical density (turbidity) of the sample with a nephelometer or photoelectric colorimeter (Fig. 22–5). The procedure is applicable to semen of all farm animals as long as the gel has been filtered out. The equipment required is as follows: photoelectric colorimeter with standardized tubes for uniform light transmission, pipets and 0.9% saline or 2.9% sodium citrate dihydrate solutions for diluting the semen at a standard rate, and a hemocytometer. The latter is a calibrated slide made for counting red blood cells. In the same manner sperm counts on a series of semen samples can be made to compare with the optical densities of the samples determined photometrically. Thereafter the calibrated photometer can be used to estimate sperm concentrations quickly and accurately. A detailed description of the procedure and a method for rechecking the photometer calibration with latex particle standards is available (Foote, 1972).

D. Morphology

Semen from most males contains some abnormally formed sperm (Table 22–1). This is not usually associated with lowered fertility until the proportion of abnormals exceeds 20%, and even then certain types may not be associated with infertility. Samples with large numbers of abnormal sperm can be detected when estimating the percentage of motile cells as described previously. A more precise estimate of the types of abnormals and their incidence can be gained from slides stained to yield live-dead ratios. Otherwise, a simple technique for studying morphology alone is to mix a high grade India ink with semen on a slide and draw out the mixture to make a thin smear. Skill is required in interpreting the slides and more than 100 sperm examined in random order should be classified per sample.

Special attention should be given to the acrosome (Saacke, 1972), as this plays an important role in fertilization. The apical ridge of the acrosome of bull and boar sperm deteriorates with aging or injury of the cell. Later the acrosome may loosen and be lost. This can only be seen with appropriate phase or interference microscopes. Acrosome changes are more highly correlated with fertility than is sperm motility (Saacke, 1972).

Improper freezing may damage the acrosome and cell membranes. Loss of enzymes from the cell, such as glutamic oxaloacetic transaminase (GOT), indicates the effectiveness of cryoprotection by the medium and freezing procedures used in preserving the integrity of sperm (Graham and Pace, 1967). Viability of the sperm when incubated at 37° C after freezing and thawing gives some indication of how long sperm will survive in the female following insemination.

E. Other Criteria

Various biochemical measurements of the DNA and associated protein of the sperm head have been researched as possible indicators of semen quality. Also metabolic tests such as oxygen utilization, fructolysis, methylene blue or resazurin color reduction time have been used. These primarily reflect the combined effects of motility and concentration of the sperm. The pH of semen and shifts in pH also were used at one time. It is clear that the complex functions of the sperm cell which contribute to its fertilizing capacity can best be monitored by a combination of carefully conducted tests.

III. PRESERVATION OF SEMEN

A. Semen Extenders

1. Principles

Ejaculated sperm do not survive for long periods unless certain agents are added. The agents, which comprise good extending media, have the following functions: (*a*) to provide nutrients as a source of energy; (*b*) to protect against the harmful effects of rapid cooling; (*c*) to provide a buffer to prevent harmful shifts in pH as lactic acid is formed; (*d*) to maintain the proper osmotic pressure and electrolyte balance; (*e*) to inhibit bacterial growth; (*f*) to increase the volume of the pure semen so that it can be used for multiple inseminations; and (*g*) to protect the sperm cells during freezing.

Pure substances and very clean equipment should be used to exclude toxic materials from the sperm environment. Extenders should be prepared aseptically and stored for less than a week unless frozen. A simple carbohydrate, such as glucose, usually is added as a source of energy for the sperm. Both egg yolk and milk are used to protect against cold-shock of the sperm cells as they are cooled from body temperature to 5° C. These substances also contain nutrients utilized by sperm. A variety of buffers in proper strength may be used to maintain a near neutral pH and proper osmotic pressure. The more common ones are citrate, phosphate, tris and other organic buffers. Semen is isosmotic with body fluids and secretions such as blood plasma and milk (approximately 300 milliosmols). Semen extenders should be formulated so as to be isosmotic with semen. To inhibit microorganisms in semen, penicillin, streptomycin, polymyxin B or other combinations of antibiotics covering a broad bacterial spectrum are added.

Glycerol usually is added to protect sperm against the otherwise lethal effects of freezing. Dimethylsulfoxide (DMSO) and a sugar such as lactose also may be beneficial, as they serve as dehydrating agents.

2. COMPOSITION OF EXTENDERS

Practically all extenders for liquid or frozen semen have either egg yolk or milk or a combination of the two as a basic ingredient. Egg yolk simply combined with sodium citrate, and milk or skim milk have been widely used for bull semen, and with modifications for ram, boar and stallion semen.

No other aspect of artificial insemination has been as intensively studied as the formulation of extenders to preserve sperm and prolong their fertility (Salisbury and Rufener, 1972; Terrill, 1972). The most remarkable discovery for freezing semen was the protective effects of glycerol by Polge and co-workers, and this has led to studies of a variety of media and freezing techniques (Salisbury and Rufener, 1972; Nagase, 1972). Countless extender formulations have been published. Selected examples of extenders currently in use for frozen semen are shown in Table 22–2. Commercially, only frozen bull semen is used extensively. It has gradually replaced refrigeration at 5° C, whereas in other species insemination is largely limited to use of freshly diluted unfrozen semen. The frozen semen extenders given for the ram, boar and stallion are experimental.

Bull semen often is frozen in skim milk and homogenized milk containing about 10% glycerol (Almquist and Wickersham, 1962). Tris-citric acid-fructose-glycerol-yolk (Foote, 1970) also has given excellent fertility results. Both the milk and tris extenders are satisfactory for liquid semen stored at 5° C. Enzymes which may alter the sperm acrosome in some way (amylase and β-glucuronidase) have increased the fertility of bull sperm frozen in yolk-citrate-glycerol (Hafs et al., 1971). Many other extenders are referred to in a recent review (Salisbury and Rufener, 1972).

The milk and tris extenders with minor modifications have been used to store semen of other species at 5° C and to freeze this semen. In the pellet method of freezing, extenders high in sugar content often are used (Table 22–2). Lactose often is substituted for raffinose on an equimolar basis.

Ram semen frozen in the extender shown in Table 22–2 (Salamon and Lightfoot, 1970), is not as fertile as when used undiluted or used at low dilutions unfrozen. Unfrozen boar semen may be stored in tris-yolk, glucose-yolk, glucose-milk or a modified carbonated yolk extender (IVT) developed at the University of Illinois. Experimentally, Graham et al. (1971) have obtained litters of pigs from semen frozen in a mixture of organic buffers, fructose, sodium citrate and egg yolk. Glycerol is detrimental to fertility and is usually excluded from the extender at the time of freezing. Commercially, swine are inseminated with liquid semen stored in glucose-bicarbonate-yolk or similar extender gassed with CO_2. Likewise, various yolk, milk and cream-gelatin extenders have been tested with stallion sperm. A preparation that gave satisfactory fertility results with semen frozen in straws is given in Table 22–2.

In areas where little refrigeration is available, semen may be stored at ambient temperature for a limited period. In New Zealand an extender called Caprogen (Shannon, 1968) is used to highly extend bull semen in the short breeding season. It consists of sodium citrate, glucose, glycine, glycerol, caproic acid and antibacterial agents, and is gassed with nitrogen before use. Egg yolk has been successfully reduced to 2% by volume. A carbonated yolk extender (IVT) and coconut milk also have given satisfactory fertility with semen stored at moderate ambient temperatures.

Antibacterial agents are especially useful in controlling microorganisms present in the extended semen stored at the higher temperatures. However, it is important that the proper antibiotics be used to control different types of organisms that may be encountered in different species. Glycerol may partially interfere with the inhibitory action of antibiotics on microorganisms.

Vegetable dyes may be included in the extender at sufficient concentrations to color it distinctly. This will not harm the sperm and facilitates identification of semen from different bulls or breeds.

B. Semen Processing

The processing of unfrozen semen will be considered first, since it has been used commercially in all species, and through the point of cooling to 5° C processing of frozen and unfrozen semen is similar.

Semen is collected at body temperature. Following collection it should be kept warm (30° C) prior to extension to avoid coldshock. This can be done by placing semen and extender in a water bath kept at 30° C. An aliquot of semen can be removed for sample evaluation and the remainder mixed with three parts of extender at 30° C. The mixture is cooled gradually to 5° C in the case of bull, ram and stallion semen and to 15° C for boar semen. Proper cooling takes at least one hour. After cooling, the semen can be further extended and packaged.

1. EXTENSION AND STORAGE OF
   LIQUID SEMEN

The objective of extending semen a prescribed amount is so the appropriate volume of semen inseminated will contain sufficient sperm to give high fertility without wasting sperm. The approximate extension rates possible with average ejaculates, appropriate sperm numbers to inseminate and other relevant data under good management conditions, are summarized in Table 22–3. An example of the calculations used to determine the final extension rates for bull semen follows:

Volume of ejaculate   = 8 ml
Concentration of sperm   =
                 1,200,000,000 per ml
Percentage of motile sperm = 70
Then, 1 ml of semen contains

$$1,200,000,000 \times \frac{70}{100} \text{ or}$$

840,000,000 motile sperm
Number of motile sperm required in
1 ml of diluted bull sperm =
                              5,000,000

$$\text{Thus dilution rate} = \frac{840,000,000}{5,000,000} =$$

168

Then 8 ml of semen can be diluted
(8 × 168) to 1344 ml.

Table 22-3. Semen Extension, Storage and Insemination Requirements

Item	Frozen Semen Cattle	Liquid Semen			
		Cattle	Sheep	Swine	Horses
Storage temperature (°C)	−196	+5	+5	+15	+5
Storage time (days)	Indefinite	4	1	2	1
Extension rate of 1 ml semen (ml)	10–75	160	10	4	2
Insemination dose					
Volume (ml)	.2–1	1	.05–.2*	50	20–50
Millions of motile sperm	15	5	50	2000	1500†
Best time to inseminate during estrus	Middle to end	Middle to end	Toward end	First and/or second day	Every one to two days
Site of semen deposition	Uterus	Uterus	Cervix	Cervix	Uterus
Number of possible females bred per male‡:					
Per ejaculate	500	1500	40	15	5
Per week	1500	4500	600	50	15
Conception on first insemination (% pregnant)	60–65	65	65	65	50–65

* Small volumes are preferable if semen is sufficiently concentrated; sperm in extended semen should be reconcentrated.
† Based on three inseminations per estrus with 500 million motile sperm each insemination.
‡ With intensive semen collection.

From these calculations it is clear that a single bull may sire many thousands of progeny per year.

Ram semen often is used within a few hours after collection without extension. When extended and stored for a day fertility appears to decrease markedly unless the sperm are reconcentrated by centrifugation. A small volume of highly concentrated sperm are preferable for insemination. Liquid boar semen usually is stored at 15° C. It can be used for two days with satisfactory fertility; boars on a three times per week collection schedule can provide a continuous supply of semen. Fertility of stallion semen declines rapidly and this should be used on the day of collection or on the following day. The extension rate used often is not more than two-fold so the total extension can be made prior to cooling.

For shipment into the field extended semen should be packaged in tubes or bottles that are nearly full. Excess air and agitation decrease the chances of sperm survival. An insulated box with ice should be used which protects the sperm against light and refrigerates the semen at about 5° C until received in the field. Subsequently semen should be stored in an ice chest or mechanical refrigerator until used.

Semen used at ambient temperatures should be wrapped to protect it from light, and handled so as to avoid temperatures approaching 40° C. Freeze-drying of semen as a means of preservation at ambient temperatures is not practical. The one successful attempt reported could not be repeated.

2. EXTENSION AND PACKAGING OF FROZEN BULL SEMEN

Most cattle inseminations today are performed with frozen semen, as freezing has permitted semen to be collected at one time and placed and used anywhere else for years, if desired (Nishikawa, 1972). Frozen bull semen usually contains two to three times more motile cells per ml after extension than liquid semen (Table 22–3) to compensate for the sperm which are killed during freezing in order to maintain maximum fertility (Pickett, 1971; Salisbury and Rufener, 1972). The number of sperm used is considerably in excess of the minimum number required. This reduces the potential number of cows which could be inseminated per ejaculate with frozen semen. However, the prolonged storage life of frozen semen permits all of an ejaculate to be used eventually, thus often more than compensating for the higher concentration of sperm used in frozen semen.

Glycerol usually is added to semen after cooling to 5° C. The amount varies from less than 5% in some yolk-sugar media to 10% milk. Some prefer to add glycerol slowly over a period of one hour and others recommend a one-step addition. Tris buffers (Foote, 1970) and the sugar buffers used with pellet freezing (Nagase, 1972) offer the advantage that glycerol can be included in the initial media used for cooling sperm. In most other extenders sperm are damaged morphologically by adding glycerol at room temperature.

The semen-extender mixture is held at 5° C for several hours before freezing to allow sperm to equilibrate in the cold (5° C). About four to six hours is optimum, depending upon the medium used. Originally this was thought to be partly necessary for equilibration with glycerol. However, it is clear that little time, if any, is required for glycerol equilibration; other changes take place at 5° C which enhance sperm survival during freezing.

Bull sperm commonly are packaged in three ways: (a) glass ampules containing 0.5–1 ml of extended semen; (b) polyvinyl chloride straws containing 0.25–0.5 ml; (c) pellets containing approximately 0.1 ml. When the smaller volumes are frozen as a unit the sperm concentration per milliliter is increased correspondingly, so that the total sperm per insemination dose is maintained. For example, semen frozen in 0.1 ml packages should have 10 × as many sperm per unit volume as semen frozen in 1.0 ml packages to contain the same total number of sperm.

Glass ampules were used nearly exclusively as the use of frozen semen developed in cattle. Ampules provided a sterile container that could be automatically labeled, filled and sealed. Each ampule contained sufficient sperm (Table 22–3) for a single insemination. Six to eight ampules are attached to a metal cane (Plate 24A) which also carries the bull's identification. Ampules of semen are frozen on the canes and subsequently stored on the canes at −196° C in liquid nitrogen (Pickett, 1971).

The ampule containers occupied considerable space in storage and many sperm were killed during freezing in ampules. Subsequently, polyvinyl chloride straws were developed by Cassou (see Pickett, 1971) patterned after a Danish straw used in earlier days of artificial insemination with liquid semen. Methods for handling various types of straws have been developed (Macpherson and Penner, 1972). The straw requires less storage space than the ampules, has slightly better freezing characteristics and can be labeled, filled and sealed automatically. Also, sperm can be transferred to the cow at the time of insemination with less loss of cells.

Pelleted semen is prepared by dropping about 0.1 ml drops of extended semen into hemispheric depressions made in a block of dry ice. Sperm survival is good following freezing. Pellets take the least amount of space when stored in bulk. Therefore they offer the cheapest storage method. The main disadvantage is the difficulty in placing a bull identification on each pellet, although this has been done by incorporating a small printed paper disc during freezing. Otherwise pellets are placed in bulk containers which are labeled and then repackaged at low temperatures into individually labeled containers prior to field distribution. Another packaging technique is to place sperm directly into catheters used eventually for insemination and to freeze these as breeding units.

3. FREEZING BULL SEMEN

Mechanical freezers and freezers using dry ice, liquid air, liquid O_2 and liquid N_2 all have been tried successfully. Liquid N_2 has increased in popularity where it is available because it is also an excellent refrigerant for low temperature long-time storage of semen. Extended semen, to be frozen as pellets or in ampules or straws, is held at about $5°$ C prior to freezing. Ampules often are placed in a liquid nitrogen freezer programmed to lower the temperature at about $3°$ C per minute to $-15°$ C. At this point the glycerolated semen is nearly frozen and the rate of freezing is increased until $-150°$ C is reached. The ampules on canes then are transferred to liquid N_2 at $-196°$ C. Straws usually are frozen in nitrogen vapor and stored at $-196°$ C (Plate 24B). Because of the large surface area of the straw and its thin wall, semen freezes more rapidly than in the ampule. Pellets start to freeze within a few seconds after being placed on a block of dry ice. In a few minutes they reach $-79°$ C and are transferred quickly into containers immersed in liquid nitrogen at $-196°$ C.

The exact nature of the freezing process which allows sperm under one set of conditions to be frozen rapidly (pellets), moderately rapidly (straws), and under other conditions requiring a slower rate (ampules) is not understood. Freezing too rapidly may cause thermal shock. At the same time slow freezing causes salt concentrations to increase as water freezes out. This increase in osmotic pressure over a prolonged period of slow freezing may be damaging to the proteins and lipoproteins of the sperm cell membranes and acrosome.

4. STORAGE OF FROZEN SEMEN

Frozen semen placed initially in dry ice and more recently in liquid nitrogen has been used successfully after nearly 20 years of storage. With the lower temperature of liquid N_2 it should be possible to store frozen semen indefinitely. Semen can be shipped to any place to be used at any time. However, it is cheaper and genetically advantageous to use semen from outstanding sires promptly, rather than to store it.

A variety of efficient vacuum sealed liquid N_2 refrigerators are available for storing frozen semen. These range in size from central storage units with a capacity of over 50,000 1.0 ml ampules and a N_2 reserve which lasts about three months, to the common field units holding up to 1500 ampules and a N_2 reservoir good for about five weeks. With the 0.5 ml ampules storage capacity can be increased by 25%. With different arrangements for storing straws and using either the 0.5 ml "midistraw" or 0.25 ml "ministraw" between 3 and 10 times as many straws as 1 ml ampules can be stored. Pellets stored in bulk occupy the least space. The large central storage units can hold up to 750,000 0.1 ml pellets, and possibly provide economical banking of semen from young bulls in sampling programs.

Special small storage units requiring only three to four refillings per year with liquid nitrogen have been developed. However, it is extremely important to check the liquid nitrogen refrigerator periodically to see that the nitrogen level is maintained. Loss of all liquid nitrogen permitting the temperature to rise considerably can result in damage to the sperm even though the semen still appears to be frozen.

5. THAWING BULL SEMEN

Frozen semen should be held continuously at low temperatures until used. Frozen sperm after thawing do not survive for as long as unfrozen sperm and they refreeze poorly. Therefore, one must be certain that the semen is going to be used soon once it is thawed. It is recommended that ampules be thawed in a container of ice water. This takes about eight minutes. Straws are usually thawed in the hand, an inside shirt pocket or water at 40° C. Pellets are best thawed by transferring to a liquid thaw media at 40° C, but under practical field conditions an ice water thawing bath is easier to maintain and is satisfactory.

6. FROZEN SEMEN OF OTHER FARM ANIMALS

The same general principles described for cattle (dairy and beef) appear to apply to handling frozen semen of other species. However, for several reasons A.I. with frozen semen has not been developed on a commercial scale in these species. The motility of frozen ram semen is similar to bull semen but fertility is lower. This is surprising in view of the fact that the closely related goat produces semen that has been frozen and used successfully for artificial insemination. Ram semen is best processed by the pellet method using the extender shown in Table 22–2. After cooling to 5° C semen can be frozen on dry ice one hour after it is cool. Pellets are thawed in a citrate-glucose solution at 37° C, centrifuged to concentrate the sperm and used promptly for insemination. The sperm-rich fraction of stallion semen or sperm after removing much of the seminal fluid (plasma) by centrifugation is frozen as pellets or in straws (Nishikawa, 1972). Storage is in liquid nitrogen. Several straws or pellets are thawed to provide sufficient sperm for each insemination.

Only recently has it been possible to freeze boar sperm and preserve some fertility. Glycerol, normally used to protect sperm of many species during freezing, reduces fertility in pigs. Sperm-rich fractions of boar semen held for an hour before cooling (Graham et al., 1971) and mixed with an organic buffer containing egg yolk or diluted in other media (Table 22–2) can be successfully frozen as pellets. Thawing is done very rapidly at body temperature or higher. Limited fertility results have been reported.

IV. INSEMINATION TECHNIQUE

Proper use of fertile semen at the time of insemination is essential for high fertility. This requires proper detection and reporting of estrus so the female may be inseminated at the correct time. Also, proper techniques of handling and inseminating the semen, and healthy females in sound breeding condition are important.

A. Detection of Estrus and Optimum Insemination Time

Acceptance of the male is the best indication of estrus, but this test may not always be practical for A.I. For cows an estrus expectancy list based on estrus 18–22 days previously will assist in detecting cows in estrus. Cows should be checked twice daily. This may not be feasible for beef cattle, but those ready for breeding can be put in separate pastures for close observation. Heat detection aids placed on the backs of cattle, or surgically sterilized bulls with marking devices can aid in estrus detection. Under the best conditions, as many as 20% of the females may not be properly detected or show behavioral symptoms of estrus.

Because of the difficulty in detecting estrus, a program of estrous cycle regulation is attractive. A number of techniques may lead to a synchronized highly fertile estrus at reasonable cost. Such programs still are largely experimental, however.

Estrus in ewes can be detected by using vasectomized rams with crayon on the brisket or with teaser rams equipped with a harness, apron and colored chalk for that purpose. Sows in estrus can be detected by using a vasectomized boar or by placing the hands firmly on the back of a sow. Those in estrus will not move and will exhibit lordosis. Acceptance of the stallion is the best indication of estrus in the mare. At the time of estrus the mare also will show frequent contractions of the vulva ("winking") and frequent urination in the presence of a stallion.

Table 22-4.

Cows First Showing Estrus	Should be Bred	Too Late for Good Results
In the morning	Same day	Next day
In the afternoon	Next day— morning or early afternoon	After 3 PM next day

The recommended time for insemination of different species is given in Table 22–3. Cows should not be inseminated before 60 days after calving for best conception rates. If a cattleman is attempting to reduce the calving interval on some cows by rebreeding sooner after calving, more services per conception should be expected. Conception rate also is lower when cows are inseminated during the first half of an average 18-hour estrus. Therefore, cows should be inseminated in the last half of estrus. A practical procedure is to check for estrus twice daily with insemination done as shown in Table 22–4.

In the ewe, insemination should take place at the middle or during the second half of estrus. Multiple inseminations during estrus, particularly with frozen semen, increase fertility. Sows come into estrus about three to five days after farrowing, but they do not ovulate at this time and should not be bred. Sows come into heat three to eight days after weaning their pigs and may be inseminated at this time. Since sows ovulate about 30–36 hours after the beginning of estrus, with rapid loss of fertility after ovulation, it is best to inseminate either late on the first day or early on the second day of estrus. Some advantage is gained by inseminating on both the first and second day of estrus. Inseminating a mare during the first heat, about nine days after foaling, is not advised because the uterus is not fully involuted and conception rate is lower. Better conception rates result when mares are inseminated about 30 days after foaling. Since mares have a long and variable estrus it is best

to inseminate every day or second day during estrus, starting on the second or third day. When mares are palpated daily, insemination can be done to coincide with ovulation, which precedes the end of estrus by one to two days.

B. Insemination Procedure

Identification of the female is the first step. The breeding record should be checked, previous breedings noted and the current insemination recorded. Basic breeding records are important for all species. Inseminating equipment used for several species is shown in Plate 24C.

Cattle. The dairy cow is usually inseminated while standing in a stanchion or stall. Beef cows in estrus should be penned and a squeeze chute provided to restrain the individual during insemination.

Semen can be deposited in the cervix with the aid of a speculum. However, the rectovaginal technique is more effective and more widely used (Sullivan et al., 1972). By manipulation of the cervix with the hand in the rectum, the inseminating catheter is passed through the spiral folds of the cow's cervix (Fig. 22–6A). Part of the semen is deposited just through the cervix into the body of the uterus and the remainder is deposited in the cervix as the catheter is withdrawn. Expulsion of the semen should be done slowly to avoid excessive sperm losses in the catheter. A common error is to penetrate beyond the body of the uterus into the uterine horns. The number of sperm required for optimum conception rate is higher for frozen semen than for liquid semen (Table 22–3). There is no advantage in depositing more than 15 million motile cells per well-timed insemination. Slightly fewer sperm per straw package than per ampule may be required because nearly all of the sperm in the straw are transferred to the cow, whereas some sperm are retained in the ampule and associated inseminating catheter. Pellets are thawed in an extender and the same equipment is used for insemination as with ampules.

In the animal which has been inseminated previously and pregnancy is a possibility, the catheter should not be forced through the cervix. Approximately 3–5% of pregnant cows show signs of estrus. The use of the dye technique in training technicians to inseminate reproductive tracts and intact cows appears to be an extremely useful method in improving the accuracy of semen placement and in improving the efficiency of technicians with low conception rates. The technique consists of dye placement in reproductive tracts of cows at specified locations and slaughter of the cattle until the inseminator has learned to deposit semen where planned. Video taping of the insemination to relate to the findings at slaughter is instructive.

Despite the small volume of semen inseminated (Table 22–3) sperm pass rapidly throughout the reproductive tract of the cow. Bull sperm apparently require little or no capacitation and are able to fertilize the egg immediately. This along with the late ovulation in the cow may explain why some cows when inseminated shortly after estrus conceive.

Sheep. The ewe can be held securely by putting the hind legs over a rail or placing her in an elevated crate during insemination. Sometimes a rotating platform arrangement (Fig. 22–6B) with the inseminator in a pit is used to inseminate more than 100 ewes per hour. This permits one man to put a ewe on the platform, another to inseminate and a third man to release the ewe.

Insemination with the aid of a speculum permits semen to be deposited into the cervix rather than the vagina. A small volume of concentrated semen (Table 22–3) is most effective in producing pregnancy. If semen has been stored diluted, sperm should be reconcentrated so that a small volume is inseminated. With normal cycling ewes, insemination of 50 million motile sperm is sufficient. With progestagen-treated ewes, to induce estrus, sperm number requirements may

be as great as 1500 million cells (Quinlivan, 1970) for maximum fertility in certain seasons. Double inseminations 12 hours apart with liquid semen increase fertility slightly but the increase is not sufficient to warrant the extra effort. However, fertility with frozen semen is less likely to approach normal levels unless double inseminations are used. The same procedures are used for goat insemination.

Swine. The sow is best inseminated without being restrained, thus avoiding a struggle that may make it more difficult for semen to flow through the cervix. With some rubbing and stroking and pressure on the back, the sow in estrus will usually stand calmly during insemination. The insemination tube is automatically guided into the cervix because the vagina tapers directly into the cervix (Fig. 22–6C). The cervix itself tapers forming tight folds. It is not possible to pass the catheter into the uterus. However, by manipulating the catheter past the first cervical folds and inflating the cuff (Plate 24C) most of the 50 ml suggested for insemination will be forced into the uterus. Otherwise much semen may be lost. A fairly large semen volume as well as high sperm numbers (Table 22–3) are required in the sow for maximum fertility. The mechanism by which the volume acts to improve fertility is unknown. Fluid volume simply may be needed to transport sperm quickly through the long uterine horns. Frozen boar sperm appear to be rapidly removed by the uterus. Even when many billions are inseminated intracervically few reach the oviducts. Surgical deposition of a few million frozen-thawed sperm in .05–.10 ml directly into the oviduct gave good fertility (Polge et al., 1970).

Horses. The inseminator should be protected from possible kicking by the mare. The mare can be restrained by hobbles, backed against baled hay or a board wall or preferably be put in a breeding chute. The technique of insemination is illustrated in Figure 22–6D.

The area around the vulva should be scrubbed before insemination to minimize contamination. The arm in a plastic sleeve, lightly lubricated, can be inserted into the vagina and the index finger inserted into the cervix. The inseminating catheter then is easily guided into the uterus, where 20–50 ml of raw or extended semen is deposited. Best results with fresh or stored unfrozen semen are obtained when 500 million motile sperm are inseminated. The minimum number of unfrozen and frozen sperm required has not been established.

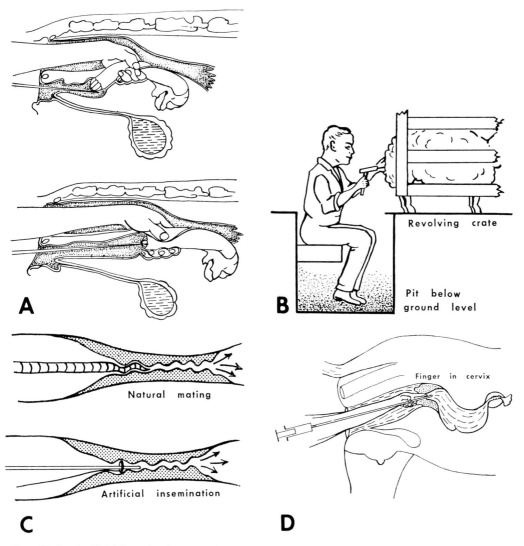

FIG. 22–6.   Artificial insemination procedures.

A, Top, wrong way of holding cervix for rectovaginal technique of inseminating cows. *Bottom,* by using correct procedure, it is relatively easy to deposit semen through the cervix. (*Redrawn from Bonadonna, 1957. Nozioni Di Fisiopathologia Della Riproduzione E Di Fecondazione Artificiale Degli Animali Domestici, Milan, courtesy of T. Bonadonna.*)

B, Pit and restraining crate facilitating ewe insemination.

C, Comparison of cervical semen deposition in natural mating and artificial insemination in swine. Arrows indicate flow of semen into the uterus.

D, For insemination of mares the index finger in the cervix assists in guiding the catheter held by the other hand into the uterus.

V. FACTORS AFFECTING CONCEPTION RATE IN ARTIFICIAL INSEMINATION

Accurate measurement of fertility is an important part of any organized artificial insemination program. The most important criteria to assess fertility are the proportion of females which produce young and the number of young produced. These statistics are often difficult to obtain commercially and the usual measure of reproductive efficiency in inseminated cattle is the 60- to 90-day nonreturn rate. This report is tabulated on a monthly basis. Two months after each month when the inseminations were performed (which is also about 90 days after the beginning of that month) the records of all inseminated cows are checked to determine what proportion of the cows returned to estrus and were reinseminated. Computerized summaries of 60- to 90-day nonreturn rates are usually prepared regularly on all bulls and technicians. With computers it would be feasible to introduce a more precise time, such as a 75-day interval. Experience with pregnancy checks shows that a 60- to 90-day nonreturn rate of 70% on first services will correspond to a 60-65% pregnancy rate. Fertility results with artificial insemination and natural service in cattle are similar. Frozen semen results tend to be slightly lower than for liquid semen.

Organizations providing artificial insemination service should periodically evaluate the reproductive efficiency of their service with spot checks of actual young born. This is particularly true in sheep and swine where manual pregnancy checks are more difficult to perform and where multiple births and large litter sizes are important. Estimates of pregnancy rates are given in Table 22–3. Litter size in swine inseminated artificially with unfrozen semen is equal to natural service, but farrowing rates to a single natural service are slightly higher. Mares should be pregnancy checked 40–60 days after breeding by an experienced person or by conducting an

immunologic blood test. Techniques for freezing ram, boar and stallion semen have not yet reached the level where frozen semen compares favorably with natural service.

The major factors determining fertility in artificial breeding are: (a) fertility of the males used to produce the semen; (b) care with which semen is collected, processed and stored; (c) skill of the inseminating technician; and (d) management of the females. Males should be carefully selected, isolated and tested before joining a bull stud. They should have completely normal testes (Foote et al., 1972), produce high quality semen and be free from disease. Regular health checks should be repeated periodically. All health codes established by the A.I. industry should be followed. Fertility of each male can be determined accurately in A.I. and those with very low fertility initially should be culled. Bulls may increase in fertility shortly after puberty, and most males show some decline in fertility as they age.

All equipment for collecting, processing and inseminating semen must be clean and sterile. No residues of cleaning compounds should be left on equipment which will come in contact with sperm. Aseptic procedures throughout are essential.

Frozen semen should be stored continuously at −196° C. *Never* let the liquid nitrogen level run low.

Finally the herdsman must have the female in optimum breeding condition and detect estrus properly if high fertility is to be achieved. Thus, realizing the potential of artificial insemination depends upon the combined skill and cooperation of people, the organization producing and distributing the semen, the inseminator and the manager of the herd or flock.

REFERENCES

Almquist, J. O. and Wickersham, E. W. (1962). Diluents for bovine semen. XII. Fertility and motility of spermatozoa in skim milk with various levels of glycerol and methods of glycerolization. *J. Dairy Sci.* 45, 782–787.

Amann, R. P. (1970). "Sperm Production Rates." In *The Testis*. A. D. Johnson, W. R. Gomes and N. L. VanDemark (eds.), Academic Press, New York, pp. 433–482.

van Duijn, C., Jr. and Verver, H. W. (1970). Sire specificity of movement characteristics and longevity of bull spermatozoa. *Ann. Biol. Anim. Bioch. Biophys. 10*, 51–58.

Foote, R. H. (1970). Influence of extender, extension rate, and glycerolating technique on fertility of frozen bull semen. *J. Dairy Sci. 53*, 1478–1782.

Foote, R. H. (1972). How to measure sperm cell concentration by turbidity. *NAAB Proc. 4th Tech. Conf. Anim. Reprod. Artif. Insem.* 57–61.

Foote, R. H., Larson, L. L. and Hahn, J. (1972). Can fertility of sires used in artificial insemination be improved? *A.I. Digest 20*(6), 6–8.

Foster, J., Almquist, J. O. and Martig, R. C. (1970). Reproductive capacity of beef bulls. IV. Changes in sexual behavior and semen characteristics among successive ejaculations. *J. Anim. Sci. 30*, 245–252.

Graham, E. F. and Pace, M. M. (1967). Some biochemical changes in spermatozoa due to freezing. *Cryobiology 4*, 75–84.

Graham, E. F., Rajamannan, A. H. J. and Schmehl, M. K. L. (1971). Fertility studies with frozen boar spermatozoa. *A. I. Digest 19*(6), 6–8.

Hafs, H. D., Boyd, L. J., Cameron, S., Johnson, W. L. and Hunter, A. G. (1971). Fertility of bull semen with added Beta-Glucuronidase. *J. Dairy Sci. 54*, 420–422.

Macpherson, J. W. and Penner, P. (1972). A modified straw technique for frozen semen. *A.I. Digest 20*(2), 6–8.

du Mesnil du Buisson, F. and Signoret, J. P. (1970). Reproductive physiology and artificial insemination in pigs. *Vet. Rec. 87*, 562–568.

Nagase, H. (1972). "Scientific and Technological Background of Pellet Freezing of Bull Semen." In *Riproduzione Animale e Fecondazione Artificiale*, Bologna, Edagricole, pp. 189–205.

Nishikawa, Y. (1972). Present state of long-term conservation of sperm and its application in various species of domestic animals. *VII Internat. Congr. Anim. Reprod. Artif. Insem.*, Munich, *I*, 143–165.

Parsonson, I. M., Hall, C. E. and Settergren, I. (1971). A method for the collection of bovine seminal vesicle secretions for microbiologic examination. *J. Amer. Vet. Med. Assn. 158*, 175–177.

Pickett, B. W. (1968). Collection and evaluation of stallion semen. *NAAB Proc. 2nd Tech. Conf. Artif. Insem. Reprod.*, 80–86.

Pickett, B. W. (1971). Factors affecting the utilization of frozen bovine semen for maximum reproductive efficiency. *A. I. Digest 19*(2), 8–23.

Polge, C., Salamon, S. and Wilmut, I. (1970). Fertilizing capacity of frozen boar semen following surgical insemination. *Vet. Rec. 87*, 424–428.

Quinlivan, T. D. (1970). The relationship between numbers of spermatozoa inseminated and fertilization rate of ova in ewes treated with fluoroprogestagen intravaginal sponges in summer and autumn. *J. Reprod. Fert. 23*, 87–102

Saacke, R. G. (1972). Semen quality tests and their relationship to fertility. *NAAB Proc. 4th Tech. Conf. Anim. Reprod. Artif. Insem.*, 22–28.

Salamon, S. and Lightfoot, R. J. (1970). Fertility of ram spermatozoa frozen by the pellet method. III. The effects of insemination technique, oxytocin and relaxin on lambing. *J. Reprod. Fert. 22*, 409–423.

Salisbury, G. W. (1968). Fertilizing ability and biological aspects of sperm storage *in vitro*. *VIth Internat. Congr. Anim. Reprod. Artif. Insem.*, Paris, 2, 1189–1204.

Salisbury, G. W. and Rufener, W. H. (1972). "Recent Research in Bovine Semen Extenders." In *Riproduzione e Fecondazione Artificiale*, Bologna, Edagricole, pp. 287–298.

Seidel, G. E., Jr. and Foote, R. H. (1969). Influence of semen collection interval and tactile stimuli on semen quality and sperm output in bulls. *J. Dairy Sci. 52*, 1074–1079.

Shannon, P. (1968). Advances in semen dilution. *Proc. N.Z. Soc. Anim. Prod. 28*, 23–31.

Sullivan, J. J., Bartlett, D. E., Elliott, F. I., Brouwer, J. R. and Kloch, F. B. (1972). A comparison of recto-vaginal, vaginal, and speculum approaches for insemination of cows and heifers. *A. I. Digest 20* (1), 6–8.

Swierstra, E. E. (1971). Sperm production of boars as measured from epididymal sperm reserves and quantitative testicular histology. *J. Reprod. Fert. 27*, 91–99.

Terrill, C. E. (1972). "Collection, Evaluation, Dilution and Conservation of Ram Semen." In *Riproduzione Animale e Fecondazione Artificiale*, Bologna, Edagricole, pp. 309–315.

15

Chapter 23

Techniques in Female Reproduction

A. Detection and Synchronization of Estrus

J. W. Lauderdale and R. G. Zimbelman

Males are inherently capable of detecting estrus, therefore estrus detection becomes a husbandry problem only if males either are not used as estrus detectors or do not do their job properly. In order to record animals accurately, some means of animal identification must be employed. Ear tags have been used extensively with cattle and sheep and ear notches with swine. Paint brands have been widely used with sheep and to a lesser degree with swine. Branding, both hot iron and freeze branding, has been used extensively with cattle and horses.

The signs of estrus in different species are summarized in Table 23–1. It is interesting to note that the estrous mare will allow the stallion to smell, bite and mount without retaliating. The mare not in estrus will not tolerate such behavior, but will either attempt to escape or retaliate with biting and kicking (Fig. 23–1). When the stallion is displaying courtship behavior the mare will frequently raise her tail and urinate.

Table 23-1. Signs of Estrus

| Species | Observable Signs | | | Mounting Behavior | |
	Swollen Vulva	Mucous Secretions	Behavior	Absence of Male	Presence of Male
Sow	Yes	No	Restless, irritable, explorative, frequent urination, vocal emissions	Immobility response to man	Stands to male
Ewe	No	No	None	None	Stands to male
Cow	Sometimes	Sometimes	Restless, nervous, off feed, mounts herdmates	Stands for herdmates	Stands for either male or herdmates
Mare	Sometimes	No	Frequent urination, "winking" vulva after urination	None	Stands to male

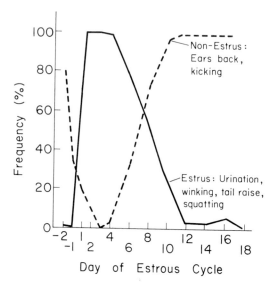

FIG. 23–1. Signs of estrus for mares. (*From Artificial Insemination of Mares. ABS Tech Manual, 1969, courtesy of Dr. John Sullivan.*)

I. DETECTION OF ESTRUS

Husbandry methods of estrus detection have been based on characteristics related to expression of estrus for each species. Of prime importance is the fact that animals are observed at regular intervals specifically for signs of estrus. Failure of the human observer to detect animals in estrus is the greatest cause of anestrus.

Ewe. Estrus can be determined most effectively by use of a ram. Vasectomized rams either wearing a marking harness containing marking material or painted with marking material on their brisket are used extensively with high degree of effectiveness. Rams are either run with the flock continuously and the ewes observed once or twice daily for marks on their rump or rams may be introduced into the flock by the observer one or more times daily and the ewes observed for estrous behavior. For the latter method, the ewe can either be separated from the flock or marked by the observer. These methods have been successful in detecting ewes in estrus. Observations of the flock in the early morning and late eve-

ning are effective, however, more frequent observations during each 24-hour interval increase effectiveness of estrus detection.

Sow. Estrus can be determined most effectively by use of either an intact or vasectomized boar. The estrus observer can introduce the boar into a pen of females and observe those females the boar attempts to mount. The boar can easily be prevented from mating with a female. Alternate methods are to allow the boar to be seen and heard by the females, but not to come in contact with the females, to play a recording of boars, or to play a recording and introduce smell of a boar. Estrous females will evidence signs of estrus and will stand to back pressure from the observer. The least efficient method of estrus detection is to observe for signs of estrus and to attempt to apply back pressure. Relatively fewer females will be detected by this latter method than with the above mentioned methods. Observations of the herd in early morning and late evening yield a high degree of estrus detection.

Cow. Cows in estrus usually stand to be mounted by herdmates, therefore observation of the herd for this type of behavior is an effective method of estrus detection. Use of vasectomized, penectomized or otherwise altered bulls, may or may not increase effectiveness of estrus detection. These types of bulls may be fitted with a marking harness, Chin-Ball marking device or have their brisket painted and run continuously with the herd. Cows are then observed for marks on their rump. An "estrus" indicator is commercially available. This device can be attached to the rump and turns red when pressure is applied such as by a mounting animal. Effectiveness of accurate estrus detection has varied, but results indicate a fully red indicator is about as accurate as human estrus detection. When properly placed, there is little chance of obtaining false negative readings (no red marker when the female is in heat). However, false positives

Table 23-2. Efficiency of Various Methods of Estrus Detection in Cattle

Method of Detection	Percent Correctly Identified In Estrus
Continuous 24 hour observation	98–100
Observed thrice daily	81–91
Observed twice daily	81–90
Observed during routine dairy activities	56
Painted bulls	98–100

(Adapted from Donaldson, 1968, *Aust. Vet. J. 44*, 496; and Williamson et al., 1972, *Vet. Rec. 91*, 50.)

(red marker) can occur if animals are crowded so that they are not able to escape from being mounted even though they are not in estrus.

During the first 60 days postpartum the percentage of cows that ovulate but are not detected in estrus may range between 10 and 60%. Thus, estrus detection may be a physiologic as well as a management problem in cattle in certain circumstances. The increase in missed estrus may reflect only a shortened period of estrus rather than an absence of estrus. Observation for estrous activity of cattle in early morning and late afternoon resulted in a high degree of accurate estrus detection, but more frequent observation resulted in more accurate estrus detection (Table 23-2).

Mare. The most effective method of estrus detection (teasing) is by observations of reactions of the mare to presence of either a stallion or gelding which has been injected with testosterone. Mares may be taken to the stallion or the stallion led through a barn or along a paddock fence. Eagerness and acceptance of the stallion as opposed to retreat from or rejection of the stallion by the mare are signs of estruslike behavior. Teasing usually is initiated prior to start of the breeding season in order to establish patterns of estrus behavior for individual mares. If a stallion is not available for estrus detection the "signs of estrus" referred to previously, must be relied on. This method has a very low degree of effectiveness.

II. ESTRUS SYNCHRONIZATION

Methods which decrease time needed to observe for estrus would be beneficial to herdsmen. Numerous investigators have attempted to develop practical methods of grouping estrus. These methods have included administration of natural or synthetic hormones by oral, injection, implant, and pessary routes, and by management of animals such as weaning programs and nutrition.

Objective of estrus synchronization is to manipulate the reproductive process so that females may be bred during a "short" predefined interval with normal fertility. Theoretically, estrus synchronization may be attained through inhibition of ovulation, induction of ovulation, or induction of or delay of corpus luteum regression. However, to attain practical usefulness, an estrus synchronizer must be effective, simple to administer, relatively foolproof, have acceptable cost/benefit ratio, and have no harmful effects. Some of the more practical methods of estrus synchronization that have been studied extensively are presented in Table 23–3.

Sow. Progesterone and progestogens have been shown to be successful for estrus grouping in cattle and sheep. However, neither daily injections of progesterone nor oral administration of various progestogens have been effective in controlling the estrous cycle of swine; this has been due to unpredictable occurrence of cystic follicle conditions. A nonsteroidal compound (ICI 33828, MATCH, methallibure or AIMAX) effectively synchronized estrus of gilts (Polge et al., 1968). However, this compound is no longer available for use.

Studies utilizing ICI 33828 feeding for 20 days in conjunction with PMS injected on day 20 followed by HCG

Table 23-3. Some Practical Methods of Estrus Synchronization
That Have Been Studied Extensively

Species	Drug	Method	Percent Synchronized	Notes
Sow	ICI 33828	Daily feeding	77	Fertility acceptable; no longer available
	ICI + PMS + HCG		100	
	Weaning + PMS + HCG		Fairly good	Available, practical, but variable results
Ewe	Progestogens	Intravaginal sponge	60–80	Fertility acceptable, available, practical
Cow	Progestogens	Daily feeding	70–90	Reduced fertility; none available
Mare	None demonstrated practically effective			

injected four days after the PMS established that precise control of estrus and ovulation could be attained in swine. In addition, insemination of pigs at a predetermined time with this control program resulted in high fertility and normal embryonic survival (Polge et al., 1968). Thus, even though this specific combination of compounds is unavailable at this time, the concept of estrus and ovulation control has been established. The task now is to define an acceptable compound or group of compounds to use commercially.

Other methods of estrus synchronization have been successfully employed. Weaning of pigs has been shown to result in some degree of estrus synchronization. Injection of PMS at time of weaning followed by HCG injected four days later resulted in estrus synchronization. For nonlactating confined gilts, a change in environment in conjunction with a PMS injection followed four days later with an HCG injection has resulted in some estrus synchronization. Degree of effectiveness appears related to prior estrus activity and age of gilts, i.e., gilts approaching puberty seem to respond better than postpuberal gilts. The obvious problem with this program is predicting when the gilts will respond.

Ewe. Daily injections of progesterone (Dutt et al., 1948), oral administration of progestogens (Lamond, 1964), and use of either implants (Dziuk et al., 1968) or intravaginal sponges (Robinson and Moore, 1967) containing a progestogen have all been reported to successfully synchronize estrus. However, the progestogen impregnated intravaginal sponge has received the greatest use. Sponges are usually placed in the vagina for 14 to 19 days. Ewes return to estrus during days 1 through 3, with the majority in estrus on day 2, after sponge removal. Fertility at this estrus has been equal to or somewhat lower than nontreated ewes.

Mare. No effective practical method of estrus synchronization has been reported to date. Progesterone injected daily will block estrus and ovulation, but this method has reduced practical merit. The synthetic progestogens studied to date have been ineffective in altering estrous cycles. HCG has been injected at time of estrus and does appear to result in a more predictable time of ovulation in

Table 23-4. Estrus Synchronization of Cattle
With Orally Effective Progestogens

Progestogen	Percent Synchronized	Percent Pregnant To	
		Synchronized Service	Subsequent Service
180 mg MAP	80	37	68
10 mg CAP	88	36	68
1 mg MGA	79	38	68

Table 23-5. Effect of Subcutaneous or Intramuscular PGF$_{2}\alpha$ on Estrous Cycle Length in Cattle

	Day After Estrus PGF$_{2}\alpha$-Tham Injected		
	2–4	6-9	13–16
Injection to estrus interval (days)			
Range	15–19	2–4	3–4
Mean	17	3	3
Percent heifers in heat	100	100	100
Percent heifers forming CL	100	100	100
Posttreatment estrous cycle length (days)	21	21	20

(Adapted from Lauderdale, 1972. *J. Anim. Sci. 35*, 246.)

individual mares. Some evidence suggests that responsiveness to HCG lessens at the second or third consecutive cycles if injected during each estrous period.

Cow. The greatest number of publications on estrus synchronization has been related to the cow (Lamond, 1964; Hansel, 1967). Injected progesterone, oral progestogens and implants, and vaginal sponges containing a progestogen have all been reported to synchronize estrus. The oral and implant routes of administration have received the greatest emphasis.

Effective estrus synchronization can be attained with each of the progestogens and routes of administration studied. Conception rate of cows inseminated at first synchronized estrus is normally somewhat lower than that for untreated herdmate controls. However, conception rate of second synchronized estrus, occurring about one normal estrous cycle later, has usually been equal to or somewhat greater than untreated herdmate controls. Administration of a progestogen for 9 to 12 days in conjunction with an injection of estrogen has been effective for estrus synchronization (Wiltbank and Kasson, 1968). Two oral progestogens have been marketed and one additional one extensively investigated for use in cattle, but none are currently available for commercial use (Table 23-4).

One exciting new method for estrus synchronization of cattle, horses, sheep, and swine is by use of compounds which cause regression of the corpus luteum. Prostaglandins have been shown to be luteolytic in ewes, cows and mares. Preliminary information available suggests prostaglandins may play a valuable role in the future for controlling time of estrus and time of breeding in farm animals (Table 23–5). However, there are wide gaps in our knowledge which must be filled before prostaglandins will be available for commercial livestock use.

REFERENCES

Dutt, R. H. and Casida, L. E. (1948). Alteration of the estrual cycle in sheep by use of progesterone and its effect upon subsequent ovulation and fertility. *Endocrinology 43*, 208–217.

Dziuk, P. J., Cook, B., Niswender, G. D., Kaltenbach, C. C. and Doane, B. B. (1968). Inhibition and control of estrus and ovulation in ewes with a subcutaneous implant of silicone rubber impregnated with a progestogen. *Am. J. Vet. Res. 29*, 2415–2417.

Hansel, W. (1967). Control of the ovarian cycle in cattle. A Review. *Aust Vet. J. 43*, 441–449.

Lamond, D. R. (1964). Synchronization of ovarian cycles in sheep and cattle. *Anim. Breed. Abstr. 32*, 269–285.

Polge, C., Day, B. N. and Groves, T. W. (1968). Synchronization of ovulation and artificial insemination in pigs. *Vet. Rec. 83*, 136–142.

Robinson, T. J. and Moore, N. W. (1967). "The Evaluation of Progesterone and SC-9880-Impregnated Intravaginal Sponges for the Synchronization of Oestrus for Large Scale Artificial Insemination of Merino Ewes in Spring." In *The Control of the Ovarian Cycle in the Sheep.* T. J. Robinson (ed.), Sydney, Sydney University Press, pp. 116–132.

Wiltbank, J. N. and Kasson, C. W. (1968). Synchronization of cattle with an oral progestational agent and an injection of an estrogen. *J. Anim. Sci. 27*, 113–116.

B. Pregnancy Diagnosis

R. ZEMJANIS

The objective of pregnancy examinations is not so much to determine those animals that have conceived as to detect the animals which have failed to do so. Nonreturns to estrus are used to measure conception rates. This measure is as accurate and reliable as the individuals responsible for observation. Often it is highly erroneous (Zemjanis, 1970b).

I. PREGNANCY DIAGNOSIS IN CATTLE

A. Pregnancy Examination by Rectal Palpation

Rectal palpation is the method of choice for clinical pregnancy diagnosis. The diagnosis is reliable and accurate as well as rapid and immediate. The ideal pregnancy examination method should detect those animals which have failed to conceive before the first estrous period following service. Rectal palpation cannot yield definite and final results at this time. If applied at a certain time of the cycle it does provide circumstantial evidence indicative of either a failure or success of the service.

1. RECTAL PALPATION 19–22 DAYS AFTER SERVICE

Findings at this time are highly suggestive. There are three possibilities to consider: (a) The uterus is quiescent and one of the ovaries contains a fully developed corpus luteum. This information suggests that the animal has probably conceived. (b) The uterus is turgid with high tone. One of the ovaries contains a follicle exceeding 15 mm in diameter. There are no palpable corpora lutea present. The diagnosis is that the animal

has not conceived and that it is in heat and should be serviced. (c) The uterus is turgid and edematous. The only palpable structure in the ovaries is a soft area with rough edges suggesting the presence of ovulation depression. The animal is nonpregnant and it has just been in heat. The diagnoses made at this time have an accuracy of 85–90%. Errors are primarily due to individual variations in cycle length. The operator should have the palpation skills needed for retraction of the uterus into the pelvic cavity and for recognition of the described uterine and ovarian changes.

2. RECTAL PALPATION BETWEEN 35 AND 40 DAYS

While the information obtained at 19–22 days is only suggestive, a definite pregnancy diagnosis can be made before the second expected estrus following service with 100% accuracy. The technique involves the following steps:

(a) Retraction of the uterus into the pelvic cavity. Uterine retraction is the key to accurate and safe early pregnancy diagnosis because gentle and thorough exploration of both horns is impossible while the uterus is in its natural semi-abdominal position. There are indeed very few animals in which the uterus is found entirely within the pelvic cavity. Retraction technique has been described (Zemjanis, 1970a).

(b) Exploration of both horns for the following general changes accompanying pregnancy:

Asymmetry of Horns. Asymmetry in pregnancy is caused by fetal fluids accumulated primarily in the pregnant horn and not by difference in wall thickness between both horns.

Fluctuation Within the Cavity of the Larger Horn. Upon intermittent digital pressure the entrapped fluid gives a typical sensation of resilience. Fluctuation is most obvious in the widest part of the pregnant horn where the wall of the uterus is considerably thinner.

(c) Exploration of uterine horns for the "positive signs" of pregnancy: while asymmetry and presence of fluid are found in all pregnant uteri, these signs are by no means limited to pregnancies. Involuting postparturient uterus and pyometra are also associated with asymmetry and presence of fluid. Therefore pregnancy diagnosis must be based on the presence or absence of signs found exclusively in pregnant uteri. They are the following:

"Fetal Membranes Slip." This involves feeling of a fold of the chorioallantoic membrane as a typical "blip." It is felt best by surrounding the pregnant horn with thumb and index finger and letting the uterine contents pass between the gently compressed fingers. The fetal membrane slip is first felt 33–34 days of pregnancy. At that time the membranes are very thin except for a connective tissue cord along the mesometrial aspect of the chorioallantoic sac. Therefore it is essential that the entire horn is surrounded and explored by gentle manipulations. Rough, hard pressure will not only prevent feeling of the fine membranes but might also even lead to trauma.

Amnionic Vesicle. The amnionic vesicle in early pregnancies is a turgid kidney-shaped structure which because of the long umbilical stalk appears to float freely within the allantoic cavity. The stalk is attached to the ventral aspect of the chorioallantoic sac at the level of or just anterior to the intercornual ligament. It can be felt as early as 30 days following breeding, when it is less than a kidney bean in size ($\frac{3}{4} \times \frac{1}{2}$ cm). At 35 days it resembles an olive in size

($1\frac{1}{2} \times \frac{3}{4}$ cm). It reaches the size of a plum ($4 \times 2\frac{1}{2}$ cm) at 40 days of pregnancy.

Since establishing of the presence or the absence of fetal membranes provides adequate and reliable basis for diagnosis there is no need to explore the pregnant horn for the amnionic vesicle. It should be palpated only in cases of suspected fetal resorptions.

3. RECTAL PALPATION OF ANIMALS 40 AND MORE DAYS AFTER BREEDING

The rapid growth of the uterus results in displacement which becomes more and more apparent as the pregnancy advances. By 60 days the uterus is entirely abdominal and cannot be retracted without extra effort. The descent is generally completed by the end of the fifth or sixth month of gestation. Further expansion is first lateral and horizontal until the seventh month when ascent begins. Hypertrophy of the middle uterine artery and fremitus begins to become apparent at 70 and 90 days of gestation respectively.

"Fetal Membrane Slip." The "fetal membrane slip" becomes more and more readily palpable as pregnancy advances. Uteri of the size up to 60-day pregnancies should be retracted for differential diagnosis.

"Cotyledons." From approximately 65 days of gestation it is possible to palpate circumscribed thickened areas in the uterine wall and the chorioallantoic membrane. They represent the joined maternal caruncles and fetal cotyledons. Cotyledons become apparent first in the pregnant horn. Since the base of the pregnant horn is accessible without retraction it should be explored for the presence or absence of cotyledons. The size of cotyledons at the base of the pregnant horn is useful in estimating the stage of pregnancy as shown in Table 23–6.

Fetus and Fetal Parts. They are palpated by ballottement of the extended uterus and feeling the rebound of solid

Table 23-6. Size of Cotyledons
at Base of Pregnant Horn

Stage of Pregnancy (Days)	Length and Width of Cotyledons (cm)
70	0.75 × 0.5
80	1.0 × 0.5
90	1.5 × 1.0
100	2.0 × 1.25
120	2.5 × 1.5
150	3.0 × 2.0
180	4.0 × 2.5
210	5.0 × 3.0
240	6.0 × 4.0
270	8.0 × 5.0

fetal structures. Other abdominal organs should not be mistaken for the uterus and its contents.

B. Other Methods of Pregnancy Diagnosis

Efforts have been made to develop biologic and chemical methods of pregnancy diagnosis. These have included hormone assays and determination of chemical and physical characteristics of vaginal and cervical secretions. So far these efforts have been fruitless and rectal palpation is still the method of choice.

II. PREGNANCY DIAGNOSIS IN SHEEP

Diagnosis by rectal palpation is physically impossible and direct abdominal ballottement yields reliable results too late in gestation to have economic value.

Vaginal biopsy is one of the few reasonably reliable methods with an accuracy of 97% in animals pregnant over 40 days (Richardson, 1972). Biopsies are obtained surgically and processed for histologic evaluation. Nonpregnant ewes have a stratified vaginal epithelium consisting of 10–12 layers of cells. The superficial cells are squamous, while the cells of the deeper layers are polygonal with bright-staining nuclei. The vaginal epithelium of the pregnant ewes is lower

and has fewer layers of cells. These are columnar and cuboidal rather than polygonal and squamous.

A. Ultrasonic Techniques

Ultrasonic techniques employed in human obstetrics have been adapted for use in pregnancy diagnosis in sheep (Lindahl, 1969; Wilson and Newton, 1969). The use of the amplitude depth ultrasonic probe and the ultrasonic fetal pulse detector has been reviewed by Richardson (1972). Ewes are restrained in a standing position or placed on their haunches. Wool is shorn from the lower abdomen. Water soluble jelly is placed on the skin to serve as a transmitting medium. The ultrasonic probe is placed on the abdomen approximately 8–10 cm anterior to the udder. Scanning is performed by moving the probe in a longitudinal direction from one flank to the other. Detection of fetal heart beat and the swishing of the umbilical vessels serves as a basis for positive diagnosis. The method is accurate in animals at least 100 days pregnant. The method has little if any value for detection of multiple pregnancies. Direct palpation of the uterus via laparotomy gives 94% accuracy in ewes five weeks pregnant (Lamond, 1964). This method is a reasonably accurate means of early pregnancy diagnosis. It is questionable, however, if any of the above methods are practical and feasible in commercial flocks.

Viscosity and arborization pattern of cervical mucus, cytology of vaginal smear, udder development, sacral reflex, anticolostrum sensitivity test, certain biochemical tests and others are generally unreliable.

Attempts to diagnose pregnancy by determination of hormone levels by chemical methods and bioassay have been discouraging. It is possible, however, that the recent rapid advances in the methodology of hormone assays may yield a practical and reliable method of early pregnancy diagnosis in sheep.

III. PREGNANCY DIAGNOSIS IN SWINE

Of the diagnostic procedures studied, only vaginal biopsy and ultrasonic probing are of reasonable practical applicability.

A. Vaginal Biopsy

This technique is based on the morphologic differences in the appearance of vaginal epithelium in pregnant and nonpregnant swine (Busch, 1963). The epithelium of the pregnant animals has no more than three layers of cells with condensed chromatin in their nuclei. The cells are arranged in rows parallel to the basement membrane. The epithelium has a regular luminal border. The nonpregnant sows have epithelium which consists of several layers of cells. Their proximal border is irregular and often invaginated. The sample is obtained with a biopsy instrument, either a biopsy punch or of any other design not less than 24 cm long. Restraint is not necessary when penned sows are sampled. Care should be taken that the region sampled represents the anterior vagina. The method is reliable in 95% of sows not less than 21 days pregnant. Reproduction records are required. The method involves sampling, processing of biopsies and morphologic examination. Therefore it is time-consuming and costly. This restricts its practical applicability.

B. Ultrasonic Pregnancy Diagnosis

Pregnancy diagnosis of 95% accuracy is obtained by the use of a particular instrument* in sows and gilts bred 30 or more days. This method is convenient to apply and yields immediate results before the second expected estrus following service.

* "ECHO-TRACE VII Ultrasonic Analyzer," METRIX, Incorp., 11122 East 47th Avenue, Denver, Colorado 80239. Distributor: P. J. Dziuk, Department of Animal Science, University of Illinois, Urbana, Illinois 61801.

C. Other Methods

Rectal palpation is found to be impractical and unreliable (Huchzermeyer and Plonait, 1960). Velle (1960) and Lunaas (1962) found high levels of estrogens in the urine of sows between the 24th and 32nd days of gestation. Estrogen tests—Cuboni's or others—are not applicable for clinical diagnosis.

IV. PREGNANCY DIAGNOSIS IN HORSES

The mare is one of the females of domestic animals in which pregnancy can be clinically diagnosed by several methods.

A. Rectal Palpation

Rectal palpation is possible in all conventional sized mares as well as in certain larger Shetland ponies. Examination of horse by rectal palpation requires restraint which must be applied in all cases. The method of restraint and its severity is determined by each individual mare's personality and behavior. Because of the limited time for conception allowed by the relatively short breeding season an early pregnancy diagnosis is even more indicated than in the cow.

1. RECTAL PALPATION AT 18–20 DAYS AFTER BREEDING

Rectal palpation yields suggestive information rather than definite final evidence for conception or nonconception. The cervix receives primary attention because the uterus and the ovaries of the mare do not exhibit changes indicative of stages of estrous cycle. The cervix of a mare which has conceived is contracted and firm. The nonpregnant mare which at the time should be returning in heat has a relaxed and rather soft cervix.

2. 30 TO 45 DAYS OF GESTATION

This is the period when pregnancy is easiest to detect. The uterus and its contents can be explored in their entirety. Pregnancy is recognized as a circumscribed spherical enlargement of the uterine horn (Fig. 23–2). The sphere represents the chorioallantoic sac which is distended so that its contents cannot be palpated. The enlargement is generally found as a ventral bulge in the lower third of the uterine horn. Most pregnancies are found in the right horn. In early gestation and primarily during the fifth week the uterus is often found to have a high degree of tone. It may respond to palpation by strong contraction above and below the enlargement making the latter more obvious. The bulge, which at 30 days is approximately 2–3 cm in diameter, grows to the size of an orange at 45 days of gestation. The technique involves finding one of the

ovaries and following the broad ligament to the uterus. The anterior aspect is then explored from one horn to the other Both horns should be compared.

3. 45 AND MORE DAYS OF GESTATION

The chorioallantoic sac grows rapidly and descends into the body portion of the uterus attaining an oval rather than spherical shape. The uterus begins its descent and by 60 days of gestation it might be impossible to surround the entire bulge. As the uterus descends, the broad ligaments and the ovaries they suspend are displaced forward, downward and medially. By the fourth month the broad ligaments are felt as tense cords extending forward, downward and medially with the bulk of the uterus presenting a fluctuating dome between them. Fetus or fetal parts can be palpated by ballottement. It must be established that the organ felt through ballottement is the uterus. From the seventh month of gestation the uterus begins its ascent.

B. Hormone Assay

The mare is unique in that high levels of hormones appear in the body fluids during pregnancy.

1. GONADOTROPIN

PMSG appears in the blood as early as 40 days following conception. Peak levels are found between 50 and 120 days. Thereafter there is a gradual decline.

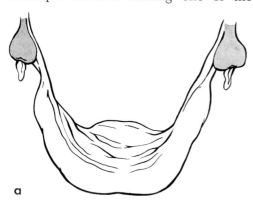

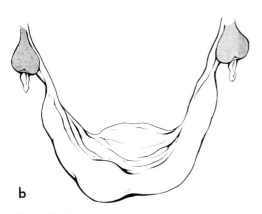

FIG. 23–2. *A*, Schematic drawing of a non-pregnant equine uterus. Anterior view. *B*, Schematic drawing of a 35-day pregnant equine uterus. Anterior view. Note distinct bulge at the base of the right horn.

Immunologic Tests. Immunologic diagnosis of pregnancy in mares (Wide and Wide, 1963) is based on the principle that pregnant mare serum gonadotropin (PMSG) when present in the blood to be tested prevents agglutination of sensitized red cells by anti-PMSG. In the United States a test kit can be obtained* providing a natural agglutinin inhibitor, rabbit anti-PMSG antibodies, neutralizer

* Obtainable as MIP-Test from Denver Chemical Manufacturing Company, Stamford, Connecticut 06904.

Table 23-7. Methods of Pregnancy Diagnosis in Farm Animals

Species	Diagnostic Method	Sample	Procedure	Time in Gestation When Applicable (Days)
Cattle	Rectal palpation	—	Tactile	30–35 to term
Sheep	Vaginal biopsy	Vaginal mucosa	Histologic	40 to term
	Ultrasonic	—	—	70 to term
Swine	Vaginal biopsy	Vaginal mucosa	Histologic	21 to term
	Ultrasonic	—	—	25–70
	Chemical	Urine	Estrogen assay	24–32
Horse	Rectal palpation	—	Tactile	30 to term
	Immunologic	Serum	Hemagglutination inhibition	50–100
	Chemical	Urine	—	120–250

for natural agglutinin inhibitor and sensitized sheep red blood cells. Four drops of serum are mixed with a drop of inhibitor. Two minutes later one drop of neutralizer is added to the mixture. Rabbit anti-PMSG antibodies are added and the reaction is read after two hours. The diagnosis is positive when a definite ring is formed by the red blood cells in the bottom of the tube. Diagnosis is negative when the red blood cells clump together in the bottom of the tube. An immunologic gel diffusion test also based on the presence of PMSG in the tested blood has been described by Wormstrand (1969). The immunologic test is most accurate between 50 and 100 days of gestation. Errors are mostly due to individual variation in PMSG levels.

Aschheim-Zondek Test. This test is a bioassay for the FSH activity of the PMSG. Immature rats are injected with whole blood and serum. The rats are killed 72 hours later. Hemorrhagic ovulation spots and edematous uterus are evidence for pregnancy. This test is most accurate between 50 and 80 days of gestation.

2. ESTROGEN

Placental estrogens appear in the urine of mares from approximately 120 days of pregnancy. From 250 until 290 days of gestation the level of estrogen declines. A chemical test for estrogen was introduced by Cuboni as early as in 1934. Lyngset (1965) compared Cuboni's test with Lunaas' test and found that both are equally reliable. According to Lunaas 10 ml of distilled water are added to 1 ml of test urine. The diluted urine is mixed with 15 ml of concentrated sulfuric acid. The mixture is examined for fluorescence. The test is reasonably accurate between 120 and 150 days. It is, however, most accurate after 150 days of gestation. Because of this its practical value is negligible.

REFERENCES

Busch, W. (1963). Beitrag zur Histologischen Diagnose der Trachtigkeit beim Schwein durch Vaginal Biopsie. *Monadsheft für Vet. Med. 18*, 813–817.

Huchzermeyer, F. and Plonait, H. (1960). Trachtigkeitdiagnose und Rectaluntersuchung beim Schwein. *Tierärztl. Umschau 15*, 339–401.

Lamond, D. R. (1964). Diagnosis of early pregnancy in the ewe. *Aust. Vet. J. 39*, 192–195.

Lindahl, I. L. (1969). Comparison of ultrasonic techniques for the detection of pregnancy in cows. *J. Reprod. Fert. 18*, 117–120.

Lunaas, T. (1962). Urinary estrogen levels in the sow during oestrous cycle and early pregnancy. *J. Reprod. Fert. 4*, 13–20.

Lyngset, O. (1965). Pregnancy diagnosis in the mare. A comparison between the chemical method of Cuboni and Lunaas. *Vet. Rec. 77*, 218–219.

Richardson, C. (1972). Pregnancy diagnosis in the ewe: A review. *Vet. Rec. 90*, 264–275.

Velle, W. (1960). Early pregnancy diagnosis in the sow. *Vet. Rec. 72*, 116–118.

Wide, M. and Wide, L. (1963). Diagnosis of pregnancy in mares by an immunological method. *Nature* (Lond.) *198*, 1017–1018.

Wilson, I. A. N. and Newton, J. E. (1969). Pregnancy diagnosis in the ewe. A method for the use on the farm. *Vet. Rec. 84*, 356–358.

Wormstrand, A. (1969). Immunological pregnancy diagnosis in the mare. *J. Amer. Vet. Med. Assn. 155*, 42–45.

Zemjanis, R. (1970a). *Diagnostic and Therapeutic Techniques in Animal Reproduction.* 2nd ed. Baltimore, Williams & Wilkins.

Zemjanis, R. (1970b). *Functional Infertility in the Cow.* Proceedings of the VI International Conference on Cattle Diseases. Anim. Assoc. Bov. Pract., Stillwater, Oklahoma, Heritage Press, pp. 215–221.

C. Egg Transfer

M. R. JAINUDEEN AND E. S. E. HAFEZ

Since Heape first succeeded in transplanting rabbit embryos in 1890, eggs have also been successfully transferred in cattle, sheep, goats, pigs and other laboratory animals. Egg transfer involves superovulation of the donor, egg recovery, manipulation of eggs, synchronization of ovulation of both donor and recipient, and technique of transfer (Fig. 23–3).

I. SUPEROVULATION

A primary requirement in egg transfer is a supply of fertilized eggs from genetically superior donors. Superovulation or an increased number of ovulations can be induced by injecting FSH, usually pregnant mare's serum gonadotropin (PMSG) during the follicular phase of the estrous cycle (Table 23–8). Ovulations occur spontaneously or after an in-

Table 23-8. A Schedule for Superovulation

Animal	PMSG Injection Dose (i.u.)	PMSG Injection Day of Cycle	HCG* (i.u.)
Cow	2000–3000	16	2000
Calf	1000–2000	—	3000
Goat	1000–1500	17	1000
Ewe	600–1000	12	—
Sow	750–1500	15	500

* Administered intravenously at onset of estrus or in the case of the calf five days after PMSG injection.

jection of LH early in estrus. Females are bred or artificially inseminated to provide fertilized eggs for transfer.

Among the problems associated with superovulation in cattle are individual variability in response, low ovulation and fertilization rates, and accelerated egg transport through the oviduct. A serious limitation is the decrease in the number of eggs obtained from successive superovulations in the same donor due to the formation of antihormones. In addition, the superovulatory response in calves has been variable and fertilization rates have also been very low.

II. COLLECTION OF EGGS

Eggs are usually collected around the time they enter the uterus. This is necessary because recovered eggs should be at a stage of development capable of surviving after transfer to the uterus of the recipient. Eggs enter the uterus faster in the sow (two days) than in the cow and ewe (three days) or in the mare (four to six days). They can be collected from the reproductive tract

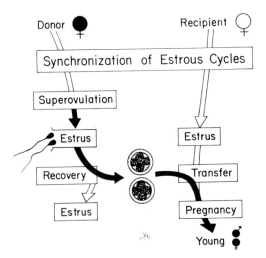

FIG. 23–3. Schematic representation of the steps involved in egg transfer in cattle, sheep and swine.

of the donor animal at slaughter (in vitro) or from the living animal (in vivo). The latter method is preferred because it permits repeated collections from valuable donors. Either surgical or nonsurgical techniques may be used.

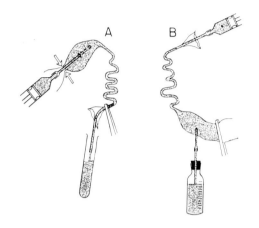

FIG. 23–4. Methods used for surgical recovery of eggs. *A*, Cattle and sheep; fluid is flushed from the uterine end toward the infundibulum; arrows indicate compression of uterine lumen to prevent backflow of fluid. *B*, Swine; fluid is flushed from the ovarian end toward the isthmus.

In the surgical method, after the donors are anesthetized and placed on their backs, the uterine horns and oviducts are exposed through a midventral laparotomy. The recovery procedure varies with the species (Fig. 23–4). In the cow and ewe, uterine and oviductal eggs are recovered by flushing each uterine horn toward the ovarian end, three to six days after ovulation. The flushings are collected by inserting a polyethylene tube into the fimbriated end of the oviduct. In the sow or mare the uterotubal junction (UTJ) offers considerable resistance to the flow of fluid in the direction of the oviduct; therefore, the oviduct is flushed toward the uterus.

With the surgical technique there is a high incidence of adhesions which interfere with ovum pick up and/or transport through the oviduct, and limits repeated collections of eggs. In cattle the nonsurgical technique of egg recovery through the cervix consists of a two-way flow system, one for admitting fluid into the uterus and the other for draining off the fluid (Fig. 23–5). The escape of any

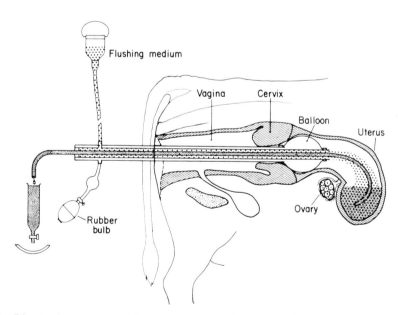

FIG. 23–5. Diagrammatic representation of a nonsurgical technique for egg recovery from the bovine uterus. After the cannula is passed through the cervix, the balloon is inflated to prevent leakage of fluid. The flushing fluid is allowed to gravitate into the uterus and the flushings are collected via the inner tube which has been previously guided into one uterine horn.

uterine flushings through the cervix is prevented by inflating a balloon fitted at the tip of the instrument.

The different media used for flushing, storing and transferring eggs are:

Sheep. Homologous blood serum alone or with an equal volume of sterile 0.9% saline. The serum is heated at 55° C for 30 minutes to destroy a thermolabile ovicidal factor.

Cattle and Swine. Commercially available tissue culture medium 199. Bicarbonate is added to this culture medium in the swine.

Addition of penicillin (1000 units/ml) and streptomycin (500 to 1000 μgm/ml) to the above media is recommended to reduce transmission of infections at time of transfer to the reproductive tract of the recipient.

III. STORAGE OF EGGS

In the interval between recovery and transfer, eggs should be held at room temperature or at body temperature (37° C) on a warm plate or in an incubator. The storage of eggs up to 10 or 20 hours in this manner does not adversely affect their survival.

A. Preservation in Vitro

The methods for the storage of mammalian embryos are described elsewhere (Hafez, 1970). Low temperature and the tissue culture technique are used for egg storage. Mammalian eggs stored below 20° C, do not undergo further development. At 10° C sheep eggs in autologous serum retain their viability for approximately 72 hours. Preservation of eggs by freezing has been unsuccessful. Only rabbit and mouse embryos can be cultured in vitro. In sheep, eggs in the early cleavage stages are difficult to culture, but some success has been achieved in culturing morula to the blastocyst stage. Bovine eggs do not develop satisfactorily in culture.

B. Preservation in Vivo

Eggs transferred between species survive for limited periods only. In the oviduct of a pseudo-pregnant or estrous rabbit, cattle and sheep eggs are viable for three to five days (Lawson et al., 1972a, b) and pig eggs for two days (Polge et al., 1972). To increase the chances of recovery of transferred eggs the oviduct of the rabbit is ligated at the uterotubal junction. The rabbit can serve as a temporary incubator to transport eggs over long distances. For example, sheep eggs that were transported in the rabbit from England developed into lambs when retransferred into the uteri of synchronized sheep in South Africa (Plate 25).

IV. EGG TRANSFER

A. Synchronization of Donor and Recipient

For successful transfers, synchrony between the egg and its environment, as related to the age of the corpora lutea is important. In cattle, sheep and swine, the estrous cycle of the recipient has to be synchronized to within ± 2 days of the donor's cycle. The reasons for pregnancy losses in asynchronous transfers are not clear but could be due to an adverse uterine environment or to an inability of the embryo to exert a luteotropic action on the corpus luteum of the recipient.

Synchronization can be achieved by one of three methods: (a) selecting a recipient which was in estrus on the same day as the donor; (b) storing eggs in vitro or in vivo until a recipient in a similar stage of the cycle as the donor is available; or (c) regulating the estrous cycles of both donor and recipient by hormonal treatment.

B. Selection of Eggs

Only morphologically normal eggs are used and all eggs showing structural abnormalities should be discarded (Plate 26). Since 4- to 32-cell stage eggs tolerate

PLATE 25

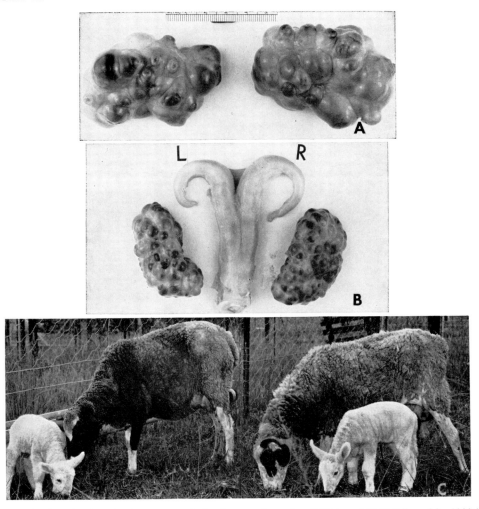

A, Ovaries from beef cattle after superovulation by gonadotropins, 3000 i.u. of PMS followed by 2000 i.u. of HCG five days later. Note the developing corpora lutea. The cow was slaughtered five days after the HCG injection.

B, Superovulated ovary from a three-month-old calf treated with 3000 i.u. of PMS followed by 300 i.u. of HCG five days later; 146 ovulation points were counted. Note the size of ovaries compared with the immature uterus.

C, Border Leicester lambs born in South Africa from embryos obtained in England, flown to South Africa in the uterus of a pseudopregnant rabbit and transferred to Dorper ewes (foster mothers) in South Africa (*Photograph by G. L. Hunter.*)

PLATE 26

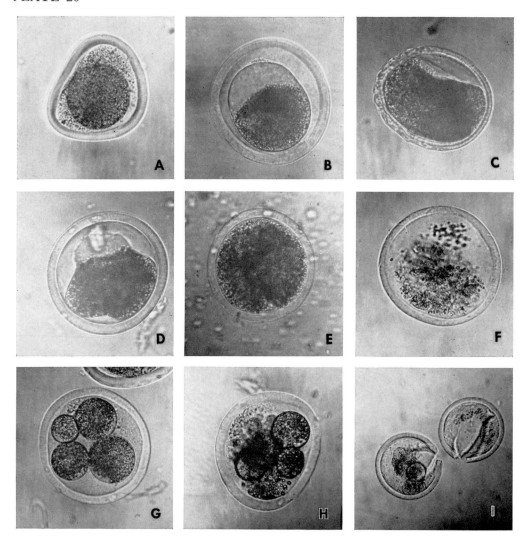

Atypical and degenerating cow eggs.

A, B, C, D, Atypical unfertilized eggs showing structural abnormalities (× 185);
E, Degenerating one-cell egg; note coarse granulation (× 185).
F, Advanced degree of degeneration (× 185).
G, H, Fragmenting eggs; note different sizes of fragments (× 185).
I, Two eggs with ruptured zona pellucida; note the escape of cytoplasm (× 63).

handling better and are more likely to survive the procedure, most transfers are performed at this stage.

The proportion of eggs which survive depends on the number transferred. The highest pregnancy rates in sheep, cattle and goats can be obtained with the transfer of a single egg into each uterine horn of the recipient. Approximately 10–12 eggs should be transferred into each uterine horn to obtain a normal-sized litter (10–12 piglets) in pigs, since only about one-half of the transferred eggs are represented by viable young at farrowing.

C. Transfer Techniques

The procedure to transfer eggs surgically is very similar for all species and involves exposure of the uterus by a midventral laparotomy under general anesthesia. The ovarian end of the uterine horn is gently picked up and is punctured with a cutting-edge surgical needle. Through this passage, a fine glass pipet containing the egg(s) is inserted and the egg(s) deposited into the uterine lumen. The volume of fluid used in egg transfer should be minimal and the deposition of large air bubbles should be avoided.

The nonsurgical transfer of eggs through the cervix (inovulation) can be performed in the cow because a pipet could easily be inserted through the cervix by rectal manipulation but successes have been limited. At least two reasons have been proposed to account for this failure. First, entry through the cervix results in severe uterine infection. This has been avoided by the addition of antibiotics to the storage and transfer media. Second, eggs are ejected through the cervix about $1\frac{1}{2}$ hours after transfer due to uterine contractions. Although these contractions are believed to be due to a reflex discharge of oxytocin during manipulation of the cervix, Rowson et al. (1972b) were unable to find significant blood levels of oxytocin following such stimulation. The premature expulsion of the egg has been overcome by a transfer technique which bypasses

the cervix (Sugie, 1965) or by distending the uterus with CO_2 immediately after the eggs are transferred through the cervix (Rowson and Moor, 1966). In the latter technique an artificial insemination pipet is passed through the cervix into one uterine horn. The egg for transfer is drawn into a length of polyethylene tubing which is connected by an adaptor to a 1 ml syringe. After the tubing is threaded through the pipet, the egg is injected directly into the uterine lumen, and carbon dioxide is passed through the same pipet to distend the uterus.

D. Conception Rates

During the last few years, a high rate of success has been achieved with the egg transfer technique. The pregnancy rates ranged from 75% for sheep (Moore, 1968) to 90% for cattle (Rowson et al., 1972a). In pigs, the pregnancy rate was almost 100% with 68% of the transferred embryos surviving in pregnant recipients (Pope et al., 1972). The higher than normal fertility rates obtained is presumably related to the selection of normal eggs for transfer and to the transfer of a maximal number of eggs compatible with intrauterine survival in the different species.

V. POTENTIAL USES AND LIMITATIONS

The transfer of embryos has both scientific and practical uses. It allows critical experimental approaches to problems in physiology and biochemistry of reproduction, genetics, cytology, animal breeding, immunology and evolution. For example, the technique can be used to study pre- and post-implantation development and to evaluate genetic-environmental interactions in relation to variability in the phenotype.

Genetic improvement via the egg has been greatly hampered by the inability to utilize the potential ova of a female. A single cow, for example, will produce one calf per year and seldom more than eight in a lifetime. In cattle, egg transfer could be utilized for genetic improvement, a change of breed in a single generation, to reduce the generation interval (transfer of calf eggs to adult cows) or facilitate progeny testing, and to obtain superior slaughter animals (beef calves) from poor yielding dairy cows. Transferring either a single fertilized egg into each uterine horn of an unmated cow or a second egg to the contralateral horn of a recipient cow which conceived a few days earlier (Rowson et al., 1971) is more effective than the superovulation technique for the production of twins in cattle.

In sheep, egg transfer has been employed to produce new types of sheep (stud poll Merino sheep), to increase litter size of certain breeds of sheep (Romney Marsh) and to obtain two lamb crops per year. In pig breeding the technique is of limited value. The transfer of supersized litters to unmated females or additional embryos to previously bred recipients does not appreciably increase litter size.

The ability of fertilized eggs to survive in the oviduct of the rabbit for a few days permits the transport of embryo over long distances and thus provides for a cheap method of exporting livestock. Already exports of sheep and pig embryos are now being made and seems certain for cattle embryos in the future.

There are three major limitations of developing egg transfer techniques to a practical scale comparable to artificial insemination. These are (a) no reliable methods for superovulation and production of fertilized eggs on a large scale;

(b) the difficulty of storing eggs in vitro in a dormant state; and (c) the lack of a simple nonsurgical technique for collecting embryos. Many of these problems are being resolved and egg transfer, just as artificial insemination, could play a significant role in animal reproduction.

REFERENCES

Hafez, E. S. E. (1970). "Egg Storage." In *Methods in Mammalian Embryology*. J. C. Daniels (ed.), San Francisco, Freeman & Co.

Lawson, R. A. S., Adams, C. E. and Rowson, L. E. A. (1972a). The development of sheep eggs in the rabbit oviduct and their viability after retransfer to ewes. *J. Reprod. Fert.* 29, 105–116.

Lawson, R. A. S., Rowson, L. E. A. and Adams, C. E. (1972b). The development of cow eggs in the rabbit oviduct and their viability after retransfer to heifers. *J. Reprod. Fert.* 28, 313–315.

Moore, N. W. (1968). The survival and development of fertilized eggs transferred between Border Leicester and Merino ewes. *Aust. J. Agric. Res.* 19, 295–302.

Polge, C., Adams, C. E. and Baker, R. D. (1972). Development and survival of pig embryos in the rabbit oviduct. *Proc. 7th Internat. Cong. Anim. Reprod. Artif. Insem.* Munich 4, 60.

Pope, C. E., Christenson, R. K., Zimmerman-Pope, V. A. and Day, B. N. (1972). Effect of numbers of embryos on embryonic survival in recipient gilts. *J. Anim. Sci.* 35, 805–808.

Rowson, L. E. A., Lawson, R. A. S. and Moor, R. M. (1971). Production of twins in cattle by egg transfer. *J. Reprod. Fert.* 25, 261–268.

Rowson, L. E. A., Lawson, R. A. S., Moor, R. M. and Baker, A. A. (1972a). Egg transfer in the cow: Synchronization requirements. *J. Reprod. Fert.* 28, 427–431.

Rowson, L. E. A., McNeilly, A. S. and O'Brien, C. A. (1972b). The effect of vaginal stimulation on oxytocin release during the luteal phase of the cow's oestrous cycle. *J. Reprod. Fert.* 30, 287–288.

Rowson, L. E. A. and Moor, R. M. (1966). Nonsurgical transfer of cow eggs. *J. Reprod. Fert.* 11, 311–312.

Sugie, T. (1965). Successful transfer of a fertilized bovine egg by nonsurgical techniques. *J. Reprod. Fert.* 10, 197–201.

Appendix I

Chromosome Numbers of Bovinae, Equinae and Caprinae Species

Common Name	Scientific Name	Chromosome Number (2N)	Fundamental Number
Domestic cattle	*Bos taurus*	60	62
Banteng	*Bos banteng*	60	62
Zebu	*Bos indicus*	60	62
Yak	*Bos grunniens*	60	62
European bison	*Bison bonasus*	60	62
American bison	*Bison bison*	60	62
Gaur	*Bos gaurus*	58	62
Nyala	*Tragelaphus angasi*	55	58
Congo buffalo	*Syncerus caffer nanus*	54	60
African buffalo	*Syncerus caffer caffer*	52	60
Asiatic buffalo	*Bubalus bubalis*	48	58
Anoa	*Anoa depressicornis*	48	60
Nilgai	*Boselaphus tragocamelus*	46	60
Four-horned antelope	*Tetracerus quadricornis*	38	38
Sitatunga	*Tragelaphus spekei*	30	58
Mongolian wild horse	*Equus przewalskii*	66	94
Domestic horse	*Equus caballus*	64	94
Donkey	*Equus asinus*	62	104
Nubian ass	*Equus asinus africans*	62	104
Mongolian wild ass	*Equus hemionus*	56	104
Tibetan wild ass	*Equus kiang*	56	104
Persian wild ass	*Equus onager*	56	104
Grevy zebra	*Equus grevyi*	46	78
African zebra	*Equus burchelli*	44	82
Grant zebra	*Equus burchelli boehmi*	44	82
Mountain zebra	*Equus zebra*	34 (?)	60
Domestic goat	*Capra hircus*	60	60
Ibex	*Capra ibex*	60	60
Markhor	*Capra falconeri*	60	60
Saiga antelope	*Saiga tatarica*	60	60
Aoudad	*Ammotragus lervia*	58	60
Afghanistan sheep	*Ovis ammon cycloceros*	58	60
Kara-Tau sheep	*Ovis ammon nigimontana*	56	60
Domestic sheep	*Ovis aries*	54	60
Mouflon	*Ovis musimon*	54	60
Red sheep	*Ovis orientalis*	54	60 (?)
Bighorn sheep	*Ovis canadensis*	54	60
Laristan sheep	*Ovis ammon laristanica*	54	60
Musk ox	*Ovibos moschatus*	48	60
Himalayan tahr	*Hemitragus jemlahias*	48	60
Rocky Mountain goat	*Oreamnos americanus*	42	60

Appendix II

Chromosome Numbers and Reproductive Ability in Equine, Bovine and Caprine Hybrids

Species and Chromosome Number (2N) Sire	Dam	Hybrids Chromosome Number (2N)	Reproductive Ability
Mongolian wild horse, 66 (*E. przewalskii*)	Domestic horse, 64 (*E. caballus*)	65	Fertile (?)
Donkey, 62 (*E. asinus*)	Domestic horse, 64 (*E. caballus*)	63 (Mule)	Sterile
Domestic horse, 64 (*E. caballus*)	Donkey, 62 (*E. asinus*)	63 (Hinny)	Males are sterile, females are fertile, only in very exceptional cases
Nubian ass, 62 (*E. asinus africanus*)	Donkey, 62 (*E. asinus*)	62	Fertile
Mongolian wild ass, 56 (*E. hemionus*)	Donkey, 62 (*E. asinus*)	59	Fertile (?)
Grevy zebra, 46 (*E. grevyi*)	Domestic horse, 64 (*E. caballus*)	55 (Zebroid)	Sterile
African zebra, 44 (*E. burchelli*)	Donkey, 62 (*E. asinus*)	53 (Zebronkey)	Sterile
Donkey, 62 (*E. asinus*)	Mountain zebra, 34 (?) (*E. zebra*)	48	Sterile
American bison, 60 (*Bison bison*)	Zebu, 60 (*Bos indicus*)	60	Females are fertile
American bison, 60 (*Bison bison*)	Domestic cattle, 60 (*Bos taurus*)	60 (Cattalo)	Male F_1 are sterile
Domestic cattle, 60 (*Bos taurus*)	American bison, 60 (*Bison bison*)	60 (Cattalo)	Male F_1 are sterile
Domestic goat, 60 (*Capra hircus*)	Barbary sheep, 58 (*Ammotragus lorvia*)	59 (?)	Full-term fetuses, but no live hybrid
Domestic goat, 60 (*Capra hircus*)	Domestic sheep, 54 (*Ovis aries*)	57	Embryos are resorbed or aborted at six weeks pregnancy
Domestic sheep, 54 (*Ovis aries*)	Mouflon, 54 (*Ovis musimon*)	54	Fertile in both sexes
Bighorn sheep, 54 (*Ovis canadensis*)	Domestic sheep, 54 (*Ovis aries*)	54	Reduced fertility

Appendix III

Reproductive Diseases of Viral, Protozoan and Bacterial Origin

Disease	Species	Etiology	Diagnosis — Clinical	Diagnosis — Other	Control
Epizootic bovine abortion	Cattle	Unknown	Abortion in late pregnancy, lymphoid tissue lesions in fetus	Histopathology	None
Granular venereal disease	Cattle	Lymphoid reaction	Nodular vulvitis, balanitis	—	None
Infectious pustular vulvovaginitis (IPV)	Cattle	Bovine herpesvirus	Respiratory disease, abortion	Isolation of virus, serum neutralization test	Isolation, cessation of breeding, vaccination
Infectious bovine rhinotracheitis (IBR)	Cattle	Bovine herpesvirus	Respiratory disease, abortion	Isolation of virus, serum neutralization test	Vaccination
Bovine viral diarrhea (BVD)	Cattle	Virus	Abortion in early pregnancy, birth defects	Serum neutralization test	Vaccination
Catarrhal vaginitis	Cattle	Virus	Catarrhal vaginitis	Isolation of virus	None
Ulcerative dermatitis	Sheep	Virus	Ulceration of lips, legs, vulva and sheath	Lamb inoculation	Inspection of sale rams or rams purchased before breeding
Hog cholera	Swine	Virulent or modified live virus vaccination	Stillborn pigs, edematous dead pigs, weak pigs	History of pregnant sow, vaccination or infection	Quarantine and slaughter of infected herds
SMEDI	Swine	Porcine picornavirus	Stillbirth mummified fetus, embryonic death, infertility	Virus isolation	Allow exposure before breeding, maintain closed herd
African swine fever	Swine	Virus	Disease resembling hog cholera, abortion in pregnant sows	Exposure to warthogs or other infected swine	Quarantine and slaughter of infected herds
Equine rhinopneumonitis	Horses	Equine herpesvirus I	Abortion in late pregnancy, respiratory disease in young	Focal necrosis of liver and edema of lungs in fetus, inclusion bodies, isolation of virus	Vaccination
Equine viral arteritis	Horses	Virus	Respiratory infection, cellulitis, abortion	Isolation of virus	Isolation of infected
Coital vesicular exanthema	Horses	Virus	Pustules on vulva, vagina, sheath and penis	None	Isolation, cessation of breeding

Disease	Species	Etiology	Diagnosis		Control
			Clinical	Other	
Enzootic abortion	Sheep	Chlamydia	Abortion, fresh (not autolyzed)	Staining elementary bodies in placenta, complement-fixation test	Vaccination
Trichomoniasis	Cattle	*Trichomonas fetus*	Infertility, pyometra and abortion in cows	Examination for trichomonads	Breeding rest, artificial insemination
Toxoplasmosis	Sheep Swine	*Toxoplasma gondii*	Encephalitis, abortion	Histopathology, dye test for antibodies	Isolation
Listeriosis	Cattle Sheep	*Listeria monocytogenes*	Nervous signs, circling, abortion	Isolation of bacterium	Avoid stress and feeding silage
Vibriosis	Cattle	*Vibrio fetus* var. *venerealis*	Infertility	Mucus agglutination test, isolation, fluorescent antibody	Artificial insemination, vaccination
	Sheep	*Vibrio fetus* var. *intestinalis*	Abortion	Isolation of bacterium	Vaccination
Leptospirosis	Cattle	*Leptospira pomona* *Leptospira hardjo*	Hemolytic anemia, abortion in late pregnancy, agalactia	Agglutination test	Vaccination, elimination of carriers with antibiotic treatment
	Swine	*Leptospira pomona* *Leptospira grippotyphosa* *Leptospira canicola*	Abortion in late pregnancy, birth of weak pigs	Isolation of *Leptospira*	
	Horses	*Leptospira pomona* *Leptospira grippotyphosa* *Leptospira icterohemorrhagiae*	Abortion in late pregnancy, periodic ophthalmia		
	Sheep	*Leptospira pomona*	Hemolytic anemia, abortion in late pregnancy		
Brucellosis	Cattle	*Brucella abortus*	Abortion in late pregnancy, sterility in bulls	Isolation of bacterium, serum and milk agglutination tests	Vaccination, test and slaughter
	Swine	*Brucella suis*	Abortion, weak pigs, sterility in boars		
	Sheep Goat	*Brucella melitensis*	Abortion		
	Sheep	*Brucella ovis*	Epididymitis in rams, abortion		
	Dog	*Brucella canis*	Abortion		

(Afshar, 1965. *Vet. Bull. 35*, 165; Blood and Henderson, 1963, *Veterinary Medicine*, 2nd ed., Baltimore, Williams & Wilkins; Howarth, 1960, *Proc. U. S. Livestock Sanit. Assoc. 64*, 401.)

Appendix IV

I. GENERAL REFERENCES

The following is a list of references recommended for specialized students conducting research studies in reproductive physiology.

1. Textbooks and Reference Books

Annals of the New York Academy of Sciences (1959). *The Uterus*. Vol. 75. New York, published by the Academy, 2 East 63rd St.

Annals of the New York Academy of Sciences (1959). *The Vagina*. New York, published by the Academy, 2 East 63rd St.

Armstrong, C. N. and Marshall, A. J. (eds.) (1964). *Intersexuality in Vertebrates Including Man*. London, Academic Press.

Arthur, G. H. (1964). *Wright's Veterinary Obstetrics*. 3rd. ed. Baltimore, Williams & Wilkins.

Asdell, S. A. (1964). *Patterns of Mammalian Reproduction*. 2nd ed. Ithaca, Cornell University Press.

Assali, N. S. (ed.) (1968). *Biology of Gestation*. Vol. II, "The Fetus and Neonate." New York, Academic Press.

Assali, N. S. (ed.) (1972). *Pathophysiology of Gestation*. New York, Academic Press. Vol. 1, "Maternal Disorders," Vol. 2, "Fetal-Placental Disorders," Vol. 3, "Fetal and Neonatal Disorders."

Austin, C. R. (1961). *The Mammalian Egg, a Study of a Specialized Cell*. Oxford, Blackwell Scientific Publications.

Austin, C. R. and Perry, J. S. (eds.) (1965). *Agents Affecting Fertility*. Boston, Little, Brown & Company.

Austin, C. R. and Short, R. V. (eds.) (1972). *Reproduction in Mammals*. 5 volumes. London, Cambridge University Press.

Baccetti, B. (ed.) (1970). *Comparative Spermatology*. New York, Academic Press.

Barcroft, J. (1947). *Researches on Pre-Natal Life*. Springfield, Charles C Thomas.

Bargmann, W. (1967). *Histologie und Mekroskopische Anatomie des Menschen*. Stuttgart, Georg Thieme Verlag.

Beatty, R. A. (1957). *Parthenogenesis and Polyploidy in Mammalian Development*. New York, Cambridge University Press.

Behrman, S. J. and Kistner, R. W. (eds.) (1968). *Progress in Infertility*. Boston, Little, Brown & Company.

Bell, E. T. and Loraine, J. A. (eds.) (1967). *Recent Research on Gonadotropic Hormones*. Edinburgh, Livingstone.

Benesch, F. and Wright, J. G. (1951). *Veterinary Obstetrics*. Baltimore, Williams & Wilkins.

Benirschke, K. (ed.) (1967). *Comparative Aspects of Reproductive Failure*. New York, Springer-Verlag.

Bielanski, W. (1972). *Rozród Zwierzat, Bydlo Owce Konie Swinie*. Warszawa, Pánstwowe Wydaronictwo Rolnicze i Leśne.

Biggers, J. D. and Schuetz, A. W. (eds.) (1972). *Oogenesis*. Baltimore, University Park Press.

Bishop, M. W. H. (ed.) (1971). *Advances in Reproductive Physiology*. Vols. V and VI. London, Academic Press.

Blandau, R. J. and Moghissi, K. (eds.) (1973). *The Biology of the Cervix*. Chicago, The University of Chicago Press.

Bloom, W. and Fawcett, D. (1962). *A Textbook of Histology*. 8th ed. Philadelphia, W. B. Saunders Company.

Boreus, L. O. (ed.) (1971). *Fetal Pharmacology*. New York, Raven.

Carey, H. M. (ed.) (1963). *Human Reproductive Physiology*. London, Butterworths.

Cole, H. H. and Cupps, P. T. (eds.) (1970) *Reproduction in Domestic Animals*. 2nd ed. New York, Academic Press.

Committee on Maternal Nutrition/Food and Nutrition Board of National Research Council (1970). *Maternal Nutrition and the Cause of Pregnancy*. Washington D.C., National Academy of Sciences.

Daniel, J. C. (ed.) (1971). *Methods in Mammalian Embryology*. San Francisco, Freeman & Co.

Desjardins, C. and Biggers, J. D. (1972). *Immunoreproduction*. New York, Academic Press.

Diamond, M. (ed.) (1968). *Perspectives in Reproduction and Sexual Behavior*. Bloomington, Indiana University Press.

Diczfalusy, E. and Standley, C. C. (eds.) (1972). *The Use of Non-Human Primates in Research on Human Reproduction*. Stockholm, Karolinska Institutet.

Diczfalusy, E. and Borell, U. (eds.) (1971). *Control of Human Fertility*. Stockholm, Almqvist & Wiksell.

Ellenberger, W. and Baum, H. (1943). *Handbuch der vergleichenden Anatomie der Haustiere*. 18th ed. O. Zietzschmann, E. Ackerknecht and H. Grau (eds.), Berlin, Springer-Verlag.

Ellenberger, W., Baum, H. and Dittrich, H. (1957). *An Atlas of Animal Anatomy for Artists*. 2nd ed. New York, Dover Publications, Inc.

Elstein, M., Moghissi, K. and Borth, R. (eds.) (1973). *Cervical Mucus in Human Reproduction*. Copenhagen, Bogtrykkeriet Forum, WHO publication, Geneva, Switzerland.

Enders, A. C. (ed.) (1963). *Delayed Implantation.* Chicago, The University of Chicago Press.

Ferguson, L. C. (1958). *Diseases of Swine.* Ames, Iowa, Iowa State University Press.

Ferreira, A. J. (1969). *Prenatal Environment.* Springfield, Charles C Thomas.

Finerty, J. C. and Cowdry, E. V. (1960). *A Textbook of Histology.* 5th ed. Philadelphia, Lea & Febiger.

Fluhmann, C. F. (1961). *The Cervix Uteri and Its Diseases.* Philadelphia, W. B. Saunders Co.

Folley, S. J. (1956). *The Physiology and Biochemistry of Lactation.* Springfield, Charles C Thomas.

Frandson, R. D. (1965). *Anatomy and Physiology of Farm Animals.* Philadelphia, Lea & Febiger.

Gibian, H. and Plotz, E. J. (eds.) (1970). *Mammalian Reproduction.* Berlin, Springer-Verlag.

Giroud, A. (1970). *The Nutrition of the Embryo.* Springfield, Charles C Thomas.

Gorbman, A. and Bern, H. A. (1962). *A Textbook of Comparative Endocrinology.* New York, Wiley.

Grady, H. G. and Smith, D. E. (eds.) (1963). *The Ovary.* Baltimore, Williams & Wilkins.

Gray, A. P. (1954). *Mammalian Hybrids.* Franham Royal, Bucks, England, Agricultural Bureaux.

Hafez, E. S. E. (ed.) (1962). *The Behaviour of Domestic Animals.* London, Balliere, Tindall & Cox.

Hafez, E. S. E. (ed.) (1971). *Comparative Reproduction of Nonhuman Primates.* Springfield, Charles C Thomas.

Hafez, E. S. E. (ed.) (1970) *Reproduction and Breeding Techniques for Laboratory Animals.* Philadelphia, Lea & Febiger.

Hafez, E. S. E. and Blandau, R. J. (eds.) (1968). *The Mammalian Oviduct.* Chicago, The University of Chicago Press.

Hafez, E. S. E. and Evans, T. N. (eds.) (1973). *Human Reproduction: Conception and Contraception.* New York, Harper & Row.

Hammond, J. (1960). *Farm Animals. Their Breeding, Growth and Inheritance.* London, Edward Arnold.

Hammond, J. (1927). *Reproduction of the Cow.* Cambridge, Cambridge University Press.

Harrop, A. E. (1960). *Reproduction in the Dog.* London, Bailliere, Tindall & Cox.

Herman, H. A. and Madden, F. W. (1964). *The Artificial Insemination of Dairy Cattle—A Handbook.* 2nd ed. New York, Lucas Bros.

Inguilla, W. and Greenblatt, R. B. (eds.) (1969). *The Ovary.* Springfield, Charles C Thomas.

Klopper, A. and Diczfalusy, E. (eds.) (1969). *Foetus and Placenta.* Oxford, Blackwell Scientific Publications.

Kon, S. K. and Cowie, A. T. (eds.) (1961). *Milk: The Mammary Gland and Its Secretion.* Vols. I and II. New York, Academic Press.

Kowlessar, M. (1961). *Physiology of Prematurity.* Madison, Madison Printing Company.

Krolling, O. and Grau, H. (1960). *Lehrbuch der Histologie und vergleichenden mikroskopischen Anatomie der Haustiere.* 10th ed. Berlin, Parey.

Lewis, D. (ed.) (1961). *Digestive Physiology and Nutrition of the Ruminant.* London, Butterworths.

Lloyd, C. W. (ed.) (1959). *Recent Progress in the Endocrinology of Reproduction.* New York, Academic Press.

McDonald, L. E. (1971). *Veterinary Endocrinology and Reproduction.* Philadelphia, Lea & Febiger.

Mack, H. C. (ed.) (1970). *Prenatal Life.* Detroit, Wayne State University Press.

McKerns, K. W. (ed.) (1969). *The Gonads.* New York, Appleton-Century-Crofts.

McLaren, A. (ed.) (1966, 1967, 1968). *Advances in Reproductive Physiology.* Vols. I, II and III. New York, Academic Press.

Mann, T. (1954). *The Biochemistry of Sperm.* London, Methuen.

Manthei, C. A. (1958). *Diseases of Swine.* Ames, Iowa, Iowa State University Press.

Marrable, A. W. (1971). *The Embryonic Pig.* London, Pitman Medical.

Marshall, A. J. (ed.) (1960). *Biology and Comparative Physiology of Birds.* New York, Academic Press.

Mathews, W. W. (1972). *Atlas of Descriptive Embryology.* New York, Macmillan.

Moghissi, K. S. and Hafez, E. S. E. (eds.) (1972). *Biology of Mammalian Fertilization and Implantation.* Springfield, Charles C Thomas.

Morrison, J. E. (1970). *Foetal and Neonatal Pathology.* 3rd ed. London, Butterworths.

Nalbandov, A. (1958). *Reproductive Physiology; Comparative Reproductive Physiology of Domestic Animals, Laboratory Animals and Man.* San Francisco, Freeman & Co.

Netter, F. H. (1958). *The CIBA Collection of Medical Illustrations.* Vol. I. Summit, N. J., CIBA.

Newth, D. R. (1970). *Animal Growth and Development.* London, Edward Arnold.

Nickel R., Schummer, A. and Seiferle, E. (1954). *Lehrbuch der Anatomie der Haustiere.* Berlin, Parey.

Nishikawa, Y. (1959). *Studies on Reproduction in Horses. Singularity and Artificial Control of Reproductive Phenomena.* Tokyo, Japan Racing Association.

Overzier, C. (ed.) (1963). *Intersexuality.* London, Academic Press.

Parkes, A. S. (ed.) (1958–60). *Marshall's Physiology of Reproduction.* Vols. I and II. London, Longmans.

Patten, B. M. (1948). *Embryology of the Pig.* 3rd ed. Philadelphia, The Blackstone Company.

Patten, B. M. (1964). *Foundations of Embryology.* 2nd ed. New York, McGraw-Hill.

Perry, J. (ed.) (1960). *The Artificial Insemination of Farm Animals.* New Brunswick, N. J., Rutgers University Press.

Perry, J. (1972). *The Ovarian Cycle of Mammals.* New York, Hafner Press.

Raven, P. (1961). *Oogenesis: The Storage of Developmental Information.* New York, Pergamon.

Reynolds, S. R. M. (1965). *Physiology of the Uterus.* New York, Hafner Press.

Rhodin, J. A. G. (1963). *An Atlas of Ultrastructure.* Philadelphia, W. B. Saunders Co.

Roberts, S. J. (1970). *Veterinary Obstetrics and Genital Diseases.* Ann Arbor, Edwards Bros.

Robinson, T. J. (ed.) (1967). *The Control of the Ovarian Cycle in the Sheep.* Sydney, Sydney University Press.

Romanoff, A. L. (1960). *The Avian Embryo.* New York, Macmillan.

Romanoff, A. L. and Romanoff, A. J. (1949). *The Avian Egg.* New York, Wiley.

Rosemberg, E. (ed.) (1968). *Gonadotropins 1968.* Los Altos, Calif., Geron-x, Inc.

Rosemberg, E. and Paulsen, C. A. (eds.) (1970). *The Human Testis.* New York, Plenum Press.

Rothchild, N. M. V. (1956). *Fertilization.* New York, Wiley.

Rowlands, I. W. (ed.) (1966). *Comparative Biology of Reproduction in Mammals.* London, Academic Press.

Rugh, R. and Shettles, L. B. (1971). *From Conception to Birth.* New York, Harper & Row.

Salisbury, G. W. and VanDemark, N. L. (1961). *Physiology of Reproduction and Artificial Insemination of Cattle.* San Francisco, Freeman & Co.

Saxena, B. B., Beling, C. G. and Gandy, H. M. (eds.) (1971). *Gonadotropins.* New York, Wiley-Interscience.

Segal, S. J., Crozier, R., Corfman, P. A. and Condliffe, P. G. (eds.) (1973). *The Regulation of Mammalian Reproduction.* Springfield, Charles C Thomas.

Shelesnyak, M. C. and Marcus, G. J. (eds.) (1969). *Ovum Implantation.* New York, Gordon and Breach.

Sherman, A. I. (ed.) (1971). *Pathways to Conception— The Role of the Cervix and the Oviduct in Reproduction.* Springfield, Charles C Thomas.

Sisson, S. (1953). *Anatomy of Domestic Animals.* 4th ed. Rev. by D. Grossman. Philadelphia, W. B. Saunders Co.

Smith V. R. (1959). *Physiology of Lactation.* Ames, Iowa, Iowa State University Press.

Stave, U. (ed.) (1970). *Physiology of the Perinatal Period.* New York, Appleton-Century-Crofts.

Sturkie, P. D. (1954). *Avian Physiology.* Ithaca, Comstock Publishing.

Swenson, M. J. (ed.) (1970). *Duke's Physiology of Domestic Animals.* 8th ed. Ithaca, Cornell University Press.

Taylor, L. W. (1949). *Fertility and Hatchability of Chicken and Turkey Eggs.* New York, Wiley.

Timiras, P. S. (1972). *Developmental Physiology and Aging.* New York, Macmillan.

Trautmann, A. and Fiebiger, J. (1952). *Fundamentals of the Histology of Domestic Animals.* Ithaca, Comstock Publishing.

Tuchmann-Duplessis, H. (ed.) (1971). *Malformations Congénitales des Mammifères.* Paris, Masson.

Turner, C. W. (1952). *The Mammary Gland. I. The Anatomy of the Udder of Cattle and Domestic Animals.* Columbia, Lucas Bros.

Van Roekel, H. (1959). *Diseases of Poultry.* Ames, Iowa, Iowa State University Press.

Van Tienhoven, A. (1968). *Reproductive Physiology of Vertebrates.* Philadelphia, W. B. Saunders Co.

Velardo, J. T. (ed.) (1958). *The Endocrinology of Reproduction.* New York, Oxford University Press.

Villee, C. A. (ed.) (1961). *Control of Ovulation.* Oxford, Pergamon,

Villee, C. A. (1960). *The Placenta and Fetal Membranes.* Baltimore, Williams & Wilkins.

Waisman, H. A. and Kerr, G. (eds.) (1970). *Fetal Growth and Development.* New York, McGraw-Hill.

Watterson, R. L. and Sweeney, R. M. (1973). *Laboratory Studies of Chick, Pig, and Frog Embryos.* 3rd ed. Minneapolis, Burgess.

Westerfield, C. (1957). *Histology and Embryology of the Domestic Animals.* 3 Vols. Ann Arbor, Edwards Bros.

Williams, W. L. (1943). *Diseases of the Genital Organs of Domestic Animals.* 3rd ed. Ithaca, N.Y., Miss Louella Williams.

Witschi, E. (1956). *Development of Vertebrates.* Philadelphia, W. B. Saunders Co.

Wolstenholme, G. E. W. and O'Connor, M. (eds.) (1969). *Foetal Anatomy.* London, Churchill.

Wolstenholme, G. E. W. and O'Connor, M. O. (eds.) (1961). *Somatic Stability in the Newly Born.* Boston, Little, Brown & Company.

Woodruff, J. D. and Pauerstein, C. J. (1969). *The Fallopian Tube.* Baltimore, Williams & Wilkins.

Wynn, R. M. (ed.) (1967). *Cellular Biology of the Uterus.* New York, Appleton-Century-Crofts.

Young, W. C. (ed.) (1961). *Sex and Internal Secretions.* 3rd ed. Vols. I & II. Baltimore, Williams & Wilkins.

Zamboni, L. (1971). *Fine Morphology of Mammalian Fertilization.* New York, Harper & Row.

Zemjanis, R. (1962). *Diagnostic and Therapeutic Techniques in Animal Reproduction.* Baltimore, Williams & Wilkins.

Zietzschmann, O. and Krölling, O. (1955). *Lehrbuch der Entwicklung geschichte der Haustiere.* 2nd ed. Berlin, Parey.

Zuckerman, S. (ed.) (1962). *The Ovary.* Vols. I and II. New York, Academic Press.

II. PROCEEDINGS OF CONGRESSES AND SYMPOSIA

The International Congress of Animal Reproduction

1. Held in 1948 in Milan, Italy
2. Held in 1952 in Copenhagen, Denmark
3. Held in 1956 in Cambridge, England
4. Held in 1961 in The Hague, Holland
5. Held in 1964 in Trento, Italy
6. Held in 1968 in Paris, France
7. Held in 1971 in Munich, Germany

The Biennial Symposia on Animal Reproduction, Held in U.S.A.

1. *Conference on Female Reproduction in Farm Animals* (1953). *Iowa State College Journal of Science 28* (1):1, 138 pages.
2. *Reproduction and Infertility* (1955). *Michigan State University Centennial Symposium Publication,* 112 pages.
3. *Reproduction and Infertility* (1958). F. X. Gassner, ed., New York, London, Pergamon Press, 273 pages

4. *The Effect of Germ Cell Damage on Animal Reproduction* (1960). Supplement to *Journal of Dairy Science*, Vol. 43. N. L. VanDemark, ed., 169 pages.
5. *Fifth Biennial Symposium on Animal Reproduction* (1961). Knoxville, University of Tennessee, proceedings not published.
6. *Gonadotropins: Their Chemical and Biological Properties and Secretory Control* (1964). H. H. Cole, ed., San Francisco and London, Freeman & Co., 250 pages.
7. *Seventh Biennial Symposium on Animal Reproduction: Environmental Influences on Reproductive Processes* (1966). *J. Anim. Sci.*, Vol. 25 (Supplement), W. Hansel and R. H. Dutt, eds., 147 pages.

Colloque de La Societe Nationale pour l'etude de la Sterilite et de la Fecondite

"Les Fonctions de Nidation Uterine et leurs Troubles." Masson et Cie, Libraire de l'Academie de Medecine, 120 Blv., Saint Germaine, Paris 6°.

CIBA Foundation Symposium

Mammalian Germ Cells (1953). Boston, Little, Brown and Company.
Preimplantation Stages of Pregnancy (1965). London, Churchill.

Symposium on Mammalian Genetics and Reproduction (1960)

Held in Oak Ridge, Tennessee, *J. Cell Comp. Physiol. 56* (Suppl. 1), 1–193.

Memoirs of the Society for Endocrinology

England, Cambridge University Press.

No. 4 *Comparative Endocrinology of Vertebrates* Part I, 1955.
No. 5 *Comparative Endocrinology of Vertebrates*, Part II, 1956.
No. 6 *Implantation of Ova*, 1959.
No. 7 *Sex Differentiation and Development*, 1960.

Transactions of the Annual Conference on "Gestation"

Starting in 1954, sponsored by the Josiah Macy, Jr. Foundation, New York.

III. SCIENTIFIC PERIODICALS AND ABSTRACTS

Acta Agricultura Scandinavica, Hovslagargatan, 2iii, Stockholm C, Sweden.
Acta Endocrinologica, Periodica, Skelmosevej 10, Copenhagen, Valby, Denmark.
American Journal of Veterinary Research, American Veterinary Medical Association, 600 S. Michigan Avenue, Chicago, Illinois.
Animal Behaviour, Bailliere, Tindall & Cox, London, W. C. 2, England.
Animal Breeding Abstracts, Commonwealth Bureaux of Animal Breeding and Genetics, Edinburgh, Scotland.
Auchthygiene, Fortpflanzungsstorungen und Besamung der Haustiere, Verlag M & H. Schaper, Hanover, Germany.
British Poultry Science, Oliver and Boyd, Ltd., London, England.
The Cornell Veterinarian, Cornell Veterinarian, Inc., Veterinary College, Ithaca, New York.
Endocrinology, Charles C Thomas, Springfield, Illinois.
Fertility and Sterility, Paul P. Hoeber, Inc., New York, New York.
International Journal of Fertility, Ben Franklin Press, Pittsfield, Massachusetts.
Journal of Agricultural Science, Cambridge University Press, Cambridge, England.
Journal of the American Veterinary Medical Association, American Veterinary Medical Association, 600 S. Michigan Avenue, Chicago, Illinois.
Journal of Animal Science, Boyd Printing Co., 49 Sheridan Avenue, Albany 10, New York.
Journal of Dairy Science, The Garrard Press, 510–522 North Hickory St., Champaign, Illinois.
Journal of Embryology and Experimental Morphology, Oxford University Press, London, E. C. 4, England.
Journal of Endocrinology, Cambridge University Press, Bentley House, 200 Euston Road, London, N.W. 1, England.
Journal of Reproduction and Fertility, Blackwell Scientific Publications, Oxford, England.
Poultry Science, The Poultry Science Association. Texas A & M College System, College Station, Texas.
The Veterinary Record, 7 Mansfield St., London, W. 1, England.

Index

Page numbers in *italics* indicate figures; page numbers followed by "t" indicate tables.